Lecture Notes in Computer Science 13836

More information about this series at https://link.springer.com/bookseries/558

Wei Qi Yan · Minh Nguyen ·
Martin Stommel (Eds.)

Image and Vision Computing

37th International Conference, IVCNZ 2022
Auckland, New Zealand, November 24–25, 2022
Revised Selected Papers

Springer

Editors
Wei Qi Yan
Auckland University of Technology
Auckland, New Zealand

Minh Nguyen
Auckland University of Technology
Auckland, New Zealand

Martin Stommel
Auckland University of Technology
Auckland, New Zealand

ISSN 0302-9743 ISSN 1611-3349 (electronic)
Lecture Notes in Computer Science
ISBN 978-3-031-25824-4 ISBN 978-3-031-25825-1 (eBook)
https://doi.org/10.1007/978-3-031-25825-1

This Springer imprint is published by the registered company Springer Nature Switzerland AG
The registered company address is: Gewerbestrasse 11, 6330 Cham, Switzerland

Preface

The 37th International Conference on Image and Vision Computing New Zealand (IVCNZ 2022) was held both online and onsite on November 24–25, 2022, in Auckland, New Zealand at the Auckland University of Technology (AUT) City Campus. The conference was endorsed by the International Association for Pattern Recognition (IAPR). IVCNZ 2022 was organized by the Centre for Robotics & Vision (CeRV), Auckland University of Technology (AUT).

Image and Vision Computing New Zealand is New Zealand's premier academic conference on all aspects of computer vision, image processing, computer graphics, virtual and augmented reality, visualization, and HCI applications related to these fields. Relevant topics include, but are not limited to: artificial Intelligence approaches to computer vision; augmented and virtual reality; automated visual surveillance; biomedical imaging and visualization; biologically inspired vision systems; biometrics; calibration techniques; computer graphics; enhancement of video and still images; face recognition; feature detection and extraction; geometric modeling in vision and graphics; graph matching; image analysis and understanding; image-based rendering; image compression and coding; machine vision applications; medical imaging applications; motion tracking and analysis; motion synthesis and control; multimedia information retrieval; novel algorithms or techniques; object recognition; pattern recognition and classification; reconstruction techniques; rendering and scientific visualization; scientific visualization; security image processing; shape recovery from multiple images; sonar and acoustical imaging; and stereo image analysis.

We invited submissions aiming either at highlighting relationships between adjacent topics within the listed areas or contributing to a particular topic within one area which is of fairly general interest.

This conference used double-blind review, which means that the authors were concealed from the reviewers, and vice versa, throughout the review process. By the end, we had 79 full papers submitted to the EasyChair conference system. We received 136 reviews in total. The average number of reviews per paper was 2.72; on average, each reviewer was assigned approximately three papers to assess. After the double-blind reviewing process, 37 papers were accepted for presentation at the conference. This resulted in a 46.8% acceptance rate. We therefore had 37 oral presentations, grouped in four sessions during the two days of the conference. From the selected papers, we chose one paper for the best paper award. Additionally, we invited four renowned keynote speakers: Mengjie Zhang, Victoria University of Wellington, New Zealand; Mohan Kankanhalli, National University of Singapore; Nikola Kasabov, Auckland University of Technology, New Zealand; and Dacheng Tao, University of Sydney, Australia.

December 2022

Wei Qi Yan
Minh Nguyen
Martin Stommel

Organization

Organising Committee

Wei Qi Yan (General Chair)	Auckland University of Technology, New Zealand
Minh Nguyen (Program Chair)	Auckland University of Technology, New Zealand
Martin Stommel (Program Chair)	Auckland University of Technology, New Zealand

Program Committee

Waleed Abdulla	University of Auckland, New Zealand
Qurrat Ul Ain	Victoria University of Wellington, New Zealand
Aisha Ajmal	Victoria University of Wellington, New Zealand
Harith Al-Sahaf	Victoria University of Wellington, New Zealand
Donald Bailey	Massey University, New Zealand
Andrew Bainbridge-Smith	University of Canterbury, New Zealand
Oliver Batchelor	University of Canterbury, New Zealand
Boris Bačić	Auckland University of Technology, New Zealand
Ying Bi	Victoria University of Wellington, New Zealand
Phil Bones	University of Canterbury, New Zealand
Will Browne	Queensland University of Technology, Australia
Stefano Cagnoni	University of Parma, Italy
Kwok-Ping Chan	University of Hong Kong, China
Joe Chen	University of Canterbury, New Zealand
Qi Chen	Victoria University of Wellington, New Zealand
Tai-Yin Chiu	University of Texas at Austin, USA
Richard Clare	University of Canterbury, New Zealand
Michael Cree	University of Waikato, New Zealand
Jinshi Cui	Peking University, China
Jeremiah D. Deng	University of Otago, New Zealand
Mansoor Ebrahim	Iqra University, Pakistan
Jaco Fourie	Lincoln Agritech Ltd., New Zealand
Xiping Fu	University of Otago, New Zealand
Chiou-Shann Fuh	National Taiwan University, Taiwan
Yongsheng Gao	Griffith University, Australia
Alfonso Gastelum Strozzi	UNAM/CCADET, Mexico
Akbar Ghobakhlou	Auckland University of Technology, New Zealand
Hamid Gholamhosseini	Auckland University of Technology, New Zealand
Andrew Gilman	PlantTech Research Institute, New Zealand

Richard Green	University of Canterbury, New Zealand
Anthony Griffin	Auckland University of Technology, New Zealand
Christopher Hollitt	Victoria University of Wellington, New Zealand
Aarij Hussaan	Iqra University, Pakistan
Xiaoyi Jiang	University of Münster, Germany
Scott King	Texas A&M University, Corpus Christi, USA
Huy Le	Auckland University of Technology, New Zealand
Andrew Lensen	Victoria University of Wellington, New Zealand
Shen Hin Lim	University of Waikato, New Zealand
Jorge Alberto Marquez Flores	Universidad Nacional Autónoma de México, Mexico
Stephen McNeill	Landcare Research, New Zealand
Brendan McCane	University of Otago, New Zealand
Yi Mei	Victoria University of Wellington, New Zealand
Steven Mills	University of Otago, New Zealand
Kourosh Neshatian	University of Canterbury, New Zealand
Jean Francois Nezan	INSA, Rennes, IETR Laboratory, France
Bach Nguyen	Victoria University of Wellington, New Zealand
Minh Nguyen	Auckland University of Technology, New Zealand
Binh P. Nguyen	Victoria University of Wellington, New Zealand
Mukku Nisanth Kartheek	Institute for Development and Research in Banking Technology & National Institute of Technology, India
Warangal Mohammad Norouzifard	University of Auckland, New Zealand
Kapila Pahalawatta	Lincoln Agritech Ltd., New Zealand
Simon Park	Open Polytechnic, New Zealand
Tuan Pham	Prince Mohammad Bin Fahd University, Saudi Arabia
Sanjoy Pratihar	Indian Institute of Information Technology, India
Kalyani Ibrahim Rahman	Open Polytechnic, New Zealand
Atiya Rana	Iqra University, Pakistan
Ramesh Rayudu	Victoria University of Wellington, New Zealand
Mariano Rivera	CIMAT, Mexico
Mahdi Setayesh	Microsoft, USA
Abubakar Siddique	Victoria University of Wellington, New Zealand
Martin Stommel	Auckland University of Technology, New Zealand
Lee Streeter	University of Waikato, New Zealand
Lech Szymanski	University of Otago, New Zealand
Rachael Tighe	University of Waikato, New Zealand
Martin Urschler	University of Auckland, New Zealand
Zhiyong Wang	University of Sydney, Australia

Stephen Weddell	University of Canterbury, New Zealand
Henry Williams	University of Auckland, New Zealand
David Wilson	Auckland University of Technology, New Zealand
Brendon J. Woodford	University of Otago, New Zealand
Burkhard Wuensche	University of Auckland, New Zealand
Bing Xue	Victoria University of Wellington, New Zealand
Wei Qi Yan	Auckland University of Technology, New Zealand
Yu-Bin Yang	Nanjing University, China
Chun Hong Yoon	SLAC National Accelerator Laboratory, USA
Mengjie Zhang	Victoria University of Wellington, New Zealand
Fanglue Zhang	Victoria University of Wellington, New Zealand
Junhong Zhao	Victoria University of Wellington, New Zealand
Shihua Zhou	Dalian University, China

Sponsors

School of Engineering, Computer & Mathematical Sciences, Auckland University of Technology

Department of Computer Science & Software Engineering, Auckland University of Technology

Centre for Robotics and Vision (CeRV), Auckland University of Technology

Contents

StencilTorch: An Iterative and User-Guided Framework for Anime Lineart Colorization

Yliess Hati[1,2](✉), Vincent Thevenin[2], Florent Nolot[1], Francis Rousseaux[1], and Clement Duhart[2]

[1] University of Reims Champagne-Ardenne, Laboratory CReSTIC, 51 100 Reims, France
{florentnolot,francisrousseaux}@univ-reims.fr

[2] Léonard de Vinci Pôle Universitaire, Research Center, 92 916 Paris La Défense, France
{yliess.hati,clement.duhart}@devinci.fr, vincent.thevenin@edu.devinci.fr

Abstract. Automatic lineart colorization is a challenging task for Computer Vision. Contrary to grayscale images, linearts lack semantic information such as shading and texture, making the task even more difficult. Modern approaches train a Generative Adversarial Network (GAN) to generate illustrations from user inputs such as color hints. While such approaches can generate high-quality outputs in real-time, the user only interacts with the pipeline once at the beginning of the process. This paper presents StencilTorch, an interactive and user-guided framework for anime lineart colorization motivated by digital artist workflows. StencilTorch generates illustrations from a given lineart, color hints, and a mask allowing for iterative workflows where the output of the first pass becomes the input of a second. Our method improves previous work on both objective and subjective evaluations.

1 Introduction

Motivation. Illustration can be summarized as the succession of four well defined tasks: sketching, inking, coloring, and post processing. Referring to Kandinsky's work [19], the colorization process can change the meaning of an entire piece of art by introducing color schemes, shading, and textures. These last three properties of the painting process turn out to be challenging for the Computer Vision task of automatic colorization. Contrary to its grayscale counterpart [8,14,45], linearts lack semantic information making the task even more difficult. Materials and 3D shapes can only be inferred from their silhouettes.

Problem. Previous work introduced the use of GAN [9] methods, one of the most widespread neural architectures for image generation. These approaches can generalize and produce perceptively qualitative illustrations. While some work focused on the use of color-based hints [5,7,12,26,31,36], others are using

W. Q. Yan et al. (Eds.): IVCNZ 2022, LNCS 13836, pp. 1–17, 2023.
https://doi.org/10.1007/978-3-031-25825-1_1

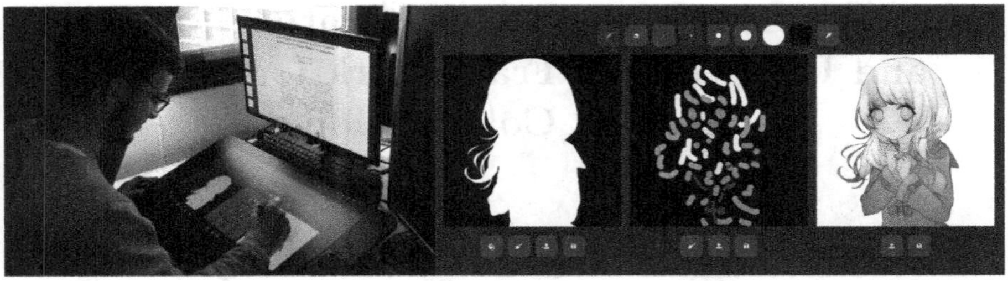

Fig. 1. The figure shows a photo of a user interacting with our model StencilTorch and a screenshot from our Web Application. The left canvas allows the user to provide or draw a mask. Hint maps can be drawn or provided in the middle canvas. Hovering over the hint section reveals a transparent lineart. In the right canvas, the user can upload his lineart and see the resulting illustration. A toolbar with basic digital painting tools is available at the very top, including a pen, an eraser, predefined brush sizes, and a color picker.

style transfer [43] or tags [21] as color cues to condition the output of the model to user intents. In such methods, feature extractors such as Illustration2Vec [35] are used to enforce semantic information on the input.

Previous methods consist of a one-step process where the user is invited to influence the generation process once, in the beginning, using information hints. This type of pipeline is not ideal in the context of creation where the artist wants to iterate and explore the design space of the illustration process. In this paper, we instead formulate the task of automatic colorization as a Human-in-the-loop process where the user collaborates with the machine in an iterative and interactive process to produce the final piece of art.

Solution. We introduce StencilTorch, an iterative and user-guided framework for anime lineart colorization. Our framework is motivated by human workflow and is inspired by previous work done by Ci et al. [5], and follow-up work by Hati et al. [12]. We train a conditional Wasserstein GAN with gradient penalty, c-WGAN-GP for short, to generate illustrations from a given lineart, natural color hints, and a mask describing the region of the image that has to be inpainted.

Our model is trained on images curated online using a similar approach to PaintsTorch [12]. The illustrations are post-processed to extract a synthetic lineart using an Extended Difference of Gaussians (xDoG) [41] and a displacement map is extracted to remove lighting and limit the amount of colors from which the color hints are sampled from.

Findings. We evaluated StencilTorch on our curated test dataset against previous work by Zhang et al. [44], PaintsChainer [31], Ci et al. [5], and PaintsTorch [12]. The models are benchmarked using both objective metrics, Fréchet Inception Distance (FID), Learned Perceptual Image Patch Similarity (LPIPS), and a subjective metric Mean Opinion Score (MOS) obtained by conducting a user study. StencilTorch improves previous work on each of those metrics.

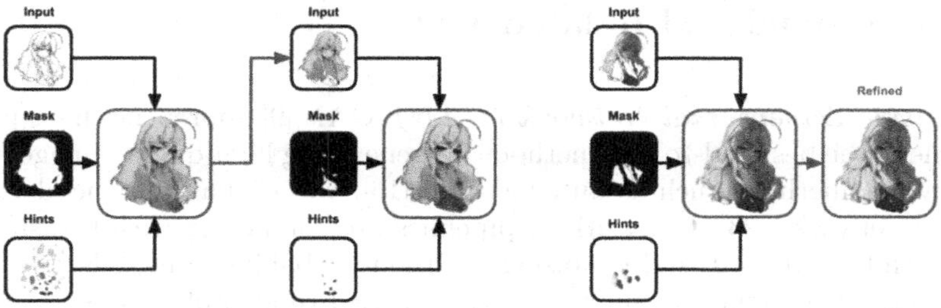

Fig. 2. StencilTorch can paint a lineart given user hints and a mask. The diagram illustrates an iterative workflow. The artist explores a potential colorization for its drawing on the left and explore an alternative colorization on the right after the introduction of shading and lighting in the input image.

Our approach not only enables iterative and collaborative workflows between the human and the computer but also captures the additional painting style introduced by the user, such as shading and illumination.

Contributions.

- A new synthetic dataset curation pipeline for producing qualitative input and output pairs for automatic anime lineart colorization. Our pipeline partially removes texture and shadow information from the illustrations, and uses semantic color segmentation to produce better synthetic hints for training.
- StencilTorch, an interactive and user-guided framework for anime lineart colorization. Given a lineart, color hints and a mask, StencilTorch generates a colored illustration. The colored output can be used as input for the next iteration enabling natural workflows between the human and the machine. Our model introduces a guide network for the generator's decoder during training and is trained for both illustration generation and inpainting.
- A study showing that StencilTorch improves previous work on our curated dataset both objectively with the FID and LPIPS metrics, and subjectively with an MOS. We also provide an ablation study to justify our design choices.
- An interactive open-source web application of the StencilTorch framework using TensorFlowJS and Web Canvases to simulate digital art programs.

Implication. Previous work focused on improving the perceptual quality of the generation process [5,12,31,45]. Our contribution is orthogonal and can be combined with such improvements enabling seamless and natural digital illustration workflows. Stenciltorch allows the artist to collaborate with the machine, as shown in Fig. 2. We believe that such approaches will allow AI-driven tools to be part of the creative environment and enable fast and broad exploration.

Reproduction. For reproducibility and transparency, we published our implementation and experimentation at www.github.com/yliess86/PaintsTorch2.

2 Background and Related Work

Generative Adversarial Network (GAN). GAN [9] approaches have proven to be one of the best end-to-end methods for generating high-quality images beating previous methods such as autoencoders [16], variational autoencoders [22], and flow networks [23]. GAN driven pipelines are competing with transformer-based architectures [40] and denoising diffusion probabilistic models [17]. They are efficient at inference, enabling closed-to or real-time applications.

Vanilla. The vanilla GAN introduced by Goodfellow et al. [9] consists of a generator $\mathcal{G}$ trained to produce images $\mathcal{G}(z)$ similar to the data distribution fed to a discriminator $\mathcal{D}$ trained to differentiate fake images from true images x. The networks are jointly trained to optimize a Min-Max objective shown in Eq. 1 and can be further split as distinct objectives as shown in Eq. 2.

$$\min_{\mathcal{D}} \max_{\mathcal{G}} \mathbb{E}_x[log(\mathcal{D}(x))] + \mathbb{E}_z[log(1 - \mathcal{D}(\mathcal{G}(z)))] \ , \tag{1}$$

$$\frac{1}{m}\sum_{i=1}^{m}[log(\mathcal{D}(x_i)) + log(1 - \mathcal{D}(\mathcal{G}(z_i)))] \ , \quad \frac{1}{m}\sum_{i=1}^{m}[log(\mathcal{D}(\mathcal{G}(z_i)))] \ . \tag{2}$$

In practice, this formulation is unstable. The generator often saturates if it does not keep up with the discriminator, which task is most of the time easier to optimize. It also suffers from vanishing gradients and mode collapse, where the generator finds a simple solution that fools the discriminator failing at generating diverse enough outputs. The literature includes techniques to overcome those issues, such as the hinge loss [25], the Wasserstein distance [2], gradient penalty [10], and the use of batch [18] and spectral [30] normalization strategies.

Wasserstein Distance. One powerful alternative to the vanilla formulation is the Wasserstein GAN or WGAN for short [2]. It resolves both the mode collapse and the vanishing gradients issues. The output activation of the discriminator is changed from a sigmoid to a linear function. This change turns the discriminator network into a critic rating the quality of the generated output rather than discriminating the fake from the real. The critic and generator objectives are shown in Eq. 3. Gradient clipping is used to satisfy the Lipschitz constraint.

$$\frac{1}{m}\sum_{i=1}^{m}[\mathcal{D}(x_i)] - [\mathcal{D}(\mathcal{G}(z_i))] \ , \qquad \frac{1}{m}\sum_{i=1}^{m}[\mathcal{D}(\mathcal{G}(z_i))] \ . \tag{3}$$

Gradient Penalty. Instead of gradient clipping, Gulrajani et al. [10] enforce a constraint on the critic such that its gradients with respect to the inputs are unit vectors. The critic loss is augmented with an additional term shown in Eq. 4 where $\hat{x}$ is sampled from a linear interpolation between real and fake samples to satisfy the critic's Lipschitz constraint.

$$\lambda \mathbb{E}_{\hat{x}}[(||\nabla_{\hat{x}} D(\hat{x})||_2 - 1)^2] \ . \tag{4}$$

User Conditioning. A GAN network can be further conditioned on user inputs such as class labels [29] and transformed into a multi-modal model. The final user is granted fine-grain control over the generated output by conditioning both the generator and the discriminator or critic networks on such input. This approach is called a conditional GAN or c-GAN for short.

Color Hints. User-guided automatic colorization through color hints is one of the leading approaches. It enables fine-grain control on the image generation process by pin-pointing colors in the regions where a specific color is intended. This type of color control has been popularized by Zhang et al. [45] and is used by the majority of the following methods [5,7,12,26,36,44].

In their work, Zhang et al. [45] claim that randomly activating pixels from the original illustration during training as color hints is enough to enable fine-grain control on the output. However, Hati et al. [12] demonstrate that this is not enough. The lack of semantic information, shading, and texture is to blame. They propose using simulated strokes to represent the user inputs more faithfully.

Both approaches employ a U-Net [33] architecture with ResNetXt blocks [13] using dilated convolution to increase the receptive field and favor speed, and Pixel Shuffling [37] to limit the generation of artifacts.

Style Transfer. An alternative input style for color hints is the use of style transfer. The user provides a target illustration from which the network has to learn the style properties while conserving the content of the given lineart using VGG16 [38] or VGG19 features as perceptual loss proxies. This approach has been extensively studied in previous work [8,14,43]. Style transfer can be combined with color palette conservation to transfer colors from the style image.

Tags. Tag-based automatic colorization has been explored by Kim et al. [21]. They introduce feature attribute vectors to condition the network for lineart colorization. By describing features such as hairstyle, hair color, eye color, and others, the network can transfer the user intent into the painting. They also introduce SECat, a neural network module to help the model focus on details.

3 Proposed Method

Our method, StencilTorch aims at generating colored illustrations from a given lineart, color hints, and a mask to favor iterative and creative workflows where the output of the first pass can become the input of a second. This section discusses the importance of the data curation process, the input generation, our model architecture, the introduction of inpainting, and our training curriculum.

3.1 Dataset Curation

The challenge of anime lineart colorization suffers from a lack of qualitative and publicly available datasets. Finding corresponding pairs of lineart and illustrations in abundance is a challenge. Online scrapping of anime drawings and synthetic lineart extraction are the preferred methods [5,12,45].

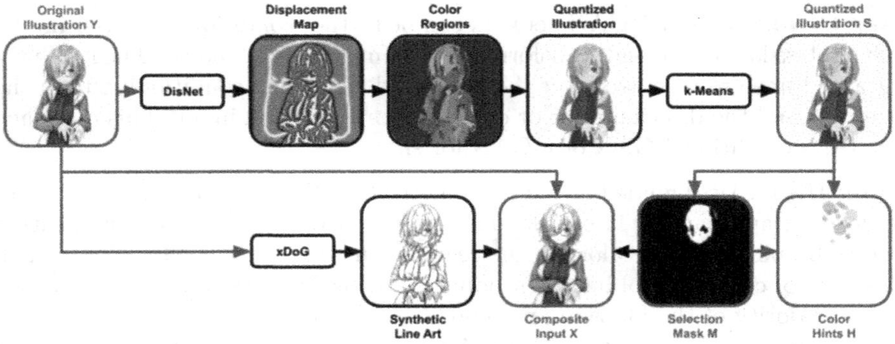

Fig. 3. StencilTorch input generation process diagram. An illustration is sampled from our curated dataset. This illustration is used to generate a synthetic lineart using the xDoG method. A displacement network DisNet is trained to generate displacement maps from colors illustrations used to produce color regions. The color regions are then quantized and reduced via k-means clustering to eliminate most of the lighting, shading, and texture-specific colors. The regions are sampled to produce a mask and color hints. The synthetic lineart, the mask, and the color hints are combined into a single input.

The few public datasets available to the community we are aware of are inconsistent in terms of perceptual quality, the nature of the images (e.g. comics pages, photos), and present illustrations from artists of different backgrounds, levels, and styles. For these reasons, we have curated a custom dataset.

Our dataset contains 21,930 scrapped anime-like illustrations for training, 3,545 for testing, and is manually filtered to ensure a perceptive quality across all samples and remove inappropriate (e.g. gore, mature, and sexual) content. This process is motivated by previous work on PaintsTorch [12] where the authors demonstrate the implications of the dataset quality in the generation process.

We find important to highlight that any dataset [1] used for the challenge of anime-like line art colorization is biased. They are reflections of the anime subculture and communities from which they are drawn. The drawings are mostly figures of female characters with visible skin. This may justify the overall salmon watercolor tone attributed to the illustrations produced by current works.

3.2 Input Generation

The lack of lineart and illustration pairs require the use of complex input generation pipelines. Previous work [5,12] proposed to generate synthetic linearts out of the curated paintings. xDoG [41] is the technique used to extract such information. It produces qualitative and clean lines from complex colored drawings.

The color hints are randomly selected by sampling parts of the illustration or averaging colors from random locations. However, we claim that this approach is

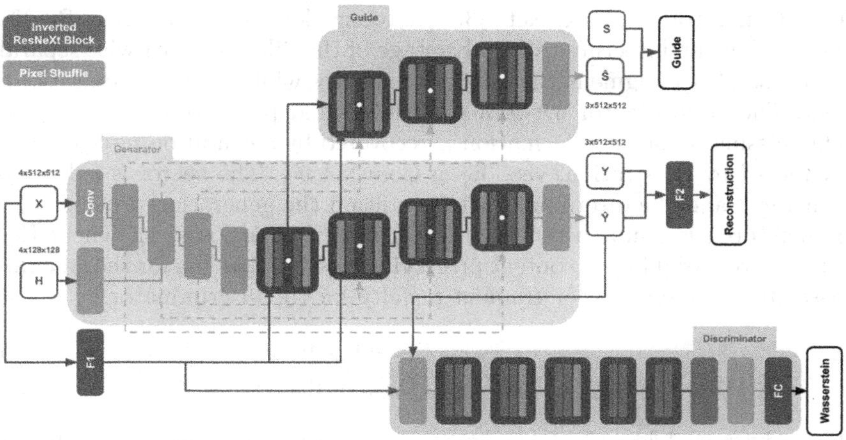

Fig. 4. StencilTorch architecture schematic. The pipeline employs five neural networks: a generator $\mathcal{G}$ responsible for generating fake illustrations from a black and white lineart, color hints, a mask, and a feature vector, a feature extractor $\mathcal{F}_1$, a VGG16 content extractor $\mathcal{F}_2$, a guide network responsible for guiding the generation process, and a discriminator network $\mathcal{D}$.

not appropriate for proper colorization. Such processes do not distinguish midtone colors from lighting, shading, nor texture, leading to misrepresentations of the user intent. Randomly activating pixels does not account for color bleeding and messy inputs. In this paper, we present a new color hint scheme selection process solving the aforementioned issues and better matching the behavior of our end users. Our pipeline is summarized in the diagram shown in Fig. 3.

The new StencilTorch input pipeline relies on color regions extraction. We train a ResNet [13] model to regress displacement maps from colored illustrations on the DanbooRegion dataset [28]. The displacement map is robust to noise and is used to extract unique color sections. We assign each region to its median color and further process the output by reducing the number of colors to 25 using k-means clustering. We empirically selected the number clusters to apply color quantization in RGB space without sacrificing too much detail. This process limits the shading, lighting, and texture information present in the input illustration. The color hints are finally sampled from this image and drawn to the hint texture as circles of various sizes to simulate a variety of user strokes.

3.3 Model Architecture

The model architecture used in StencilTorch is a c-WGAN-GP inspired by previous work by Ci et al. [5] and Hati et al. [12]. Our adversarial framework consists of five neural networks: a generator, a discriminator, also called a critic, a feature extractor, a style network, and a guide network. A schematic of the entire architecture is shown in Fig. 4.

The Generator is a U-Net [33] autoencoder with inverted ResNeXt blocks [42]. The architecture takes advantage of the dilated depth-wise separable convolutions [4] to augment the network capacity while being computationally efficient. The discriminator network presents similar properties.

The missing semantic information is recovered by a conditioning feature vector extracted by Illustration2Vec [35], a ConvNet $\mathcal{F}_1$. The vector is fed to both the generator and the discriminator to condition the generation process.

StencilTorch is train to optimize a weighted mix of losses shown in Eqs. 5 and 6: an adversarial loss, a content preserving loss, a guide loss for the generator, a Wasserstein distance and a gradient penalty for the discriminator.

$$\mathcal{L}_{\mathcal{D}} = \mathcal{L}_w + \mathcal{L}_{gp} \ , \tag{5}$$

$$\mathcal{L}_{\mathcal{G}} = \lambda_{adv}\mathcal{L}_{adv} + \mathcal{L}_{cont} + \mathcal{L}_{guide} \ . \tag{6}$$

The adversarial loss $\mathcal{L}_{adv}$ shown in Eq. 7 trains the generator to fool the discriminator. We choose to optimize the hinge loss [25] as it stabilizes training while improving the perceptual quality. The loss is weighted by $\lambda_{adv} = 1e-4$.

$$\mathcal{L}_{adv} = \mathbb{E}_{\hat{Y}}[max(1 - \mathcal{D}(\hat{Y}, \mathcal{F}_1(X)), 0)] \ . \tag{7}$$

The content loss $\mathcal{L}_{cont}$ shown in Eq. 8 ensures the preservation of the original content present in the lineart. The content information is provided by a VGG16 model we refer to as $\mathcal{F}_2$ pretrained on ImageNet [34].

$$\mathcal{L}_{cont} = \frac{1}{chw}||\mathcal{F}_2(\hat{Y}) - \mathcal{F}_2(Y)||_2 \ . \tag{8}$$

The guide loss $\mathcal{L}_{guide}$ shown in Eq. 9 helps the decoder disentangling feature maps to recover flat colorization in early stages. The loss is inspired by previous work by Zhang et al. [43]. The guide network is used during training, removed at inference time, and has the same architecture as the generator decoder. It produces an output $\bar{S}$ we optimize to be close to the quantized illustration S obtained in the input generation process.

$$\mathcal{L}_{guide} = \frac{1}{chw}||\hat{S} - S||_2 \ . \tag{9}$$

The Wasserstein loss $\mathcal{L}_w$ shown in Eq. 10 is responsible for training the discriminator. For the adversarial loss, we optimize its hinge formulation.

$$\mathcal{L}_w = \mathbb{E}_{\hat{Y}}[max(1 + \mathcal{D}(\hat{Y}, \mathcal{F}_1(X)), 0)] + \mathbb{E}_Y[max(1 - \mathcal{D}(Y, \mathcal{F}_1(X)), 0)] \ . \tag{10}$$

StencilTorch optimizes an additional gradient penalty term $\mathcal{L}_{gp}$ shown in Eq. 11. As proposed by Karras et al. [20] the gradient penalty is augmented with a drifting term to further enforce stabilization during the training of the discriminator. The gradient penalty is weighted by the hyperparameter $\lambda_{gp} = 10$ and a drifting weight $\epsilon_{drif} = 1e-3$.

$$\mathcal{L}_{gp} = \lambda_{gp}\mathbb{E}_{\hat{Y}}[(||\lambda_{\hat{Y}}\mathcal{D}(\hat{Y}, \mathcal{F}_1(X))||_2 - 1)^2] + \epsilon_{drift}\mathbb{E}_Y[\mathcal{D}(Y, \mathcal{F}_1(X))^2] \ . \tag{11}$$

3.4 Mask Inpainting

Selection masks are part of the tools used by digital artists to select regions of the illustration and avoid bleeding artifacts. Inspired by this kind of workflow, we introduce the use of selection masks to limit the GAN generation process to specific areas of the input. This process allows for partial inpainting of the illustration. It is responsible for our iterative workflow where the output of a first pass can be used as the input of a second pass, with or without intermediate modification. As shown in Fig. 2, it allows the generator to pick insights from the artist's style and naturally fits the standard digital artist workflow.

During training, masks are generated by sampling and merging the color regions computed in the input generation process. The selection mask is first used for the synthetic lineart and its corresponding colored illustration compositing. The lineart is drawn where the mask is white, the illustration where it is black. The composite input is stacked with the mask and is used instead of the black-and-white lineart to feed the generator. The mask is finally reused for inpainting when computing the final generated output illustration and acts as attention weights during the entire process.

3.5 Curriculum Learning

Curriculum leaning [3,6] is a type of pipeline in which a model is trained on tasks that gradually increase in difficulty. It has been shown to increase models' performances in convolutional neural networks [11].

In StencilTorch, we apply curriculum learning by progressively increasing the number of regions used for inpainting during training. In early stages, the surrounding pixels are used to provide context for inpainting. At the end of training, the model is forced to paint the entire black-and-white lineart. This feature is implemented by sampling the proportion of regions to inpaint $p = max(\epsilon, u),\ u \sim \mathcal{U}(0, 1)$ where ϵ is linearly annealed over time $\epsilon = 0.9 \rightarrow 0$.

4 Implementation

StencilTorch is trained using the PyTorch library for fast prototyping and iterations. Pytorch is, however, not suited for building client-side Web Applications. More actions have to be taken.

Model Export. We exported the PyTorch model as an ONNX model using the TorchScript intermediate representation. Some layers must be adapted to account for supported operations between the two frameworks. The ONNX model is then transpiled into a TensorFlow model that is further processed to be exported as a TensorFlowJS instance. This final instance of the model can be used in a client-side Web App.

Web Application. We also implement a Web Application that provides similar tooling to digital painting software, such as a brush, an eraser, a color picker, a color wheel with different brush sizes, and a canvas. A screenshot of the Web App is provided in Fig. 1. The user is invited to draw or import a lineart, a mask, and color hints. The inputs are processed by the TensorFlowJS model and outputted on a result canvas in real-time. The user can download every input and output for reproduction or integration with their favorite digital art tool.

5 Evaluation Setup

Data. We evaluate and train our method on our custom dataset containing illustrations scrapped from the web and filtered manually, 21,930 for training, and 3,545 for test. The images are resized to 512 on their smallest side and randomly cropped to 512×512 during training, and center cropped at test time.

Metrics. To measure the perceptual quality of the generated images, the GAN literature reports evaluations of both objective and subjective metrics. Art is a subjective matter, human evaluation is required.

Concerning the objective metrics, we measure two neural feature-based scores in our test set, the FID [15], and the LPIPS [44]. Those metrics uses an ImageNet [34] pretrained neural network, respectively InceptionNet [39], and AlexNet [24], to measure similarities between a pair of images, a fake one, the generated image, and a real one, the target, in feature space.

To assess subjective preferences, we conducted a user study to evaluate an MOS. The study consists of showing colorized linearts to the user whose task is to rate the quality of the illustration on a scale from 1 (bad) to 5 (excellent). Our study consists of 20 images for each model. Our study population comprises 46 individuals aged from 16 to 30, with 26% women and 35% experienced in drawing or colorization.

Baseline. We evaluate StencilTorch against previous works: PaintsChainer [31], Ci et al. [5], Zhang et al. [45], and PaintsTorch [12].

Training. The models are trained end-to-end using the AdamW optimizer [27] with a learning rate $\alpha = 1e - 4$ and beta parameters $\beta_1 = 0.5$ and $\beta_2 = 0.9$. They are trained for 40 epochs using a batch size of 32 on each of the four GPUs during 24 hours straight.

Measurements. All experiments and measurements are realized on a DGX1-station from NVidia equipped with an Intel Xeon E5-2698 20-Core Processor, 512 Go of DDR4 RAM, and four V100 GPUs with 32 Go of VRAM each.

Table 1. Benchmark and ablation of StencilTorch against previous works [5,12,31,45]. The table reports both FID and LPIPS evaluations. BN stands for Batch Normalization, G for Guide network, and C for Curriculum Learning. In average StencilTorch improves previous work. The evaluation use different amount of color hints, No Hints, regular Hints, and Full Hints.

Model	No Hint	Hints	Full Hints	Mean
FID ↓				
Zhang et al.	134.06	274.87	242.58	245.33
PaintsChainer	54.97	99.63	112.16	93.02
Ci et al.	52.48	96.22	106.73	85.14
PaintsTorch	51.54	95.71	98.37	81.87
StencilTorch	**51.16**	94.40	106.05	**81.63**
StencilTorch + BN	70.50	118.82	106.33	96.78
StencilTorch + G	85.00	**91.60**	**93.80**	89.98
StencilTorch + G + C	103.12	159.27	153.42	136.03
LPIPS ↓				
Zhang et al.	0.28	0.46	**0.26**	0.37
PaintsChainer	0.28	0.71	0.60	0.54
Ci et al.	0.23	0.62	0.59	0.48
PaintsTorch	0.18	0.59	0.56	0.44
StencilTorch	**0.16**	0.51	0.58	0.40
StencilTorch + BN	0.19	0.50	0.56	0.41
StencilTorch + G	0.31	0.50	0.58	0.46
StencilTorch + G + C	0.21	**0.30**	0.55	**0.36**

Table 2. Mean Opinion Scores, and the 95% confidence t-test p-values for the mean comparing StencilTorch to previous works [5,12,31,45] on our curated test set. StencilTorch improves previous contributions.

Model	MOS ↑	STD ↓	p-value ↓
PaintsChainer	1.79	0.51	$6.04e^{-23}$
Ci et al.	2.18	0.56	$7.72e^{-18}$
Zhang et al.	2.83	0.67	$9.84e^{-08}$
PaintsTorch	3.05	0.42	$9.15e^{-09}$
StencilTorch	**3.71**	**0.28**	

6 Results

We quantitatively in Table 1, 2, and qualitatively in Fig. 2, 5, and 7 show that StencilTorch outperforms previous works [5,12,31,45] in FID, LPIPS and MOS.

Fig. 5. Mosaic of inputs and output pairs generated with one StencilTorch pass. The top image represent the colorization, bottom right the lineart, bottom center the inpainting mask and bottom right the color hints.

Benchmark. StencilTorch is benchmarked against previous contributions in FID, LPIPS and MOS evaluations on our curated test set. One challenge for such evaluation is the generation of consistent color hints for every lineart of the test set. We propose to evaluate every approach using no hints, regular hint sampling, and full hints, meaning dense color sampling from the color region extracted during the input generation. On average, our approach improves previous work by Zhang et al. [45], PaintsChainer [31], Ci et al. [5], and PaintsTorch [12].

Ablation. The ablation study evaluates the impact of our design choices. We report the FID and the LPIPS on our test set. Our curriculum strategy helps the generator handle color hints on every of the three hint quantities we evaluated.

Limitations. Although our approach allows the generation of illustrations given a black-and-white lineart, a mask, and a hint map, the generated output often lacks shadows and textures compared to previous work by Hati et al. [12]. Such phenomena can be observed in Fig. 7. The flat coloring produced by our method is certainly due to the guide network involved in the entire decoding process during training. It seems, however, that the problem fades away when our model is used for inpainting as it was intended (see Fig. 2).

Fig. 6. Impact of the hint concentration on StencilTorch output. The lineart and mask are shown in the first column. Column 2 to 4 demonstrate different concentration, from no hint to a sufficient amount. Top rows represent the outputs, and bottom rows the hints. The last illustration is manually refined by an artist.

Fig. 7. Comparison of StencilTorch, PaintsTorch [12], and PaintsChainer [31] from left to right given the same lineart and hint map. The illustration generated by StencilTorch is the result of a single pass without any user in the loop. StencilTorch and PaintsTorch provide cleaner outputs. The colors of StencilTorch appear flat but do not present artifacts in comparison to PaintsTorch. The output of our model can be refined in collaboration with the user using inpainting mode.

7 Conclusion

Our work StencilTorch addresses the need for AI-driven tools that naturally integrate into the artist workflow and allow fast prototyping and iteration in collaboration with the machine. While current approaches have focused on improving the generation of user-guided anime lineart colorization, we explored the use of inpainting masks. This reformulation of the colorization process allows natural workflows to emerge. The output of a first pass is a potential input for a second. Our study demonstrates that our approach beats previous work on subjective metrics, FID, LPIPS, and objective metrics, MOS.

Future Work. The use of curriculum learning to progressively inpaint the lineart does not seem to generate good quality illustrations when no hints are provided as shown in Fig. 6. Recent advances in the domain of Denoising Diffusion Probabilistic Model (DDPM) such as DALL-E 2 [32] demonstrate unprecedented performance in image generation while providing natural conditioning. In future work, we want to explore the use of such a technique for automatic anime lineart colorization in the continuity of our quest for human-computer collaboration.

References

1. Anonymous, community, D., Branwen, G.: Danbooru 2020: A large-scale crowd-sourced and tagged anime illustration dataset, January 2021. https://www.gwern.net/Danbooru2020
2. Arjovsky, M., Chintala, S., Bottou, L.: Wasserstein generative adversarial networks. In: Proceedings of the 34th International Conference on Machine Learning. Proceedings of Machine Learning Research, vol. 70, pp. 214–223. PMLR, International Convention Centre, Sydney, Australia, 06–11 Aug 2017. https://proceedings.mlr.press/v70/arjovsky17a.html
3. Bengio, Y., Louradour, J., Collobert, R., Weston, J.: Curriculum learning. In: Proceedings of the 26th Annual International Conference on Machine Learning, pp. 41–48. ICML 2009, Association for Computing Machinery, New York, NY, USA (2009). https://doi.org/10.1145/1553374.1553380
4. Chollet, F.: Xception: Deep learning with depthwise separable convolutions. In: 2017 IEEE Conference on Computer Vision and Pattern Recognition (CVPR), pp. 1800–1807 (2017). https://doi.org/10.1109/CVPR.2017.195
5. Ci, Y., Ma, X., Wang, Z., Li, H., Luo, Z.: User-guided deep anime line art colorization with conditional adversarial networks. In: Proceedings of the 26th ACM International Conference on Multimedia, pp. 1536–1544. MM 2018, Association for Computing Machinery, New York, NY, USA (2018). https://doi.org/10.1145/3240508.3240661
6. Elman, J.L.: Learning and development in neural networks: the importance of starting small. Cognition **48**(1), 71–99 (1993)
7. Frans, K.: Outline colorization through tandem adversarial networks. CoRR abs/1704.08834 (2017). arxiv:1704.08834
8. Furusawa, C., Hiroshiba, K., Ogaki, K., Odagiri, Y.: Comicolorization: semi-automatic manga colorization. In: SIGGRAPH Asia 2017 Technical Briefs, SA 2017, Association for Computing Machinery, New York, NY, USA (2017). https://doi.org/10.1145/3145749.3149430

9. Goodfellow, I.J., et al.: Generative adversarial nets. In: Proceedings of the 27th International Conference on Neural Information Processing Systems, vol. 2, pp. 2672–2680. NIPS 2014, MIT Press, Cambridge, MA, USA (2014). https://dl.acm.org/doi/10.5555/2969033.2969125
10. Gulrajani, I., Ahmed, F., Arjovsky, M., Dumoulin, V., Courville, A.C.: Improved training of Wasserstein GANs. In: Advances in Neural Information Processing Systems, vol. 30. Curran Associates, Inc. (2017), https://proceedings.neurips.cc/paper/2017/file/892c3b1c6dccd52936e27cbd0ff683d6-Paper.pdf
11. Hacohen, G., Weinshall, D.: On the power of curriculum learning in training deep networks. In: Proceedings of the 36th International Conference on Machine Learning, ICML 2019, 9–15 June 2019, Long Beach, California, USA. Proceedings of Machine Learning Research, vol. 97, pp. 2535–2544. PMLR (2019). https://proceedings.mlr.press/v97/hacohen19a.html
12. Hati, Y., Jouet, G., Rousseaux, F., Duhart, C.: PaintsTorch: a user-guided anime line art colorization tool with double generator conditional adversarial network. In: European Conference on Visual Media Production. CVMP 2019, Association for Computing Machinery, New York, NY, USA (2019). https://doi.org/10.1145/3359998.3369401
13. He, K., Zhang, X., Ren, S., Sun, J.: Deep residual learning for image recognition. In: Proceedings of the IEEE Conference on Computer Vision and Pattern Recognition (CVPR), June 2016
14. Hensman, P., Aizawa, K.: CGAN-based manga colorization using a single training image. In: 2017 14th IAPR International Conference on Document Analysis and Recognition (ICDAR), vol. 3, pp. 72–77. IEEE Computer Society, Los Alamitos, CA, USA, Nov 2017. https://doi.org/10.1109/ICDAR.2017.295, https://doi.ieeecomputersociety.org/10.1109/ICDAR.2017.295
15. Heusel, M., Ramsauer, H., Unterthiner, T., Nessler, B., Hochreiter, S.: GANs trained by a two time-scale update rule converge to a local Nash equilibrium. In: Proceedings of the 31st International Conference on Neural Information Processing Systems, pp. 6629–6640. NIPS'17, Curran Associates Inc., Red Hook, NY, USA (2017)
16. Hinton, G.E., Salakhutdinov, R.R.: Reducing the dimensionality of data with neural networks. Science **313**(5786), 504–507 (2006)
17. Ho, J., Jain, A., Abbeel, P.: Denoising diffusion probabilistic models. Adv. Neural. Inf. Process. Syst. **33**, 6840–6851 (2020)
18. Ioffe, S., Szegedy, C.: Batch normalization: accelerating deep network training by reducing internal covariate shift. In: Bach, F., Blei, D. (eds.) Proceedings of the 32nd International Conference on Machine Learning. Proceedings of Machine Learning Research, vol. 37, pp. 448–456. PMLR, Lille, France, 07–09 July 2015. https://proceedings.mlr.press/v37/ioffe15.html
19. Kandinsky, W., Sadleir, M.: Concerning the Spiritual in Art. Dover Publications, New York (1977). (oCLC: 3042682)
20. Karras, T., Aila, T., Laine, S., Lehtinen, J.: Progressive growing of GANs for improved quality, stability, and variation. CoRR abs/1710.10196 (2017). arxiv.org:1710.10196
21. Kim, H., Jhoo, H.Y., Park, E., Yoo, S.: Tag2pix: line art colorization using text tag with Secat and changing loss. In: 2019 IEEE/CVF International Conference on Computer Vision (ICCV), pp. 9055–9064 (2019). https://doi.org/10.1109/ICCV.2019.00915

22. Kingma, D.P., Welling, M.: Auto-encoding variational bayes. In: 2nd International Conference on Learning Representations, ICLR 2014, Banff, AB, Canada, April 14–16, 2014, Conference Track Proceedings (2014). arxiv.org:1312.6114
23. Kingma, D.P., Dhariwal, P.: Glow: Generative flow with invertible 1×1 convolutions. In: Advances in Neural Information Processing Systems, vol. 31 (2018)
24. Krizhevsky, A., Sutskever, I., Hinton, G.E.: ImageNet classification with deep convolutional neural networks. Commun. ACM **60**(6), 84–90 (2017)
25. Lim, J.H., Ye, J.C.: Geometric GAN (2017)
26. Liu, Y., Qin, Z., Wan, T., Luo, Z.: Auto-painter: cartoon image generation from sketch by using conditional Wasserstein generative adversarial networks. Neurocomputing **311**, 78–87 (2018)
27. Loshchilov, I., Hutter, F.: Fixing weight decay regularization in adam. CoRR abs/1711.05101 (2017). arxiv.org:1711.05101
28. Zhang, L., Ji, Y., Liu, C.: DanbooRegion: an illustration region dataset. In: Vedaldi, A., Bischof, H., Brox, T., Frahm, J.-M. (eds.) ECCV 2020. LNCS, vol. 12358, pp. 137–154. Springer, Cham (2020). https://doi.org/10.1007/978-3-030-58601-0_9
29. Mirza, M., Osindero, S.: Conditional generative adversarial nets. arXiv preprint arXiv:1411.1784 (2014)
30. Miyato, T., Kataoka, T., Koyama, M., Yoshida, Y.: Spectral normalization for generative adversarial networks. CoRR abs/1802.05957 (2018). arxiv.org:1802.05957
31. Pixiv: Pelica Paint. https://petalica-paint.pixiv.dev/index_en.html (2017)
32. Ramesh, A., Dhariwal, P., Nichol, A., Chu, C., Chen, M.: Hierarchical text-conditional image generation with clip LATENTs. arXiv preprint arXiv:2204.06125 (2022)
33. Ronneberger, O., Fischer, P., Brox, T.: U-net: convolutional networks for biomedical image segmentation. In: Navab, N., Hornegger, J., Wells, W.M., Frangi, A.F. (eds.) MICCAI 2015. LNCS, vol. 9351, pp. 234–241. Springer, Cham (2015). https://doi.org/10.1007/978-3-319-24574-4_28
34. Russakovsky, O., et al.: ImageNet large scale visual recognition challenge. Int. J. Comput. Vision (IJCV) **115**(3), 211–252 (2015). https://doi.org/10.1007/s11263-015-0816-y
35. Saito, M., Matsui, Y.: Illustration2vec: a semantic vector representation of illustrations. In: SIGGRAPH Asia 2015 Technical Briefs, pp. 5:1–5:4. SA 2015, ACM, New York, NY, USA (2015). https://doi.org/10.1145/2820903.2820907
36. Sangkloy, P., Lu, J., Fang, C., Yu, F., Hays, J.: Scribbler: controlling deep image synthesis with sketch and color. In: 2017 IEEE Conference on Computer Vision and Pattern Recognition, CVPR 2017, Honolulu, HI, USA, 21–26 July 2017, pp. 6836–6845. IEEE Computer Society (2017). https://doi.org/10.1109/CVPR.2017.723
37. Shi, W., et al.: Real-time single image and video super-resolution using an efficient sub-pixel convolutional neural network. In: Proceedings of the IEEE Conference on Computer Vision and Pattern Recognition, pp. 1874–1883 (2016)
38. Simonyan, K., Zisserman, A.: Very deep convolutional networks for large-scale image recognition. In: Bengio, Y., LeCun, Y. (eds.) 3rd International Conference on Learning Representations, ICLR 2015, San Diego, CA, USA, 7–9 May 2015, Conference Track Proceedings (2015). arxiv.org:1409.1556
39. Szegedy, C., et al.: Going deeper with convolutions. In: Proceedings of the IEEE Conference on Computer Vision and Pattern Recognition, pp. 1–9 (2015)

40. Vaswani, A., et al.: Attention is all you need. In: Guyon, I., et al. (eds.) Advances in Neural Information Processing Systems, vol. 30. Curran Associates, Inc. (2017). https://proceedings.neurips.cc/paper/2017/file/3f5ee243547dee91fbd053c1c4a845aa-Paper.pdf
41. Winnemöller, H., Kyprianidis, J.E., Olsen, S.C.: XDOG: an extended difference-of-Gaussians compendium including advanced image stylization. Comput. Graph. **36**(6), 740–753 (2012). https://doi.org/10.1016/j.cag.2012.03.004, www.sciencedirect.com/science/article/pii/S009784931200043X, 2011 Joint Symposium on Computational Aesthetics (CAe), Non-Photorealistic Animation and Rendering (NPAR), and Sketch-Based Interfaces and Modeling (SBIM)
42. Xie, S., Girshick, R.B., Dollár, P., Tu, Z., He, K.: Aggregated residual transformations for deep neural networks. In: 2017 IEEE Conference on Computer Vision and Pattern Recognition (CVPR), pp. 5987–5995 (2017)
43. Zhang, L., Ji, Y., Lin, X., Liu, C.: Style transfer for anime sketches with enhanced residual u-net and auxiliary classifier GAN. In: 2017 4th IAPR Asian Conference on Pattern Recognition (ACPR), pp. 506–511 (2017). https://doi.org/10.1109/ACPR.2017.61
44. Zhang, R., Isola, P., Efros, A.A., Shechtman, E., Wang, O.: The unreasonable effectiveness of deep features as a perceptual metric. In: CVPR (2018)
45. Zhang, R., et al.: Real-time user-guided image colorization with learned deep priors. ACM Trans. Graph. **36**(4), 1–11 (2017). https://doi.org/10.1145/3072959.3073703

UnseenNet: Fast Training Detector for Unseen Concepts with No Bounding Boxes

Asra Aslam(✉) and Edward Curry

Data Science Institute, University of Galway, Galway, Ireland
asra.aslam.7@gmail.com, edward.curry@nuigalway.ie

Abstract. Training of object detection models using less data is currently the focus of existing N-shot learning models in computer vision. Such methods use object-level labels and takes hours to train on unseen classes. There are many cases where we have large amount of image-level labels available for training and cannot be utilized by few shot object detection models for training. There is a need for a machine learning framework that can be used for training any unseen class and can become useful in real-time situations. In this paper, we proposed an "Unseen Class Detector" that can be trained within a short time for any possible unseen class without bounding boxes with competitive accuracy. We build our approach on "Strong" and "Weak" baseline detectors, which we trained on object detection and image classification datasets, respectively. Unseen concepts are fine-tuned on the strong baseline detector using only image-level labels and further adapted by transferring the classifier-detector knowledge between baselines. We use semantic as well as visual similarities to identify the source class (i.e. Sheep) for the fine-tuning and adaptation of unseen class (i.e. Goat). Our model (UnseenNet) is trained on the ImageNet classification dataset for unseen classes and tested on an object detection dataset (OpenImages). UnseenNet improves the mean average precision (mAP) by 10% to 30% over existing baselines (semi-supervised and few-shot) of object detection. Moreover, training time of proposed model is < 10 min for each unseen class.

Keywords: Weakly supervised learning · Object detection · Transfer learning · Domain adaptation · Computer vision

1 Introduction

Detection of objects is one of the significant challenges in computer vision. Multiple object detection models have been proposed to date, including Fast R-CNN [7], Faster R-CNN [25], SSD [21], RetinaNet [19], and YOLO [23]. The conventional trend is to train such high-performance models on images with bounding

https://github.com/Asra-Aslam/UnseenNet.

W. Q. Yan et al. (Eds.): IVCNZ 2022, LNCS 13836, pp. 18–32, 2023.
https://doi.org/10.1007/978-3-031-25825-1_2

box annotations for weeks and detect objects only for certain classes present in object detection datasets (like Pascal VOC [6]: 20 classes, MCOCO [20]: 80 classes, OpenImages (OID) [13]: 600 classes). The problem arrive not because we are highly dependent on annotated object bounding boxes based datasets but we ignore the classification data available on the web, which can be accumulated on request using labels of classes. Current improvements are based on few shot approaches [3,11,18,27,32–35], and where existing image-level labels cannot be utilized. Moreover training time of these approaches is increasing to days to increase performance. Thus, such approaches are not suitable for real-time applications.

Several works [9,17,29,31] are bringing a lot of potential in the area of *unseen* concepts by converting classifiers to detectors using image-level labels. However, accuracy of these approaches are quite low as current improvements are focused on few shot learning based models. In this work, we attempt to make detectors using classification data while providing better accuracy and responding to unseen classes in real-time.

Our work aims to answer the question: Can we train detectors for any possible unseen concept (with only image-level labels) within a limited amount of time while providing competitive accuracy?

We divide the proposed framework "UnseenNet" into two parts; first, we train two detectors "Strong Baseline" and "Weak Baseline" (shown in Figs. 1 and 2) on 100 classes for weeks (∼300 epochs). Here, we train *strong baseline* on object detection datasets (MCOCO and OID), and *weak baseline* on image classification dataset (ILSVRC [26]). We use pipeline of YOLO [24] with MobileNet [10] backbone to train detectors for fast detection and classification.

The second part is designated for the training of unseen classes. Here, we collect images (with no bounding boxes) from the web using only names of unseen classes. Then we choose a class (i.e., seen class) from *strong baseline detector* similar to *unseen* class, fine-tune it on collected classification data, then finally adapt classifier to detector. We conduct experiments to evaluate the performance of proposed framework (shown in Fig. 3) in low response time for unseen concepts. We show that UnseenNet improves the mean average precision (mAP) by 10 to 30% over existing approaches within 10 min of training time.

2 Task Definition

In the case of our training detector for unseen concepts without bounding boxes, we assume that we have access to the object detection datasets (i.e., training images with bounding box annotations for the small number of classes) and image classification datasets (i.e., training images with only image-level labels for the large but finite number of classes). Our objective is to train detectors for any unseen concept (i.e., an infinite number of classes) without bounding box annotations within a short time to make them useful for real-time applications.

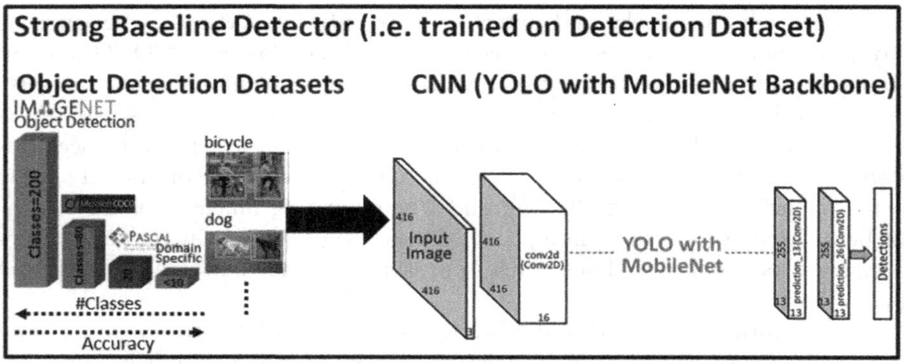

Fig. 1. Strong baseline detector trained on object detection datasets (i.e. images with bounding boxes) which contains limited number of classes to provide high accuracy [1].

3 Related Work

Training of Object Detection Models: Existing object detection models (YOLO [23], RetinaNet [19], SSD [21]) require images with bounding box annotations for training, thus falls in the category of fully-supervised models. Such models require training for weeks for a limited number of classes, thus cannot process any new/unseen class. We attempt to overcome this limitation by providing training on unseen concepts without bounding boxes in a short time.

Use of Datasets: Most popular object detection datasets include Pascal VOC [6], MCOCO [20], OID [13], and ILSVRC detection challenge [26]. Though these datasets show promising results on the training of object detection models, indeed fail to fulfill the data requirements of object detection models. On the other side, image classification datasets (ImageNet [5]: 1000, MNIST [16]: 10, CIFAR-10 [14]: 10 classes) could be useful for training weak detectors regardless of having only image-level labels. Presently, significance of detection/classification datasets only appears in the comparison of deep learning models. We propose settings of utilizing all existing (object-detection, domain-specific, image-classification) datasets to train detectors for unseen classes of real-world scenarios.

Weakly Supervised Detection: Weakly supervised learning [2,12] is an emerging solution for large-scale unseen concepts. Weakly supervised learning also formulated as a Multiple Instance Learning (MIL) problem. $WSOD^2$ [36] uses bottom-up object evidences and top-down classification output with an adaptive training mechanism. Multiple MIL based methods appear in literature [4,8,28] for the weakly supervised object localization. Most of these methods use pre-trained ImageNet [5] for initialization. However, our model uses the fastest MobileNetv3 [10] as backbone and own trained "Strong Baseline Detector" for initialization. Such methods are evaluated on Pascal VOC [6], which is known in computer vision for a long time; its classes shouldn't be considered unseen.

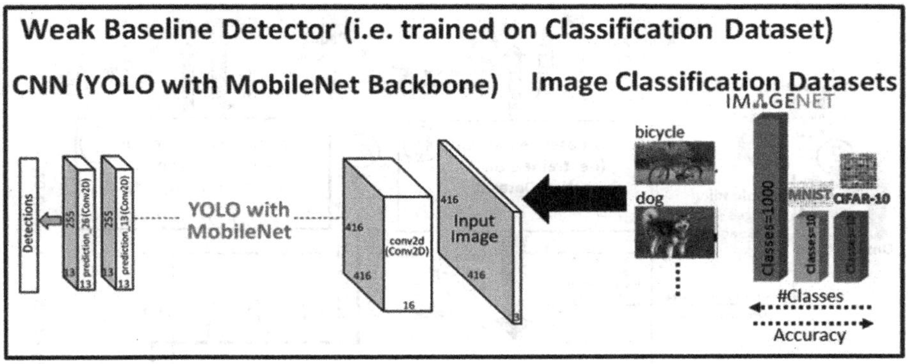

Fig. 2. Weak baseline trained on image classification datasets i.e. images with only labels (no bounding boxes).

Another area of related work includes the knowledge transfer from source to target domain. Large Scale Detection through Adaptation (LSDA) [9] based approaches transforms classifiers into object detectors using only image-level labels. Tang et al. [29] improve LSDA by incorporating informed visual knowledge and semantic similarities. Uijlings et al. [31] proposed a revisit knowledge transfer for detectors training in the weakly supervised settings and outperformed all the baselines. We utilize the concept of transferring knowledge between classifier and detector in our work. Some of the significant advantages of our work includes better accuracy, short training time, flexibility for training any unseen (new) concept and no requirement of bounding boxes.

N-Shot Learning: It is a branch of machine learning which handles the challenge of training models with only a small amount of data. In terms of terminology, we refer to it as N-way K-Shot-classification, where N is the number of classes and K is the number of labeled training samples from each class. Recently few-shot learning based approaches [3,11,18,27,32–35] are achieving promising results for task of object detection. Such approaches requires multiple stages and more training time. Moreover, these methods does not utilize image-level labels which can be easily acquired rather than object-level labels.

4 UnseenNet Framework

We propose an "UnseenNet" detector (shown in Fig. 3) which allow user to construct detectors for unseen classes without the need for detection data (no bounding boxes) within the short training time. Our model is based on making use of existing object detection datasets of bounded vocabulary (consists of *seen* concepts) to construct detectors for *unseen* concepts (i.e. unbounded vocabulary) by using the differences between a weak detector (trained on image classification dataset) and a strong detector (trained on object detection datasets). We

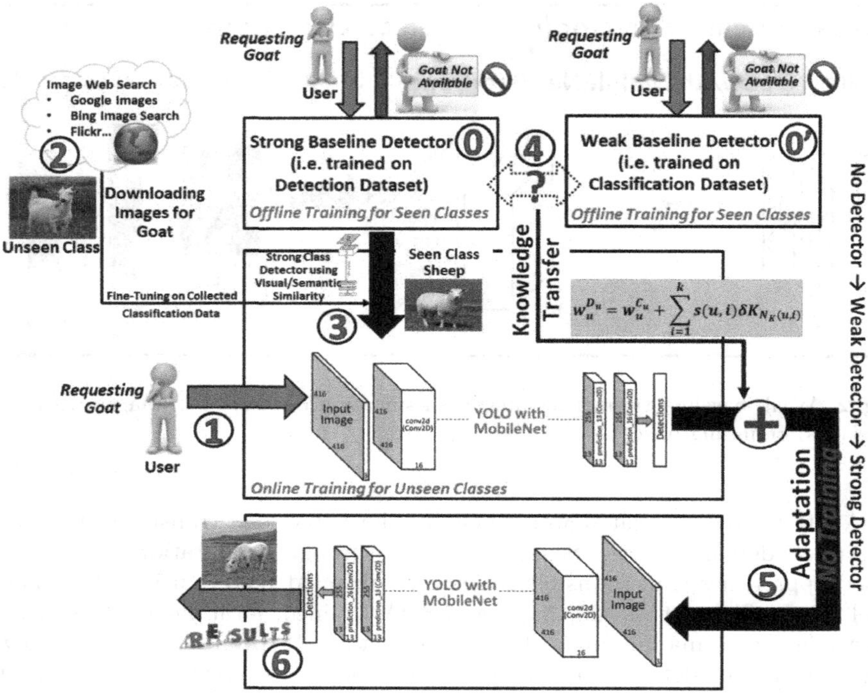

Fig. 3. An illustration of "UnseenNet" model. An illustration of our "UnseenNet" model. We make use of existing object detection datasets (have bounding box annotations) and image classification datasets (have only image-level labels) and train two separate detectors "Strong Baseline Detector (0)" and "Weak Baseline Detector (0')" respectively, in advance. (1) On request of any *unseen* concept, (2) we download images using only image-level labels (like goat). (3) *Strong baseline detector* is then fine-tuned on collected images of unseen concepts (like a goat) on the most semantically similar class (like sheep). (4) At this stage, we also compute the visual similarity of the constructed unseen class detector (trained on classification data) with seen classes of *weak baseline detector*, combine it with semantic similarities, and select top-k classes ranked on comprehensive similarities. (5) Finally, we transfer the knowledge of classifier-detector differences of top classes to the constructed unseen class detector, adapt it into the stronger detector without further training, and (6) return detection results. Note: Baseline detectors (0, 0') are trained offline while other parts (1–6) gets train on unseen class request.

describe below the construction of our strong and weak baseline detectors offline for *seen* concepts and training of detectors online for *unseen* concepts while investigating the object detection model's training time, which we are referring to as the *response-time* of our model on *unseen* concepts.

4.1 Training Baseline Detector Offline for Seen Concept (with Bounded Vocabulary)

First, we setup an architecture of YOLO with MobileNet backbone and construct two baseline detectors as follows:

Strong Baseline Detector (D_S) shown in Fig. 1 is a $|K|$ class detector trained on existing object detection datasets. It is a detector that is trained on strong labels (i.e., bounding box annotations). Presently we have taken 100 classes (like LSDA) by considering all classes of Microsoft COCO (80 classes [20]) and 20 classes of OID [13]. Please note that 20 classes of Pascal VOC [6] are also present in Microsoft COCO.

Weak Baseline Detector (D_W) shown in Fig. 2 is another $|K|$ class detector trained on image classification dataset. We trained it on weak labels (i.e., images-level labels). In this detector, we consider the same classes on which we trained the previous *Strong Baseline*, but we use ILSVRC [26] classification data.

4.2 Training Online Detector for Unseen Concept (for Unbounded Vocabulary)

On request of an unseen class (u), say *goat*, first our model (shown in Fig. 3) provides an environment to collect images for 'goat" from the Web using Google Images[1], Flickr[2], or Bing Image[3] search. Second, it use the "Strong Baseline Detector", and set up detector for *unseen* class (i.e. goat) using similar *seen* class (like sheep). It is important to note that a similar class (like sheep for goat) can be chosen only using semantic similarity at this stage as visual features of an unseen class cannot be computed before training. Next, we fine-tune the detector on images collected for "goat". Now we have a new detector having $|K|$ classes that can detect *goat*. Since *goat* class is trained only on image-level labels, we call it a *weak detector* or simply a classifier "C_u" for unseen class. At this point, our model's response time for *unseen* concepts is equal to the time for fine-tuning.

We presume that fine-tuning induce a *specific category* bias transformation in the detection network towards class "goat". Moreover, this network already encodes a *generic "background" category* due to previously trained on detection data (because of strong baseline), which is positive, as this will automatically make the new detector much more effective in localizing the new class without detection data. Finally, the previous classifier C_u adapts into a corresponding detector D_u. This assumes that "difference between classification and detection of a target object category has a positive correlation with similar categories" detailed in large scale detection approaches [9,29].

Suppose weights of the output layer of D_S (Strong Baseline Detector) and D_W (Weak Baseline Detector) are w^{D_S} and w^{D_W} respectively. We know that

[1] https://github.com/hardikvasa/google-images-download.
[2] https://www.flickr.com/services/api/.
[3] https://pypi.org/project/bing-image-downloader/.

for any *seen* category $i \in K$, final detection weights should be computed as $w_i^{D_S} = w_i^{D_W} + \delta_{K_i}$, where δ_{K_i} is the difference in weights for the seen category.

By using this knowledge difference and denoting the kth nearest neighbor in set K of category u as $N_K(u,k)$, we adapt the final output detection weights for categories u as:

$$w_u^{D_u} = w_u^{C_u} + \sum_{i=1}^{k} s(u,i)\delta K_{N_K(u,i)} \tag{1}$$

where $k \leq |K|$, and $s(u,i)$ denotes the similarity of seen class (i) with unseen class (u). Equation 1 uses the *weighted nearest neighbor* scheme ([29,30], where weights are assigned to seen categories based on how similar they are to the unseen category. We select top-k weighted nearest neighbor categories $(s(u,i))$ using Eq. 2. Other than the semantic similarity, we also compute the visual similarity at this stage by using the minimal Euclidean distance between the detection parameters of the last layers of detectors D_W and C_u. Suppose K_v is the set of visually similar (s_v) categories and K_s is the set of semantically similar (s_s) categories, then comprehensive similarity $s(u,i)$ for unseen category with seen categories is evaluated as:

$$s(u,i) = \alpha s_v(u,i) + (1-\alpha)s_s(u,i),\ i \in \{K_v \cap K_s\} \tag{2}$$

where $\alpha \in [0,1]$ is a parameter introduced in literature [29,30] to control the influence of the two similarity measures. We use minimal Euclidean distance between feature distributions of the last layers as visual similarity [9] and naive *path-based* semantic similarity measure of WordNet [22] along with a weighted average scheme to compute the comprehensive similarity $(s(u,i))$ scores.

Finally, we call this adapted detector "D_u" a *strong detector* for unseen class. We analyze the response-time of our model in Sect. 5 from the stage of *no detector* to *weak detector* (C_u), and eventually to a *strong detector* (D_u).

5 Experiments

5.1 Implementation Details

Data Preparation We trained Strong and Weak Baseline Detectors on seen classes offline and performed experiments on unseen classes while having training time constraints

Seen Classes
Strong Baseline Detector Training: In this case, We consider all 80 classes of Microsoft COCO [20] and 20 classes of Open Images OID [13] to train a strong baseline detector with bounding box annotations. We select 20 classes from OID by sorting its 600 classes on the basis number of images per class and considering the top 20 with the highest number of images available for training.

Weak Baseline Detector Training: Here, We take the same 100 seen classes, retrieve images with labels from the ISLVRC [26] dataset (i.e., images have no

Table 1. The mean average precision (mAP) while using ILSVRC for Weak Level labels and MCOCO & OID for Strong Level labels. First, we show the performance for existing weakly-supervised methods. We also include the performance of semi-supervised LSDA. Row 5–7 shows our model's results on classification network, class invariant adaptation while fine-tuning specific class, then including Classifier to Detector Adaptation. We show the training time (10 min) our model takes to provide similar detection mAP. It is important to note that, inference time of UnseenNet is 9.2fps.

Method			mAP	Response time
LSDA [9]			16.33	5.5 h
Visual knowledge transfer [29]			20.03	> 5.5 h
ZSDTR [37]			20.16	–
Knowledge transfer MI [31]			23.3	–
UnseenNet (Ours)	(Classification network with No Adapt)		22.82	5 min
			27.92	10 min
			27.04	50 min
	(Class Invariant Adapt & Specific Class Fine-Tuning)		33.36	5 min
			42.03	10 min
			39.77	50 min
	(Class Invariant Adapt, Specific Class Fine-Tuning, & Adapt)	Weighted Avg NN–5	36.17	5 min
			42.24	**10 min**
			39.79	50 min
		Weighted Avg NN–10	38.55	5 min
			42.36	**10 min**
			39.80	50 min
		Weighted Avg NN-100	39.91	5 min
			43.07	**10 min**
			39.88	50 min

bounding boxes), and train *weak baseline detector* by giving full image size in place of annotations.

Unseen Classes

We chose classes from the ILSVRC [26] that are also present in Open Images OID [13] (consist of 600 classes) and consist of reasonable number of testing images (>100). So that we can evaluate the model on an object detection dataset, which gets trained on image classification dataset. That is, we use the testing dataset of OID for unseen classes to serve as groundtruth in the evaluations. We also perform qualitative evaluations on additional 16 unseen classes that we downloaded from the web using Google Images API[4]. Such classes are not present in any dataset (Pascal VOC, MCOCO, OID, ImageNet etc.) to-date. This clearly proves our model's significance for unseen concepts (known or unknown).

In our experiments, we consider the pipeline of YOLOv3 [23,24] and MobileNetv3 [10] for fast detection and classification. Specifically, we used the three layers (38, 117, 165) from the MobileNetv3 (Small) within YOLO to make the prediction[5]. We trained our baseline detectors first on learning rate of 10^{-3}

[4] https://github.com/hardikvasa/google-images-download.

[5] https://github.com/david8862/keras-YOLOv3-model-set.

Table 2. Comparison with few-shot detection methods

Method	mAP on Splits		Shots
	5-Class	20-Class	
LSTD [3]	35.27	3.2	10
FSRW [11]	44.53	5.6	
Meta-RCNN [35]	48.33	8.7	
MetaDet [33]	44.4	7.1	
MPSR [34]	53.1	9.8	
TFA [32]	48.73	10.0	
FSCE [27]	57.37	11.9	
cos-FSOD [18]	54.77	20.3	
UnseenNet	**68.78**	**51.29**	0

till 100 epochs, then we used the decay type exponential till 200 epochs; finally, we used the 10^{-4} till 300 epochs as validation loss stopped decreasing near this point. However, for the training of our unseen classes, we used the constant learning rate of 10^{-4}, which could be increased in future experiments for faster results. We kept the slowest possible learning rate, as our model should serve as the base-work for handling dynamic unseen concepts in short training time. Finally, we utilize the benchmark object detection metrics project[6] to evaluate our detections with IOU = 0.5.

We assume it is essential to specify that ImageNet and Object detection datasets use different name for the same classes, so we are using the vocabulary of WordNet to give a single name to each class and also provide mappings of different datasets with our model. We used the path vector of WordNet for the semantic similarity measure. Visual similarity is simply computed using the minimal Euclidean distance of weights of the unseen class detector (trained on classification data) and weights of weak baseline detector, which is the same as described in LSDA [9]. To complete the weighted average scheme's evaluations over the simplified (visual and semantic) similarity measures, we also verify the value of parameter $\alpha = 0.6$, which is responsible for computation of the degree of similarity of the unseen category with seen categories.

We estimated the total number of epochs required to train the model for the designated training time by considering the batch size, number of available training images, and speed of our GPU for the completion of one step. The total number of epochs computed as:

$$epochs = \frac{ResponseTime}{((Num\ of\ Images/Batch\ Size) * t)} \tag{3}$$

where, "response time" denotes the total training time allowed, "Num of Images" is the number of available training images, and t is time GPU takes to complete one step, which is 0.465 s in our case. Here, "Num of Images/Batch-Size" is the

[6] https://github.com/Adamdad/Object-Detection-Metrics.

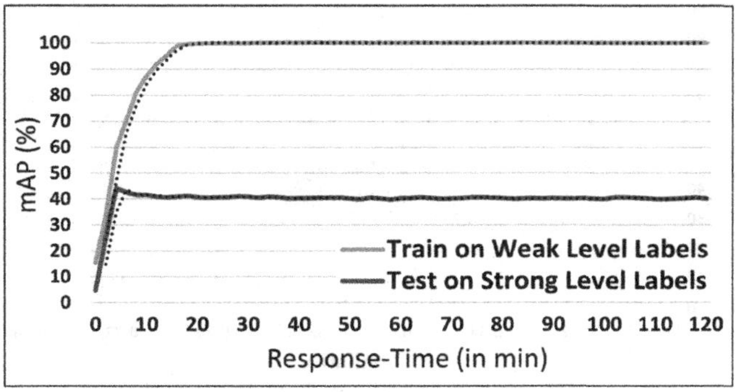

Fig. 4. Examples of mAP with Response-Time, For each "Unseen" category, we use the top-10 weighted average nearest neighbor "Seen" categories for adaptation. This shows that maximum mAP could be achieved in 10 min.

number of steps. We used default batch-size 16. We conducted experiments on NVIDIA TITAN Xp GPU (8 Core Processor×16), Driver 440.1.

5.2 Quantitative Evaluation on Unseen Categories

Comparative Analysis with Existing Models. We compare the performance of the UnseenNet in Table 1 against weakly supervised object detection models. We show mean average precision (mAP) for unseen categories along with required training time. We evaluate our model by considering different number (5, 10, and 100) of nearest neighbors of "unseen" categories with "seen" categories while using weighted average nearest neighbor scheme (Eq. 1).

The first 4 rows show the results of existing approaches including LSDA [9], its improved version with visual knowledge transfer [29], zero shot learning ZSDTR [37] (as we are also not giving any shots of bounding boxes), and revisiting knowledge transfer MI based approach [31]. We can observe that mean Average Precision (mAP) of existing weakly supervised approaches are low while there training time is very high (>5.5 h) or days and weeks in existing scenarios.

It is necessary to evaluate our model first by training only on classification data because we are using YOLOv3–MobileNetv3 [10,24] in contrast to R-CNN–AlexNet [15,25]. We show that this amendment improves the performance from 16.33 to 22.82. Here, we show the mAP for different response time (5 min, 10 min, 50 min). We choose these response times using testing and training (shown in Fig. 4) detail in Sect. 5.2.

Second, we show the mAP using Class Invariant Adapt (Strong Baseline Detector) and fine-tuning the nearest "seen" class on target "unseen" class classification data. Finally, we apply the specific class adaptation by using the weighted average of "N" nearest neighbor classes, where N could be 5, 10, and 100. This step does not require training. We show the final detection performance

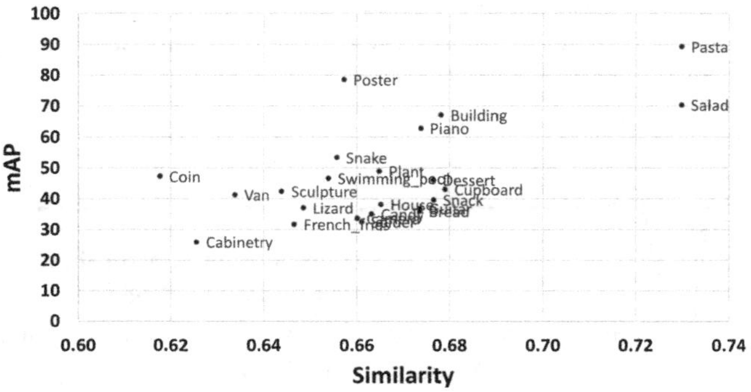

Fig. 5. mAP of unseen classes with similarity scores

(average on 100 classes) by indicating our model's total time. Best results indicate that we can reach from stage of **no** detector for unseen concepts to weak detector (mAP **42.03**) and strong detector (mAP **43.07**) within 10 min of training. Lastly, the testing/inference of proposed model is 9.2fps (Table 1).

In present case, UnseenNet does not require any shots of bounding box based annotations, and use only image-level labels for finetuning. However, existing few-shot object detection models are showing great promise by providing competitive performance with only few shouts of annotated bounding boxes. Thus, we compared our model performance with recent few-shot detection approaches [3,11,18,27,32–35]. Presently these models perform experiments by considering base classes of Pascal VOC for training and then novel classes also of Pascal VOC; and also in case of Microsoft COCO, base and novel classes belongs to same dataset. However, in our case we used Pascal VOC and Microsoft COCO classes in our strong baseline detector, thus we chose novel classes from OID dataset and still we get better performance than existing few-shot detection methods. Existing approaches used 5-class and 20-class novel splits, so we randomly generated 5-class pairs and 20-class pairs in our dataset of unseen classes. We can observe in Table 2 for case of 5-class splits we are getting mAP 68.78 which is >10% improvement over existing models. Similarly in case of 20 class splits our performance is mAP 51.09 which is greatest among existing approaches. Again, we trained on classification dataset (ImageNet) and tested on images Open Images dataset which consist of multiple objects in single image. It is important to note that we consider here performance of 10 shots of existing approaches which was best among all shots. Moreover we use only image-level labels thus UnseenNet does not need any shot.

Experimental Results with Response-Time. To retrieve the range of response-time effective in our model, we train each category until the point testing accuracy starts to decrease (to avoid over-fitting). We show average performance of all unseen concepts with training time in Fig. 4. We first train our

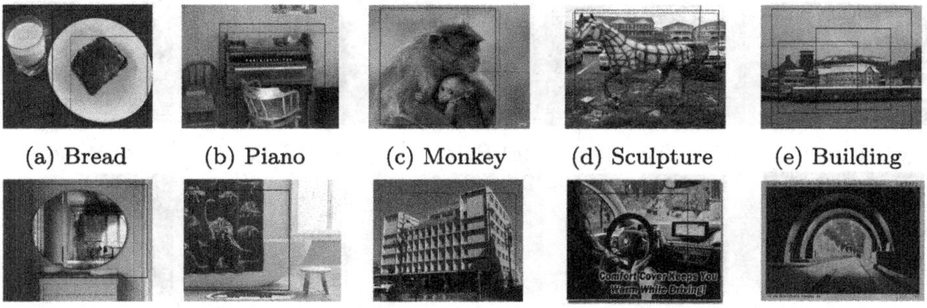

(a) Bread (b) Piano (c) Monkey (d) Sculpture (e) Building

(f) Mirror (g) Curtain (h) Hospital (i) Steering (j) Tunnel

Fig. 6. Examples of correct detections of our model on "Unseen" categories are shown in red color and groundtruth (taken from OID) in green. Second row unseen classes are downloaded online, and no groundtruth available to date. (Color figure online)

model on weak level labels (i.e., without bounding boxes) and then test on strong labels (i.e., with bounding boxes). Here weak labels are taken from ImageNet classification data and strong labels are taken from OID dataset. We observe that the maximum mAP of each class could be achieved within 10 min of training. After that, mAP decreases and remain constant. However, we recommend 10 min of training to attain maximum mAP 43.07 to avoid any unexpected reduction in mAP due to over-fitting. It is worth noting that mAP at 0 min training time is not zero due to the strong baseline detector for initialization.

Experimental Results with Unseen Concepts. We present an analysis (Fig. 5) mAP with similarities of unseen categories with *seen* categories (top-10) for few examples of our unseen classes. The simple average similarity score:

$$s_j = \frac{\sum_{i=1}^{m} s(j,i)}{m} \tag{4}$$

where m is 10 presently and $s(j,i)$ is the comprehensive similarity (shown in Eq. 2) between unseen (j) and seen (i) category computed using $\alpha = 0.6$. It shows if we have *unseen* classes (like building, pasta, salad) more similar to *seen* classes, then our model have high probability of giving high performance with the exception for small size objects or availability of less training data.

5.3 Qualitative Evaluation on Unseen Categories

We show visual examples of our model detections in Fig. 6. Examples of correct detections of our model on "Unseen" categories shown in red color and groundtruth (taken from OID) in green. Here Fig. 6 (a) – (e) includes classes of ILSVRC, and Fig. 6 (f)–(j) consist of additional unseen classes which not present in any object detection or image classification dataset to date. Correct detections

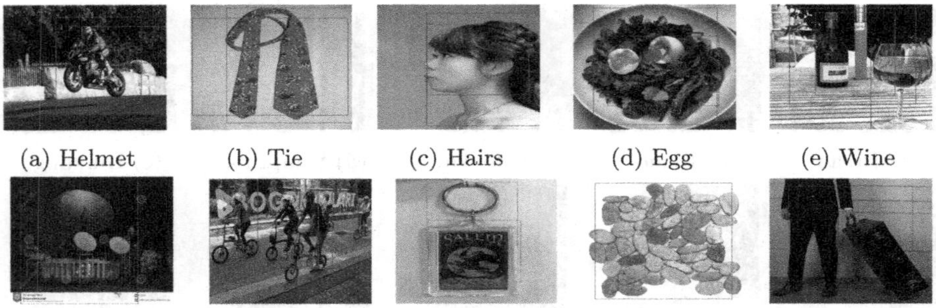

(a) Helmet (b) Tie (c) Hairs (d) Egg (e) Wine

(f) SatelliteDish (g) Pedestrian (h) Key (i) Stone (j) Handle

Fig. 7. Examples of incorrect detections (Label Object Correctly but Incorrect Localization) are shown in red and groundtruth in green. Second row unseen classes are downloaded online, and no groundtruth available to date. (Color figure online)

of unseen concepts verify that UnseenNet can be trained on any class within a 10 min of training. It also reduces the need to create large object detection datasets.

Some examples of incorrect detections are shown in Fig. 7, where Fig. 7 (a)–(e) includes classes of ILSVRC, and Fig. 7 (f)–(j) consist of additional unseen classes that our model downloaded online and are not present in any object detection dataset to date. This demonstrates that if we have "unseen" classes less similar to "seen" classes, then UnseenNet could label them *correctly* because of the training on classification data with *incorrect* localization due to absence of detection data.

6 Conclusions and Future Work

We presented an "UnseenNet" model that has the ability to construct a detector for any unseen concept without bounding boxes while training in a short time and providing competitive accuracy. We found that starting from a "strong baseline detector" trained on existing object detection datasets speed up the training rather than using only ImageNet pre-trained model to train unseen concepts. Moreover, in conjunction with semantic and visual similarity measures, classifier-detector conversion methods make our model more robust. Our evaluations demonstrate that *UnseenNet* outperforms the baseline approaches in terms of training time for any unseen class and improves the mAP from 10% to 30% over existing detection based methods.

In the future, UnseenNet could be improved with more effective detectors and classifiers. Presently, we provided the size of images as bounding boxes for training weak baseline detector. This could be improved by background extraction or segmentation approaches and provide better annotations for training. Lastly, Strong and Weak baseline detectors could include large number of *seen* classes to obtain more similar classes.

References

1. Aslam, A., Curry, E.: A survey on object detection for the internet of multimedia things (IOMT) using deep learning and event-based middleware: Approaches, challenges, and future directions. Image Vis. Comput. **106**, 104095 (2021)
2. Bilen, H., Vedaldi, A.: Weakly supervised deep detection networks. In: Proceedings of the IEEE CVPR, pp. 2846–2854 (2016)
3. Chen, H., Wang, Y., Wang, G., Qiao, Y.: LSTD: a low-shot transfer detector for object detection. In: Proceedings of the AAAI conference, vol. 32 (2018)
4. Cinbis, R.G., Verbeek, J., Schmid, C.: Weakly supervised object localization with multi-fold multiple instance learning. IEEE Trans. Pattern Anal. Mach. Intell. **39**(1), 189–203 (2016)
5. Deng, J., Dong, W., Socher, R., Li, L.J., Li, K., Fei-Fei, L.: Imagenet: a large-scale hierarchical image database. In: IEEE Conference on Computer Vision and Pattern Recognition, 2009. CVPR 2009, pp. 248–255. IEEE (2009)
6. Everingham, M., Van Gool, L., Williams, C.K.I., Winn, J., Zisserman, A.: The pascal visual object classes (VOC) challenge. Int. J. Comput. Vision **88**(2), 303–338 (2010)
7. Girshick, R.: Fast R-CNN. In: Proceedings of the IEEE International Conference on Computer Vision, pp. 1440–1448 (2015)
8. Gokberk Cinbis, R., Verbeek, J., Schmid, C.: Multi-fold mil training for weakly supervised object localization. In: Proceedings of the IEEE Conference on Computer Vision and Pattern Recognition, pp. 2409–2416 (2014)
9. Hoffman, J., et al.: LSDA: large scale detection through adaptation. In: Advances in Neural Information Processing Systems, pp. 3536–3544 (2014)
10. Howard, A., et al.: Searching for mobilenetv3. In: Proceedings of the IEEE International Conference on Computer Vision, pp. 1314–1324 (2019)
11. Kang, B., Liu, Z., Wang, X., Yu, F., Feng, J., Darrell, T.: Few-shot object detection via feature reweighting. In: Proceedings of the IEEE/CVF International Conference on Computer Vision, pp. 8420–8429 (2019)
12. Kolesnikov, A., Lampert, C.H.: Improving weakly-supervised object localization by micro-annotation. arXiv preprint arXiv:1605.05538 (2016)
13. Krasin, I., et al.: Openimages: A public dataset for large-scale multi-label and multi-class image classification. Dataset. **2**, 3 (2017) https://github.com/openimages
14. Krizhevsky, A., Hinton, G., et al.: Learning multiple layers of features from tiny images (2009)
15. Krizhevsky, A., Sutskever, I., Hinton, G.E.: Imagenet classification with deep convolutional neural networks. In: Advances in Neural Information Processing Systems, pp. 1097–1105 (2012)
16. LeCun, Y., Cortes, C., Burges, C.: MNIST handwritten digit database. ATT Labs. **2** (2010). http://yann.lecun.com/exdb/mnist
17. Li, Y., Zhang, J., Huang, K., Zhang, J.: Mixed supervised object detection with robust objectness transfer. IEEE Trans. Pattern Anal. Mach. Intell. **41**(3), 639–653 (2018)
18. Li, Y., et al.: Few-shot object detection via classification refinement and distractor retreatment. In: Proceedings of the IEEE/CVF CVPR, pp. 15395–15403 (2021)
19. Lin, T.Y., Goyal, P., Girshick, R., He, K., Dollár, P.: Focal loss for dense object detection. In: Proceedings of the IEEE ICCV, pp. 2980–2988 (2017)

20. Lin, T.-Y., et al.: Microsoft COCO: common objects in context. In: Fleet, D., Pajdla, T., Schiele, B., Tuytelaars, T. (eds.) ECCV 2014. LNCS, vol. 8693, pp. 740–755. Springer, Cham (2014). https://doi.org/10.1007/978-3-319-10602-1_48
21. Liu, W., et al.: SSD: single shot multibox detector. In: Leibe, B., Matas, J., Sebe, N., Welling, M. (eds.) ECCV 2016. LNCS, vol. 9905, pp. 21–37. Springer, Cham (2016). https://doi.org/10.1007/978-3-319-46448-0_2
22. Pedersen, T., Patwardhan, S., Michelizzi, J.: Wordnet: similarity: measuring the relatedness of concepts. In: Demonstration papers at HLT-NAACL 2004, pp. 38–41. Association for Computational Linguistics (2004)
23. Redmon, J., Divvala, S., Girshick, R., Farhadi, A.: You only look once: unified, real-time object detection. In: Proceedings of the IEEE Conference on Computer Vision and Pattern Recognition, pp. 779–788 (2016)
24. Redmon, J., Farhadi, A.: Yolov3: An incremental improvement. arXiv preprint arXiv:1804.02767 (2018)
25. Ren, S., He, K., Girshick, R., Sun, J.: Faster R-CNN: towards real-time object detection with region proposal networks. In: Advances in Neural Information Processing Systems, pp. 91–99 (2015)
26. Russakovsky, O., et al.: ImageNet large scale visual recognition challenge. Int. J. Comput. Vision **115**(3), 211–252 (2015). https://doi.org/10.1007/s11263-015-0816-y
27. Sun, B., Li, B., Cai, S., Yuan, Y., Zhang, C.: FSCE: few-shot object detection via contrastive proposal encoding. In: Proceedings of the IEEE/CVF CVPR (2021)
28. Tang, P., et al.: Weakly supervised region proposal network and object detection. In: Ferrari, V., Hebert, M., Sminchisescu, C., Weiss, Y. (eds.) ECCV 2018. LNCS, vol. 11215, pp. 370–386. Springer, Cham (2018). https://doi.org/10.1007/978-3-030-01252-6_22
29. Tang, Y., Wang, J., Gao, B., Dellandréa, E., Gaizauskas, R., Chen, L.: Large scale semi-supervised object detection using visual and semantic knowledge transfer. In: Proceedings of the IEEE Conference on Computer Vision and Pattern Recognition, pp. 2119–2128 (2016)
30. Tang, Y., et al.: Visual and semantic knowledge transfer for large scale semi-supervised object detection. IEEE Trans. Pattern Anal. Mach. Intell. **40**(12), 3045–3058 (2017)
31. Uijlings, J., Popov, S., Ferrari, V.: Revisiting knowledge transfer for training object class detectors. In: Proceedings of the IEEE CVPR, pp. 1101–1110 (2018)
32. Wang, X., Huang, T.E., Darrell, T., Gonzalez, J.E., Yu, F.: Frustratingly simple few-shot object detection. In: ICML (2020)
33. Wang, Y.X., Ramanan, D., Hebert, M.: Meta-learning to detect rare objects. In: Proceedings of the IEEE/CVF ICCV, pp. 9925–9934 (2019)
34. Wu, J., Liu, S., Huang, D., Wang, Y.: Multi-scale positive sample refinement for few-shot object detection. In: Vedaldi, A., Bischof, H., Brox, T., Frahm, J.-M. (eds.) ECCV 2020. LNCS, vol. 12361, pp. 456–472. Springer, Cham (2020). https://doi.org/10.1007/978-3-030-58517-4_27
35. Yan, X., Chen, Z., Xu, A., Wang, X., Liang, X., Lin, L.: Meta R-CNN: towards general solver for instance-level low-shot learning. In: Proceedings of the IEEE/CVF International Conference on Computer Vision, pp. 9577–9586 (2019)
36. Zeng, Z., Liu, B., Fu, J., Chao, H., Zhang, L.: Wsod2: learning bottom-up and top-down objectness distillation for weakly-supervised object detection. In: Proceedings of the IEEE International Conference on Computer Vision, pp. 8292–8300 (2019)
37. Zheng, Y., Cui, L.: Zero-shot object detection with transformers. In: 2021 IEEE International Conference on Image Processing (ICIP), pp. 444–448. IEEE (2021)

Person Detection Using an Ultra Low-Resolution Thermal Imager on a Low-Cost MCU

Maarten Vandersteegen[1(✉)], Wouter Reusen[2], Kristof Van Beeck[1], and Toon Goedemé[1]

[1] KU Leuven EAVISE, Jan Pieter De Nayerlaan 5, Sint-Katelijne-Waver, Belgium
{Maarten.Vandersteegen,KristofVan.Beeck,Toon.Goedeme}@kuleuven.be
[2] Melexis Technologies NV, Transportstraat 1, Tessenderlo, Belgium
wre@melexis.com

Abstract. Detecting persons in images or video with neural networks is a well-studied subject in literature. However, such works usually assume the availability of a camera of decent resolution and a high-performance processor or GPU to run the detection algorithm, which significantly increases the cost of a complete detection system. However, many applications require low-cost solutions, composed of cheap sensors and simple microcontrollers. In this paper, we demonstrate that even on such hardware we are not condemned to simple classic image processing techniques. We propose a novel ultra-lightweight CNN-based person detector that processes thermal video from a low-cost 32×24 pixel static imager. Trained and compressed on our own recorded dataset, our model achieves up to 91.62% accuracy (F1-score), has less than 10k parameters, and runs as fast as 87ms and 46ms on low-cost microcontrollers STM32F407 and STM32F746, respectively.

Keywords: Person detection · Low-resolution · Thermal · Neural networks · Microcontrollers · Pruning · Quantization

1 Introduction

Bounding-box based person detection using neural networks is one of the most covered topics in computer vision literature [15,22,29,30] which is a strong indication of its importance for many applications. However, most studies assume moderate to high quality sensors and powerful processors like GPUs with lots of resources to work with, which comes with bulky hardware, high power consumption and a big price tag in the order of thousands of dollars. Moreover, high quality sensors such as color cameras may reveal privacy-sensitive information about persons, leading to privacy concerns and mistrust in technology. In many real-life applications, such problems are unacceptable and therefore low-resolution sensors and embedded GPUs, low-power accelerators or even standard microcontrollers (MCUs) are preferred, costing rather hundreds of dollars, tens of dollars or even a few dollars, respectively. With an exponential decrease in

W. Q. Yan et al. (Eds.): IVCNZ 2022, LNCS 13836, pp. 33–47, 2023.
https://doi.org/10.1007/978-3-031-25825-1_3

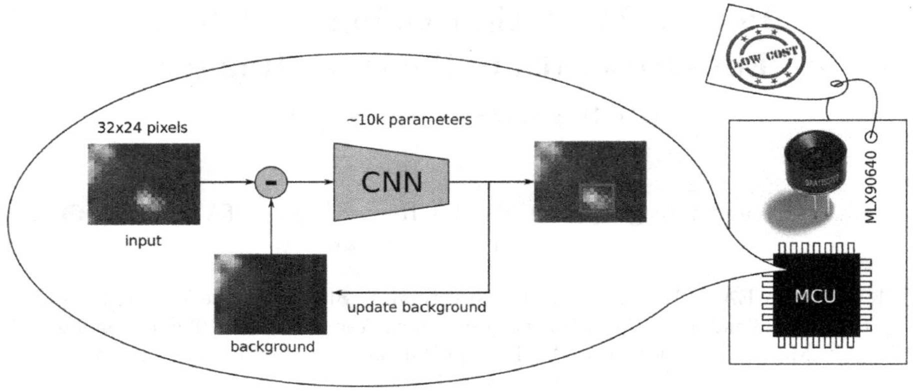

Fig. 1. Our proposed system is built upon a low-cost thermal imager (MLX90640) and a low-cost MCU, which runs a custom designed and compressed CNN object detector with a unique background subtraction mechanism.

price, comes an exponential decrease in computational resources, which makes deployment of neural networks increasingly complex, especially for MCUs. This requires smart solutions that work with low-resolution sensors, sparse model architectures and state-of-the-art compression technologies to squeeze out every last redundant computation, which in turn poses challenges in retaining sufficient prediction accuracy.

Nonetheless, this work succeeds in building a highly accurate person detection system with a low-cost MCU, the cheapest category of processors, a low-cost sensor and a highly optimized deep-learning based detection algorithm. Due to the thermal spectrum, combined with ultra low-resolution, it must be noted that the identity of persons in front of the camera is guaranteed to be preserved [12].

Our methodology, depicted in Fig. 1, consists of a custom CNN-based single-stage object detector and a unique video background subtraction algorithm to help distinguish static person-like objects from real persons. Given the fact that deep learning is skilled at extracting useful context from the background of an image, the combination of deep learning and background subtraction might seem odd at first glance. However, this is not the case when working with such low-resolution (thermal) images, since background features from our sensor are often indistinguishable from a person, even for a human observer. Figure 2 depicts a few of such difficult examples, with and without our background subtraction applied.

Our contributions are summarized as follows:

- This work proposes a truly low-cost and privacy preserving detection system for stationary setups, featuring a novel detection algorithm.
- We introduce a benchmarking system to compare the accuracy of our model against the baseline – the standard person detection software from Melexis, the manufacturer of the MLX90640 thermal sensor – and prove that our models outperform the former, achieving up to 15.6% higher in F_1-score.

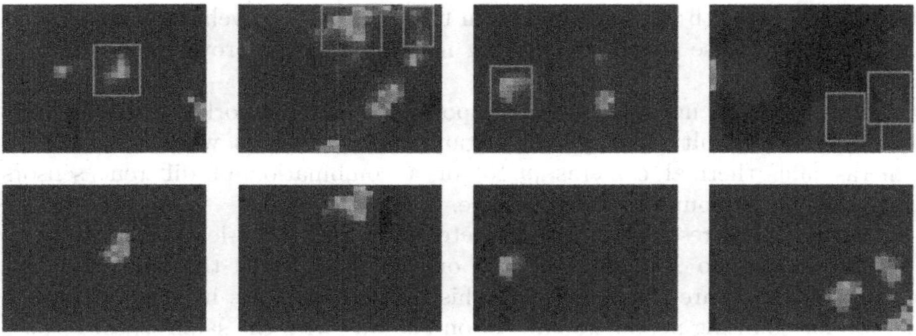

Fig. 2. Example images from the dataset with annotated persons (top row) and the same images with our proposed background subtraction applied (bottom row), which masks out static objects that seem indistinguishable from a real person.

- We are the first to prove the advantage of background-subtraction in combination with a CNN for ultra low-resolution thermal object detection.
- We present extensive compression experiments and a benchmark on inference time and memory utilization of our models deployed with Tensorflow Lite or microTVM on two MCU targets.
- We release our dataset in order to support future research[1].

Our paper is structured as follows: Sect. 2 discusses related work on person detection and neural network compression, given their direct relevance to our application, Sect. 3 elaborates on our detection algorithm, the proposed compression method and how we collected our dataset, Sect. 4 discusses our results while conclusions are made in final Sect. 5.

2 Related Work

2.1 Person Detection

Since the dawn of the deep-learning area, person detection has been addressed mostly with neural network based object detectors such as (Fast(er)) R-CNN [30], YOLO [29] and SSD [22]. In a constant battle to improve the accuracy and reducing the computational budget, many new designs and improvements emerged in later years [15,20,33,38]. Other attempts for improving the accuracy focus on (1) exploiting temporal queues in video object detection and (2) using different sensor modalities.

Video object detection or recognition methods exploit the temporal domain to improve the detection accuracy. Liu et al. [21] for example, propose to use a convolutional LSTM module into their SSD object detector, Simonyan et al. [31] use an additional optical-flow input image in their two-stream approach, Li et al.

[1] Our dataset: https://iiw.kuleuven.be/onderzoek/eavise/mldetection/home.

[18] experiment with 3D convolutions in their video-based vehicle detector, and Kang et al. [11] use a tubelet tracking mechanism to improve their detection results.

Different sensor modalities are incorporated in several works for more robust detection in difficult viewing circumstances. Experiments with time-of-flight cameras [39], thermal cameras [9,10] or a combination of different sensors [28,36,37] can be found in the literature.

Although low-resolution object detection with deep-learning has been attempted down to 96×96 pixels [2], or even lower with the help of super-resolution [4], we are the first to try this on 32×24 pixels in a direct way. A method for counting people in a detection-like way with the same sensor as ours [12] exists, however their dataset seems less challenging (no person-like background objects) and their model is much heavier (130k params) compared to ours (10k params).

Constrained by the very limited resources of regular MCUs, we base our model on tiny YOLOv2 [29], which is small and much more scalable compared to two-stage detectors or multi-headed detectors like SSD or later YOLO versions. Even though we process video, spatio-temporal building blocks with a large memory footprint like 3D convolutions are to be avoided, together with convolutional LSTMs, given that the latter are difficult to tune and make compression extremely complex. We however included experiments with a motion image [31] in Sect. 4.2, because it's cheap to calculate.

2.2 Model Compression

Although several efficient CNN architectures like MobileNet [8] or EfficientNet [32] emerged from the need for lighter models, additional compression is often required to meet model size, computational budget, energy, or time constraints. The most popular compression techniques can be divided in three categories: (1) model quantization, (2) model pruning and (3) Network-Architecture-Search (NAS).

Quantization approaches come in two flavors: (1) Post-Training Quantization (PTQ) and (2) Training-Aware Quantization (QAT). PTQ is the easiest and most widely used quantization approach for obtaining 8-bit CNNs and is a standard feature in most deployment frameworks like Tensorflow Lite [34] or TensorRT [25], while QAT is the preferred choice for sub-8-bit precision [24].

Pruning methods can be divided in two categories: (1) unstructured pruning and (2) structured pruning. Unstructured pruning methods [3,14,16] aim to remove unnecessary neural connections by setting individual weight values to zero. However, this introduces weight sparsity which requires special libraries to support the acceleration of such models. Structured pruning methods like filter pruning [6,17,23,27] on the other hand, remove entire convolution filters at once and avoid the need for special acceleration libraries or hardware. Different methods are proposed to identify good filter candidates for pruning, ranging from methods that remove filters with the smallest L_1 or L_2 filter norms [6,17,27], to more complex ones that for example require changes in the loss function [23].

NAS has been adopted by several works [1,7,19] to find small architectures that can be directly deployed on resource constrained devices. Although such techniques show remarkable performance, their complexity is not needed in our work. Instead, we prefer iterative channel pruning with simple norm-based saliency, which has been proved by Ophoff et al. [27] to work surprisingly well for constrained object detection problems. We believe that our approach can be defined as a constrained object detection problem because: (1) we only have a single class, (2) we target fixed camera viewpoints and (3) most importantly, our background subtraction method greatly reduces the background variance in the input. These constraints allow simple, but well proven structured pruning methods, to compress our model by more than a factor ×100. In addition, we also quantize our models to 8-bit through regular PTQ.

3 Approach

In order to train a CNN-based person detector, a large scale dataset is needed. Since no person detection datasets of our target sensor (MLX90640) are publicly available, we recorded and annotated our own dataset, which is described in more detail in Sect. 3.1. Section 3.2 describes our CNN-based detector and its background subtraction technique, and Sect. 3.3 elaborates on the proposed model compression method.

3.1 Dataset

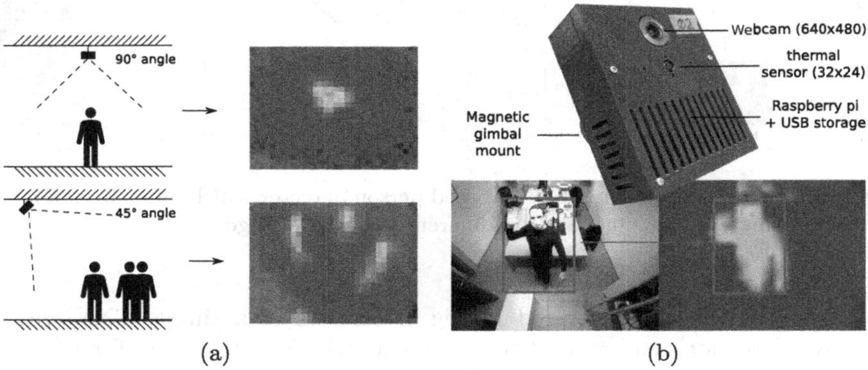

Fig. 3. (a) Illustration of the 45-degree and 90-degree recording angles maintained in our dataset and (b) our recorder featuring the MLX90640 thermal sensor and a webcam. After running existing tracking software on the visible video, the bounding boxes can be used as annotations for the thermal images after manual inspection.

A large bounding-box annotated dataset is constructed of 190 lengthy video clips at 8FPS, containing 96k thermal video frames in total. Recordings are made using a static camera setup that is mounted on the ceiling in 12 different locations including offices, residential rooms and laboratory rooms. Each video clip is labeled with a *45-degree* or *90-degree* tag, indicating that the setup is pointed straight down or at an angle of approximately 45 degrees, respectively, as shown in Fig. 3a. Video frames are stored as 16-bit .tiff images containing the raw temperature measurements in degrees Celsius, multiplied by 100. We split our dataset in train, validation and test set, where the videos of the validation and test sets are recorded at different locations compared to the videos of the training set, in order to avoid overfitting.

To help automate the annotation process, existing person detection software, based on Mask-RCNN [5] and an object tracker, is used to track persons in a high resolution color video feed, coming from a webcam that is included in the recorder setup. Since the recording angle and field-of-view of both cameras have been made equal, the generated tracking data from the webcam can be used as annotation data for the thermal video, after careful manual inspection and correction of the tracking data using the CVAT [26] annotation tool. Figure 3b depicts our recorder and an observation example of both sensors.

3.2 Model

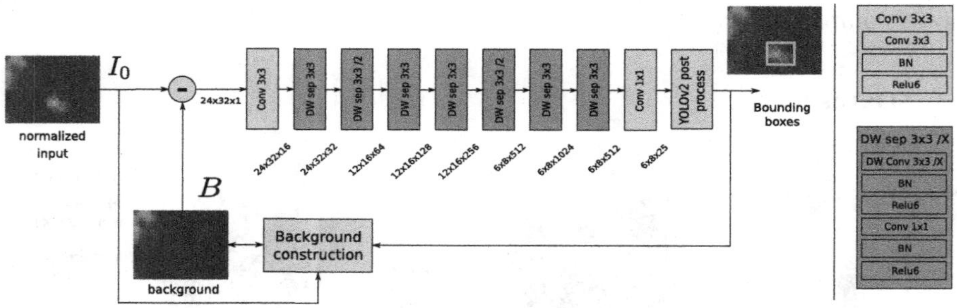

Fig. 4. Our proposed model is a CNN based person detector which infers from a background subtracted input image and a differential motion image.

Our proposed model, illustrated in Fig. 4, consists of a simple CNN with a YOLOv2 detection head and a background subtraction algorithm. First, a constructed background image $B \in \mathbb{R}^{H \times W}$ is subtracted from the current normalized input image $I_0 \in \mathbb{R}^{H \times W}$. Second, the result is sent through the CNN backbone, consisting of a single 3×3 convolution followed by 7 depthwise-separable convolutions. Finally, the YOLOv2 head, a 1×1 convolution followed by its typical post-processing steps, produces the bounding boxes. The following subsections (1) motivate the addition of a background subtraction algorithm, while discussing its implementation and (2) elaborate on the design choices of our CNN architecture.

Background Subtraction. This allows the model to make a clear distinction between stationary objects with a person-like heat signature and actual persons. Distinguishing a person from a computer for example on a single 32×24 pixel thermal image, is often impossible without further context, even for a human observer. The top row images from Fig. 2 depict a few of these difficult samples from our dataset to illustrate this. In such cases, the only difference between both objects is that a person moves from time to time, while a stationary object does not. Based on this assumption, a robust distinction can be made by maintaining an up-to-date background image of stationary features. Our experiments in Sect. 4.2 prove that this method gives an additional boost of up to 28% in AP detection accuracy.

Our proposed background updating algorithm, given by Algorithm 1, uses the current input image I_0, the current background image B and the detections D from the current input image to produce the updated background image B^*. First, all bounding boxes in D are enlarged by one pixel in all directions. Second, a binary mask M is constructed that contains white pixels at locations that overlap with one or more bounding boxes from D, and black pixels otherwise. Third, a background candidate $\hat{B}$ is created from background pixels of I_0 and foreground pixels of B, and finally the new background B^* is created through an exponential-moving-average filter with decay factor $\alpha = 0.99$. Figure 2 shows the qualitative effect of our background subtraction algorithm on a few samples from our dataset.

Algorithm 1: Background update

Input: input image $I_0 \in \mathbb{R}^{H \times W}$, current background image $B \in \mathbb{R}^{H \times W}$ and detections bounding boxes D

Output: updated background image B^*

Procedure:

$D \leftarrow \text{enlarge\_boxes}(D)$

$M \in \{0,1\}^{H \times W}$

$\forall i \in \{1, ..., H\}, \forall j \in \{1, ..., W\}$

$$M_{i,j} = \begin{cases} 1 & \text{if } i, j \text{ overlaps with a bounding box from } D \\ 0 & \text{otherwise} \end{cases}$$

$\hat{B} \leftarrow B \odot M + I_0 \odot (1 - M)$

$B^* \leftarrow \alpha B + (1 - \alpha)\hat{B}$

During initialization, the first background image is constructed by averaging three consecutive frames, in the assumption that no persons are present at startup. Additional computations for calculating the background are negligible, since this happens only once every 25 frames. During training, ground truth annotations are used to construct the background image instead of detections.

Network Architecture. Our network architecture, is inspired by tiny YOLOv2 because we value some of its design choices: (1) a single headed output, which simplifies the network graph and its post-processing, (2) a small amount of layers, which significantly reduces the time overhead for calling the layers, and (3) no residual connections, other branching structures or exotic layer types to simplify compression and deployment. Moreover, the number of layers and number of output channels per layer of our uncompressed model are the same as in tiny YOLOv2 (details in Fig. 4). We however prefer depthwise-separable convolutions in order to create a sparser model to start with, therefore reducing the burden on the compression stage. ReLU6 is preferred over leaky ReLU since it can be fused into its preceding convolution layer in most deployment frameworks. To retain sufficient spatial output resolution, our model down-samples the input resolution by a factor four, in contrast to the more commonly used factor 32 in other higher resolution object detectors. Our model has $num\_anchors \times 5 = 5 \times 5 = 25$ output channels with anchor box sizes that are adjusted to our dataset.

3.3 Compression

We compress our model in two stages: first, iterative channel pruning is applied, followed by post-training quantization to 8-bit for both weights and activations.

Channel Pruning. We propose structured channel pruning, which avoids the need for libraries with sparsity support on the target MCU and compresses the activation tensors as well to reduce RAM memory. Each iteration, the 5% filters with the lowest normalized L_2-norm, as proposed by Ophoff et al. [27], are removed, followed by a fine-tuning step. When fine-tuning is finished, the weights with the lowest validation loss are selected to initialize the model for the next iteration. To speed up the pruning process, we adopt the following early stopping criteria: given the validation loss of the current pruned model L_{curr} and the validation loss of the unpruned model L_{start}, we skip or stop fine-tuning when $L_{curr} \leq 1.03 \times L_{start}$ and immediately continue with the next iteration.

Quantization. After channel pruning is completed, PTQ is applied to convert both the model's weights and activations to 8-bits. We use the Tensorflow Lite model converter for this task and provide a calibration set of 100 images to estimate the scales and zero-points of the activation tensors.

4 Experiments

In the first section of our experiments (Sect. 4.1), we compare our models accuracy-wise against the baseline, which is the proprietary in-house software from Melexis. Section 4.2 studies the added value of our background subtraction method, Sect. 4.3 elaborates on the compression ratios and accuracy losses and final Sect. 4.4 compares the time performance and resource occupation of our models, deployed with two different run-time frameworks on two different MCU targets.

4.1 Comparison

In this section we compare the accuracy of our detection model against the accuracy of our baseline, the in-house developed detection software by Melexis, the manufacturer of the MLX90640. Melexis's proprietary detection software is built and perfected by an in-house development team, over the course of two years. Their algorithm uses dynamic noise suppression, contrast and edge enhancement in pre-processing, and calculates a dynamic threshold to identify local maxima and minima after background subtraction. A multi-object tracker is used to further improve the accuracy. Compared to a 90-degree or overhead camera view, detecting persons in video from a 45-degree camera angle turns out to be much more difficult. Therefore, accuracy results are reported separately for both views throughout the whole experiments section.

Table 1. F_1-score accuracy results of our baseline – Melexis's proprietary detection software – compared to our models with and without compression applied.

Model	F1 45-degree	F1 90-degree
Baseline	64.31%	89.02%
Ours	**83.52%**	**91.76%**
Ours compressed	79.9%	**91.62%**

Table 1 presents the F_1-scores on the test set of our model in uncompressed and compressed state, and the baseline. For both camera angles, our models outperform Melexis's software by 19.2% and 2.7% on the 45-degree and 90-degree test sets, respectively. In compressed state, where our model is heavily pruned and quantized to 8-bit, this still results in a performance boost of 15.6% and 2.6% on the 45-degree and 90-degree test sets, respectively, making our approach the preferred solution.

The Average-Precision (AP) metric, which is calculated from a Precision-Recall curve, is popular in most object detection papers, but requires a model to be able to produce a prediction probability for each produced bounding-box. Since Melexis's software is not capable of producing such probabilities, it does not make sense to use this metric here. We therefore prefer F_1-score instead, which is a percentile number produced from a single precision and a single recall value. Selecting a single precision-recall point for a detector like ours requires selecting a single threshold value. We find the optimal threshold by maximizing the F_1-score through adjustment of the threshold value. We use a standard 0.5 overlap threshold for matching detections to ground-truth boxes and set the NMS overlap threshold to 0.3.

4.2 Background Subtraction

This section presents results on the added value of our proposed background subtraction method. Figure 5a and 5b depict the PR-curve charts of a number of models tested on 45-degree and 90-degree labeled dataset images, respectively. Models that are tested on the 90-degree test set are trained on the 90-degree training set, while model tested on the 45-degree test set are trained on both 90-degree and 45-degree training data.

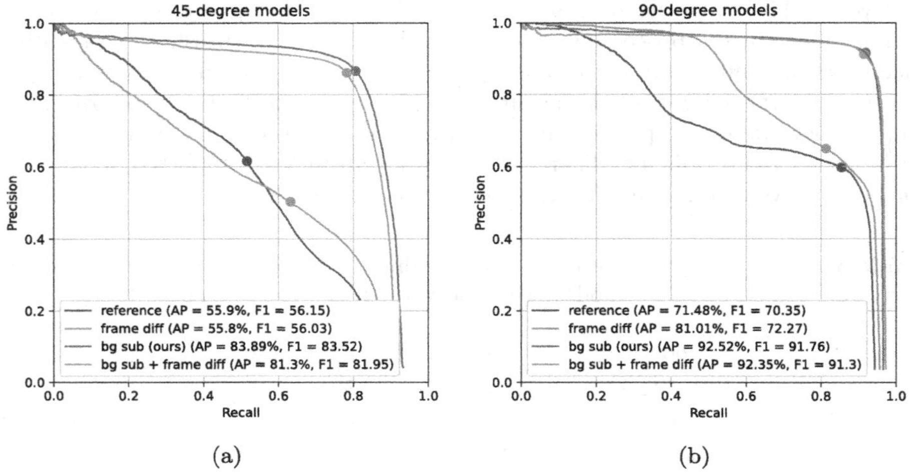

Fig. 5. PR-curves of (a) 45-degree and (b) 90-degree models. For each curve, AP and F_1 scores are presented (best viewed in color).

Each chart presents results of four different models: (1) *reference*, which is our model without background subtraction, (2) *frame diff*, which is the reference model, but with an additional difference image on a second input channel, (3) *bg sub*, which is our proposed model with background subtraction and (4) *bg sub + frame diff*, which is our model with background subtraction and the additional difference image. Inspired by the optical-flow input from Simonyan et al. [31], we experiment with a difference image, which is a computationally cheap alternative to optical-flow and a way to inject short-term motion information into the network. We construct the difference image by subtracting a previous image from the current image, where we set the frame stride between both images equal to five. From our results, it can be concluded that this difference image seems to help improve the accuracy a little when applied to the *reference* model on the 90-degree dataset. However, the added value of the difference image diminishes in comparison to our proposed background subtraction method, which boosts the accuracy by 28% and 21% AP on the 45-degree and 90-degree datasets, respectively. Since the background subtraction method eliminates the confusion between a person and an imposter background object, the network gets much less confusing samples during training, resulting in such a tremendous performance boost. The *bg sub + frame diff* models however did not improve on the *bg sub*, so the difference image is left out from our proposed method.

All models in Fig. 5 are directly trained on our dataset from a random initialization for up to 50k iterations (≤3 h on a GTX1080), using the SGD optimizer with a high weight decay $d = 0.03$ to reduce overfitting. The first 1000 iterations, a warm-up stage exponentially increases the learning rate up to the base learning rate $lr = 0.001$. Subsequently, a reduce-on-plateau learning rate scheduler decays the learning rate with factor 10 whenever the validation loss did not drop in the previous 5000 iterations. For data augmentation, we use random contrast, brightness, horizontal and vertical flipping manipulations.

4.3 Model Compression

The compression results of our 45-degree and 90-degree models are presented in Table 2. For each model, the AP and F_1 accuracies are given for the original model, the pruned model and the pruned + quantized model. We start with a 1.26M parameter model and prune that down to around a 10k parameters model, with negligible loss in accuracy for the 90-degree model and acceptable loss in accuracy for the 45-degree model. This results in pruning rates of up to ÷136 and ÷52 for the parameters and MACS, respectively. Since standard PTQ to 8-bit weights and activations hardly influences the accuracy, there is no need for QAT.

Table 2. Accuracies, model size, number of Multiply-Accumulates and number of bits per weights/activation for our original, pruned and pruned + quantized models.

Model	AP	F_1	#params	#MAC	W/A
45-degree	83.89%	83.52%	1.26 M	68.36 M	32/32
45-degree-pruned	80.48%	80.5%	13.86 k (÷91)	1.95 M (÷35)	32/32
45-degree-pruned-quant	79.47%	79.9 %	13.86 k (÷91)	1.95 M (÷35)	8/8
90-degree	92.52 %	91.76%	1.26 M	68.36 M	32/32
90-degree-pruned	92.49%	91.75%	9.23 k (÷136)	1.32 M (÷52)	32/32
90-degree-pruned-quant	92.29%	91.62 %	9.23 k (÷136)	1.32 M (÷52)	8/8

Both models are pruned for up to 60 iterations on a GTX1080, which completes in about 1.5 days. We then select the model with the highest test accuracy that is close to 10k parameters. The 45-degree and 90-degree models that we selected are pruned for 44 and 49 iterations, respectively. Each fine-tuning step takes at most 10k iterations, which is $1/5^{\text{th}}$ of a regular training time. The learning rate is set to 0.0001 and lowered by a factor 10 after 5k iterations.

4.4 Deployment

After pruning and quantization, we deploy both 45-degree and 90-degree models on two popular low-cost MCUs: (1) the STM32F746 with a Cortex-M7 core and (2) the STM32F407 with a Cortex-M4 core. We compare the inference time and memory utilization between two open-source neural-network deployment frameworks for our application: Tensorflow Lite for microcontrollers (TFLite) [34] and microTVM [35]. We configure both frameworks to use the CMSIS-NN [13] microkernel library as a backend, because it supports SIMD instructions that can execute four 8-bit multiply-accumulate operations in a single cycle, which significantly boosts the performance. For our benchmark, we compile the generated C-code with all optimizations for speed enabled (`GCC -O3`) and set the clock speeds of both platforms to their maximum, which is 216MHz for the STM32F746 and 168MHz for the STM32F407.

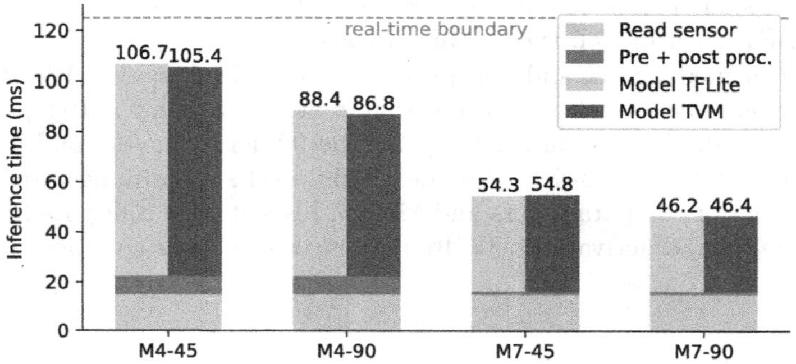

Fig. 6. Inference times of both 45-degree and 90-degree models on an STM32F407 (M4) and an STM32F746 (M7) MCU, using either Tensorflow Lite (TFLite) or microTVM (TVM).

Figure 6 presents the inferences times of our models (*Model TFLite/Model TVM*), sensor acquisition time (*Read sensor*) and other processing (*Pre + post proc*), which includes pixel temperature calculation, input normalization, yolo post-processing, NMS and background calculation. All our models are capable of running in real-time ($\geq$ 8 FPS) on both microcontrollers with sufficient headroom for doing other tasks.

Figure 7a and 7b report the RAM memory and flash memory usage, respectively. The utilization of RAM memory in our microTVM experiments is noticeably lower compared to that of Tensorflow Lite, but the biggest difference can be seen in flash-memory usage, where microTVM uses less than half the size compared to Tensorflow Lite. This is because Tensorflow Lite's model interpreter engine is a fixed size component that becomes significantly large when working with very small models like ours. In contrast, microTVM generates direct function calls to the CMSIS-NN micro kernels without the need for interpreter code, which tremendously reduces the flash usage. Note that the statistics only report the used resources of the model.

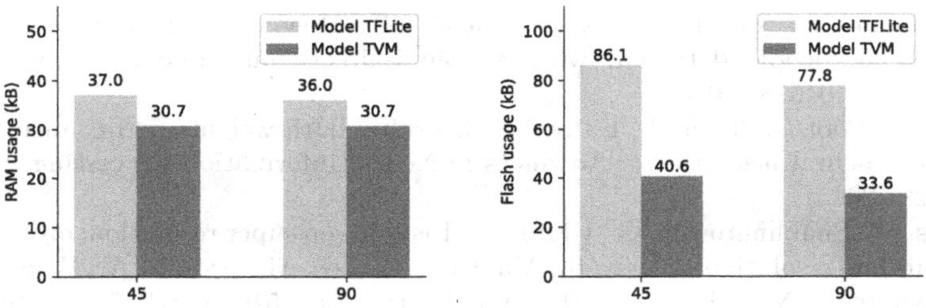

(a) RAM memory usage (b) Flash memory usage

Fig. 7. RAM and flash memory usage statistics of both 45-degree and 90-degree models, measured when using either Tensorflow Lite (TFLite) or microTVM (TVM).

5 Conclusion

Although person detection is a wanted asset in many real-life applications, it is not always feasible due to (1) low-cost constraints and (2) privacy related issues if a regular camera is involved.

This work solves both problems by proposing a low-cost detection system in the order of magnitude of tens of dollars, based on a privacy-preserving ultra low-resolution thermal imager and a low-cost microcontroller. Even though it is extremely difficult to create an accurate deep-learning model that works with such limited resources, our compressed models achieve an accuracy of 79.9% and 91.62% (F1-score) on our 45-degree and 90-degree benchmark test, respectively, outperforming the standard person detection software of Melexis by a significant margin. We achieve our goal by proposing a smart background-subtraction mechanism that eliminates confusion between person-like objects and real persons, boosting our model's performance up to 28%. Our processing pipeline with 90-degree model is running at 46ms/image and 87ms/image on an STM32F746 and STM32F407, respectively, and only requires 34kB of flash memory and 31kB of RAM.

In future work, an additional spatio-temporal module could be added to potentially further improve the accuracy, or more extreme compression could be used to further down-size the model.

Acknowledgement. This work is supported by VLAIO and Melexis via the AI@EDGE TETRA project.

References

1. Banbury, C., et al.: MicroNets: neural network architectures for deploying TinyML applications on commodity microcontrollers. Proc. Mach. Learn. Syst. **3**, 517–532 (2021)

2. Callemein, T., Van Beeck, K., Goedemé, T.: How low can you go? Privacy-preserving people detection with an omni-directional camera. arXiv preprint arXiv:2007.04678 (2020)
3. Han, S., Pool, J., Tran, J., Dally, W.: Learning both weights and connections for efficient neural network. In: Advances in Neural Information Processing Systems, vol. 28 (2015)
4. Haris, M., Shakhnarovich, G., Ukita, N.: Task-driven super resolution: object detection in low-resolution images. In: Mantoro, T., Lee, M., Ayu, M.A., Wong, K.W., Hidayanto, A.N. (eds.) ICONIP 2021. CCIS, vol. 1516, pp. 387–395. Springer, Cham (2021). https://doi.org/10.1007/978-3-030-92307-5_45
5. He, K., Gkioxari, G., Dollár, P., Girshick, R.: Mask R-CNN. In: Proceedings of the IEEE International Conference on Computer Vision, pp. 2961–2969 (2017)
6. He, Y., Kang, G., Dong, X., Fu, Y., Yang, Y.: Soft filter pruning for accelerating deep convolutional neural networks. arXiv preprint arXiv:1808.06866 (2018)
7. Hendrickx, L., Van Ranst, W., Goedemé, T.: Hot-started NAS for task-specific embedded applications. In: Proceedings of the IEEE/CVF Conference on Computer Vision and Pattern Recognition, pp. 1971–1978 (2022)
8. Howard, A.G., et al.: MobileNets: efficient convolutional neural networks for mobile vision applications. arXiv preprint arXiv:1704.04861 (2017)
9. Ippalapally, R., Mudumba, S.H., Adkay, M., HR, N.V.: Object detection using thermal imaging. In: 2020 IEEE 17th India Council International Conference (INDICON), pp. 1–6. IEEE (2020)
10. Jiang, C., et al.: Object detection from UAV thermal infrared images and videos using Yolo models. Int. J. Appl. Earth Obs. Geoinf. **112**, 102912 (2022)
11. Kang, K., et al.: T-CNN: tubelets with convolutional neural networks for object detection from videos. IEEE Trans. Circuits Syst. Video Technol. **28**(10), 2896–2907 (2017)
12. Kraft, M., Aszkowski, P., Pieczyński, D., Fularz, M.: Low-cost thermal camera-based counting occupancy meter facilitating energy saving in smart buildings. Energies **14**(15), 4542 (2021)
13. Lai, L., Suda, N., Chandra, V.: CMSIS-NN: efficient neural network Kernels for arm Cortex-M CPUs. arXiv preprint arXiv:1801.06601 (2018)
14. Laurent, C., Ballas, C., George, T., Ballas, N., Vincent, P.: Revisiting loss modelling for unstructured pruning. arXiv preprint arXiv:2006.12279 (2020)
15. Law, H., Deng, J.: CornerNet: detecting objects as paired keypoints. In: Ferrari, V., Hebert, M., Sminchisescu, C., Weiss, Y. (eds.) Computer Vision – ECCV 2018. LNCS, vol. 11218, pp. 765–781. Springer, Cham (2018). https://doi.org/10.1007/978-3-030-01264-9_45
16. LeCun, Y., Denker, J., Solla, S.: Optimal brain damage. In: Advances in Neural Information Processing Systems, vol. 2 (1989)
17. Li, H., Kadav, A., Durdanovic, I., Samet, H., Graf, H.P.: Pruning filters for efficient convnets. arXiv preprint arXiv:1608.08710 (2016)
18. Li, S., Chen, F.: 3D-DETNet: a single stage video-based vehicle detector. In: Third International Workshop on Pattern Recognition, vol. 10828, pp. 60–66. SPIE (2018)
19. Lin, J., Chen, W.M., Lin, Y., Gan, C., Han, S., et al.: MCUNet: tiny deep learning on IoT devices. Adv. Neural. Inf. Process. Syst. **33**, 11711–11722 (2020)
20. Lin, T.Y., Goyal, P., Girshick, R., He, K., Dollár, P.: Focal loss for dense object detection. In: Proceedings of the IEEE International Conference on Computer Vision, pp. 2980–2988 (2017)

21. Liu, M., Zhu, M.: Mobile video object detection with temporally-aware feature maps. In: Proceedings of the IEEE Conference on Computer Vision and Pattern Recognition, pp. 5686–5695 (2018)
22. Liu, W., et al.: SSD: single shot MultiBox detector. In: Leibe, B., Matas, J., Sebe, N., Welling, M. (eds.) ECCV 2016. LNCS, vol. 9905, pp. 21–37. Springer, Cham (2016). https://doi.org/10.1007/978-3-319-46448-0_2
23. Liu, Z., Li, J., Shen, Z., Huang, G., Yan, S., Zhang, C.: Learning efficient convolutional networks through network slimming. In: Proceedings of the IEEE International Conference on Computer Vision, pp. 2736–2744 (2017)
24. Nagel, M., Fournarakis, M., Amjad, R.A., Bondarenko, Y., van Baalen, M., Blankevoort, T.: A white paper on neural network quantization. arXiv preprint arXiv:2106.08295 (2021)
25. Nvidia TensorRT, September 2022. https://developer.nvidia.com/tensorrt
26. Computer Vision Annotation Tool (CVAT), September 2022. https://github.com/opencv/cvat
27. Ophoff, T., Gullentops, C., Van Beeck, K., Goedemé, T.: Investigating the potential of network optimization for a constrained object detection problem. J. Imaging **7**(4), 64 (2021)
28. Ophoff, T., Van Beeck, K., Goedemé, T.: Exploring RGB+ depth fusion for real-time object detection. Sensors **19**(4), 866 (2019)
29. Redmon, J., Farhadi, A.: Yolo9000: better, faster, stronger. In: Proceedings of the IEEE Conference on Computer Vision and Pattern Recognition, pp. 7263–7271 (2017)
30. Ren, S., He, K., Girshick, R., Sun, J.: Faster R-CNN: towards real-time object detection with region proposal networks. Adv. Neural. Inf. Process. Syst. **28**, 91–99 (2015)
31. Simonyan, K., Zisserman, A.: Two-stream convolutional networks for action recognition in videos. arXiv preprint arXiv:1406.2199 (2014)
32. Tan, M., Le, Q.: EfficientNet: rethinking model scaling for convolutional neural networks. In: International Conference on Machine Learning, pp. 6105–6114. PMLR (2019)
33. Tan, M., Pang, R., Le, Q.V.: EfficientDet: scalable and efficient object detection. In: Proceedings of the IEEE/CVF Conference on Computer Vision and Pattern Recognition, pp. 10781–10790 (2020)
34. Tensorflow lite, September 2022. https://www.tensorflow.org/lite
35. microTVM, September 2022. https://tvm.apache.org/docs/topic/microtvm/index.html
36. Vandersteegen, M., Van Beeck, K., Goedemé, T.: Real-time multispectral pedestrian detection with a single-pass deep neural network. In: Campilho, A., Karray, F., ter Haar Romeny, B. (eds.) ICIAR 2018. LNCS, vol. 10882, pp. 419–426. Springer, Cham (2018). https://doi.org/10.1007/978-3-319-93000-8_47
37. Wolpert, A., Teutsch, M., Sarfraz, M.S., Stiefelhagen, R.: Anchor-free small-scale multispectral pedestrian detection. arXiv preprint arXiv:2008.08418 (2020)
38. Wong, A., Famuori, M., Shafiee, M.J., Li, F., Chwyl, B., Chung, J.: YOLO Nano: a highly compact you only look once convolutional neural network for object detection. In: 2019 Fifth Workshop on Energy Efficient Machine Learning and Cognitive Computing-NeurIPS Edition (EMC2-NIPS), pp. 22–25. IEEE (2019)
39. Xiang, H., Zhou, W.: Real-time people detection based on top-view TOF camera. In: Twelfth International Conference on Graphics and Image Processing (ICGIP 2020), vol. 11720, pp. 11–19. SPIE (2021)

A Real-Time Kiwifruit Detection Based on Improved YOLOv7

Yi Xia(✉), Minh Nguyen, and Wei Qi Yan

Auckland University of Technology, 1010 Auckland, New Zealand
xiayi.shawn.001@gmail.com

Abstract. In New Zealand (NZ), agriculture is an essential industry, Kiwifruits contribute significantly to the country's overall exports. Traditionally Kiwifruits require manually picking up and heavily relies on human resources, which result in Kiwifruit yields often being affected by human labours. With the rapid development of deep learning in agriculture, agricultural automation has become an efftive way for the industry. Accurate and fast Kiwifruit detection can accelerate the process in the industry. In this paper, we propose an improved Kiwifruit detection model based on YOLOv7. We collected digital images from natural Kiwifruit orchards and produced a manually labelled, data-augumented Kiwifruit image dataset. We add the attention module to YOLOv7 and increase the weight of visual features while suppressing the weight of invalid features. The results show that our proposed method has higher detection accuracy than the original YOLOv7 model, while the detection speed is sufficient for real-time usage. The results of our experiments provide a technical reference for automated picking in modern Kiwifruit supply chain.

Keyword: Deep learning · YOLOv7 · Attention mechanism · Real-time detection · CBAM

1 Introduction

Kiwifruits appeare in worldwide markets and have become one of the most iconic namecards of New Zealand [5]. However, the rapidly growing industry has also brought significant challenges, including labor shortages that have led to industry losses. Therefore, developing an efficient supply chain related to picking, sorting, cleaning, and packaging is an effective way to improve efficiency in the current Kiwifruit industry. We are use of the state-of-the-art artificial intelligence, in particular, deep learning, to increase the Kiwifruit yield estimation, improve picking efficiency and reduce the costs of human labors [18].

Fast and accurate models are the foundation of Kiwifruit counting [1]. Traditionally, fruit detection algorithms [9, 19, 36] extracted key feature parameters such as color and shape of visual objects through digital image processing. The image segmentation algorithms are harnessed in order to detect the visual objects. However, the conventional algorithms are less robust. The factors such as lighting conditions and fruit location

W. Q. Yan et al. (Eds.): IVCNZ 2022, LNCS 13836, pp. 48–61, 2023.
https://doi.org/10.1007/978-3-031-25825-1_4

affected the results of the model detection. Therefore, the practicality of fruit yelds estimation in orchards under natural conditions is low.

However, recent advances in deep learning [10] and computer vision [34] have combined more fields with artificial intelligence and computer vision [22, 25]. Mainly, digital image processing based on deep learning has been extensively applied to modern agriculture, such as plant pest control, fruit ripeness detection [33], and fruit freshness grading [6].

Visual object detection has been conducted based on deep learning [20, 21]. Wang and Yan proved that the one-stage YOLOv5 algorithm outperformed the two-stage Faster R-CNN algorithm in the leaf detection task through comparative experiments, especially in the speed of object detection [30]. Bazame et al. proposed the best-performing YOLOv4-based algorithm for detecting and classifying coffee beans on tree branches by comparing YOLOv3, YOLO4 and YOLOv4-tiny algorithms [2]. The mAP of the model is 81%. Lawal propounded an accurate and fast algorithm for fruit detection-YOLOMuskmelon, which combines the ReLU-activated ResNet-43 Backbone with residual block alignment, SPP, CIoU loss, FPN, and DIoU-NMS to improve detection performance, the average accuracy of this model is 89.6% [13]. Liu et al. put forward SE-Mask R-CNN algorithm for detecting apples in complex environments [15]. The method improves the resource allocation of the model to the effective feature maps by adding SENet to the backbone network. Liu et al. offered TomatoDet [14], an anchorless frame algorithm for tomato detection, which was applied to solve the detection of tomatoes under complex environmental conditions, such as uneven lighting, leaf or branch occlusion, and overlap between fruits. The algorithm incorporates an attention mechanism in CenterNet and introduces a circular representation to optimize the detector. The average precision of this algorithm is 98.16%. Jilbert et al. proffered an algorithm based on the YOLOv5 model for detecting coconut fruits using UAVs [12]. The accuracy of this algorithm is 88.4%. The mAP of the improved model is improved by 3.5%, and the model size is compressed by 62.77%. Although a plethora of studies have obtained better model performance, with the rapid development of deep learning, the better performing YOLOv7 [28] algorithm can obtain more accurate results under faster conditions.

Kiwifruit detection based on deep learning algorithm proposed in this paper is accurate, fast, and is able to be generalized to accommodate the complex conditions in natural orchards. The algorithm for visual object detection is based on the YOLOv7 model, the one-stage object detection algorithm does not need to generate candidate frames. This algorithm directly converts the problem of visual object localization into a regression problem. Therefore, YOLO model is superior to computing speed, this fast detection capability can be better applied to visual object detection.

In this paper, we propose an improved YOLOv7 algorithm combined with the Convolutional Block Attention Module (CBAM) [32]. This method is able to improve the detection accuracy of small, overlapping, and multiple Kiwifruits. A CBAM module is added to the backbone of YOLOv7 network and assign weights to channel features and spatial features in the feature map to increase the model sensitivity while reducing attention to invalid features, thereby improving the model's detection of Kiwifruit in orchards. In this paper, we also have other contributions: (1) We created a new Kiwifruit dataset, (2) conducted data augmentation, (3) verified our YOLOv7 model by using ablation experiments.

This paper is organized as follows: In Sect. 2, we introduce the related work, our method is detailed in Sect. 3, our results are demonstrated in Sect. 4. In Sect. 5, we conclude this paper and envision our future work.

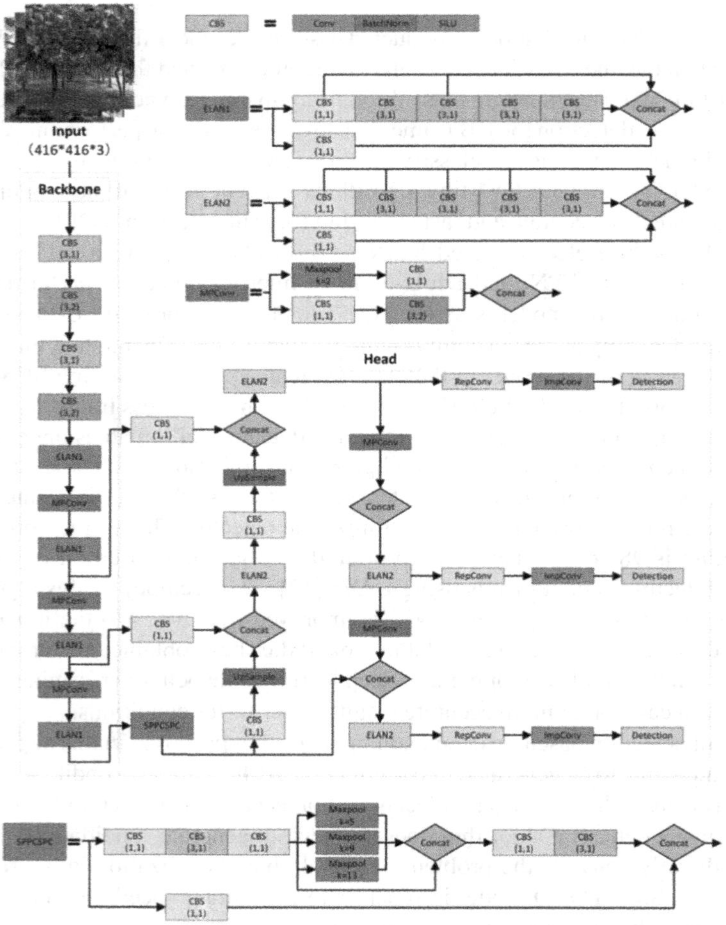

Fig. 1. YOLOv7 neural network architecture.

2 Related Work

2.1 YOLOv7

Visual object detection is a computer vision problem that locates and labels visual objects by drawing a bounding box around the object, and determines the class label to which the given box belongs to. Visual object detection is an important research topic in computer vision and broadly employed to the areas such as face recognition, car plate number recognition, intelligent transportation and autonomous vehicles. The YOLO family has witnessed visual object detection in the era of deep learning. Since the publication of YOLOv1 [23] in 2015, YOLO has been updated iteratively. The lightweight and high accuracy of YOLO models have set the benchmark for the state-of-the-art methods of visual object detection.

YOLOv4 [3], Scaled-YOLOv4 [27], and YOLOR [29] were proposed in 2020 and 2021. The latest object detection model - YOLOv7 [28] was proffered in 2022. It outperforms most of well-known object detectors such as R-CNN [8], YOLOv4 [3], YOLOR [29], YOLOv5, YOLOX [7], PPYOLO [17], and DETR [4] etc. YOLOv7 reduces about 40% of the number of parameters and 50% of the computational costs of real-time object detection. It is split into two main areas of optimization: Model architecture optimization and training process optimization. YOLOv4 improved the accuracy at the cost of training but did not increase inference cost [28]. However, YOLOv7 takes use of a re-parameterized approach to replace the original modules. It adopts dynamic label assignment, which has the effect of assigning labels to output layers more efficiently [37]. The YOLOv7 structure is similar to YOLOv5, the main improvement is the replacement of internal components of the network structure.

Figure 1 shows the overall structure of YOLOv7. In Fig. 1, we see that YOLOv7 consists of three components: Input, backbone, and head. The backbone layer extracts feature maps, the head layer is employed for prediction. As shown in Fig. 1, firstly the processed image is input into the backbone network in YOLOv7, a feature map is output with three layers of different sizes through the head layer network, and finally the prediction result is exported through Rep convolution and Imp convolution.

2.2 Attention Mechanism

Attention mechanisms [26] have made significant achievements in image processing in recent years. The essence of attention mechanism is to detect the information which is interested and suppressed the useless information. Three main attention mechanisms are based on how weights are applied to feature spatial and channel: Spatial attention mechanisms [38], channel attention mechanisms [11], mixed spatial and channel attention mechanisms [24]. The attention mechanisms have different effects on different computer vision tasks.

Squeeze-and-Excitation Network (SENet) [11] is a channel-based attention model that models the importance of each feature channel and then enhances or suppresses different channels for different tasks. A bypass branch is branched out after the regular convolution operation. Firstly, the squeeze operation is performed, compressing the spatial dimension with features so that each 2D feature map becomes an actual number, and the number of feature channels remains unchanged. Then the excitation operation generates weights for each feature channel, which is applied to show the correlation between the modelled feature channels. Once the model gets the weights for each feature channel, it can show the importance of different channels by applying that weight to each original feature channel. The model achieves a more significant performance improvement with a minor increase in computation. Efficient Channel Attention Net

(ECA-Net) [31] improved on the SENet module. The module indicates a way for local cross-channel interaction without dimensionality reduction, which effectively avoids the effect of dimensionality reduction on the learning effect of channel attention. The experimental results show that ECA-Net has low complexity while obtaining excellent performance.

Convolutional Block Attention Module (CBAM) [27] is a simple and effective attention module for feedforward convolutional neural networks that connects the spatial attention module after the channel attention module. The CBAM structure is shown in Fig. 2. The focus of spatial attention is on the position of objects in the image, while channel attention focuses on the objects in the image. Instead of using a single maximum pooling or average pooling, the attention module harnesses the summation or stacking of the maximum and average pooling.

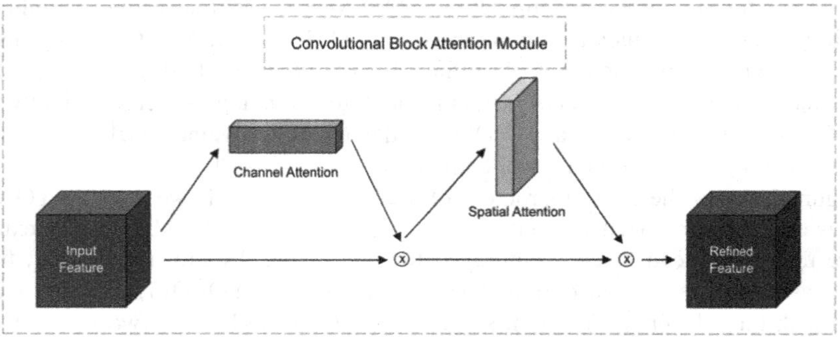

Fig. 2. The main structure of convolutional block attention module.

The channel attention module structure is shown in Fig. 3. The input feature maps are subjected to global max pooling and global average pooling based on width and height respectively to obtain two $1 \times 1 \times C$ feature maps, which are then fed into a two-layer MLP with the number of neurons in the first layer as C/r, where r is the reduction rate and the activation function as ReLU, the number of neurons in the second layer is C. This two-layer neural network is shared. The channel attention feature and the input feature map are multiplied elementwise to generate the final channel attention feature. Finally, the input features are multiplied elementwise to generate the input features required by the spatial attention module.

The spatial attention module structure is shown in Fig. 4. The feature map output from the channel attention module is employed as the input feature map. The module firstly conducts channel-based global max pooling and global average pooling to obtain two feature maps. A concatenation operation (channel splicing) is undergone on the two feature maps based on the channels. A convolution is then performed to reduce the dimensionality to one channel. Then it goes through a sigmoid to generate a spatial attention feature.

Finally, the feature is multiplied by the input feature of the module to obtain the final generated feature. The module is shown to provide improvements in both classification and detection performance.

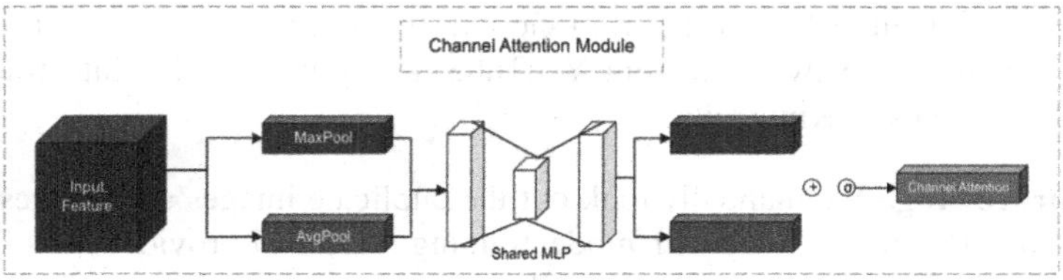

Fig. 3. The structure of channel attention module.

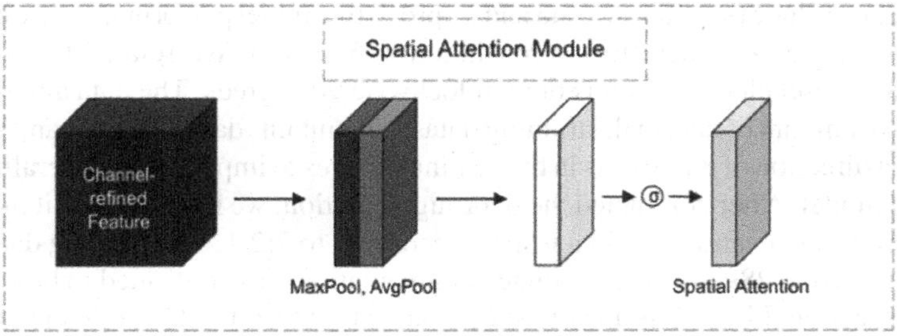

Fig. 4. The structure of spatial attention module.

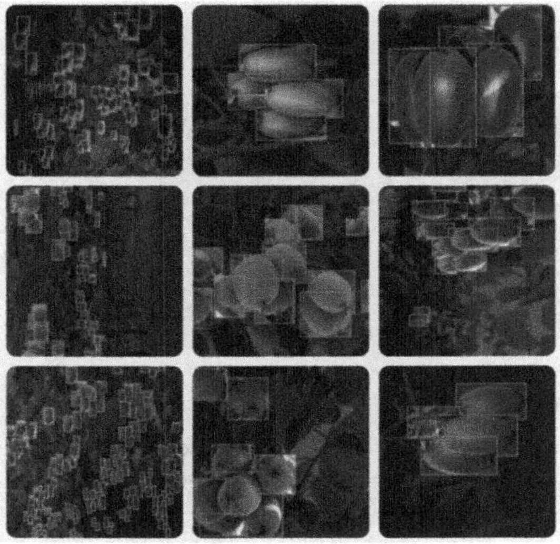

Fig. 5. The image dataset labeled on Roboflow.

3 Our Methods

3.1 Dataset

Data Collection. There is not publicly labelled dataset available for Kiwifruits. In this paper, two methods were employed to collect the Kiwifruit dataset. Firstly, we collected digital images of natural Kiwifruits from Google Images. Secondly, we retrieved and downloaded videos of Kiwifruits from YouTube, we split the video into frames. We collected 117 images of Kiwifruits.

Data Preprocessing. We manually took out the duplicate images and images without Kiwifruits to reduce redundancy with model training. YOLOv7 provides Roboflow tool, which can label the images and automatically export the custom dataset. Therefore, we uploaded the filtered images to Roboflow for manually labelling, there are 7,114 labels in this dataset. The labelling results are shown in Fig. 5.

In order to reduce the training time and improve the model performance, we collected the images and resized them to the resolution 416 × 416. We rotated the images in the training dataset clockwise and counterclockwise 90 degrees. The data augmentation increases the amount of data in the training dataset, maintains data diversity, and alters the distribution direction of Kiwifruits in the original images to improve the generalization of the trained model. After completed the data augmentation, we randomly split the dataset into a training set, a valid set, and a test set according to 7:2:1. The training dataset was finally increased to 289 images, the image augmentation was confirmed to be correct by manual inspection. Figure 6 is an attribute visualization result of the augmented dataset. The number of labels in the dataset is shown in Fig. 6 (a), the location of the labels in the dataset is shown in Fig. 6 (b), the width and height of the labels in the dataset are shown in Fig. 6 (c).

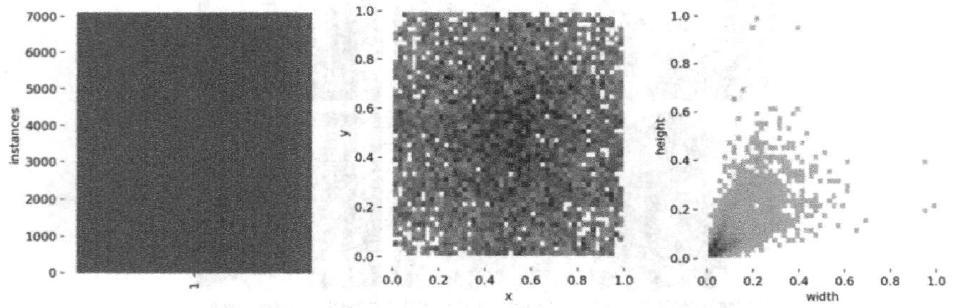

(a) (b) (c)

Fig. 6. (a) The number and the class of labels in the dataset. (b) The location of the labels in the images of the dataset. (c)The size of the labels in the dataset.

3.2 Modelling

In this paper, we were use of a manually collected, preprocessed, and labelled Kiwifruit image samples. In this paper, a CBAM model was added to the front of the backbone in YOLOv7 net to deal with the dense nature of objects, the high overlap rate and the small size of the objects in the Kiwifruit videos. The method improves the overall accuracy of the object detection model by integrating CBAM and YOLOv7 together to assign weights of channel features and spatial features of visual objects in the feature map, increase the attention to detect visual objects and suppresse attention to non-objects. The structure of the improved model is shown in Fig. 7.

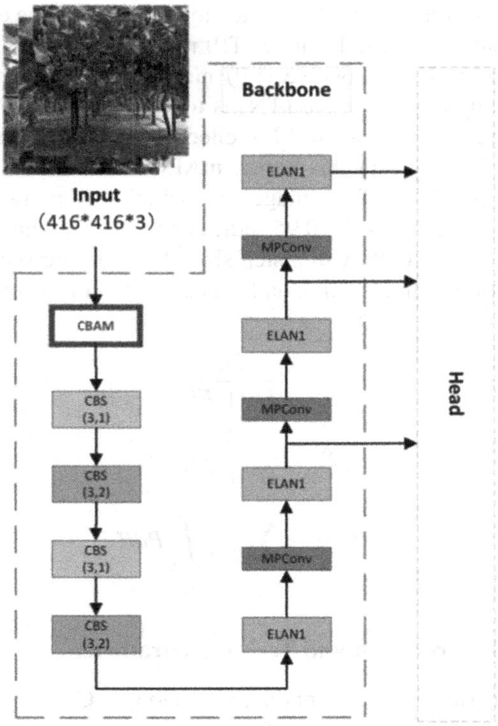

Fig. 7. The location of the CBAM in the improved model.

As shown in Fig. 7, a pre-processed image of size 416 × 416 × 3 is input into the backbone. The output feature map is firstly processed through the global max pooling and global average pooling in CBAM, then through a multilayer perceptron with shared weights, which conducts an addition operation based on the two feature maps through the sigmoid activation function. After the channel attention module is completed, the feature maps are input into the spatial attention module. The two feature maps are combined by using global max pooling and global average pooling. Then the number of channels is reduced to 7 × 7 convolution [29]. The sigmoid activation function obtains the spatial attention feature maps. Finally, the outputs of channel attention module and the spatial attention module are multiplied to obtain the output feature map of CBAM. The feature maps from the CBAM are fed into the CBS module in the original Backbone, the final predictions are output to implement the object detection by the model.

4 Our Results

4.1 Evaluation Metrics for Kiwifruit Detection

In this paper, we are use of precision (P), recall (R), and mean average precision (mAP) as the evaluation metrics for the Kiwifruit detection algorithm. The experimental results encapsulated four outcomes, True Positive (TP) refers to manually marked Kiwifruits being detected correctly, False Positive (FP) means the object that was incorrectly detected as a Kiwifruit, True Negative (TN) is to the negative samples with negative system prediction, and False Negative (FN) reflects to Kiwifruits that are missed. Two mAP indicators are employed in this paper, mAP@0.5 and mAP@0.95. mAP@0.5 refers to the average precision of all images in each class if assigned IoU to 0.5, and then all classes are averaged. mAP@0.95 indicates to the average mAP over different IoU thresholds (from 0.5 to 0.95 with a step size of 0.05). Intersect over Union (IoU) reflects to the proportion of intersection and concatenation of the object prediction box and the true box.

$$P = \frac{TP}{TP + FP} \tag{1}$$

$$R = \frac{TP}{TP + FN} \tag{2}$$

$$mAP = \frac{1}{N}\sum_{i=1}^{N}\int_{0}^{1} PdR \tag{3}$$

4.2 Experimental Environment and Training Parameters

In this paper, the experiments were conducted in Google Collaboratory platform. We were use of Python 3.7 (version 3.7.14), Pytorch (version 1.12.1), and CUDA (version 11.2) for the YOLOv7 training. The Tesla T4 (16G) GPU was utilized for the detection model training. The size of all images used for training in this experiment is 416 $\times$ 416, batch-size is 16 and epochs are 150 times.

4.3 Experimental Results and Comparisons

The results of the improved model for the detection of Kiwifruits are shown in Fig. 8, which shows that the model is better for detecting high density, overlapping objects and small objects.

In order to verify the effectiveness of the improved YOLOv7 model, we compared the original YOLOv7 model. We also compared the more popular YOLOv5s model. In addition, we compared six improved YOLOv5s models. We took use of two models to insert three attention mechanisms, SE, ECA and CBAM, into different positions of the YOLOv5s model. In the first method, we inserted these three attention mechanisms in front of the SPFF module of Backbone in YOLOv5 model. We named the improved algorithm YOLOv5_SE, YOLOv5_ECA, and YOLOv5_CBAM. In the second method, we replaced these three attention mechanisms with the C3 module within

Backbone in the original model and named it YOLOv5_C3SE, YOLOv5_C3ECA, and YOLOv5_C3CBAM. Figure 9 (a) shows the first method of inserting the attention module. Figure 9 (b) indicates the second method. The training parameters of the comparison experimental models are consistent. The results of the different models are compared in Table 1.

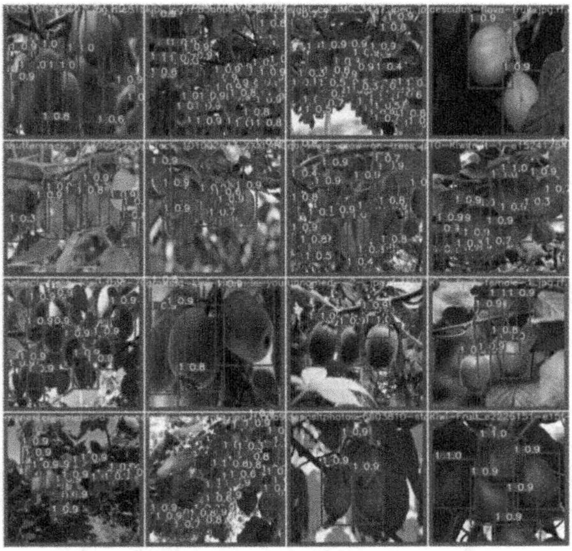

Fig. 8. Test results using the improved model.

In Table 1, YOLOv7_CBAM model attained the best performance with the same experimental parameters. YOLOv7 and YOLOv7_CBAM models outperformed YOLOv5s and six attention mechanisms addition models based on YOLOv5s in the Kiwifruit detection experiments. The improved YOLOv7 model increased the precision by 1.1%, recall by 3.8%, the mAP@0 .5 value by 0.8% and the mAP@0.95 value by 0.5% in comparison to the original YOLOv7 model. YOLOv5s model with the C3 module, replaced by the attention mechanism, has the smallest size due to the small number of model layers and the small number of model parameters. We also see from Table 1 that the approach of adding the attention mechanism to the YOLO algorithm produced better performance than replacing the original CBS module. From the comparison, we see that the addition of CBAM can help YOLOv7 model to improve the performance of object detection, which affirms the effectiveness of the approach proposed in this paper.

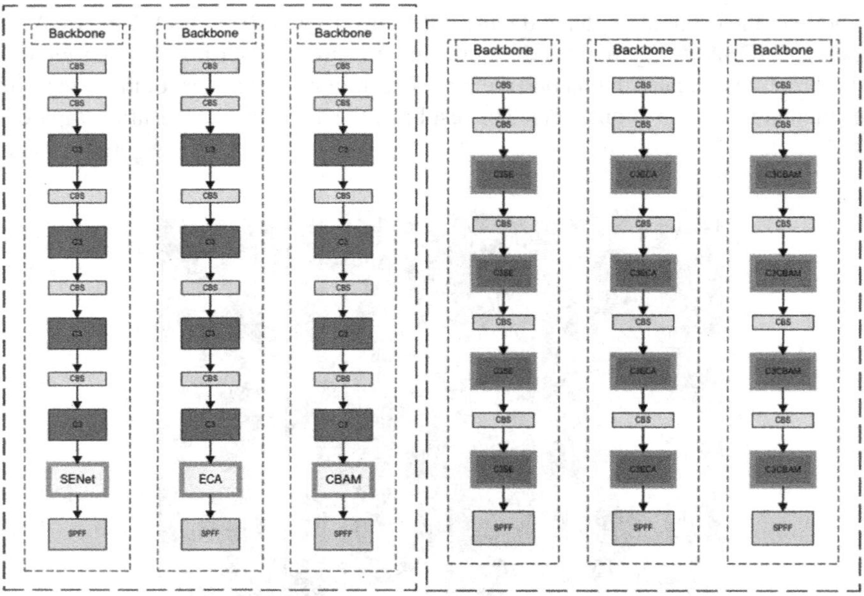

(a) (b)

Fig. 9. (a) The method for inserting three attention mechanisms ahead of the SPFF module in the Backbone of the YOLOv5s algorithm. (b) The method for replacement of the C3 module in Backbone of YOLOv5s algorithm with three attention mechanisms.

Table 1. Comparison of the improved model with other models.

Model	Precision	Recall	mAP@0.5	mAP@0.95	Model size
YOlOv5s	92.2%	86.0%	94.1%	63.1%	14.3MB
YOLOv5_SE	91.8%	87.7%	94.3%	61.2%	15.4MB
YOLOv5_C3SE	90.6%	85.6%	92.4%	58.6%	**13.1MB**
YOLOv5_ECA	91.4%	86.1%	94.2%	60.0%	14.3MB
YOLOv5_C3ECA	90.9%	85.1%	92.5%	60.6%	**13.1MB**
YOLOv5_CBAM	**94.1%**	85.1%	93.4%	60.5%	14.5MB
YOLOv5_C3CBAM	89.8%	85.3%	93.0%	60.1%	**13.1MB**
YOLOv7	92.1%	88.1%	95.3%	66.7%	74.8MB
YOLOv7_CBAM	93.1%	**91.9%**	**96.1%**	**67.2%**	74.8MB

5 Conclusion

Fruit detection and yield estimation have a significant impact on agricultural automation. Fastly and highly accurate fruit detection algorithms can aid harvesting robots in performing their picking tasks efficiently. In this paper, we presented a YOLOv7-based method for detecting Kiwifruits in orchards, which combines CBAM mechanism with the original YOLOv7 model. The method improved the current YOLOv7 model, the best-performed model in the YOLO family. The addition of the CBAM module assists the model in increasing attention to the object and reducing attention to useless features. Our experimental results show that the improved YOLOv7 model performed better than the original YOLOv7 in Kiwifruit detection. The results showcase the effectiveness of the improved algorithm. The future work in this research project is to reduce the model size and propose a highly accurate and lightweight improved Kiwifruit detection model. Meanwhile, tracking and counting of Kiwifruits in orchards will be achieved by combining multiobject tracking together to provide theoretical and technical references for further applications in practical scenarios [1, 16, 35].

References

1. An, N., Yan, W.: Multitarget tracking using Siamese neural networks. ACM Trans. Multimed. Comput. Commun. App. **17**, 1–6 (2021)
2. Bazame, H., Molin, J., Althoff, D., Martello, M.: Detection, classification, and mapping of coffee fruits during harvest with computer vision. Comput. Electron. Agric. **183**, 106066 (2021)
3. Bochkovskiy, A., Wang, C., Liao, H.: YOLOv4: Optimal speed and accuracy of object detection, https://arxiv.org/abs/2004.10934
4. Carion, N., Massa, F., Synnaeve, G., Usunier, N., Kirillov, A., Zagoruyko, S.: End-to-end object detection with transformers. In: Vedaldi, A., Bischof, H., Brox, T., Frahm, J.-M. (eds.) Computer Vision – ECCV 2020. LNCS, vol. 12346, pp. 213–229. Springer, Cham (2020). https://doi.org/10.1007/978-3-030-58452-8_13
5. Ferguson, A.: 1904—the year that Kiwifruit (Actinidia deliciosa) came to New Zealand. N. Z. J. Crop. Hortic. Sci. **32**, 3–27 (2004)
6. Fu, Y., Nguyen, M., Yan, W.Q.: Grading methods for fruit freshness based on deep learning. SN Comput. Sci. **3**, 264 (2022)
7. Ge, Z., Liu, S., Wang, F., Li, Z., Sun, J.: YOLOX: Exceeding YOLO series in 2021 (2021). https://arxiv.org/abs/2107.08430
8. Girshick, R., Donahue, J., Darrell, T., Malik, J.: Rich feature hierarchies for accurate object detection and semantic segmentation. In: IEEE Conference on Computer Vision and Pattern Recognition, pp. 580–587 (2014)
9. Gongal, A., Karkee, M., Amatya, S.: Apple fruit size estimation using a 3D machine vision system. Inf. Process. Agric. **5**, 498–503 (2018)
10. Goodfellow, I., Bengio, Y., Courville, A.: Deep Learning. MIT Press, Cambridge (2016)
11. Hu, J., Shen, L., Sun, G.: Squeeze-and-excitation networks. In: IEEE Conference on Computer Vision and Pattern Recognition, pp. 7132–7141 (2018)
12. Jilbert, M. N., Jennifer, C.D.: On-tree mature coconut fruit detection based on deep learning using UAV images. In: IEEE International Conference on Cybernetics and Computational Intelligence, pp. 494–499 (2022)
13. Lawal, O.: YOLOMuskmelon: quest for fruit detection speed and accuracy using deep learning. IEEE Access **9**, 15221–15227 (2021)
14. Liu, G., Hou, Z., Liu, H., Liu, J., Zhao, W., Li, K.: TomatoDet: anchor-free detector for tomato detection. Front. Plant Sci. **13**, 942875 (2022)
15. Liu, Y., Yang, G., Huang, Y., Yin, Y.: SE-Mask R-CNN: an improved Mask R-CNN for apple detection and segmentation. J. Intell. Fuzzy Syst. **41**, 6715–6725 (2021)

16. Liu, Z., Yan, W., Yang, B.: Image denoising based on a CNN model. In: IEEE ICCAR (2018)
17. Long, X., et al.: PP-YOLO: An effective and efficient implementation of object detector. https://arxiv.org/abs/2007.12099
18. Massah, J., AsefpourVakilian, K., Shabanian, M., Shariatmadari, S.: Design, development, and performance evaluation of a robot for yield estimation of Kiwifruit. Comput. Electron. Agric. **185**, 106132 (2021)
19. Olaniyi, E., Oyedotun, O., Adnan, K.: Intelligent grading system for banana fruit using neural network arbitration. J. Food Process Eng. **40**, e12335 (2016)
20. Pan, C., Liu, J., Yan, W., et al.: Salient object detection based on visual perceptual saturation and two-stream hybrid networks. IEEE Trans. Image Process. **30**, 4773–4787 (2021)
21. Pan, C., Yan, W.: A learning-based positive feedback in salient object detection. In: IEEE IVCNZ (2018)
22. Pan, C., Yan, W.Q.: Object detection based on saturation of visual perception. Multimed. Tools App. **79**(27–28), 19925–19944 (2020). https://doi.org/10.1007/s11042-020-08866-x
23. Redmon, J., Divvala, S., Girshick, R., Farhadi, A.: You only look once: unified, real-time object detection. In: IEEE CVPR, pp. 779–788 (2016)
24. Shan, T., Yan, J.: SCA-Net: a spatial and channel attention network for medical image segmentation. IEEE Access. **9**, 160926–160937 (2021)
25. Shen, D., Xin, C., Nguyen, M., Yan, W.: Flame detection using deep learning. In: IEEE ICCAR (2018)
26. Vaswani, A., et al.: Attention is all you need. In: Advances in Neural Information Processing Systems, vol. 30 (2017)
27. Wang, C., Bochkovskiy, A., Liao, H.: Scaled-YOLOv4: Scaling cross stage partial network. https://arxiv.org/abs/2011.08036
28. Wang, C., Bochkovskiy, A., Liao, H.: YOLOv7: Trainable bag-of-freebies sets new state-of-the-art for real-time object detectors. https://arxiv.org/abs/2207.02696
29. Wang, C., Yeh, I., Liao, H.: You Only Learn One Representation: Unified network for multiple tasks. https://arxiv.org/abs/2105.04206
30. Wang, L., Yan, W.Q.: Tree leaves detection based on deep learning. In: Nguyen, M., Yan, W.Q., Ho, H. (eds.) Geometry and Vision. CCIS, vol. 1386, pp. 26–38. Springer, Cham (2021). https://doi.org/10.1007/978-3-030-72073-5_3
31. Wang, Q., Wu, B., Zhu, P., Li, P., Zuo, W., Hu, Q.: ECA-Net: Efficient channel attention for deep convolutional neural networks. https://arxiv.org/abs/1910.03151
32. Woo, S., Park, J., Lee, J.-Y., Kweon, I.S.: CBAM: convolutional block attention module. In: Ferrari, V., Hebert, M., Sminchisescu, C., Weiss, Y. (eds.) Computer Vision – ECCV 2018. LNCS, vol. 11211, pp. 3–19. Springer, Cham (2018). https://doi.org/10.1007/978-3-030-01234-2_1
33. Xiao, B., Nguyen, M., Yan, W.Q.: Apple ripeness identification using deep learning. In: Nguyen, M., Yan, W.Q., Ho, H. (eds.) Geometry and Vision. CCIS, vol. 1386, pp. 53–67. Springer, Cham (2021). https://doi.org/10.1007/978-3-030-72073-5_5
34. Yan, W.:Computational Methods for Deep Learning: Theoretic, Practice and Applications Texts in Computer Science. TCS. Springer, Cham (2021).https://doi.org/10.1007/978-3-030-61081-4
35. Yan, W.: Introduction to Intelligent Surveillance: Surveillance Data Capture, Transmission, and Analytics. 2nd Edn. Springer, Cham (2019). https://doi.org/10.1007/978-3-319-60228-8
36. Zhao, K., Yan, W.Q.: Fruit detection from digital images using CenterNet. In: Nguyen, M., Yan, W.Q., Ho, H. (eds.) Geometry and Vision. CCIS, vol. 1386, pp. 313–326. Springer, Cham (2021). https://doi.org/10.1007/978-3-030-72073-5_24

37. Zheng, K., Yan, W., Nand, P.: Video dynamics detection using deep neural networks. IEEE Trans. Emerg. Top. Comput. Intell. **25**, 223–234 (2017)
38. Zhu, X., Cheng, D., Zhang, Z., Lin, S., Dai, J.: An empirical study of spatial attention mechanisms in deep networks. IEEE CVPR, pp. 6688–6697 (2019)

A Novel Explainable Deep Learning Model with Class Specific Features

Deepthi Praveenlal Kuttichira[1](✉), Basim Azam[1], Brijesh Verma[1], Ashfaqur Rahman[2], and Lipo Wang[3]

[1] Griffith University, Brisbane, Australia
[2] Data 61/CSIRO, Canberra, Australia
[3] Nanyang Technological University, Singapore, Singapore

Abstract. The predictive accuracy of any machine learning model is highly depended on the features used to train the model. For this reason, it is important to extract good discriminative features from the raw data. This extraction of good features from raw data is a challenging task. Deep learning models like Convolutional Neural Networks (CNNs) have the ability to automatically extract features from raw data and also have excellent predictive capabilities. This excellent predictive capability and ability to extract good features from raw data have made CNN very popular, especially in the field of computer vision. Even though CNN is popular, like many other deep learning models it is also notoriously black-box model. The predictions made by a CNN model cannot be explained based on features that influenced the given predictions. In our work we put forth an architecture that has convolutional layers to extract features automatically and the predictions made by this model can be explained based on specific features/neurons that resulted in the prediction. The model put forth in this paper has accuracy that is on par with the state-of-the-art models. Also, its predictions are explainable with target class specific feature importance.

Keywords: Feature extraction · Explainable deep learning · Convolutional neural networks · Logistic regression

1 Introduction

The digital revolution of the current era has resulted in surplus data. Analyzing this data can help in retrieving useful information, but process this big data is a challenging task for a human. Machine Learning (ML) has a key role in processing and extracting useful information from this data. One such ML model that has gained wide popularity in the field of computer vision is convolutional neural network (CNN). This popularity of CNN can be attributed manly two reasons. First is the excellent predictive accuracy of the CNN model and second is its ability to learn good discriminative features from the raw input data.

W. Q. Yan et al. (Eds.): IVCNZ 2022, LNCS 13836, pp. 62–74, 2023.
https://doi.org/10.1007/978-3-031-25825-1_5

Extraction of good features is pertinent to the predictive accuracy of any ML model [1]. For this reason, feature engineering is an important part of the model training pipeline. Feature engineering is non-trivial and is often time consuming. Traditional feature extraction involved manual hand-engineered feature extraction techniques. This involved usage of complex mathematical formulas and required domain knowledge.

Some of the traditional feature extraction method used for extraction of discriminative features from image data are Histogram of Gradients (HOG) [2], Scale Invariant Feature Transform (SIFT) [3] and Local Binary Patterns (LBP) [4]. Selecting which of these methods is the best to extract good features from a given raw data requires domain expertise. This is because each of this method works the best for certain types of data. For example, HOG is generally used in the contest of human detection tasks whereas LBP is best applied to extract texture information from data. So, a model that can automatically learn discriminative features from raw data is advantageous and simplifies the need of manual feature engineering. This is where model like CNNs are gaining its popularity. CNNs have been known to learn features as good as hand crafted features when trained with sufficiently large train data [5].

A basic CNN architecture comprises of a feature extractor and a classifier. The feature extractor of CNN is build using convolutional and max-pooling layers. These layers learn discriminative features from the raw data [6]. This feature extractor is followed by a flatten layer which essentially vectorizes the feature learned by the feature extractor. This feature vector is used for classification by the classifier part of the CNN. The classifier part of a CNN is essentially a feed-forward neural network. This classifier part learns the inter-dependencies in the feature vector and maps it to the output. The weights learned by the feature extractor and the classifier are updated and optimized using back-propagation algorithm. The ability of the feature extractor to automatically learn features eliminates the need of heavy manual feature engineering. Though the automatic feature learning ability of CNNs is extremely advantageous, the mapping between the learned feature and the prediction made is complex and cannot be explained. This lack of explainability of the predictions made, makes these models back-box models. Even the engineers who built these models are unable to explain the predictions made by these models.

Recently, there has been a rise in the interest to explain the predictions made by the ML models. This is because in many fields where the stakes are high like in the medical field where the predictions made by these models can affect the life and treatment plan for a patient, it is pertinent to know the reasons for a certain prediction. There are many studies that tries to tackle the issue of explainability of ML models. One approach to explaining features is to visualize the features learned by a model in order find explanations. These methods aim to visualize the features learned by each of the layers of a CNN model to understand the final prediction made by the model [7, 8]. Other approach is to explain the predictions of a model by calculation the significance of each pixel in the input image in making a certain prediction. Saliency maps is an example for this line of approach [9]. In Saliency maps, each pixel in the input image is weighted in accordance to the significance it has on the prediction of a target class. Saliency maps gives insights on which parts of the input image are the most influential in target class prediction. Grad-CAM is another technique that computes the gradient of the target class and traces it back to the input space to identify the regions in input space that influenced the predictive decision the most [10]. Another approach is to project the features learned by the layers of a CNN in feature space to the input space. This is done using Deconvolutional neural networks [11]. Trying to calculate the variable importance is another general approach used to explain the predictions of many ML models [12].

There are many works that uses the convolutional layers of a CNN to automatically extract discriminative features from raw data [13–16]. Many works have been proposed to combine the feature extraction capabilities of a CNN model with other classifiers. Some works combine these automatically extracted features with SVM [17, 18], while others use these features with clustering algorithms like K-Nearest Neighbours [19, 20]. The goal of these works is to improve predictive accuracy of the proposed model.

In our work the goal is to not just have good accuracy but to put forth an architecture where the predictions of the model can be explained based on the features that influenced the decision most. For this we propose to extract features from raw data and use an inherently interpretable classifier like logistic regression to make predictions. We extract the weights of this logistic regression model to extract the features that influenced the prediction of target class the most. We also provide the visualization of the neurons that impacted the prediction the most by plotting the activation functions of these target class specific neurons. The main contributions of this paper are as follows:

1. Introduction and investigation of an architecture that automatically extract features from raw data and has predictive accuracy on par with other state-of-the-art methods.
2. The predictions made by the model are explained using the selection of class-specific neurons and visualizing its activation function.
3. Detailed evaluation and comparative analysis of the proposed method on benchmark image data sets.

The rest of the paper is organized as follows. Section 2 describes the proposed method in detail while Sect. 3 deals with the explanation of experiments and description of dataset. Section 4 highlights the results formulated from experiments and presents analysis based on outcomes of the experimental work. Section 5 finally concludes the work.

2 Proposed Method

The goal of our work is to put forth a novel classifier that has state-of-the art accuracy whose predictions can be explained in terms of the features/neurons that influences the predictions the most and by visualizing the activation functions of these most influential neurons for the predictions. Our proposed method involves feature learning/extraction, training of logistic regression model, extracting influential neuron for specific target class and finally visualizing the activation functions of these influential neurons.

We begin with the input data $D = \{x_i, y_i\}_{i=1}^{n}$ where n the number of instances in the input data and x_i is a $m-$ dimensional image vector of the form $x_i = \{x_{i1}, x_{i2}, \ldots x_{im}\}$. Entries in the vector x_i corresponds to each pixel in the input image. The input data can be split into training and testing set as $D^{trainf} = \{x_i^{train}, y_i^{train}\}_{i=1}^{n-k}$ and $D^{testf} = \{x_i^{test}, y_i^{test}\}_{i=1}^{n-k}$, where k is the number of images selected for training.

Our initial step involves training a CNN with D^{train}. A CNN model comprises of a feature extractor and a classifier. The convolutional and max-pooling layers form the feature extractor, and the fully connected layer forms the classifier of a CNN. In a convolutional layer a filter/kernel is moved over an input image to learn important features from it. A kernel/filter is a matrix that has weights in it.

The features are learned by taking the element wise dot product of the weights in the filter with the corresponding input patch in the image and summing the products together. The weights in the filter decide the kind of features learned. Some filters act as edge detectors and others captures more abstract features from the input.

Let the input size of the image I be (I_{width}, I_{height}) and the size of the kernel K be (K_{width}, K_{height}). The size of the output O obtained after convolution function (O_{width}, O_{height}), where $O_{width} = I_{width} - K_{width} + 1$ and $O_{height} = I_{height} - K_{height} + 1$. This obtained output is called the feature/activation map. Each layer has multiple filters. So, the output of a convolutional layer will be of the size $(O_{width} \times O_{height} \times kl)$ where kl is the number of filters in that layer. Pooling layer is used for dimensionality reduction. For pooling a square window is run over the input image and either an average value or the maximum value in the window corresponding to the input image is retained. The output of the layer before the flatten layer contains the features extracted by the CNN. These extracted features are flattened by the flatten layer before being fed to the classifier.

The classifier of a CNN is essentially a fully connected feed-forward neural network. The filters and pooling layer learn the important features from the input image. The neurons in the hidden layers in the feed-forward neural network essentially models the dependencies between these extracted features. In the first hidden layer we can understand the modelling of these dependencies better as the activation of a neuron is directly proportional to the weighted features. As the hidden layers increase, these dependencies become more complex and understanding these dependencies become difficult.

In our proposed work, we extract the output of the neurons in the first hidden layer of the feed-forward neural network part of the CNN network as the features. Let these extracted features for input x_i be z_i, where $z_i = f_{cnnf}(x_i)$ where f_{cnnf} denotes the function of convolution neural network up to the first hidden layer of the CNN model. The function f_{cnnf} acts as our automatic feature extractor. The vector z_i is of the size r where r = number of neurons in first hidden layer.

Using the feature extractor of the trained CNN we create the new training and testing data $D^{trainf} = \{z_i^{train}, y_i^{train}\}_{i=1}^{k}$, and $D^{testf} = \{x_i^{test}, y_i^{test}\}_{i=1}^{n-k}$. We use this new training data D^{trainf} to train logistic regression model. A logistic regression model makes prediction as follows (Fig. 1).

$$P\left(y_i^{pred} = 1\right) = \frac{1}{1 + e^\wedge(\beta_o + \beta_1 z_{i1} + \cdots + \beta_p z_{ir})} \tag{1}$$

Taking the natural log of Equation (1) will reduce the right-hand side as $\beta_o + \beta_1 z_{i1} + \cdots + \beta_p z_{ir}$, which is in the form of linear regression. β represents the weights corresponding to the features. The weights indicate the significance of the corresponding features.

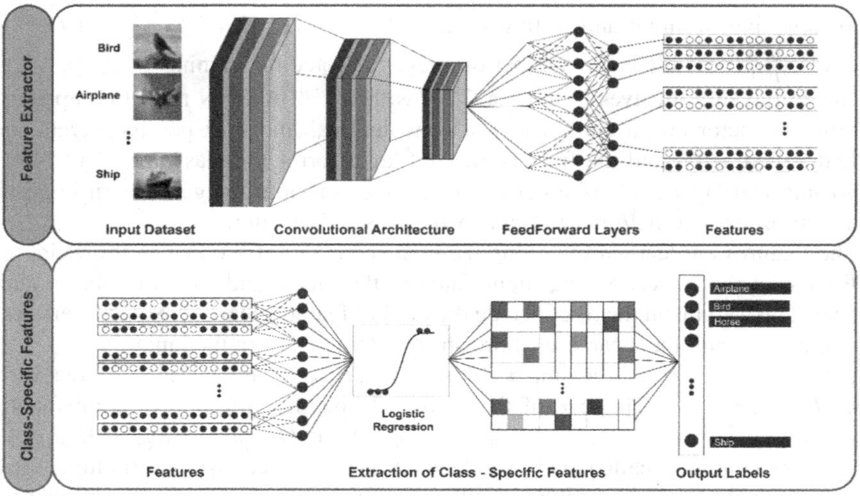

Fig. 1. Our proposed model has a feature extractor that extracts feature from the raw input data. Our feature extractor consists of convolutional and max pooling layers of neurons. We extract the features after the first hidden layer of the feedforward neural network part of the architecture. These extracted features are used to train a logistic regression classifier. Once the logistic regression classifier is trained, we extract the significant weights (weights that are above 0) for each class in the target class set T. Each class ti will have a set of weights significant to it. These significant weights for each class are represented using various colored boxes in the weight matrix of logistic regression classifier. The features corresponding to these significant weights are extracted as the class specific weights

The features that correspond to the highest weights are the most important features. In our proposed method the features correspond to the output of the neurons in the hidden neuron. We sort the weights in ascending order for each class in the target class set T. We select the weights that are above 0, which are essentially the weights that have a positive correlation between the feature and the prediction. For each class we extract these weights and corresponding features. These features are the class specific features. In order to visualize these class specific features, we plot the activation functions of these selected class specific neurons. The algorithms used for our proposed method are given below.

Algorithm 1: Training and Feature Extraction

Input: $D^{trainf} = \{x_i^{train}, y_i^{train}\}_{i=1}^{k}$

Output: *Feature Extractor* f_{cnnf} *and trained logistic regressor* f_{lr}.

1. Train the CNN using D^{train} until training loss reaches an acceptable point.
2. Extract the features from feature extractor f_{cnnf}
3. Transform the input x_i^{train} to z_i^{train} using feature extractor as $z_i = f_{cnnf}(x_i)$.
4. Create new training set as $D^{trainf} = \{z_i^{train}, y_i^{train}\}_{i=1}^{k}$.
5. Train the logistic regression model using D^{trainf}.

Algorithm 2: Testing

Input: *Feature extractor* f_{cnn}, *Trained logistic regression model* f_{lr} *and testing dataset* $D^{testf} = \{x_i^{test}, y_i^{test}\}_{i=1}^{n-k}$

Output: *Testing Accuracy, Extracted Features, Weights corresponding to extracted features.*

1. Pass the testing input $\{x_i^{test}\}_{i=1}^{n-k}$ to Feature Extractor f_{cnnf} to obtain $\{z_i^{test}\}_{i=1}^{n-k}$.
2. Pass the extracted features to the logistic regression model.
3. Obtain the predictions and evaluate the outputs using accuracy metrics.
4. Extract the weights corresponding to the feature from *logistic regression model* f_{lr} .

Algorithm 3: Extracting Class Specific Neurons and Feature Visualization

Input: *Feature extractor* f_{cnn}, *Trained logistic regression model* f_{lr} *and testing dataset* $D^{testf} = \{x_i^{test}, y_i^{test}\}_{i=1}^{n-k}$

Output: *Testing Accuracy and Feature Visualization.*

1. Pass the testing input $\{x_i^{test}\}_{i=1}^{n-k}$ to Feature Extractor f_{cnnf} to obtain $\{z_i^{test}\}_{i=1}^{n-k}$.
2. Plot the extracted features of each input against the number of neurons in hidden layer.
3. Observe the highest peak value obtained for each class.
4. Record the neuron for which the highest peak value obtained class-wise.
5. Plot the output of the neuron for all inputs to observe its activation for different classes.

3 Experiments

We have conducted our experiments on two benchmark data sets MNIST and CIFAR-10. The description of the data set are as follows:

- MNIST: This data set is a collection of handwritten images of digits from 0–9. The train set contains 60,000 images and the test set contains 10,000 images. All the images are grayscale images of the size 28 × 28. The number of target classes is 10.
- .CIFAR-10: This data set is a collection of RGB images of size 32 × 32 × 3. There are 10 target classes such as frog, horse, ship etc. This data set has 50,000 train images and 10,000 test images

For our experiments, first we train a CNN model to classify the MNIST and CIFAR-10 data sets respectively. The CNN model used to classify MNIST data set has the architecture as, the input layer followed by a convolutional layer that contains 32 filters of size 3 × 3, followed by a max-pooling layer that uses a pooling window of size 2 × 2. This layer is followed by 2 convolutional layers that has 64 and 32 filters in each of the respective layers. The filters are of the size 3 × 3. This is followed by a flatten layer and then the first hidden layer of the feed-forward neural network part of the CNN. The CNN for MNIST has only 1 such hidden layer which is followed by the output layer. The number of neurons in the hidden layers of the feed-forward neural network part of the CNN has been varied from 2 to 20 to analyze the impact of varying feature size on classification accuracy.

The architecture of CNN used for classification of CIFAR-10 data has 4 convolutional layers. The filters used in all convolutional layers are of the size 3 × 3.The first 2 convolutional layer has 32 filters each and the last 2 convolutional layers has 64 filters each. The second and third convolutional layer is followed by a max-pooling layer with the pooling wing window of size 2 × 2. After the final convolutional layer, there is a dropout and batch normalization layer before the flatten layer. The feed forward neural network part of this architecture has 2 hidden layers before the output layer. The number of neurons in the first hidden layer of the feed forward part is again varied from 20 to 200, to analyze the impact of feature length on classification accuracy.

The extracted features/neuron outputs are used to train a logistic regression classifier. For a comparative study of the accuracy, we also train other ML models such as decision tree (DT), Support vector machine (SVM) and Multi-layer perceptron (MLP) on the extracted features. We call the model from which the features are extracted the baseline model.

Once the logistic regression model is trained, we extract the class-wise weights. The features that have a weight greater than zero positively impacts the classification of the given class. We select these features as the most important features for the target class. To demonstrate the significance of these features, we again train a logistic regression model only with features extracted for each class. The class-wise accuracy is calculated to analyze the effect of these features in identifying the images belonging to the target class. We also plot the output of the selected neurons to observe the pattern it has for each class in the data set.

All the experiments are implemented using python. For the implementation of the CNN architectures in the proposed method TensorFlow and Keras package was used.

4 Results and Analysis

Our first set of experiments involves extracting features from the first hidden layer of the feed forward part of the baseline model and training a logistic regression model. As mentioned, before we also show the results on DT, SVM and MLP models for comparative analysis. The results obtained for train data set and test data set for MNIST data is shown in Tables 1 and 2 respectively. The results obtained for train and test data for CIFAR-10 data set is shown in Table 3 and 4. It can be seen from Table 1, 2, 3 and 4 that our proposed method of training a logistic regression model with extracted features performs the best in terms of accuracy. It can also be seen that in case of MNIST data set, the best accuracy is obtained when neurons in the hidden layer is set to 20. For CIFAR-10, the best accuracy is obtained when trained with the feature length of 200.

Table 1. Accuracy (%) on test set of benchmark MNIST dataset

# Neurons in first hidden layer	Neuron features + Logistic regression	Neuron features + DT	Neuron features + MLP	Neuron features + SVM
2	**99.29**	98.19	87.56	64.89
20	**99.31**	98.96	99.34	99.35
40	**99.23**	98.77	99.32	99.22
80	**99.27**	98.71	99.32	99.23
100	**99.16**	98.82	99.21	99.20
200	**99.14**	98.63	99.33	99.19

Table 2. Accuracy (%) on train set of benchmark MNIST dataset

# Neurons in first hidden layer	Neuron features + Logistic regression	Neuron features + DT	Neuron features + MLP	Neuron features + SVM
2	**99.09**	99.97	88.02	99.63
20	**99.91**	100	99.95	99.99
40	**99.89**	100	99.92	99.99
80	**99.92**	100	99.93	99.99
100	**99.95**	100	99.97	99.99
200	**99.96**	100	99.99	99.99

We extracted the weights of the best performing logistic regression model for both data sets respectively. Based on the highest weights, the features/neurons that is most significant for a particular class is extracted. A logistic regression model is trained for each set of these features extracted and the accuracy of the target class for which the features were extracted is calculated. For MNIST data the result is shown in Table 5 and for CIFAR-10 data, it is shown in Table 6. From the Tables 5 and 6, some neuron/feature is significant for multiple neurons, but each target class has a unique set of features/neurons. For MNIST data, the significance of neuron 1 can be observed for both class 1 and class 4. Figure 2 visualizes the activation function of neuron 1 when the instances of target class 1, 4, 0 and 5 is passed to it.

Table 3. Accuracy (%) on test set of benchmark CIFAR-10 dataset

# Neurons in first hidden layer	Neuron features + Logistic regression	Neuron features + DT	Neuron features + MLP	Neuron features + SVM
2	**79.04**	77.46	80.57	78.37
20	**89.56**	84.89	88.38	89.61
40	**89.90**	85.42	88.57	89.82
80	**89.80**	82.71	86.49	89.12
100	**89.93**	81.62	85.09	89.22
200	**90.45**	81.53	82.04	89.00

Table 4. Accuracy (%) on train set of benchmark CIFAR-10 dataset

# Neurons in first hidden layer	Neuron features + Logistic regression	Neuron features + DT	Neuron features + MLP	Neuron features + SVM
2	**80.77**	99.94	84.33	88.77
20	**93.52**	100	93.24	94.12
40	**94.06**	100	92.64	94.05
80	**94.77**	100	91.62	94.94
100	**94.82**	100	90.70	94.10
200	**95.70**	100	89.93	93.84

Tables 7 and 8 shows the comparative analysis of the accuracy obtained by the proposed method with other state-of-the-art methods for MNIST and CIFAR-10 data. It can be seen from the Tables 7 and 8 that our proposed method either outperforms or performs on par with other state-of-the-art methods.

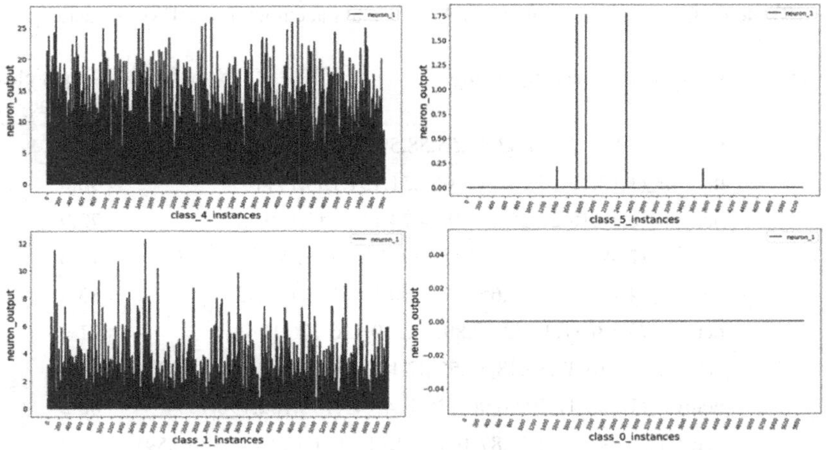

Fig. 2. The figure shows the activation function of neuron 1 for the input instances of classes 4, 5, 1 and 0 of MNIST data. It can be seen from the above figure that the neuron 1 activates the most for classes 1 and 4 and hence it is significant in determining the classes 1 and 4

Table 5. Class-specific features and target class accuracy for MNIST data.

Target class	Class specific neuron/feature	Target class accuracy
0	neuron 5, neuron 8, neuron 10, neuron 18	97.85
1	neuron 1, neuron 7, neuron 10, neuron 13, neuron 18	96.56
2	neuron 0, neuron 8–11	98.35
3	neuron 9, neuron 11–13	97.22
4	neuron 1, neuron 8–9, neuron 14, neuron 17	95.92
5	neuron 5, neuron 10, neuron 12, neuron 17	95.85
6	neuron 8, neuron 13, neuron 17	96.03
7	neuron 11, neuron 14	94.55
8	neuron 0, neuron 6–7, neuron 9, neuron 13	90.04

Table 6. Class-specific features and target class accuracy for CIFAR-10 data.

Target class	Class specific neuron/feature	Target class accuracy
0	neuron_(12,21,25,27,43,47,55,58,59,82,95,98)	82.4
1	neuron_(1,14,15,17,25,34,38,50,51,151,155,171)	93.0
2	neuron_(6,10,18,23,26,36,40,54,74,113,143,189,19)	72.9
3	neuron_(2–4,11,20,37,39,44,46,66,106,107,124,136,175)	72.6
4	neuron_(4,28,41,56,60,65,71,74,80,116,133,134,139)	83.9
5	neuron_(3,5,6,31,39,42,79,89,93,119,136,176,194)	71.9
6	neuron_(13,16,27,36,48,52,57,87,101,150,164,169,178,193)	88.9
7	neuron_(18,35,41,70,86,90,136,144,149,152,182,190,197)	88.2
8	neuron_(0,19,25,72,75,87,101,104,112,121,122,127,135,159)	88.8
9	neuron_(4,8,25,36,62,105,125,152,162,171)	88.1

Table 7. Accuracy comparison for benchmark MNIST data with other state-of-the-art methods.

Model	Accuracy (%)
Baseline CNN model	99.07
Extracted feature + Logistic regression	99.30
Tsetlin machine [21]	98.20
Park et al. [22]	98.10
CBoF & DSH [23]	99.45

Table 8. Accuracy comparison for benchmark CIFAR-10 data with other state-of-the-art methods.

Model	Accuracy (%)
Baseline CNN model	90.01
Extracted feature + Logistic regression	90.04
CBoF & DSH [23]	88.70
kMobileNet 16ch [24]	89.81

5 Conclusion

In our work we have proposed a method by which we can achieve excellent classification accuracy and also explain the prediction made on the basis of class specific neurons that impacts the prediction the most. Visualization of these class-specific neurons also sheds insights into the influence of these features in the classification of an instance into the target class. The experimental evaluation of the proposed method on two benchmark datasets MNIST and CIFAR-10 showed excellent performance. In our future work, we will extend the evaluation of the proposed model on more benchmark data sets. In future the visualization in the feature space will be projected on to the input space for better human interpretation. We will also explore better techniques for updating the weights of the proposed model.

Acknowledgments. This research was supported under Australian Research Council's Discovery Projects funding scheme (project number DP210100640).

References

1. Bengio, Y., Courville, A., Vincent, P.: Representation learning: a review and new perspectives. IEEE Trans. Pattern Anal. Mach. Intell. **35**(8), 1798–1828 (2013)
2. Dalal, N., Triggs, B.:Histograms of oriented gradients for human detection. In: 2005 IEEE Computer Society Conference on Computer Vision and Pattern Recognition (CVPR'05), vol. 1, pp. 886–893. IEEE (2005)
3. Lowe, D.G.: Distinctive image features from scale-invariant keypoints. Int. J. Comput. Vision **60**(2), 91–110 (2004)
4. Ojala, T., Pietikainen, M., Harwood, D.:Performance evaluation of texture measures with classification based on kullback discrimination of distributions. In: Proceedings of 12th International Conference on Pattern Recognition, vol. 1. pp. 582–585. IEEE (1994)
5. Lin, W., Hasenstab, K., Moura Cunha, G., Schwartzman, A.: Comparison of hand-crafted features and convolutional neural networks for liver MR image adequacy assessment. Sci. Rep. **10**(1), 1–11 (2020)
6. LeCun, Y., Bengio, Y., et al.: Convolutional networks for images, speech, and time series. Handb. Brain Theory Neural Netw. **3361**(10), 1995 (1995)
7. Olah, C., Mordvintsev, A., Schubert, L.: Feature visualization. Distill **2**(11), e7 (2017)
8. Wang, Z.J., et al.: CNN explainer: Learning convolutional neural networks with interactive visualization. IEEE Trans. Visual. Comput. Graph. **27**(2), 1396–1406 (2020)
9. Mundhenk, T.N., Chen, B.Y., Friedland, G.: Efficient saliency maps for explainable AI. arXiv preprint arXiv:1911.11293 (2019)
10. Selvaraju, R.R., Cogswell, M., Das, A., Vedantam, R., Parikh, D., Batra, D.: Grad-cam: Visual explanations from deep networks via gradient-based localization. In: Proceedings of the IEEE International Conference on Computer Vision, pp. 618–626 (2017)
11. Zeiler, M.D., Fergus, R.: Visualizing and understanding convolutional networks. In: Fleet, D., Pajdla, T., Schiele, B., Tuytelaars, T. (eds.) ECCV 2014. LNCS, vol. 8689, pp. 818–833. Springer, Cham (2014). https://doi.org/10.1007/978-3-319-10590-1_53
12. Dunn, I., Mingardi, I., Zhuo, Y.D.: Comparing interpretability and explainability for feature selection. *arXiv preprint* arXiv:2105.05328 (2021)
13. Zhang, S., *et al.* A convolutional neural network based auto features extraction method for tea classification with electronic tongue. Appl. Sci. **9**(12), 2518 (2019)
14. Zheng, Y., Li, X., Si, Y., Qin, W., Tian, H.: Hybrid deep convolutional neural network with one-versus-one approach for solar flare prediction. Mon. Not. R. Astron. Soc. **507**(3), 3519–3539 (2021)
15. Sinha, T., Verma, B.: Auto-associative features with non-iterative learning-based technique for image classification. In: 2021 International Joint Conference on Neural Networks (IJCNN), pp. 1–6. IEEE (2021)

16. Xue, G., Liu, S., Ma, Y.: A hybrid deep learning-based fruit classification using attention model and convolution autoencoder. Complex Intell. Syst. 1–11 (2020)
17. Niu, X.-X., Suen, C.Y.: A novel hybrid cnn–svm classifier for recognizing handwritten digits. Pattern Recogn. **45**(4), 1318–1325 (2012)
18. Tang, Y.: Deep learning using linear support vector machines. *arXiv preprint* arXiv:1306.0239 (2013)
19. Huang, F.J., LeCun, Y.: Large-scale learning with SVM and convolutional for generic object categorization. In: 2006 IEEE Computer Society Conference on Computer Vision and Pattern Recognition (CVPR'06), vol. 1. pp. 284–291 IEEE (2006)
20. Notley, S., Magdon-Ismail, M.: Examining the use of neural networks for feature extraction: a comparative analysis using deep learning, support vector machines, and k-nearest neighbor classifiers. *arXiv preprint* arXiv:1805.02294 (2018)
21. Abeyrathna, K.D., et al.: Massively parallel and asynchronous tsetlin machine architecture supporting almost constant-time scaling. In: International Conference on Machine Learning. pp. 10–20. PMLR (2021)
22. Park, J., Lee, J., Jeon, D.: 7.6 A 65 nm 236.5 nJ/classification neuro morphic processor with 7.5% energy overhead on-chip learning using direct spike-only feedback. In: 2019 IEEE International Solid-State Circuits Conference-(ISSCC), pp. 140–142. IEEE (2019)
23. Passalis, N., Tefas, A.: Training lightweight deep convolutional neural networks using bag-of-features pooling. IEEE Trans. Neural Netw. Learn. Syst. **30**(6), 1705–1715 (2018)
24. Schuler, J.P.S., Romani, S., Abdel-Nasser, M., Rashwan, H., Puig, D.: Grouped pointwise convolutions reduce parameters in convolutional neural networks. Mendel **28**(1), 23–31 (2022)

Correcting Charge Sharing Distortions in Photon Counting Detectors Utilising a Spatial-Temporal CNN

Aaron Smith[1,2](✉), James Atlas[1,2], and Ali Atharifard[2]

[1] University of Canterbury, Christchurch, New Zealand
aaron.smith@pg.canterbury.ac.nz, james.atlas@canterbury.ac.nz
[2] MARS Bioimaging Limited, Christchurch, New Zealand
ali.atharifard@marsbioimaging.com

Abstract. Charge sharing induces spectral and spatial distortions on photon counting detectors which must be corrected using methods such as charge summing circuitry. We propose a method of correction using a spatial-temporal convolutional neural network based on the CycN-Net design. Our results were compared to an analytical scalar matrix correction and a U-Net. We show improvements in two energy channels set to 50 and 60 kev with a mean absolute percentage error reduced from 4.84% and 7.46% to 3.95% and 5.14% respectively when compared to the scalar matrix approach. We believe this shows the potential viability of utilising the spatial-temporal CNN approach for correcting charge sharing distortions in higher energy ranges, where photon counts tend to be lower for photon counting detectors.

Keywords: Neural network · Photon counting · Spectral CT

1 Introduction

Photon counting computerised tomography (CT) scanners are becoming an increasingly useful area of medical imaging [5]. The ability of the photon counting CT scanner to generate medical scans with high spatial resolutions and with lower doses when compared to traditional CT scanners [9] enables patient care benefits for medical imaging. Photon counting detectors' (PCD) discretised energy values enable enhanced material decomposition [13]. However, the decreased pixel size used to detect individual photons at increased resolutions introduces distortion issues such as charge sharing. Charge sharing introduces spectrum distortion via a single incident photon being recorded as two or more lower energy photons. This also results an increase in noise present in the scan [12].

Some photon counting chips, such as the Medipix3RX, have extra built-in circuitry to combat the effects of charge sharing by enabling charge summing mode (CSM) [3]. This charge summing circuitry consumes valuable space on the board and as such is only implemented in four out of the seven regular energy channels. The other three channels are currently unusable in reconstruction due to the artifacts they introduce. Finding an accurate correction method for these

W. Q. Yan et al. (Eds.): IVCNZ 2022, LNCS 13836, pp. 75–90, 2023.
https://doi.org/10.1007/978-3-031-25825-1_6

three channels would increase spectral resolution and the ability to differentiate between materials in material decomposition. Other approaches to correcting the charge sharing problem have also been suggested at the hardware level, such as the implementation of multi-energy inter-pixel coincidence counters [14], which are used to count the charge sharing occurrences and subsequently correct the scan afterwards. Some other work has developed analytical methods in correcting spectrum distortions in photon counting detectors [4].

Machine learning is increasingly becoming an area of interest for tomographic image reconstruction in the medical imaging domain [16]. Some machine learning applications to challenging medical imaging problems include reducing motion artefacts in 4D cone-beam CT [18], increasing small feature preservation on CT reconstruction [15] and enabling operators to manage the trade-offs of denoising CT images with user driven denoisers [2].

Previous methods to correct charge sharing effects include neural networks trained on simulated data [10], and the use of U-Nets to correct spectral distortion issues encountered in single pixel mode when using energy integrated images as targets [8]. This paper proposes what we believe to be a novel solution to the charge sharing problem which is based on the unique design of the Medipix3RX. Our solution is to use a U-Net based on the CycN-Net design [18], which is fed spatial-temporal correlated data. The data contains projection images with the three single pixel mode (SPM), channels and the arbitration counter as the prior with an N times series window to generate charge sharing corrected SPM channels. The corrected SPM channels are compared to the target charge summing mode channels, set to the same energy band. The aim is to correct the charge sharing issues while still maintaining the high level of spatial resolution made possible with PCD, since image features are present in the time series scan. This would allow the CSM channels to be set to different thresholds on a future scan with the net result being four CSM channels, three SPM corrected channels and one arbitration counter channel. The rest of this paper is organised as follows: In Sect. 2, we describe the charge sharing problem and the corrections performed at the hardware level by the Medipix3RX chip. Section 3 provides an outline of the proposed framework and implementation. Section 4 provides an outline of the experimental setup. Section 5 presents results obtained and comparisons to current state of the art methods when performed on a variety of datasets. Section 6 analyses current issues with our implementation, followed by the conclusion of the paper in Sect. 7 and an overview of future work in Sect. 8.

2 Background

2.1 Charge Sharing

Charge sharing occurs in photon counting detectors due to a combination of factors. Firstly, the pixel size is small when compared to other integrated energy CT detectors, which provides high spatial resolution. Secondly, photon counting detectors measure photons through an electron cloud, which are produced when the photon strikes the surface of the detector. This means that the electron cloud that is produced has a high probability of overlapping significant areas of multiple pixels. This may lead to multiple pixels registering a lower energy photon, from a single higher energy incident photon. If the charge shared between these pixels is less than the lowest energy threshold of the chip, the pixel will not register the photon event, thereby missing a count. When image reconstruction algorithms such as filtered back projection are performed later down the pipeline, these lower energy, multiple photon ray paths are computed through paths that never existed. This may lead to a increase in noise present, and to incorrect material decomposition. Equation 1 [4] describes the expected width at the anode σ_t of a charge cloud with a Gaussian charge density distribution from an initial electron cloud width σ_i, taken to be 5 µm [7],

$$\sigma_t = \sqrt{2\frac{k_B T z d}{qU} + (\frac{z d N q}{10\pi\epsilon U})\frac{1}{\sqrt{5\sigma_i}} + \sigma_i^2}, \tag{1}$$

where d is the depth of the pixel's crystal layer, U is the bias voltage and z is the electron cloud's height above the anode. The number of chargers per photon is given as $N = E_e/\Delta_e$ where Δ_e is the energy per electron hole pair, which for a CdTe detector is 4.43eV/ehp [11]. The variables defined here should not change greatly between projections, leaving the dominate factor in the variation of the charge cloud dependant on the original photon collision point.

Charge sharing is not a uniform occurrence across all sensors. The charge clouds produced can vary due to manufacturing variation that effects the pixel electric fields and how the charge is dissipated for the electron cloud.

2.2 Medipix3RX

The Medipix3RX is a photon counting detector which contains a 256×256 matrix of pixels, which each have a squared pixel space of 55 µm × 55 µm [3]. When spectroscopic mode is enabled, every group of 2×2 pixels forms a single, larger pixel which reduces the matrix to 128×128 pixels with an area of 110 µm × 110 µm. The Medipix3RX detector has seven normal energy channels, plus an arbitration counter. Four of these channels are capable of mitigating the charge sharing issue when set to charge summing mode. This mode utilises summing circuits located in the corner boundaries between pixels in a 2×2 cluster to detect charge sharing events. If a given pixel has the largest charge deposition with respect to its neighbours in the grid and the summing circuit recorded value exceeds a given threshold, the photon count is recorded for that pixel. Figure 1 provides a diagram of this charge summing circuit.

Charge summing allows the Medipix3RX chip to provide a better approximation of the photon collision point when compared to the original single pixel mode and, by collecting all the charge for a single photon, also corrects the spectrum distortion. Due to hardware space limitations only four out of seven channels have this feature. The remaining three channels are limited to single pixel mode and cannot be directly used in reconstruction without inducing errors. This means that the remaining channels must cover larger energy ranges, decreasing the ability to distinguish between materials via material decomposition.

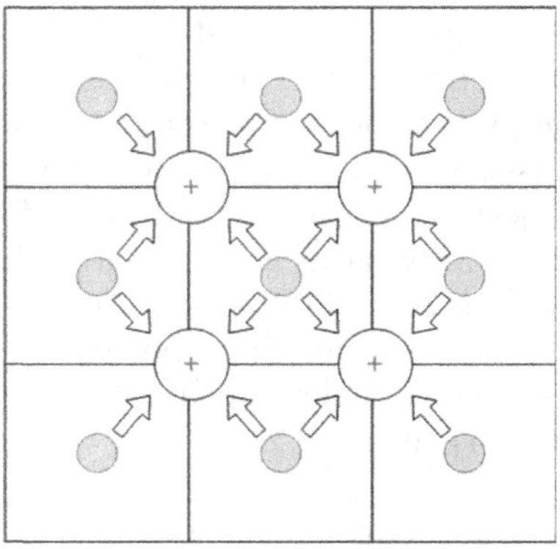

Fig. 1. Charge summing circuit: Charge is deposited into the material, which is connected to a comparator. This is used to determine which of the charges are the largest in each 2×2 grid so that the charges can be assigned to a single pixel.

[6] provides a detailed analysis of the Medipix3's charge summing correction circuits for the later Medipix3RX version, analysing its effects and showing that the algorithm is effective in suppressing charge sharing events.

3 Proposed Framework

3.1 Network

The goal of our machine learning model is to correct the acquired single-pixel mode PCD data to have the non-distorted, ideal spectral and spatial data. From above, we have determined that the Medipix3RX hardware charge summing algorithm is an approximate correction of the charge sharing distortions to the spectrum and thus is an approximate inverse of equation one, assigning all the energy of the cloud to the pixel which had the majority of the initial electron cloud σ_i in its region. As a result, we can use this data as the target for training the model. Our model should therefore be a function that maps our distorted input x to the target charge sharing corrected data y. To do this, the following minimisation loss function is applied to a neural network:

$$\min_f mean \frac{||f(x) - y||}{y + \epsilon} * 100, \tag{2}$$

where f is the network forward pass and ϵ is a small value to prevent zero division. This loss function defines a mean absolute percentage error (MAPE) and is used instead of mean squared error because higher energy threshold channels will count the same or fewer photons than the lower energy counterpart for the same exposure. This is because the energy threshold is set so all photons with an energy level equal to or greater than the threshold are recorded. This leads to a reduced mean square error for higher energy bins, causing the network to favour improvements in the lowest energy bin.

A typical neural network approach to this type of correction problem is to implement a U-Net. However, a U-Net fails to fully utilise all the features in the dataset. The first issue is the utilisation of the arbitration counter, which contains a total count of all photons recorded above the lowest energy threshold. This channel is not an ideal input to the U-Net with the single pixel mode channels, since we are not trying to correct it. It also follows its own distributions different from the SPM data. However, for comparison to the later network, we implement a standard U-Net outlined in Fig. 2.

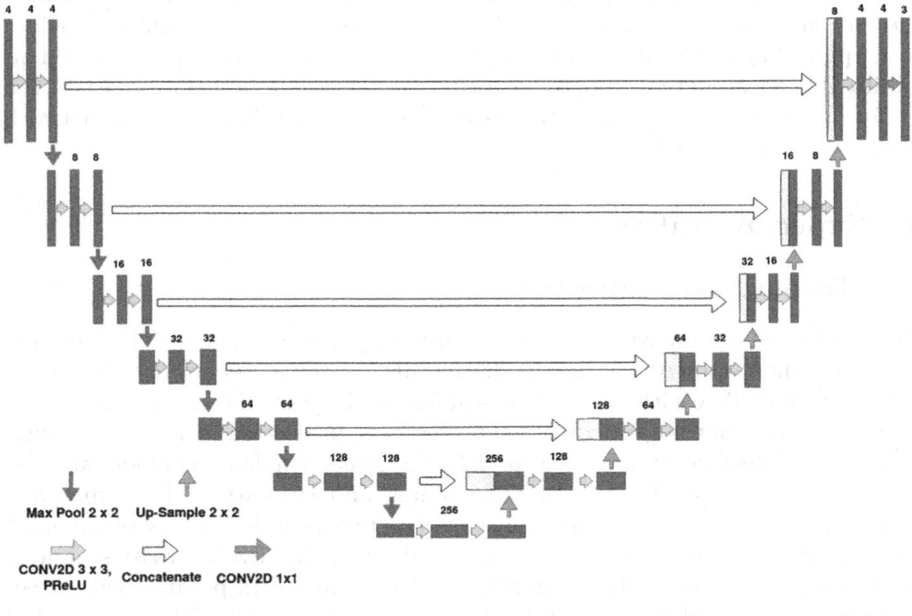

Fig. 2. Comparison U-Net Architecture: Input is the three single pixel mode images and the arbitration counter. The network contains blocks of Conv2D operations with a 3×3 window per level, followed by a 2×2 max pool for the compression process. Upsampling pathways utilise Conv2D, also with 3×3 windows. All Conv2D operations with 3×3 windows are followed by a PReLu activation function. Batch normalisation is also used before each Conv2D operation. The beginning 2D convolution block for each upsampling layer is provided with the previous upsampled result concatenated with the adjacent down-sampling result. The final upsampling block contains an extra Conv2D operation with a 1×1 window.

Because of the issues outlined with the utilisation of the arbitration counter, we modify the standard U-Net. We add another compression pathway for the arbitration counter data and connect it to the up-sampling pathway, giving us a total of two different compression pathways. The arbitration counter in this way is treated as a prior. If we then consider a stack of projections from frame $n - i$ to $n + i$ distorted by a charge sharing distribution, we have a case where the objects observed by the scanner will appear shifted based on a function of n. However, since charge sharing distribution has no time-dependent variables between multiple frames, we can utilise a stack of these images to increase the total number of counts in order to measure from the distribution and stabilise the corrected projection. This is done by allowing the network to account for object movement and reduce the variance in the charge sharing distribution.

From this, we derive a network structure that can process the time series data, based on CycN-Net [18] which was used for stabilising images distorted by respiratory cycles in 4D cone beam CT scans. We do make a key change to the authors' design in the up-sampling pathway. The CycN-Net design summed the current level of concatenated blocks with the past TConv output, before being passed into the current TConv block. We removed this step and added an additional concatenation step, to keep all the filters and found this increased the performance of the network on the datasets studied in this paper. The network design is described in Fig. 3.

4 Experimental Setup

4.1 Training and Testing Data

The purpose of the following experiments is to evaluate the ability of the modified CycN-Net model to correct charge sharing distortions at set energy thresholds for the Medipix3RX with different crystal materials. We will compare the corrections to the original single pixel mode projections, projections corrected through an analytical correction approach we define as scaled single pixel mode, and the U-net we defined previously. For the training and evaluation of the modified CycN-Net model and the comparison U-Net, two complete scans of different whole lamb legs were conducted, with two different MARS V6 wrist scanners. Both scanners contained 12 Medipix3RX PCD arranged in parallel, with spectroscopic mode enabled, for a total detector size of 128×1536. The first scanner utilised Medipix3RX detectors with CdZnTe as the crystal material with a thickness of 2 mm. The second scanner utilised Medipix3RX detectors with a CdTe crystal material with a thickness of 0.75 mm. These sensors have different detection properties, including their absorption efficiency, with 0.75 mm CdTe being transparent to some higher energy photons which the 2 mm CdZnTe absorbs. For both scanners, the energy bin thresholds were set to 40, 50, 60 and 70 kev for the charge summing mode enabled channels. The single-pixel mode channels were paired with the 40, 50 and 60 kev channels. The arbitration counter was set to a threshold of 7 kev. The tube current was set to 30 μA, the tube voltage was set to 120 kVp and the exposure time was set to 160 ms.

The lamb leg was used as a training phantom for a few key reasons. Firstly, it comprises biological materials and contains a variety of materials typically found

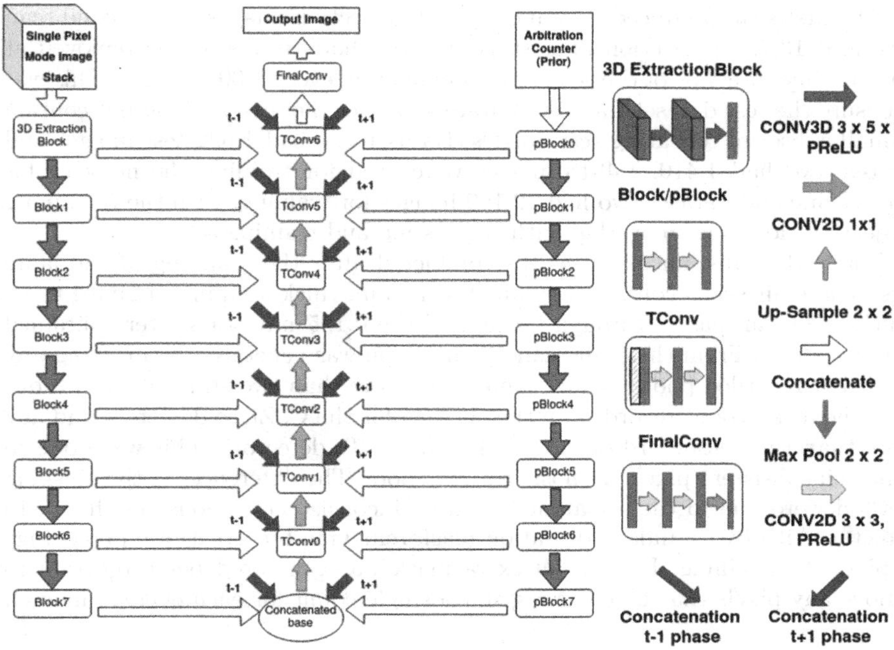

Fig. 3. Modified CycN-Net Architecture: The network consists of three copies stacked together. Each copy receives a block of 5 projections, with the middle projection being the projection to be corrected. The extra copies of the network are used to provide additional time stabilisation for the target frame. The 3D extraction block utilises two Conv3D blocks, each with a $3 \times 5 \times 5$ window. Blocks and pBlocks contain the same operations described in the U-net, including batch normalisation and PReLu activation functions. The concatenated base consists of the concatenated results of Block7, pBlock7 and the results from the t-1 and t+1 versions of these blocks. TConv blocks are again the same as the U-Net's, but with additional filters concatenated at the start of each block. These concatenated filters are the results from the previous TConv block joined with the result of the corresponding p/block and the t-1 and t+1 p/blocks. The FinalConv block is the same as the described U-Net.

in human wrist CT scans, such as muscle, fat, and bone, all in approximately the right locations. Secondly, it is large and has a varied, asymmetrical shape. This means different materials will appear in multiple regions of the scan. This is of note because, in a more homogeneous sample, such as the Pyrex cylinder we originally used, the network tends to memorise the sample scan pattern and fails to generalise. Image morphology methods fail on this dataset due to differences in the sensor geometry and internal electric fields for individual pixels, meaning that variance in the scanned object had more importance, as artificially shifting it in post-processing induced errors.

The first scan produced a total of 16,832 projection images. The second scan produced 13,795 projection images. Both scans had groups of 100 removed at varying time intervals of the scan at approximately every 1,000 images. This was to ensure the test dataset was an accurate subset of the scan and did not contain a small subset of repeating scan angles. Dataset one had 1,692 test images and dataset two had 1,410. 1,4918 images were used for training the network for dataset one and dataset two had 12,199 images for training, with the remaining projections lost due to overlap with the testing and training set.

For both scans, an air scan was conducted after the scanning of the lamb legs. These air scans contain only air, the 1.6 mm thick aluminium guard tube, which the beam passes through twice, and the 0.125 mm brass filter contained in the scanner. From these air scans, a mask file was generated by masking any pixel that recorded photon counts that were more than two standard deviations from the mean count recorded on the air scan for the CdZnTe detectors and any more than three standard deviations for the CdTe detectors. This was done to remove inconsistent pixels from the training pool. The differences in the masking method were due to the apparent increased inconsistent response of the CdTe detector, with two standard deviation mask removing too great of a percentage of pixels for training. Finally, an extra mask was generated per projection to remove any pixels that failed to register a single count for each energy channel.

4.2 Network Training and Testing

For training both the U-Net and the modified CycN-Net an RTX 3080 10Gb was used with TensorFlow version 2.7.0. The loss function used was the previously described MAPE and was minimised using the ADAM optimiser. The starting learning rate was set to 10^{-3} and decreased by half after four epochs and then decreased by half again after another four epochs. Both the U-Net and the modified CycN-Net were trained for 15 epochs on dataset one and 20 epochs on dataset two, with the epoch with the best score on the test metrics being used for the results. The batch size for the U-Net training was set to 16 however due to increased network and input data size the batch size for CycN-Net was limited to only four.

5 Results

5.1 Metrics and Evaluation

To evaluate our approach to correcting charge sharing distortions, we measure the distortion induced in the single pixel mode channels by calculating mean absolute percentage error (MAPE) against the corresponding charge summed channels. We also utilise the peak signal to noise ratio (PSNR) as a metric for performance as defined in Eq. 3:

$$PSNR = 10\log_{10}(\frac{Maxv^2}{MSE}), \tag{3}$$

where $Maxv$ is the maximum pixel value for the projection in the charge summed image and MSE is the mean squared error. Mean squared error is defined in Eq. 4:

$$MSE = \frac{1}{N}\sum_{i=1}^{N}(x_i - y_i)^2, \tag{4}$$

where N is the number of pixels in our projection image, x_i is the corrected photon count for pixel index i and y is the original photon counts for pixel index i.

We also chose to use structural similarity [17] (SSIM) as an additional metric for analysis as defined in Eq. 5:

$$SSIM(x, y) = \frac{(2\mu_x\mu_y + C_1)(2\sigma_x y + C_2)}{(\mu_x^2 + \mu_y^2 + C_1)(\sigma_x^2 + \sigma_y^2 + C_2)}, \tag{5}$$

where x is defined as our corrected projection, y is our ground truth charge summed mode channel, with μ_x and μ_y being the window means. σ_x and σ_y are the variances of the x and y sample, with σ_{xy} being their covariance. C_1 and C_2 are small values to stabilise division. We calculate the SSIM in a two dimensional window, calculating the SSIM for each energy channel independently.

We generate comparison results for three approaches to correct charge sharing distortions. To provide a baseline analytical approach for scaled single pixel mode (SSPM), we calculate a mean scaling transformation matrix for the original single pixel mode channels per pixel, to match the mean value. This is described in Eq. 6:

$$z_{jc} = \frac{1}{M}\sum_{i=1}^{M}\frac{(x_{jc} + \epsilon)}{(y_{jc} + \epsilon)}, \tag{6}$$

where z is the scalar matrix, M is the size of air scan dataset, x is the original single pixel mode projection and y is the charge summed mode projection. The pixel index is j, c is the energy channel and ϵ is a small value to prevent zero division.

Following this, the matrix is then used to scale the single pixel mode counts to be the approximate correct value in the following equation:

$$sspm_{jc} = \frac{x_{jc}}{z_{jc}}, \tag{7}$$

where $sspm$ is our scaled single pixel mode projection. This is to account for the consistent reduced counts detected by the single pixel mode channels by artificially increasing the counts to be closer to the corresponding charge summed channels for each projection. We also take into account that different pixels have different charge sharing distributions, so this matrix attempts to approximate that distribution. The result should be an improvement for the metrics defined over the direct comparison of the single pixel mode channels; however it will still contain errors due to the fact that some counts will sometimes be assigned to the wrong pixel. We then calculate the metrics for projections corrected by a U-Net described earlier. Finally we calculate our metrics for the projections corrected by the modified CycN-Net design.

5.2 Results for Experiment One

Figure 4 displays our results for experiment one. We display a heat-map of the MAPE for the 60 kev energy channel, thresholded to 20% to better show the differences between the output and desired channel.

We can see that, in general, both the SSPM and U-Net have high errors within the denser materials, around the bone and tissue. The U-Net also has high error at material edges. However, the CycN-Net has a consistently lower error in this region. Table 1 outlines our results for the MAPE for experiment one.

Table 1. Mean absolute percentage error

Energy channel	SPM	SSPM	U-Net	Modified CycNnet
40 kev	26.78272	**3.34848**	4.99544	3.61354
50 kev	41.49372	4.83571	5.38180	**3.95082**
60 kev	50.80966	7.45931	6.47575	**5.13708**
Average	39.69537	5.21450	5.06868	**4.23381**

We see that the U-Net offers some improvement over the SSPM approach, but only in the highest energy channel. CycN-Net results in some improvement in the 50 kev threshold and it significantly reduced MAPE in the highest energy channels. However, the SSPM approach is more consistent in the lowest energy channel. Table 2 displays our results for the computed PSNR on the test dataset.

Table 2. PSNR

Energy channel	SPM	SSPM	U-Net	Modified CycN-Net
40 kev	22.6004	**41.38833**	36.86978	38.73384
50 kev	17.76955	37.07403	35.84707	**37.64204**
60 kev	15.77009	33.19969	34.43373	**35.95446**
Average	18.71335	37.22068	35.71686	**37.44345**

Again, we see the same trend, with the scaled approach working the best in the lowest energy bins and the CycN-Net performing the best in the remaining two, higher energy bins, although the 50 kev performance is comparable. The U-Net still performs better than the SSPM approach at the 60 kev threshold. Table 3 shows our computed SSIM results.

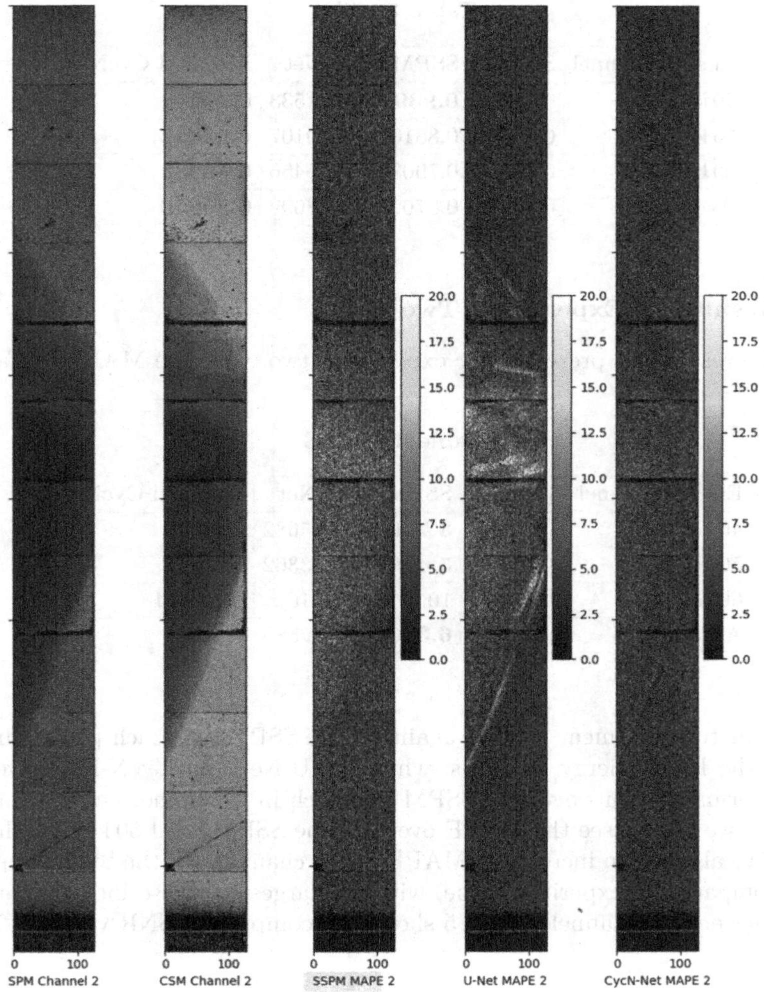

Fig. 4. MAPE comparison of the three correction methods at 60 kev threshold: From left original single pixel mode channel on the, followed by the corresponding ground truth charge summed mode channel, the MAPE heat map for the SSPM approach, the U-Net MAPE heat map and the CycN-Net MAPE heat map. Here we see that all methods do a reasonable job of predicting the air, but both the SSPM and the U-Net struggle with the darker regions

Here, the CycN-Net performs equivalently to the SSPM approach in the lowest energy channel and significantly better in the remaining channels. However, the U-Net is worse than the SSPM approach in all channels.

Table 3. SSIM

Energy channel	SPM	SSPM	U-Net	Modified CycN-Net
40 kev	0.70738	**0.93941**	0.81533	0.93825
50 kev	0.5144	0.88105	0.79107	**0.92092**
60 kev	0.38553	0.79035	0.75456	**0.86632**
Average	0.53577	0.87027	0.78699	**0.90850**

5.3 Results for Experiment Two

Here, the results are presented for experiment two, with the MAPE in Table 4.

Table 4. MAPE

Energy channel	SPM	SSPM	U-Net	Modified CycN-Net
40 kev	31.21718	**3.70097**	4.95682	5.02096
50 kev	33.47278	**5.4297**	5.83362	5.803
60 kev	21.05811	10.01085	8.6161	**8.43344**
Average	28.58269	**6.38050**	6.46885	6.41913

Similar to experiment one, we again see the SSPM approach performing the best in the lower energy channels, while the U-Net and CycN-Net have relative performance gains over the SSPM approach in the higher energy channels. However, we do not see the MAPE overtake the SSPM until 60 kev for the networks. We also see an increase in MAPE for all channels for the SSPM approach when compared to experiment one, with the largest of these increase being in the higher energy channels. Table 5 shows the computed PSNR values.

Table 5. PSNR

Energy channel	SPM	SSPM	U-Net	Modified CycN-Net
40 kev	18.18138	**37.85414**	34.52062	34.57402
50 kev	17.30738	**34.08984**	33.38491	33.33244
60 kev	20.49841	29.34949	30.55597	**30.61967**
Average	18.66239	**33.76449**	32.82050	32.84204

The best PSNR values are seen in the SSPM channel at the 40 kev and 50 kev range with the CycN-Net performing better at the 60 kev. The overall computed PSNR values are lower than the results obtained from experiment one. Table 6 displays our calculated SSIM.

Table 6. SSIM

Energy channel	SPM	SSPM	U-Net	Modified CycN-Net
40 kev	0.49543	**0.87897**	0.6717	0.66845
50 kev	0.44756	**0.81241**	0.70577	0.70642
60 kev	0.37888	0.70297	0.71571	**0.71833**
Average	0.44062	**0.79812**	0.69773	0.69773

For the SSIM values, the trend repeats, however there is a notable exception. For all past metrics the CycN-Net has performed better or on par with the U-Net, however for energy channel zero this is not the case.

6 Discussion

As shown by our results, the implementation of a U-Net reduces the impact of charge sharing in single-pixel mode only energy channels for energy bins at the 60 kev threshold. We can further improve the results by reducing the charge sharing distortions through the use of the CycN-Net. This allows the CycN-Net to outperform the SSPM approach consistently at the 50 kev and above energy range on the 2 mm CdZnTe detector dataset. These performance gains are made possible through a network structure allowing for the utilisation of spatial-temporal features present in projection images, which stabilises the charge sharing corrections. This approach tends to provide the most improvement over the single-pixel mode images at the higher energy ranges and in regions with reduced photon counts. We believe this to be because with reduced counts, a single charge sharing event has greater distortion. As such, there is increased variance in charge sharing distribution per projection, so a single scalar matrix which is calculated on the mean of the larger distribution is only able to reduce charge sharing distortions in a limited capacity. Figure 4 supports this hypothesis, showing that most of the MAPE is concentrated in the darker regions of the projection. It also shows that CycN-Net is able to produce a result with reduced error in this region.

Our ultimate purpose is to extract useful spectral information at later stages in the image processing pipeline, so more materials can be differentiated with increased accuracy in a single scan. To the best of our knowledge, we believe we have outlined a novel method of correcting charge sharing distortions in post-processing by utilising time series information.

An area of interest is the performance differences between the two different datasets seen by all explored correction methods. Although the CycN-Net improves the image quality by all used metrics when compared to the SSPM images at the 60 kev range, there is a large difference in the average metrics for all correction methods. Further testing will need to be done to determine the cause of these performance differences. One such potential cause of this is the

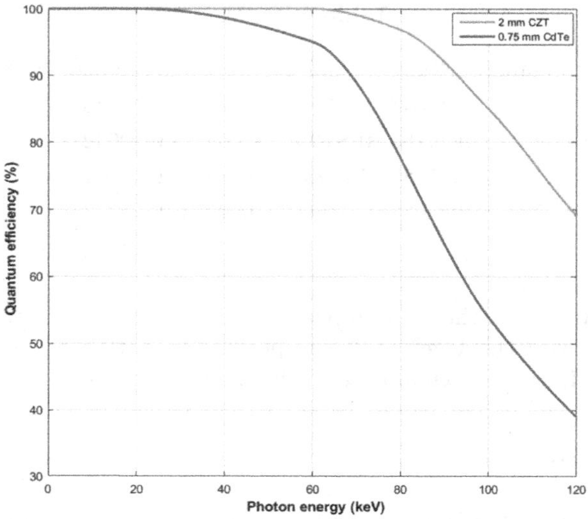

Fig. 5. Quantum efficiency for detectors 2 mm CdZnTe and 0.75 mm CdTe over photon energy: The red line shows the CdZnTe quantum efficiency over increasing photon energy and the blue line showing the quantum efficiency of the 0.75 mm CdTe over photon energy (Color figure online)

difference in detector response curve observed between the 2 mm CdZnTe and the 0.75 mm CdTe sensor displayed in Fig. 5 [1].

We see that the CdTe sensor has a response curve that begins to decrease in the 30 kev range, compared to the 2 mm CdZnTe which has 100% quantum efficiency until a photon energy of approximately 65 kev. This reduced counting of higher energy photons could be distorting the charge sharing distribution, making it more difficult to predict.

7 Conclusion

In conclusion, we introduce a novel approach to correct charge sharing distortions by customizing a CycN-Net architecture and training it to correct charge sharing distortions on the Medipix3RX photon counting detector single pixel mode channels in post processing. Our study suggests the viability of correcting charge sharing distortions in this manner, which will potentially allow the single pixel mode energy channels to be utilised in tomographic reconstruction and the material decomposition processes. This is expected to increase the accuracy of material identification for larger groups of materials. Our network design reduces the mean absolute percentage error, increases the peak signal to noise ratio and the structural similarity index measurement on the 2 mm CdZnTe dataset at the 50 and 60 kev thresholds. The 0.75 mm CdTe dataset showed the network was capable of improving these metrics at the 60 kev range.

8 Future Work

In the near future, we plan to conduct a material identification and quantification study on eight materials in a single scan, utilising both the charge summing mode channels and the single pixel mode channels. This will be performed to test whether the addition of these corrected channels can improve the accuracy of the material identification and the quantification when compared to the cases where only the charge-summed mode channels are utilised or when charge-summed mode channels and the uncorrected single pixel mode channels are used. We also plan to investigate a variation on the CycN-Net architecture where we utilise the SSPM result as a prior for the network as the SSPM is computationally quick to calculate. This should enable the CycN-net to at least be equivalent to the SSPM on the lower energy channels and may enable further improvements.

Acknowledgement. This project was funded by the Ministry of Business, Innovation and Employment (MBIE), New Zealand under contract number UOCX1404, by MARS Bioimaging Ltd and the Ministry of Education through the MedTech CoRE. The authors would like to acknowledge the Medipix2, Medipix3 and Medipix4 collaborations. Also, we would like to take this opportunity to acknowledge the generous support of the MARS Collaboration. European MARS Collaboration: S. A. Adebileje, S. D. Alexander, M. R. Amma, M. Anjomrouz, F. Asghariomabad, S. T. Bell, R. Senzing, F. O. Bochud, A. P. H. Butler, P. H. Butler, P. Carbonez, C. Chambers, K. M. Chapagain, A. I. Chernoglazov, J. A. Clark, J. S. Crighton, S. Dahal, T. Dapamede, A. Denys, N. J. A. deRuiter, D. Dixit, R. M. N. Doesburg, K. Dombroski, N. Duncan, S. P. Gieseg, A. Gopinathan, B. P. Goulter, J. L. Healy, L. Holmes, K. Jonker, T. Kirkbride, C. Lowe, V. B. H. Mandalika, A. Matanaghi, M. Moghiseh, M. Nowak, B. Paulmier, D. Racine, P. Renaud, D. Rundle, N. Schleich, E. Searle, J. S. Sheeja, L. Vanden Broeke, F. R. Verdun, V. Vitzthum, Vivek V. S., E. P. Walker, M. Wijesooriya, W. R. Younger.

References

1. Atharifard, A., et al.: Per-pixel energy calibration of photon counting detectors. J. Instrum. **12**(03), C03085 (2017)
2. Bai, T., et al.: Deep interactive denoiser (did) for x-ray computed tomography. IEEE Trans. Med. Imaging **40**(11), 2965–2975 (2021)
3. Ballabriga, R., et al.: The medipix3rx: a high resolution, zero dead-time pixel detector readout chip allowing spectroscopic imaging. J. Instrum. **8**(02), C02016 (2013)
4. Dreier, E.S., et al.: Spectral correction algorithm for multispectral CDTE x-ray detectors. Opt. Eng. **57**(5), 054117 (2018)
5. Flohr, T., et al.: Photon-counting CT review. Physica Med. **79**, 126–136 (2020)
6. Gimenez, E., et al.: Study of charge-sharing in medipix3 using a micro-focused synchrotron beam. J. Instrum. **6**(01), C01031 (2011)
7. Hamel, L., et al.: Optimization of single-sided charge-sharing strip detectors. In: 2006 IEEE Nuclear Science Symposium Conference Record. vol. 6, pp. 3759–3761 (2006). https://doi.org/10.1109/NSSMIC.2006.353811

8. Holbrook, M., Clark, D., Badea, C.: Deep learning based spectral distortion correction and decomposition for photon counting CT using calibration provided by an energy integrated detector. In: Medical Imaging 2021: Physics of Medical Imaging, vol. 11595, p. 1159520. International Society for Optics and Photonics (2021)
9. Leng, S., et al.: Dose-efficient ultrahigh-resolution scan mode using a photon counting detector computed tomography system. J. Med. Imaging **3**(4), 043504 (2016)
10. Li, M., Rundle, D.S., Wang, G.: X-ray photon-counting data correction through deep learning. arXiv preprint arXiv:2007.03119 (2020)
11. McGregor, D., Hermon, H.: Room-temperature compound semiconductor radiation detectors. Nucl. Instrum. Methods Phys. Res. Sect. A Acceler. Spectro. Detect. Assoc. Equip. **395**(1), 101–124 (1997). https://doi.org/10.1016/S0168-9002(97)00620-7, www.sciencedirect.com/science/article/pii/S0168900297006207
12. Pellegrini, G., et al.: Direct charge sharing observation in single-photon-counting pixel detector. Nucl. Instrum. Methods Phys. Res. Sect. A **573**(1–2), 137–140 (2007)
13. Symons, R., et al.: Photon-counting CT for simultaneous imaging of multiple contrast agents in the abdomen: an in vivo study. Med. Phys. **44**(10), 5120–5127 (2017)
14. Taguchi, K.: Multi-energy inter-pixel coincidence counters for charge sharing correction and compensation in photon counting detectors. Med. Phys. **47**(5), 2085–2098 (2020)
15. Tao, X., Wang, Y., Lin, L., Hong, Z., Ma, J.: Learning to reconstruct CT images from the VVBP-tensor. IEEE Trans. Med. Imaging **40**, 3030–3041 (2021)
16. Wang, G., Jacob, M., Mou, X., Shi, Y., Eldar, Y.C.: Deep tomographic image reconstruction: yesterday, today, and tomorrow-editorial for the 2nd special issue machine learning for image reconstruction. IEEE Trans. Med. Imaging **40**(11), 2956–2964 (2021)
17. Wang, Z., Bovik, A., Sheikh, H., Simoncelli, E.: Image quality assessment: from error visibility to structural similarity. IEEE Trans. Image Process. **13**(4), 600–612 (2004). https://doi.org/10.1109/TIP.2003.819861
18. Zhi, S., Kachelrieß, M., Pan, F., Mou, X.: CYCN-net: a convolutional neural network specialized for 4D CBCT images refinement. IEEE Trans. Med. Imaging **40**(11), 3054–3064 (2021)

Vehicle-Related Distance Estimation Using Customized YOLOv7

Xiaoxu Liu(✉) and Wei Qi Yan

Auckland University of Technology, Auckland 1010, New Zealand
xiliu@aut.ac.nz

Abstract. With the popularity of autonomous driving, the development of ADAS (Advanced Driver Assistance Systems), especially collision avoidance systems, has become an important branch in the field of deep learning. In the face of complex traffic environments, collision avoidance systems need to detect vehicles quickly and accurately in traffic distance to the vehicle in front. Against this background, in this paper, we aim at investigating how to build a fast and robust model for vehicle distance estimation. The theoretical insights are synthesized in the context of odometry and customized YOLOv7 based on what a conceptual framework is proposed. In this paper, KITTI is employed as the dataset for model training and testing. Being one of the pioneer works on distance estimation based on KITTI, the unique value of this research work lies in the first time using YOLOv7 with attention model as a distance estimation model and getting 4.253 on RMSE.

Keyword: Autonoumous vehicles · YOLOv7 · Vehicle detection · Distance estimation · Scene understanding

1 Introduction

Advanced Driver Assistance Systems (ADAS) provide a safe and automated driving experience that will reshape our relationship with automobile. In the near future, autonomous vehicles will allow passengers to experience a personalized and interconnected driving experience, given vehicles the ability to sense, act seamlessly and intelligently handle real-time road conditions [36, 37]. Amongst them, vision and radar systems play an important role in ADAS. The vision system is responsible for sensing the surroundings and taking the necessary measures to ensure the safety of all road users. At the same time, the radar systems continuously sense the distance between vehicles in real time, improving driving efficiency and safety [54–58]. For decades, one of the most popular ideas in the literature for solving distance detection problem pertaining to ADAS is visual object detection of current traffic environment and the distance to surrounding obstacles by means of deep learning methods [27, 28]. An important breakthrough in deep learning-based neural networks is that visual tasks do not have to be coded manually [38–41]. Deep learning neural networks allow various features to be extracted automatically from training examples [34, 47–52]. A neural network is considered to have "deep" learning capability if it has input and output layers with at least one implicit intermediate layer [29, 30, 35].

W. Q. Yan et al. (Eds.): IVCNZ 2022, LNCS 13836, pp. 91–103, 2023.
https://doi.org/10.1007/978-3-031-25825-1_7

Recent theoretical developments have revealed that the YOLO series is currently one of the most advanced methods for efficient implementation of deep neural networks for vision processing [42–44]. The YOLO series are much efficient, there is no complex detection process and only the image needs to be fed into the neural network to obtain the detection results. Moreover, YOLO series are very good at avoiding background errors and generating false positives.

For distance detection, a strategy to obtain distance information is through laser detecting and ranging [31–33]. Laser-based distance measurement is widely considered at the time of developing the Collision Warning System [1–3]. However, LiDAR is very complex, expensive and low yielding, which can only be used for testing vehicles at present. In addition, ultrasound, infrared and microwave radar can also be employed for vehicle detection and ranging, but the range of ultrasound and infrared is narrow, microwave radar is susceptible to interference and the reliability of the detection results is weak, while these methods cannot distinguish between detection targets. This poses a few problems while carrying out that the hardware devices such as radar and infrared are expensive and complex to integrate with the camera and have limitations in terms of measurement accuracy. Moreover, few studies have revealed on abandoning costly distance measuring hardware equipment and inferring the distance information from the detected 2D video frames.

Therefore, deep learning-based detection has a promising application prospect and can be combined with the methods to achieve better results. By calibrating the internal and external parameters of the camera, the distance to the vehicle in front is estimated by using the visual projection model principle and geometric ranging methods to warn of a possible collision.

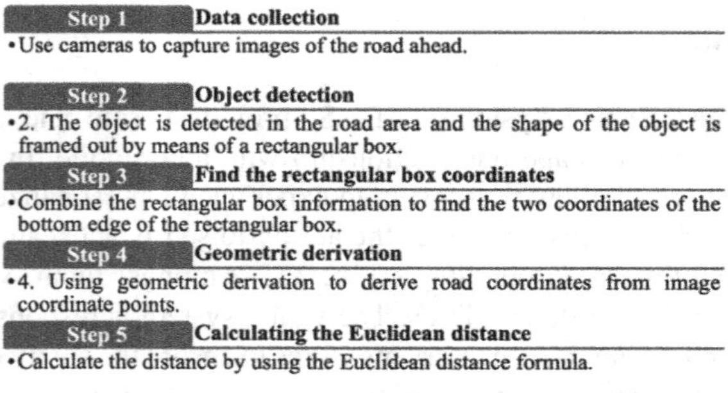

Fig. 1. The steps of monocular ranging method

The existing ranging methods based on visual information comprise two branches: Monocular camera-based ranging methods and binocular camera-based ranging methods. The general principle of monocular camera ranging is to firstly identify the target by image matching and then to estimate the target distance by its size in the image. The general steps shown in Fig. 1 are data collection, object detection, finding the rectangular box coordinates, geometric derivation and calculating Euclidean distance. Among them, the circumferential ranging method has a larger fisheye lens distortion, and the circumferential camera is generally employed for low-speed scenes, mainly for detecting ground markings, so the camera lens faces down; the other is the front-view camera ranging, which is characterized by the other is forward-looking camera ranging, which is featured by a smaller aberration of the front-view lens, and the camera is mounted under the rear-view mirror of the car, which can be harnessed in low-speed and high-speed vehicle scenes for detecting vehicles, pedestrians and obstacles in front, so the camera lens have to face forward [4].

Compared to the front view camera ranging method, the circumferential fisheye camera, because the lens faces downwards and the aberration coefficient is large, based on the camera model, the usage of mathematical geometry for ranging is no longer tried and will result in a larger error; the idea is based on the single strain matrix and affine transformation for ranging. The core knowledge is to get the four points of aberration correction map to customize the solution of the single strain matrix corresponding to the four points of the image, which is extremely relevant to the accuracy of the calibration. The front view camera is a normal camera with low aberrations based on the camera model by deriving the relationship between the pixel coordinates and the world coordinates.

Binocular vision imitates human eye structure and takes use of two or more cameras to collect images of different orientations of the same target. The 3D information of target can be accurately calculated by the matching image points between the left and right images under the binocular camera model. Binocular ranging is split into four steps as shown in Fig. 2: Camera calibration, binocular calibration, binocular matching, and calculation of the distance.

The advantage of binocular ranging is that there is no limit to the recognition rate, because in principle there is no need for recognition before measurement, but rather all obstacles are measured directly; and there is no need to maintain a sample database for binocular ranging, because there is no concept of a sample for binoculars. The advantages of monocular ranging, on the other hand, are the low cost, the low requirement for computing resources and the relatively simple system architecture.

Step 1 **Camera calibration**

- **The binocular cameras need to be calibrated to obtain the internal and external parameters, the single response matrix of both cameras.**

Step 2 **Binocular calibration**

- **The original image is corrected according to the calibration results and the two corrected images are in the same plane and parallel to each other.**

Step 3 **Binocular matching**

- **Pixel point matching is performed on the two images after correction.**

Step 4 **Calculating of distance information**

- **The distance information for each pixel is calculated from the matching results to obtain the final distance value.**

Fig. 2. The main steps of binocular method

It is of interest to know whether high precision and high-speed object detection and ranging can be still achieved while keeping costs low. Therefore, the aim of this study is to develop a more sophisticated model for vehicle detection and ranging using monocular camera. The contributions of this paper are listed as follows:

(1) A proven object detection and ranging model based on monocular camera will be implemented.
(2) To our knowledge, this is the first time using YOLOv7 with attention model as the basic architecture for distance estimation.
(3) A higher detection and ranging performance will be generated.

2 Literature Review

In this section, we review the recent literature on machine vision-based vehicle detection and vehicle ranging. Vehicle detection is the basis for vehicle ranging, while distance estimation from the vehicle ahead provides important data to support vehicle collision avoidance systems. Therefore, more and more computer vision research publications focus on vehicle detection and ranging tasks.

We review two streams of literature: In the first stream of relevant work, we study vehicle detection and ranging with binocular camera. The most difficult problem to overcome for a precise assessment of vehicle speed is measuring distance in a real-world coordinate system, which is a straightforward operation when employing stereo vision. The parallax values of the pixels contained in the vehicle region may be used to directly calculate the relative distance for each detected vehicle. The basic principle of stereo vision is to observe the same scene from multiple viewpoints (usually two). This enables the obtention of digital images of the three-dimensional scene. By using epipolar geometric principles, three-dimensional shape and position of the surrounding scene is rebuilt [6]. Chui et al., proposed an algorithm that includes four modules: pre-processing, edge detection, line segment matching, and vehicle search and distance estimation for their multi-resolution stereovision system. The coarse-to-fine detection algorithm is employed for vehicle detection task. In coarse-to-fine detection algorithm, tedious pre-processing operations are inevitable. The pre-processing performs down-sampling and low-pass filtering processes that increase computation and slow down detection speed. In the vehicle search and distance estimation module, the front vehicles and their distances will be found and estimated using the average distances of the horizontal and vertical line segments based on the right image [5].

Brojeshwar et al., proposed a more advanced method to deal with the object detection and distance estimation problem. The detection is based on Viola-Jones algorithm where Haar-like features are applied to train a cascaded classifier using the well known AdaBoost algorithm [7–9]. By adopting a revolutionary method called stereo vision, the distance is determined by finding corners with high eigen values in segmented areas of both images. Using the left image as a guide, a sub-image is created for each corer identified there, transferred to the right image using homography, and a match is discovered using colour correlation. The best matching point near the homography-obtained point is located via a spiral search. A collection of coordinate pairs from both photos is the outcome. The fact that related points will lie on epipolar lines is used to perform a more accurate match between this set of points, and the end result is a set of points that have exact correspondence in both images. This step was taken to avoid inaccurate matches due to thresholds and differences in the color resolution capabilities of the two cameras [10].

However, the binocular distance estimated by using traditional machine learning methods that require a lot of time for pre-processing and have low accuracy. Moreover, for binocular ranging itself, the computational effort is very high and the performance of the computing unit is very demanding, which makes the productization and miniaturization of binocular systems difficult and costly. Furthermore, the alignment of the binoculars has a direct impact on the accuracy of the distance measurement. The binocular stereo vision method relies on the natural light in the environment to capture images, due to environmental factors such as changes in light angle and light intensity, the brightness of the two images can vary considerably, which can pose a great challenge to the matching algorithm. The binocular ranging method is also not suitable for scenes that are monotonous and lack texture. Since the binocular stereo vision method matches images based on visual features, it can be difficult to match scenes that lack visual features (e.g., sky, white walls, desert, etc.), resulting in large matching errors or even matching failure.

The second stream of literature relevant to our work concerns detection and ranging with a monocular camera. The knowledge of dimensions in the real coordinates of certain features, objects or road sections is a fundamental problem in estimating distances by using monocular systems. This is often referred to as the scale factor for converting pixels to real-world coordinates. Another requirement is to consider the flat road assumption [11–13]. Firstly, based on indicator lines, augmentation lines or areas [17–19], these methods do not require calibration of the camera system, but rather measure the actual distance between two or more virtual lines on the road, or the actual size of the road area. The distance estimation is then posed as a detection problem in which all vehicles are detected at the same distance whenever they cross a predefined virtual line or area. Since the virtual line or area is located on the road, an accurate distance estimation involves the precise location of the contact point of a part of the vehicle. This part of the vehicle should be identical at the second position to obtain a consistent estimate of speed [20].

Based on the true size of the object, including the license plate and the vehicle [14–16, 20], inverse perspective mapping (IPM) was used by Jong to develop a technique of object detection and a way of calculating a vehicle's distance from a bird's eye view. ACFs were used in the suggested technique to create the AdaBoost-based vehicle detector. The LUV colour, edge gradient, and orientation (histograms of oriented gradients) of the input picture were used to extract the ACFs. The distance between the detected vehicle and the autonomous vehicle was then determined by implementing IPM and creating a 3D picture projected in three dimensions from a 2D input image [21].

Arabi et al. offered a thorough approach for calculating the distance to a vehicle that is following just using visual information from a monocular camera with low resolution. To do this, a pair of cars were outfitted with real-time kinematic (RTK) GPS, and the lead vehicle had specially made gadgets that captured video of the trailing vehicle. The process was then repeated with a pickup truck in the following position after 40 trials were recorded with a sedan as the following vehicle. The video data was then used to perform vehicle recognition and distance estimate using the DeepStream streaming analytics toolbox and ANN [22].

In contrast to the works of Jong [21] and Arabi et al. [22], the work was based on the AdaBoost-based vehicle detector in traditional machine learning, which may yield lower detection accuracy than the models based on advanced deep learning. However, the work takes use of less monetary cost because the model is based on a 2D to 3D transformation to estimate distances rather than relying on a hardware device. In comparison, Arabi costs more money, but invokes more advanced deep learning techniques to detect objects and estimate distance [22].

Following our research and review of the extensive literature, we find that the existing methods on monocular camera-based vehicle detection and ranging are often based on conventional machine learning methods or high monetary cost filming devices, few studies have focused on using deep learning methods to address vehicle detection and ranging at a reduced monetary cost. For the task of estimating distances based on images, deep neural networks have high performance for image processing and modern deep learning frameworks will be by far the best choice for processing image data. Without relying on expensive ranging equipment, deep learning models with improved vehicle detection accuracy can make vehicle distances calculated based on detection frame information more accurate. Therefore, research into vehicle detection and ranging implemented using monocular cameras and deep learning frameworks is urgent and necessary.

3 Our Methods

We are use of YOLOv7 as the underlying architecture for our deep learning models. The purpose of YOLOv7 is to solve two problems. For model structural re-referencing, a planned model structural re-referencing is proposed by using the concept of gradient propagation paths to analyse the structural re-referencing strategies applicable regarding each layer in different networks.

Whilst using a dynamic label assignment strategy, new problems arise in the training of models with multiple output layers, such as how to better assign dynamic targets to the outputs of different branches. To address this issue, the authors proposed a new approach to label assignment called the coarse-to-fine guided label assignment strategy.

YOLOv7 also provided "extend" and "compound scaling" methods for real-time detectors, which allow for a more sophisticated approach. The methods for more efficient use of parameters and computational effort. At the same time, this method can effectively reduce the parameters of a real time detector by up to 50% and offers faster inference and higher detection accuracy [23].

Generally, YOLOv7 firstly resizes the input image to 640 × 640 and inputs it into the backbone network, then outputs a feature map with three layers of different sizes through the head layer network, and outputs the prediction results through the REP module and the conv module.

Different from the original YOLOv7 model, our proposed model replaces the conv module in backbone with the Convolutional Block Attention Module (CBAM) [24]. The Convolutional Block Attention Module consists of two modules, the Channel Attention (CAM) and the Spatial Attention Module (SAM). The CAM enables the network to focus on the foreground of an image, allowing the network to pay more attention to meaningful regions, while the SAM enables the network to focus on locations in the whole image that are rich in contextual information. To create two feature maps, the input feature maps were subjected to global max pooling and global average pooling depending on width and height. They were then input into an MLP that had a large number of neurones in the first layer and used ReLU for activation. There was a common two-layer neural network. The final channel attention feature is created by first applying element-wise summing to the MLP output features, followed by sigmoid activation. The spatial attention module is produced after element-wise multiplication of the channel attention feature and the input feature map. The function is,

$$M_C(F) = \sigma(MLP(AvgPool(F)) + MLP(MaxPool(F))) = \sigma\left(W_1\left(W_0\left(F^c_{avg}\right)\right)\right. \\ \left.+W_1\left(W_0\left(F^c_{max}\right)\right)\right) \tag{1}$$

where σ denotes the sigmoid function, $W_0 \in \mathbb{R}^{C/r \times C}$, and $W_1 \in \mathbb{R}^{C \times C/r}$. Note that the MLP weights, W_0 and W_1, are shared for both inputs and the ReLU activation function is followed by W_0.

The channel attention module's feature map output is used as this module's input feature map. First, based on the channel, we do global maximum and global average pooling, yielding two feature maps. Based on the channel, these two feature maps are then combined. The spatial attention feature is then produced by the sigmoid, and to create the final feature, the feature is multiplied by the input feature of the module. The function is [45]:

$$M_s(F) = \sigma\left(f^{7\times7}([AvgPool(F); MaxPool(F)])\right) = \sigma\left(f^{7\times7}\left(\left[F^s_{avg}; F^s_{max}\right]\right)\right) \tag{2}$$

where σ denotes the sigmoid function and $f^{7\times7}$ represents a convolution operation with the filter size of 7 × 7.

After the detection task is completed, we get four numbers in the bounding box, namely (x_0, y_0, *width, height*). Where x_0 and y_0 are applied to tile or adjust the bounding box. The width and height are adopted for measuring the object and actually describing the detected object and details. The width and height will vary depending on the distance of the object from the camera.

The monocular camera generates a one-to-one relationship between the object and the image, the relationships of variables. Using this principle, we can deduce a relationship between known parameters. Using the principle of similar triangles, we can obtain the final formulas as follows [25, 26, 46, 53]:

$$\text{distance} = (2 * 3.14 * 180) \div (\text{w} + \text{h} * 360) * 1000 + 3 \tag{3}$$

where w is the weight of the detection bounding box, and h is the height of the detection bounding box.

4 Experimental Results

In this paper, we present an advanced deep learning-based vehicle detection and distance estimation model for low-spend monocular cameras. This was experimentally investigated by PYTHON 2.7, RTX5000 GPU and 32 GB RAM. Our data samples are from the KITTI dataset. The KITTI dataset contains the internal and external parameters of the in-car camera, as well as the coordinates, width and height of the detection frame. We randomly chose 4,000 samples for developing our deep learning model and divided them into 7:3 for training and test. The outputs demonstrated in this section match state of the art methods. The result in Fig. 3 provides evidence that our modified YOLOv7 with the distance estimation algorithm [25] is able to generate satisfying performance of vehicle detection and distance estimation.

We set appropriate parameter values (*epochs* = 3500, *batch_size* = 1, *learning_ rate* = 0.01) to train our modified YOLOv7. The network training process in Fig. 4 shows that losses of both training and validation decrease between 0 epochs and 1,000 epochs, until after 1,000 epochs the loss curves decrease slightly and flattens out around 0.068.

We present a quantitative comparison in the constructed KITTI dataset for all the evaluation metrics in Table 1. We compare several advanced YOLO models and the transformer model. The results show that YOLOv7 performs significantly better than the other YOLO models and transformer. Furthermore, our modified YOLOv7 that is added the convolutional block attention module performs even better than the original YOLOv7. It may indicate that the delivers significantly better results due to the convolutional block attention module is combined with YOLOv7.

We also grouped the distances into three categories: 0–10 m, 10–20 m and >20 m. For each category, we calculated the average RMSE in Table 2. The results indicate that our modified YOLOv7 outperforms the original YOLOv7 in all the three distance categories. To summarize the findings in Table 1 and Table 2, this may raise concerns about object detection and distance estimation tasks that can be successfully addressed by YOLOv7 model and improve performance using the attention module.

Fig. 3. The example of vehicle detection and distance estimation using the modified YOLOv7

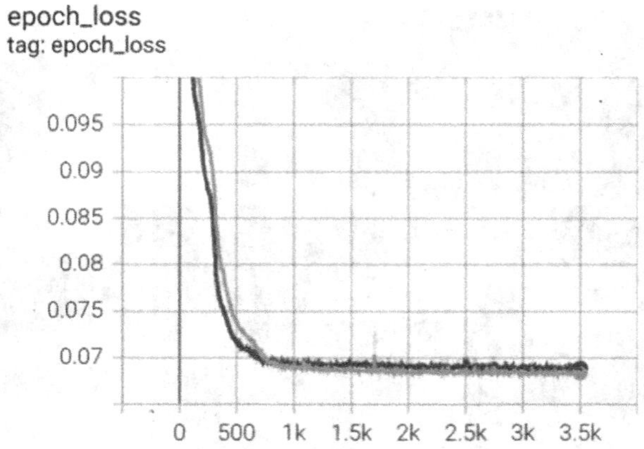

Fig. 4. The training process of the modified YOLOv7. The blue curve indicates the training loss while the orange curve indicates the validation loss.

Table 1. Quantitative comparisons of multiple deep neural networks

Training modality	RMSE
YOLOv5	4.138
YOLOv6	4.516
YOLOv7	4.275
Modified YOLOv7	**4.253**
Detection transformer (DETR)	4.833

Table 2. Average RMSE of different neural networks in different distance categories

Training modality	0–10 m	10–20 m	>20 m
YOLOv7	4.533	4.374	3.918
Modified YOLOv7	4.519	3.742	4.498

5 Conclusion

A combination of YOLOv7 and attention module could enable a low-cost monocular camera-based vehicle detection and distance estimation task with satisfactory results. This project is the first comprehensive investigation of distance estimation by using customized YOLOv7 and the performance of our customized YOLOv7 obtain 4.253 of RMSE. This would be a fruitful area for further work to add other attention mechanisms and receive further enhancements.

References

1. Tinchev, G., Penate-Sanchez, A., Fallon, M.: Learning to see the wood for the trees: deep laser localization in urban and natural environments on a CPU. IEEE Robot. Autom. Lett. **4**(2), 1327–1334 (2019)
2. Kuznietsov, Y., Stuckler, J., Leibe, B.: Semi-supervised deep learning for monocular depth map prediction. In: IEEE Conference on Computer Vision and Pattern Recognition, pp. 6647–6655, (2017)
3. Liao, Y., Huang, L., Wang, Y., Kodagoda, S., Yu, Y., Liu, Y.: Parse geometry from a line: Monocular depth estimation with partial laser observation. In: IEEE International Conference on Robotics and Automation (ICRA), pp. 5059–5066, (2017)
4. Zhang, J., Hu, S., Shi, H.: Deep learning based object distance measurement method for binocular stereo vision blind area. Int. J. Adv. Comput. Sci. Appl. **9**(9) (2018)
5. Chiu, C.C., Chung, M.L., Chen, W.C.: Real-time front vehicle detection algorithm for an asynchronous binocular system. J. Inf. Sci. Eng. **26**(3), 735–752 (2010)
6. Zhao, M., Mammeri, A., Boukerche, A.: Distance measurement system for smart vehicles. In: International Conference on New Technologies, Mobility and Security (NTMS), pp. 1–5 (2015)
7. Paul, V., Michael, J.: Rapid object detection using a boosted cascade of simple features. In: International Conference on Computer Vision and Pattern Recognition (2001)
8. Goncalo, M., Paulo, P., Urbano, N.: Vision-based pedestrian detection using HAAR-LIKE features, Robotica **46**, 321–328 (2006)
9. Rainer, L., Alexander, K., Vadim, P.: An empirical analysis of boosting algorithms for rapid objects with an extended set of Haar-like features. Intel Technical Report MRL-TR (2002)
10. Bhowmick, B., Bhadra, S., Sinharay, A.: Stereo vision based Pedestrians detection and distance measurement for automotive application. In: International Conference on Intelligent Systems, Modelling and Simulation, pp. 25–29 (2011)
11. Gunawan, A.A.S., et al.: Detection of vehicle position and speed using camera calibration and image projection methods. Procedia Comp. Sci. **157**, 255–265 (2019)
12. Kim, J.-H., et al.: Reliability verification of vehicle speed estimate method in forensic videos. Foren. Sci. Int. **287**, 195–206 (2018)
13. Huang, T.: Traffic speed estimation from surveillance video data. In: 2018 IEEE/CVF Conference on Computer Vision and Pattern Recognition Workshops (CVPRW), pp. 161–165 (2018)
14. Vakili, E., et al.: Single-camera vehicle speed measurement using the geometry of the imaging system. Mult. Tools Apps. **79**, 19307–19327 (2020)
15. Llorca, D.F., et al.: Two-camera based accurate vehicle speed measurement using average speed at a fixed point. In: The IEEE International Conference on Intelligent Transportation Systems, pp. 2533–2538 (2016)
16. Wu, W., et al.: Vehicle speed estimation using a monocular camera. In: Proceedings of SPIE 9407, Video Surveillance and Transportation Imaging Applications. SPIE (2015)
17. Dahl, M., Javadi, S.: Analytical modeling for a video-based vehicle speed measurement framework. Sensors **20**, 160 (2020)
18. Javadi, S., et al.: Vehicle speed measurement model for video-based systems. Comp. Elec. Eng. **76**, 238–248 (2019)
19. Czapla, Z.: Vehicle speed estimation with the use of gradient-based image conversion into binary form. In: 2017 Signal Processing: Algorithms, Architectures, Arrangements, and Applications (SPA), pp. 213–216 (2017)
20. Fernández Llorca, D., Hernández Martínez, A., García Daza, I.: Vision-based vehicle speed estimation: a survey. IET Intel. Transport Syst. **15**(8), 987–1005 (2021)

21. Kim, J.: Efficient vehicle detection and distance estimation based on aggregated channel features and inverse perspective mapping from a single camera. Symmetry **11**(10) 1205 (2019)
22. Arabi, S., Sharma, A., Reyes, M., Hamann, C., Peek-Asa, C.: Farm vehicle following distance estimation using deep learning and monocular camera images. Sensors **22**(7), 2736 (2022)
23. Wang, C.Y., Bochkovskiy, A., Liao, H.Y.M.: YOLOv7: trainable bag-of-freebies sets new state-of-the-art for real-time object detectors (2022). https://doi.org/10.48550/arXiv.2207.02696
24. Woo, S., Park, J., Lee, J.-Y., Kweon, I.S.: CBAM: convolutional block attention module. In: Ferrari, V., Hebert, M., Sminchisescu, C., Weiss, Y. (eds.) ECCV 2018. LNCS, vol. 11211, pp. 3–19. Springer, Cham (2018). https://doi.org/10.1007/978-3-030-01234-2_1
25. Khan, M., Paul, P., Rashid, M., Hossain, M., Ahad, M.: An AI-based visual aid with integrated reading assistant for the completely blind. In: IEEE Transactions on Human-Machine Systems, pp. 91–99 (2017)
26. Liu, X. Yan, W.: Depth estimation of traffic scenes from image sequence using deep learning PSIVT (2022)
27. Liu, X., Yan, W.: Traffic-light sign recognition using Capsule network. Multim. Tools Appl. **80**, 15161–15171 (2021)
28. Liu, X., Yan, W.: Vehicle-related scene segmentation using CapsNets. In: IEEE IVCNZ (2020)
29. Liu, X., Nguyen, M., Yan, W.: Vehicle-related scene understanding using deep learn. In: Asian Conference on Pattern Recognition (2019)
30. Liu, X.: Vehicle-related scene understanding using deep learning. Master's Thesis, Auckland University of Technology, New Zealand (2019)
31. Mehtab, S., Yan, W.: FlexiNet: fast and accurate vehicle detection for autonomous vehicles-2D vehicle detection using deep neural network. In: ACM ICCCV: (2021)
32. Mehtab, S., Yan, W.: Flexible neural network for fast and accurate road scene perception. Multim. Tools Appl. **81,** 7169–7181 (2021)
33. Mehtab, S., Yan, W., Narayanan, A.: 3D vehicle detection using cheap LiDAR and camera sensors. In: IEEE IVCNZ (2021)
34. Yan, W.: Computational Methods for Deep Learning: Theoretic Practice and Applications. Springer, Cham (2021)
35. Yan, W.: Introduction to Intelligent Surveillance: Surveillance Data Capture, Transmission, and Analytics. Springer, Cham (2019). https://doi.org/10.1007/978-3-030-10713-0
36. Gu, Q., Yang, J., Kong, L., Yan, W., Klette, R.: Embedded and real-time vehicle detection system for challenging on-road scenes. Opt. Eng. **56**(6), 06310210 (2017)
37. Ming, Y., Li, Y., Zhang, Z., Yan, W.: A survey of path planning algorithms for au-tonomous vehicles. Int. J. Commer. Veh. **3**, 448-468 (2021)
38. Shen, D., Xin, C., Nguyen, M., Yan, W.: Flame detection using deep learning. In: In-ternational Conference on Control, Automation and Robotics (2018)
39. Xin, C., Nguyen, M., Yan, W.: Multiple flames recognition using deep learning. In: Handbook of Research on Multimedia Cyber Security, pp. 296–307 (2020)
40. Luo, Z., Nguyen, M., Yan, W.: Kayak and sailboat detection based on the improved YOLO with Transformer. In: ACM ICCCV (2022)
41. Le, R., Nguyen, M., Yan, W.: Training a convolutional neural network for transportation sign detection using synthetic dataset. In: IEEE IVCNZ (2021)
42. Alexey, B., ChienYao, W., Mark, L.: YOLOv4: optimal speed and accuracy of object detection. Image and Video Processing (2020)
43. Chuyi, L. et al.: YOLOv6: a single-stage object detection framework for industrial applications. In: Proceedings of the IEEE Conference on Computer Vision and Pattern Recognition (2022)

44. Chienyao, W., Alexey, B., Mark, L.: YOLOv7: Trainable bag-of-freebies sets new state-of-the-art for real-time object detectors. In: Proceedings of the IEEE Conference on Computer Vision and Pattern Recognition (2022)
45. Hu, J., Shen, L., Sun, G.: Squeeze-and-excitation networks. In: Proceedings of the IEEE Conference on Computer Vision and Pattern Recognition, pp. 7132–7141 (2018)
46. Cao, Y.T., Wang, J.M., Sun, Y.K., Duan, X.J.: Circle marker based distance measurement using a single camera. Lect. Notes Softw. Eng. **1**(4), 376 (2013)
47. Pan, C., Liu, J., Yan, W., Zhou, Y.: Salient object detection based on visual perceptual saturation and two-stream hybrid networks. In: IEEE Transactions on Image Processing (2021)
48. Pan, C., Yan, W.: Object detection based on saturation of visual perception. Multim. Tools Appl. **79**(27–28), 19925–19944 (2020)
49. Pan, C., Yan, W.: A learning-based positive feedback in salient object detection. In: IEEE IVCNZ (2018)
50. Shen, Y., Yan, W.: Blind spot monitoring using deep learning. In: IEEE IVCNZ (2018)
51. Zheng, K., Yan, W., Nand, P.: Video dynamics detection using deep neural networks. IEEE Trans. Emerg. Top. Comput. Intell. **2**, 24–234(2017)
52. An, N., Yan, W.: Multitarget tracking using Siamese neural networks. In: ACM Transactions on Multimedia Computing, Communications and Applications (2021)
53. Leslie, M., et al.: Identification of the MuRF1 skeletal muscle ubiquitylome through quantitative proteomics. Function **192**(4), zqab029 (2021)
54. Xinyu, Z., Hongbo, G., Jianhui, Z.H.A.O., Mo, Z.H.O.U.: Overview of deep learning intelligent driving methods. J. Tsinghua Univ. (Sci. Technol.) **58**(4), 438–444 (2018)
55. Grigorescu, S., Trasnea, B., Cocias, T., Macesanu, G.: A survey of deep learning techniques for autonomous driving. J. Field Robot. **37**(3), 362–386 (2020)
56. Muhammad, K., Ullah, A., Lloret, J., Del Ser, J., de Albuquerque, V.H.C.: Deep learning for safe autonomous driving: current challenges and future directions. IEEE Trans. Intell. Transp. Syst. **22**(7), 4316–4336 (2020)
57. Mozaffari, S., Al-Jarrah, O.Y., Dianati, M., Jennings, P., Mouzakitis, A.: Deep learning-based vehicle behavior prediction for autonomous driving applications: a review. IEEE Trans. Intell. Transp. Syst. **23**(1), 33–47 (2020)
58. Li, Y., et al.: Deep learning for lidar point clouds in autonomous driving: a review. IEEE Trans. Neural Netw. Learn. Syst. **32**(8), 3412–3432 (2020)

Extending Temporal Data Augmentation for Video Action Recognition

Artjoms Gorpincenko(✉) and Michal Mackiewicz

School of Computing Sciences, University of East Anglia, Norwich, England
{a.gorpincenko,m.mackiewicz}@uea.ac.uk

Abstract. Pixel space augmentation has grown in popularity in many Deep Learning areas, due to its effectiveness, simplicity, and low computational cost. Data augmentation for videos, however, still remains an under-explored research topic, as most works have been treating inputs as stacks of static images rather than temporally linked series of data. Recently, it has been shown that involving the time dimension when designing augmentations can be superior to its spatial-only variants for video action recognition [34]. In this paper, we propose several novel enhancements to these techniques to strengthen the relationship between the spatial and temporal domains and achieve a deeper level of perturbations. The video action recognition results of our techniques outperform their respective variants in Top-1 and Top-5 settings on the UCF-101 [55] and the HMDB-51 [38] datasets.

Keywords: Data augmentation · Temporal domain · Action recognition

1 Introduction

Deep convolution neural networks (CNNs) have become the standard approach for a large number of computer vision tasks, by virtue of their unique ability to learn the most useful features from the data in the unmanned manner. However, large amounts of diverse labeled training imagery are usually required to guarantee models' high accuracy, which are often unavailable. Acquiring and annotating new data is generally expensive, time-consuming, and sometimes even impossible, resulting in networks underfitting or overfitting, depending on the training set variance. In recent years, several deep learning areas have been explored to tackle the aforementioned problems, such as domain adaptation [15,18,45,51,61], network regularization [37,44,50,56], data generation [1,17,21,63], and data augmentation [9,11,47,65,66], all showing significant performance gains over their respective baselines.

Due to its ability of expanding and populating the training distribution through synthetically created samples, pixel space augmentation was successfully used as the main driver in a number of semi-supervised [47,54,65,66], self-supervised [4,25,46], and domain adaptation [15,45,51] studies. The use of

W. Q. Yan et al. (Eds.): IVCNZ 2022, LNCS 13836, pp. 104–118, 2023.
https://doi.org/10.1007/978-3-031-25825-1_8

feature space augmentation was also explored for both static and sequential imagery [5,10,22,42], yielding improvements in models' accuracy. Data augmentation for videos, however, still remains an under-explored research area, as most works have been treating inputs as stacks of static images rather than temporally linked series of data. A recent study has shown that the time domain consideration while designing augmentations can be superior to its spatial-only variants for video recognition [34].

In this paper, we expand on the previous work [34]. We argue that some of the proposed techniques can be extended even further to fully utilise the time domain and achieve a deeper level of temporal perturbations, which results in more accurate and robust classifiers. The contributions of this paper can be summarised as follows:

1. We expand the list of available augmentations in RandAugment-T [34] by adding VideoReverse, FrameFadeIn, and VideoCutMix, augmentations that are video-specific and are done within a single sample;
2. We increase the amount of magnitude checkpoints for all augmentation techniques to allow for non-linear temporal perturbations;
3. We propose to linearly change the bounding pox positions for cut-and-paste algorithms, such as CutOut [11], CutMix [65], and CutMixUp [64], and their extensions, as well as the mixing ratio in MixUp [66] and CutMixUp [64] extensions;
4. The recognition results of the aforementioned techniques on the UCF-101 [55] and the HMDB-51 [38] datasets either maintain competitive or exceed performance achieved by the previous work [34].

2 Related Work

2.1 Spatial Augmentation

The earliest experiments that demonstrate the effectiveness of data augmentation are based on basic image modifications, such as axis flipping, rotations, translations, random cropping, and colour space alterations [6,7,36,52]. These techniques are easy to implement, bear minimum computational overhead and are very likely to preserve the label after transformation. However, combining the aforementioned operations together can result in heavily inflated datasets and high risk of label warp. Therefore, a number of studies has been done on search algorithms that aim to find the optimal subset of augmentations for a particular task [8,10,40,60]. Finally, RandAugment [9] presents an efficient framework that works out of the box for applying operations sequentially and without a separate search phase.

Image mixing is an approach that involves blending a pair of samples into one, enforcing the classifier to behave linearly in-between training data points. Performance gains can be observed even by averaging pixel values of two random images and retaining only one out of the two labels [27]. This idea was further extended to more sophisticated techniques which proposed mixing at different ratios and working with soft labels [16,65,66], as well as their non-linear derivatives [57].

Adding small amounts of noise to the input images during training encourages CNNs to have smoother and stronger decision boundaries on the data manifold and results in learning more robust features [49]. The concept was thoroughly studied in the field of adversarial attacks, where the rival network's objective is to learn augmentations that result in misclassifications in the classification model [20,47,48].

Creating synthetic data with the help of generative adversarial networks (GANs) [19] is yet another way to augment a dataset. With recent advancements in the field, GANs are now able to generate images that look real to human observers, in spite of illustrating entities that are not present in the training set [2,21,30,31]. The GAN framework also can be extended to improve the quality of samples created by variational auto-encoders [12] or perform style transfer to map existing imagery to the domain of interest [39,59,67].

2.2 Video Recognition

A clear-cut approach to video classification using CNNs is to include the temporal domain by extending the dimensionality of convolutional operations. 3D filters achieved superior results when compared to 2D, proving that the time domain has a lot of value [28]. The inclusion of the temporal axis opened up a whole research area that is aimed at exploring its various fusion techniques. The most popular ones are slow fusion to improve the time awareness of the model [29], late fusion, where temporal features are blended at the last layer [29], longer fusion, which explores the benefit of extending the temporal depth [62], and ensembling networks with different temporal awareness [62]. Finally, a combination of 2D an 1D kernels is proposed to substantially reduce the amount of learnable parameters without any loss in performance [58].

Motivated by the fact that humans use different streams to process appearance and motion data, multiple stream models were proposed [53]. The aim is to have separate spatial and temporal tracks, hence making it easier to encode relevant features in the respective streams. This is further enhanced by supplying different inputs - whereas the spatial path takes RGB frames, which contain appearance information, the temporal path receives optical flow frames that contain motion data. Later work shows that earlier fusion of the streams allows to retain the performance while halving the amount of learnable parameters [14].

2.3 Temporal Augmentation

Although a substantial amount of work has been done on spatial augmentation, the field of temporal augmentation remains under-explored. Random Mean Scaling [33] stochastically varies the low-frequency feature components to regularize classifiers, whereas FreqAug [32] experiments with randomly removing them. RandAugment-T [34] extends the spatial-only framework to the time dimension and presents a set of modifications on cut-and-paste and blend algorithms, such as CutOut [11], CutMix [65], MixUp [66], and CutMixUp [64], to produce temporally localisable features. Our work expands on the latter and proposes a set of modifications that can be used to make video classifiers more robust and accurate.

3 Methods

3.1 Single Video Augmentation

RandAugment [9] is an automated data augmentation framework that randomly selects a number of transformations for a given image. From a list of K operations, RandAugment takes N augmentations with the magnitude of M. Each transformation has a probability of $\frac{1}{K}$ to be chosen. A total of $K = 14$ operations are presented: Identity, Rotate, Posterise, Equalise, Sharpness, Translate-X, Translate-Y, Colour, AutoContrast, Solarise, Contrast, Brightness, Shear-X, and Shear-Y.

RandAugment-T [34] introduces M_1 and M_n, two magnitude points that are placed at the start and the end of each video. This allows for smooth augmentation transitions across the frames and brings the temporal component to the equation, where possible. The work also extends the list of available transformations by including ColourInvert, albeit it having static magnitude. All the operations mentioned above are taken directly from image augmentation, and are applied to a single video. Operations such as Identity, AutoContrast, Equalise, and ColourInvert do not have varying M, and hence are applied evenly across the sample.

Although previous work sticks to the aforementioned list of transformations [8,9,26,41], the purpose of this paper is to propose temporal augmentations, rather than suggest a new augmentation policy. Therefore, we expand the list of available operations by introducing VideoReverse, FrameFadeIn, and VideoCutMix (Fig. 1) - transformations that are designed specifically for video samples. VideoReverse turns the video backwards, creating a rewind effect, yet maintaining the semantics and integrity of the sample. FrameFadeIn is inspired by FadeMixUp [34], with the main difference being the use of a single sample and a simpler mixing ratio calculation:

$$\tilde{x}_t = (1 - \lambda_t)x_t + \lambda_t x_{n-t}, \tag{1}$$

where $\tilde{x}$, x, n, and λ indicate the mixed data, original data, total number of frames, and mixing ratio, respectively. Unlike FadeMixUp, we do not sample

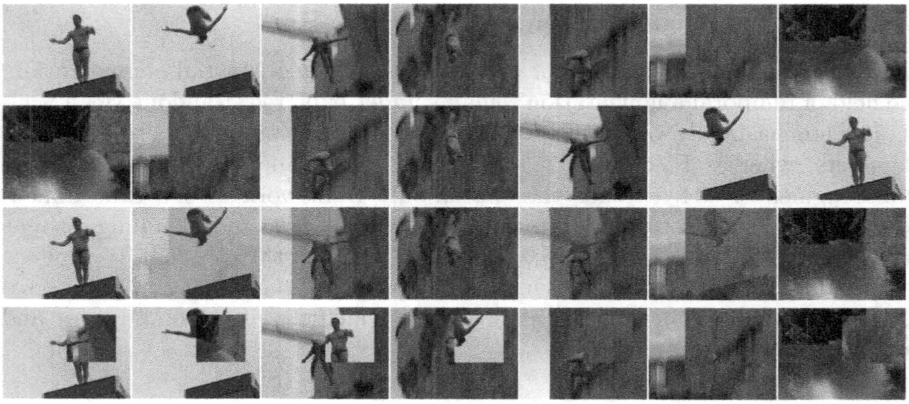

Fig. 1. Unaugmented video, VideoReverse, FrameFadeIn, and VideoCutMix in the 1st, 2nd, 3rd, and 4th row, respectively.

start and end points for λ interpolation. Instead, we gradually increase it from 0 to 0.5 until the middle of the video, then decrease it back to 0:

$$\lambda_t = \begin{cases} \frac{t}{n}, & \text{if } t \leq \frac{n}{2}, \\ \frac{n-t}{n}, & \text{otherwise.} \end{cases} \quad (2)$$

This maintains a healthy trade-off between spatial and temporal perturbations - when the distance between frames is large, the mixing ratio is small, and vice versa. Although it is possible to use sampled magnitudes instead, it significantly increases the risk of breaking temporal consistency. VideoCutMix is a temporal extension of CutMix [34,65] that can be applied to a single video:

$$\tilde{x_t} = M \odot x_t + (1 - M) \odot \hat{x_t}, \quad (3)$$

where M, $\hat{x}$, and $\odot$ denote the binary region mask indicating where to drop out or fill in from two separate frames, video with randomly shuffled frames, and element-wise multiplication, respectively. Although cut-and-pasting happens within the same sample, the nondeterministic nature introduces a certain risk of altering data to the point where semantics may be significantly damaged or lost. To keep it at minimum, we set the region ratio to 0.2 of the original frame size and keep the position of the bounding box static. As with all single sample augmentations, labels remain unchanged in VideoReverse, FrameFadeIn, and VideoCutMix.

3.2 MagAugment

RandAugment-T [34] implements augmentation transitions across frames by putting two magnitude checkpoints, M_1 and M_n, at the start and the end of

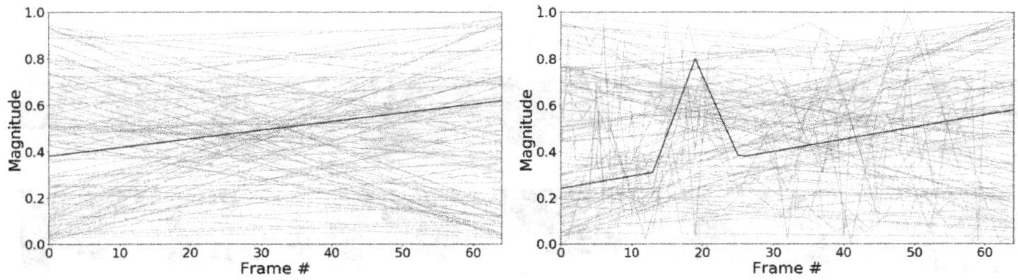

Fig. 2. Visual comparison of magnitude manipulations proposed in RandAugment-T [34] (left) and MagAugment (right), comprising of 100 randomly sampled M arrays. Blue lines - randomly highlighted samples. For MagAugment, β was set to 8.

samples, and calculating the other M_t via linear interpolation. The introduced change in magnitude leads to better video action recognition performances, when compared to its static variant [34]. Our hypothesis is that having more magnitude checkpoints placed along the sample results in greater generalisation performance, as they are more likely to mimic perturbations observed in real-life conditions. Phenomena such as flashes, sudden camera shaking and/or movement, loss of focus, and exposure adjustments tend to happen in much shorter time periods than the length of the entire video. In this subsection, we propose MagAugment (Fig. 2) - a framework designed to increase the magnitude diversity even further, without interrupting the temporal consistency.

We start with the linear signal connecting the two ends of the magnitude array. To introduce short and sporadic magnitude swings, we sample a point from the uniform distribution, $M_p \sim U(M_{min}, M_{max})$, where the parameters represent the minimum and maximum magnitude values for a given transformation. The duration of the perturbations in frames is set to $j \sim U(1, \beta)$, where β is the MagAugment parameter. Finally, the location of the point is drawn from $p \sim U(1 + j, n - j)$, where n is the total amount of frames. The process can be repeated to model several fluctuations. To incorporate the magnitude swings into the original signal, we linearly interpolate from M_{p-j} to M_p, then back to M_{p+j}. As a result, the overall augmentation direction is maintained, while allowing for occasional, more aggressive changes in pixel space that do not necessarily follow the general trend.

3.3 Temporal Deleting, Cut-and-Pasting, and Blending

The temporal adaptations of CutOut [11] and CutMix [65] apply a bounding box, B, to every frame of a given sample, without changing its position, r. In CutMix, the frame sequences are also aligned with the video used for mixing. The concept of static location is practiced in the algorithms' extensions as well - CubeCutOut, CubeCutMix, CutMixUp, and CubeCutMixUp [34]. The temporal version of MixUp [66] has a fixed mixing ratio, λ, and remains so in its extensions too - CutMixUp, FrameCutMixUp, CubeCutMixUp [34,64]. Such an

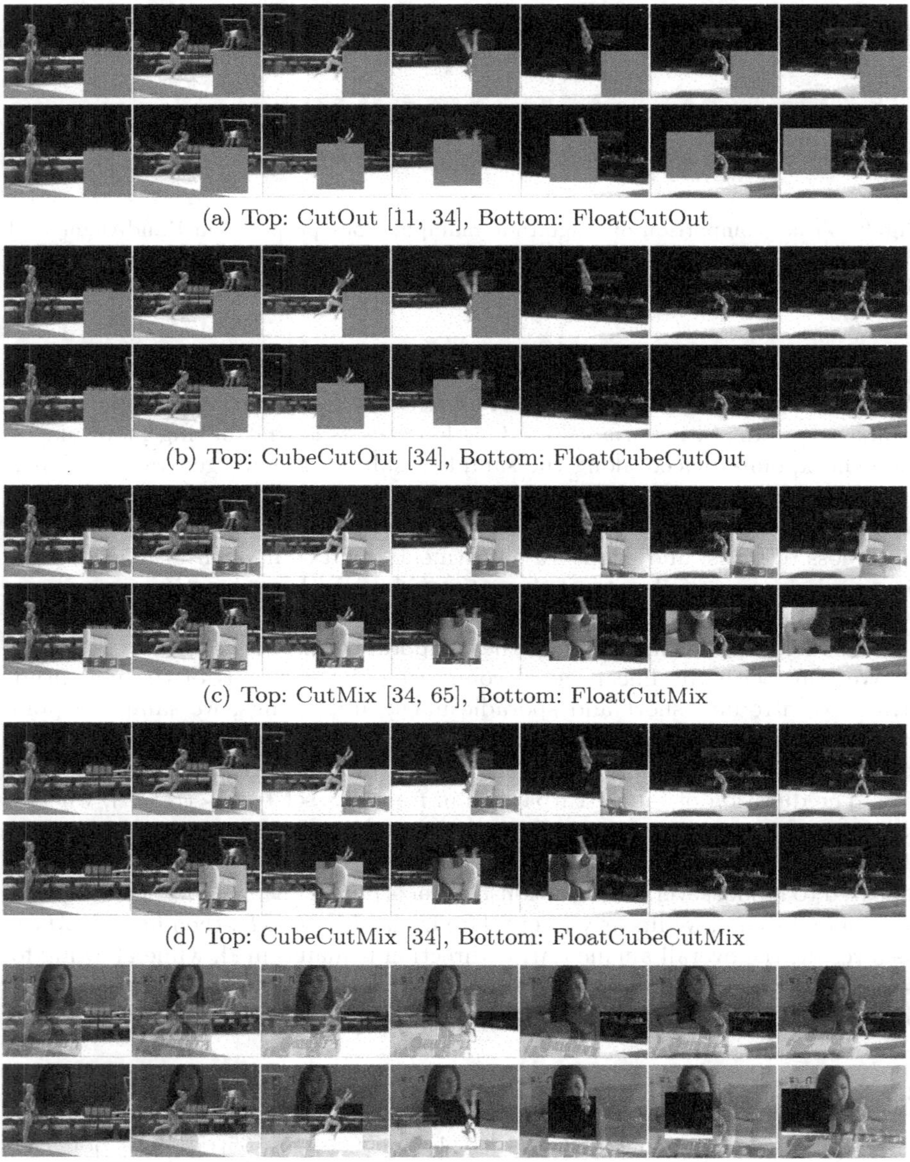

(a) Top: CutOut [11, 34], Bottom: FloatCutOut

(b) Top: CubeCutOut [34], Bottom: FloatCubeCutOut

(c) Top: CutMix [34, 65], Bottom: FloatCutMix

(d) Top: CubeCutMix [34], Bottom: FloatCubeCutMix

(e) Top: CutMixUp [34], Bottom: FloatCutMixUp

Fig. 3. Visual comparison of temporal deleting, cut-and-pasting, and blending algorithms with static and dynamic r and λ.

(f) Top: CubeCutMixUp [34], Bottom: FloatCubeCutMixUp

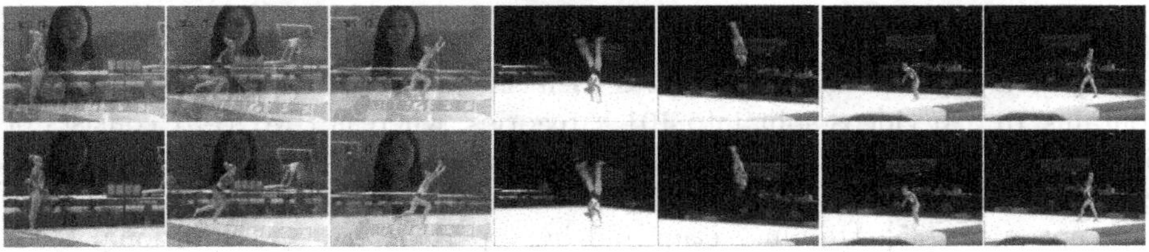

(g) Top: FrameCutMixUp [34], Bottom: FloatFrameCutMixUp

Fig. 3. (*continued*)

idea removes the stochastic behaviour that would be introduced if the aforementioned augmentations were applied to frames separately, without acknowledging them as a part of data series. However, the regularisation techniques themselves can be temporally varied too. By taking a deterministic approach, we are able to enhance the level of spatiotemporal augmentations and involve more bounding box positions and mixing ratios within a batch.

In this subsection, we propose dynamic r and λ, by linearly changing them across the time dimension. The concept is similar to RandAugment-T, only this time we generate r_1/r_n or λ_1/λ_n instead of magnitude points for the start and the end of a training sample. Therefore, for delete and cut-and-paste algorithms r becomes:

$$\begin{aligned} r_{1_x} &\sim U(r_w, W - r_w), r_{n_x} \sim U(r_w, W - r_w), r_w = W\sqrt{1-I}, \\ r_{1_y} &\sim U(r_h, H - r_h), r_{n_y} \sim U(r_h, H - r_h), r_h = H\sqrt{1-I}, \\ I &\sim \text{Beta}(\alpha, \alpha), \end{aligned} \tag{4}$$

where W, H, U, and α are the frame width, frame height, uniform distribution, and beta distribution parameter, respectively. Please note that unlike the previous implementations [11,34,64,65], we ensure that the bounding box is fully within the frame at all times, therefore guaranteeing label consistency across the time dimension and omitting the label recalculation step. r_w and r_h are calculated once per video. When start and end points are found, the rest is computed via linear interpolation between the two. Although the authors of FadeMixUp [34] introduced dynamic λ, CutMixUp, FrameCutMixUp, CubeCutMixUp still used the static one. To make the three algorithms temporally varied when it comes to blending, we substitute MixUp with FadeMixUp. All of the above results in seven new regularisation approaches: FloatCutOut, FloatCubeCutOut, FloatCutMix, FloatCubeCutMix, FloatCutMixUp, FloatCubeCutMixUp, and FloatFrameCutMixUp (Fig. 3).

4 Experiments

We train and test the approaches mentioned in this paper on the UCF-101 [55] and HMDB-51 [38] datasets to assess their effectiveness. The UCF-101 dataset contains 13 320 videos split into 101 categories, whereas HMDB-51 consists of 6 766 videos split into 51 categories. To keep the comparison with the previous work fair [34], we use the same training and testing splits, network architecture [13], optimiser [35], training setup and hyperparameters, and additional techniques, such as learning rate warm-up [23], cosine learning rate scheduling [43]. Please note that for methods proposed by Kim et al. [34], we report results achieved by running the published code[1] ourselves. For all tables, **bold text** indicates the highest accuracy. For the UCF-101, the displayed numbers represent the results on the 1st VIPriors action recognition challenge split. For the HMDB-51, we report the average results obtained from 3 different splits [38].

4.1 Single Video Augmentation

In this subsection, we evaluate VideoReverse, FrameFadeIn, VideoCutMix, and MagAugment, the results can be found in Table 1. RandAugment indicates static magnitude, applied evenly to all the frames of a given video. For RandAugment-T+, $M1$ and M_n are set to $M-\delta$ and $M+\delta$, respectively, where $\delta = U(0, 0.5*M)$, and M comes from the values used by RandAugment. RandAugment-T++ stands for the extended version, which includes VideoReverse, FrameFadeIn and VideoCutMix (abbreviated as VR, FFI, and VCM, respectively). We also include an ablation study by disabling each of the transformations. For MagAugment, a grid search of $\beta \in [2, 4, 8, 16]$ and the amount of magnitude checkpoints, $P \in [1, 2, 3, 4]$, was used to obtain the highest accuracy, with $\beta = 8$ and $P = 2$ demonstrating the best performance. We apply MagAugment to all transformations present in RandAugment-T++, apart from the ones that cannot facilitate varying magnitude - Identity, Reverse, AutoContrast, Equalise, ColourInvert, FrameFadeIn, and VideoCutMix.

The results show that including more single video augmentations provides a benefit with no added computational overhead, thanks to the nature of RandAugment. However, since the improvements in performance are rather small, it is unclear whether all the proposed augmentations are useful. By enabling MagAugment, we obtain 2.36% and 1.46% accuracy increases over spatial-only RandAugment, compared to 0.65% and 0.24% achieved by RandAugment-T, in UCF-101 Top1 and HMDB-51 Top1 settings, respectively.

[1] https://github.com/taeoh-kim/temporal_data_augmentation.

Table 1. Video action recognition results on the UCF-101 and HMDB-51 datasets for single video augmentation techniques.

Method	UCF Top-1	UCF Top-5	HMDB Top-1	HMDB Top-5
Baseline [13]	54.93	77.43	39.12	69.89
RandAugment [9,34]	69.82	88.57	49.24	79.94
RandAugment-T+ [34]	70.47	89.94	49.48	80.17
RandAugment-T++	**70.74**	**90.04**	**49.60**	**80.21**
RandAugment-T++ - VR	70.52	89.94	49.50	80.10
RandAugment-T++ - FFI	70.58	89.98	49.58	80.06
RandAugment-T++ - VCM	70.76	90.12	49.54	80.17
MagAugment	**72.18**	**93.78**	**50.70**	**81.12**

Table 2. Video action recognition results on the UCF-101 and HMDB-51 datasets for temporal deleting, cut-and-pasting, and blending techniques.

Method	UCF Top-1	UCF Top-5	HMDB Top-1	HMDB Top-5
Baseline [13]	54.93	77.43	39.12	69.89
CutOut [11,34]	51.16	74.25	36.93	68.07
CubeCutOut [34]	51.82	76.73	37.50	68.53
*F*CutOut	54.24	76.23	39.02	69.89
*F*CubeCutOut	**54.68**	**77.19**	**39.25**	**70.00**
CutMix [34,65]	53.03	76.78	34.69	65.67
CubeCutMix [34]	54.91	77.34	36.75	67.23
*F*CutMix	55.25	77.27	37.32	68.24
*F*CubeCutMix	**55.66**	**78.00**	**39.42**	**69.98**
CutMixUp [34,64]	60.08	82.14	43.13	74.19
CubeCutMixUp [34]	60.16	82.14	43.15	74.24
FrameCutMixUp [34]	61.02	82.97	42.88	74.08
*F*CutMixUp	62.41	84.59	45.06	75.78
*F*CubeCutMixUp	62.38	84.70	45.12	75.85
*F*FrameCutMixUp	**63.04**	**85.64**	**45.98**	**76.90**

4.2 Temporal Deleting, Cut-and-Pasting, and Blending

We present the results of Cutout [11], CutMix [65], and CutMixUp [64], and their temporal extensions, which can be found in Table 2. We prefix our methods with F to save space and indicate floating bounding box positions and mixing ratios. Single video augmentation is turned off in this experiment. Although the CutOut variants struggle to beat the baseline and the CutMix spin-offs demonstrate a rather small boost in accuracy, it is clear that having dynamic r and

λ helps the model to consistently achieve better performance - when compared side by side, the floating extensions demonstrate an average gain of 2.45%, when compared to their static variants in the Top-1 settings. FloatFrameCutMixUp scores the highest accuracy, improving over the baseline by 8.11% and 6.86% in UCF-101 Top-1 and HMDB-51 Top-1 settings, respectively. FloatCubeCutOut, FloatCubeCutMix, and FloatFrameCutMixUp perform the best in their respective groups, suggesting that retaining some of the frames of a video unaffected might yield additional benefits (Fig. 3b, Fig. 3d, Fig. 3g).

5 Conclusions

In this paper, we introduced several novel temporal data augmentation methods. We showed that developing video-specific transformations and including more aggressive magnitude transitions is beneficial for networks that aim to solve video action recognition. We extended temporal versions of CutOut, CutMix, and CutMixUp further by changing their nature from static to dynamic, and observed an improvement performance. Future work includes combining single video augmentations with delete, cut-and-paste, and blend techniques to expand the total amount of possible augmentation combinations, covering more baseline models to analyse applicability and versatility of the proposed methods, and testing the framework on larger datasets, such as Kinetics [3] and Something-Something-v2 [24].

Acknowledgement. The authors are grateful for the support from the Natural Environment Research Council and Engineering and Physical Sciences Research Council through the NEXUSS Centre for Doctoral Training (grant #NE/RO12156/1).

References

1. Antoniou, A., Storkey, A., Edwards, H.: Data augmentation generative adversarial networks. arXiv preprint arXiv:1711.04340 (2017)
2. Brock, A., Donahue, J., Simonyan, K.: Large scale gan training for high fidelity natural image synthesis. arXiv preprint arXiv:1809.11096 (2018)
3. Carreira, J., Zisserman, A.: Quo vadis, action recognition? A new model and the kinetics dataset. In: Proceedings of the IEEE Conference on Computer Vision and Pattern Recognition, pp. 6299–6308 (2017)
4. Chen, T., Kornblith, S., Norouzi, M., Hinton, G.: A simple framework for contrastive learning of visual representations. In: International Conference on Machine Learning, pp. 1597–1607. PMLR (2020)
5. Chu, P., Bian, X., Liu, S., Ling, H.: Feature space augmentation for long-tailed data. In: Vedaldi, A., Bischof, H., Brox, T., Frahm, J.-M. (eds.) ECCV 2020. LNCS, vol. 12374, pp. 694–710. Springer, Cham (2020). https://doi.org/10.1007/978-3-030-58526-6_41
6. Cireşan, D., Meier, U., Masci, J., Gambardella, L.M., Schmidhuber, J.: High-performance neural networks for visual object classification. Computing Research Repository - CORR (2011)

7. Cireşan, D., Meier, U., Schmidhuber, J.: Multi-column deep neural networks for image classification. In: Proceedings/CVPR, IEEE Computer Society Conference on Computer Vision and Pattern Recognition. IEEE Computer Society Conference on Computer Vision and Pattern Recognition (2012). https://doi.org/10.1109/CVPR.2012.6248110
8. Cubuk, E.D., Zoph, B., Mane, D., Vasudevan, V., Le, Q.V.: Autoaugment: learning augmentation policies from data. arXiv preprint arXiv:1805.09501 (2018)
9. Cubuk, E.D., Zoph, B., Shlens, J., Le, Q.V.: Randaugment: practical automated data augmentation with a reduced search space. In: Proceedings of the IEEE/CVF Conference on Computer Vision and Pattern Recognition Workshops, pp. 702–703 (2020)
10. DeVries, T., Taylor, G.W.: Dataset augmentation in feature space. arXiv preprint arXiv:1702.05538 (2017)
11. DeVries, T., Taylor, G.W.: Improved regularization of convolutional neural networks with cutout. arXiv preprint arXiv:1708.04552 (2017)
12. Doersch, C.: Tutorial on variational autoencoders. arXiv preprint arXiv:1606.05908 (2016)
13. Feichtenhofer, C., Fan, H., Malik, J., He, K.: Slowfast networks for video recognition. In: Proceedings of the IEEE/CVF International Conference on Computer Vision, pp. 6202–6211 (2019)
14. Feichtenhofer, C., Pinz, A., Zisserman, A.: Convolutional two-stream network fusion for video action recognition. In: Proceedings of the IEEE Conference on Computer Vision and Pattern Recognition, pp. 1933–1941 (2016)
15. French, G., Mackiewicz, M., Fisher, M.: Self-ensembling for visual domain adaptation. In: International Conference on Learning Representations (2018)
16. French, G., Oliver, A., Salimans, T.: Milking cowmask for semi-supervised image classification. arXiv preprint arXiv:2003.12022 (2020)
17. Frid-Adar, M., Diamant, I., Klang, E., Amitai, M., Goldberger, J., Greenspan, H.: Gan-based synthetic medical image augmentation for increased CNN performance in liver lesion classification. Neurocomputing **321**, 321–331 (2018)
18. Ganin, Y., Lempitsky, V.: Unsupervised domain adaptation by backpropagation. In: International Conference on Machine Learning, pp. 1180–1189. PMLR (2015)
19. Goodfellow, I., et al.: Generative adversarial nets. In: Advances in Neural Information Processing Systems, vol. 27 (2014)
20. Goodfellow, I.J., Shlens, J., Szegedy, C.: Explaining and harnessing adversarial examples. arXiv preprint arXiv:1412.6572 (2014)
21. Gorpincenko, A., French, G., Knight, P., Challiss, M., Mackiewicz, M.: Improving automated sonar video analysis to notify about jellyfish blooms. IEEE Sens. J. **21**(4), 4981–4988 (2021). https://doi.org/10.1109/JSEN.2020.3032031
22. Gorpincenko, A., French, G., Mackiewicz, M.: Virtual adversarial training in feature space to improve unsupervised video domain adaptation (2020)
23. Goyal, P., et al.: Accurate, large minibatch SGD: training imagenet in 1 hour. arXiv preprint arXiv:1706.02677 (2017)
24. Goyal, R., et al.: The "something something" video database for learning and evaluating visual common sense. In: Proceedings of the IEEE International Conference on Computer Vision, pp. 5842–5850 (2017)
25. He, K., Fan, H., Wu, Y., Xie, S., Girshick, R.: Momentum contrast for unsupervised visual representation learning. In: Proceedings of the IEEE/CVF Conference on Computer Vision and Pattern Recognition, pp. 9729–9738 (2020)

26. Ho, D., Liang, E., Chen, X., Stoica, I., Abbeel, P.: Population based augmentation: efficient learning of augmentation policy schedules. In: International Conference on Machine Learning, pp. 2731–2741. PMLR (2019)
27. Inoue, H.: Data augmentation by pairing samples for images classification. arXiv preprint arXiv:1801.02929 (2018)
28. Ji, S., Xu, W., Yang, M., Yu, K.: 3D convolutional neural networks for human action recognition. IEEE Trans. Pattern Anal. Mach. Intell. **35**(1), 221–231 (2013). https://doi.org/10.1109/TPAMI.2012.59
29. Karpathy, A., Toderici, G., Shetty, S., Leung, T., Sukthankar, R., Fei-Fei, L.: Large-scale video classification with convolutional neural networks. In: 2014 IEEE Conference on Computer Vision and Pattern Recognition, pp. 1725–1732 (2014). https://doi.org/10.1109/CVPR.2014.223
30. Karras, T., Laine, S., Aila, T.: A style-based generator architecture for generative adversarial networks. In: Proceedings of the IEEE/CVF Conference on Computer Vision and Pattern Recognition, pp. 4401–4410 (2019)
31. Karras, T., Laine, S., Aittala, M., Hellsten, J., Lehtinen, J., Aila, T.: Analyzing and improving the image quality of StyleGAN. In: Proceedings of the IEEE/CVF Conference on Computer Vision and Pattern Recognition, pp. 8110–8119 (2020)
32. Kim, J.Y., Ha, J.E.: Spatio-temporal data augmentation for visual surveillance. IEEE Access (2021). https://doi.org/10.1109/ACCESS.2021.3135505
33. Kim, J., Cha, S., Wee, D., Bae, S., Kim, J.: Regularization on spatio-temporally smoothed feature for action recognition. In: Proceedings of the IEEE/CVF Conference on Computer Vision and Pattern Recognition, pp. 12103–12112 (2020)
34. Kim, T., Lee, H., Cho, M.A., Lee, H.S., Cho, D.H., Lee, S.: Learning temporally invariant and localizable features via data augmentation for video recognition. In: Bartoli, A., Fusiello, A. (eds.) ECCV 2020. LNCS, vol. 12536, pp. 386–403. Springer, Cham (2020). https://doi.org/10.1007/978-3-030-66096-3_27
35. Kingma, D.P., Ba, J.: Adam: a method for stochastic optimization. arXiv preprint arXiv:1412.6980 (2014)
36. Krizhevsky, A., Sutskever, I., Hinton, G.E.: Imagenet classification with deep convolutional neural networks. In: Pereira, F., Burges, C., Bottou, L., Weinberger, K. (eds.) Advances in Neural Information Processing Systems, vol. 25. Curran Associates, Inc. (2012)
37. Krogh, A., Hertz, J.: A simple weight decay can improve generalization. In: Advances in Neural Information Processing Systems, vol. 4 (1991)
38. Kuehne, H., Jhuang, H., Garrote, E., Poggio, T., Serre, T.: HMDB: a large video database for human motion recognition. In: 2011 International Conference on Computer Vision, pp. 2556–2563 (2011). https://doi.org/10.1109/ICCV.2011.6126543
39. Lee, S., Park, B., Kim, A.: Deep learning based object detection via style-transferred underwater sonar images. IFAC-PapersOnLine **52**(21), 152–155 (2019). https://doi.org/10.1016/j.ifacol.2019.12.299
40. Lemley, J., Bazrafkan, S., Corcoran, P.M.: Smart augmentation learning an optimal data augmentation strategy. IEEE Access **5**, 5858–5869 (2017)
41. Lim, S., Kim, I., Kim, T., Kim, C., Kim, S.: Fast autoaugment. In: Advances in Neural Information Processing Systems, vol. 32 (2019)
42. Liu, B., Wang, X., Dixit, M., Kwitt, R., Vasconcelos, N.: Feature space transfer for data augmentation. In: Proceedings of the IEEE Conference on Computer Vision and Pattern Recognition (CVPR) (2018)
43. Loshchilov, I., Hutter, F.: SGDR: stochastic gradient descent with warm restarts. arXiv preprint arXiv:1608.03983 (2016)

44. Loshchilov, I., Hutter, F.: Decoupled weight decay regularization. arXiv preprint arXiv:1711.05101 (2017)
45. Mao, X., Ma, Y., Yang, Z., Chen, Y., Li, Q.: Virtual mixup training for unsupervised domain adaptation (2019)
46. Misra, I., Maaten, L.V.D.: Self-supervised learning of pretext-invariant representations. In: Proceedings of the IEEE/CVF Conference on Computer Vision and Pattern Recognition, pp. 6707–6717 (2020)
47. Miyato, T., Maeda, S., Koyama, M., Ishii, S.: Virtual adversarial training: a regularization method for supervised and semi-supervised learning. IEEE Trans. Pattern Anal. Mach. Intell. **41**(8), 1979–1993 (2019). https://doi.org/10.1109/TPAMI.2018.2858821
48. Moosavi-Dezfooli, S.M., Fawzi, A., Frossard, P.: Deepfool: a simple and accurate method to fool deep neural networks. In: Proceedings of the IEEE Conference on Computer Vision and Pattern Recognition, pp. 2574–2582 (2016)
49. Moreno-Barea, F.J., Strazzera, F., Jerez, J.M., Urda, D., Franco, L.: Forward noise adjustment scheme for data augmentation. In: 2018 IEEE Symposium Series on Computational Intelligence (SSCI), pp. 728–734 (2018). https://doi.org/10.1109/SSCI.2018.8628917
50. Prechelt, L.: Early stopping - but when? In: Orr, G.B., Müller, K.-R. (eds.) Neural Networks: Tricks of the Trade. LNCS, vol. 1524, pp. 55–69. Springer, Heidelberg (1998). https://doi.org/10.1007/3-540-49430-8_3
51. Shu, R., Bui, H.H., Narui, H., Ermon, S.: A DIRT-T approach to unsupervised domain adaptation. arXiv preprint arXiv:1802.08735 (2018)
52. Simard, P., Steinkraus, D., Platt, J.: Best practices for convolutional neural networks applied to visual document analysis. In: Seventh International Conference on Document Analysis and Recognition, 2003, pp. 958–963 (2003). https://doi.org/10.1109/ICDAR.2003.1227801
53. Simonyan, K., Zisserman, A.: Two-stream convolutional networks for action recognition in videos. In: Proceedings of the 27th International Conference on Neural Information Processing Systems - Volume 1, NIPS 2014, pp. 568–576. MIT Press, Cambridge (2014)
54. Sohn, K., et al.: Fixmatch: simplifying semi-supervised learning with consistency and confidence. Adv. Neural. Inf. Process. Syst. **33**, 596–608 (2020)
55. Soomro, K., Zamir, A.R., Shah, M.: UCF101: a dataset of 101 human actions classes from videos in the wild. arXiv preprint arXiv:1212.0402 (2012)
56. Srivastava, N., Hinton, G., Krizhevsky, A., Sutskever, I., Salakhutdinov, R.: Dropout: a simple way to prevent neural networks from overfitting. J. Mach. Learn. Res. **15**(56), 1929–1958 (2014)
57. Summers, C., Dinneen, M.J.: Improved mixed-example data augmentation. In: 2019 IEEE Winter Conference on Applications of Computer Vision (WACV), pp. 1262–1270. IEEE (2019)
58. Sun, L., Jia, K., Yeung, D., Shi, B.E.: Human action recognition using factorized spatio-temporal convolutional networks. In: 2015 IEEE International Conference on Computer Vision (ICCV), pp. 4597–4605. IEEE Computer Society, Los Alamitos (2015). https://doi.org/10.1109/ICCV.2015.522
59. Terayama, K., Shin, K., Mizuno, K., Tsuda, K.: Integration of sonar and optical camera images using deep neural network for fish monitoring. Aquacult. Eng. **86**, 102000 (2019). https://doi.org/10.1016/j.aquaeng.2019.102000

60. Tran, T., Pham, T., Carneiro, G., Palmer, L., Reid, I.: A Bayesian data augmentation approach for learning deep models. In: Proceedings of the 31st International Conference on Neural Information Processing Systems, NIPS 2017, pp. 2794–2803. Curran Associates Inc., Red Hook (2017)
61. Tzeng, E., Hoffman, J., Saenko, K., Darrell, T.: Adversarial discriminative domain adaptation. In: Proceedings of the IEEE Conference on Computer Vision and Pattern Recognition, pp. 7167–7176 (2017)
62. Varol, G., Laptev, I., Schmid, C.: Long-term temporal convolutions for action recognition. IEEE Trans. Pattern Anal. Mach. Intell. **40**(6), 1510–1517 (2017)
63. Wang, Y.X., Girshick, R., Hebert, M., Hariharan, B.: Low-shot learning from imaginary data. In: Proceedings of the IEEE Conference on Computer Vision and Pattern Recognition, pp. 7278–7286 (2018)
64. Yoo, J., Ahn, N., Sohn, K.A.: Rethinking data augmentation for image super-resolution: a comprehensive analysis and a new strategy. In: Proceedings of the IEEE/CVF Conference on Computer Vision and Pattern Recognition, pp. 8375–8384 (2020)
65. Yun, S., Han, D., Chun, S., Oh, S.J., Yoo, Y., Choe, J.: Cutmix: regularization strategy to train strong classifiers with localizable features. In: 2019 IEEE/CVF International Conference on Computer Vision (ICCV), pp. 6022–6031 (2019). https://doi.org/10.1109/ICCV.2019.00612
66. Zhang, H., Cisse, M., Dauphin, Y.N., Lopez-Paz, D.: mixup: beyond empirical risk minimization. arXiv preprint arXiv:1710.09412 (2017)
67. Zhu, J.Y., Park, T., Isola, P., Efros, A.A.: Unpaired image-to-image translation using cycle-consistent adversarial networks. In: Proceedings of the IEEE International Conference on Computer Vision, pp. 2223–2232 (2017)

Pre-text Representation Transfer for Deep Learning with Limited and Imbalanced Data: Application to CT-Based COVID-19 Detection

Fouzia Altaf[1(✉)], Syed M. S. Islam[1], Naeem K. Janjua[1], and Naveed Akhtar[2]

[1] Edith Cowan University Joondalup, Joondalup, Australia
faltaf@our.ecu.edu.au
[2] University of Western Australia Crawley, Perth, Australia

Abstract. Annotating medical images for disease detection is often tedious and expensive. Moreover, the available training samples for a given task are generally scarce and imbalanced. These conditions are not conducive for learning effective deep neural models. Hence, it is common to 'transfer' neural networks trained on natural images to the medical image domain. However, this paradigm lacks in performance due to the large domain gap between the natural and medical image data. To address that, we propose a novel concept of Pre-text Representation Transfer (PRT). In contrast to the conventional transfer learning, which fine-tunes a source model after replacing its classification layers, PRT retains the original classification layers and updates the representation layers through an unsupervised pre-text task. The task is performed with (original, not synthetic) medical images, without utilizing any annotations. This enables representation transfer with a large amount of training data. This high-fidelity representation transfer allows us to use the resulting model as a more effective feature extractor. Moreover, we can also subsequently perform the traditional transfer learning with this model. We devise a collaborative representation based classification layer for the case when we leverage the model as a feature extractor. We fuse the output of this layer with the predictions of a model induced with the traditional transfer learning performed over our pre-text transferred model. The utility of our technique for limited and imbalanced data classification problem is demonstrated with an extensive five-fold evaluation for three large-scale models, tested for five different class-imbalance ratios for CT based COVID-19 detection. Our results show a consistent gain over the conventional transfer learning with the proposed method.

Keywords: Transfer learning · Imbalanced data · COVID-19

1 Introduction

In the medical imaging domain, data labelling requires medical experts, who must carefully analyse the samples to provide the correct annotation. Not only

W. Q. Yan et al. (Eds.): IVCNZ 2022, LNCS 13836, pp. 119–130, 2023.
https://doi.org/10.1007/978-3-031-25825-1_9

that this process is tedious, expensive and strongly reliant on the availability of medical experts, the data itself suffers from plenty of challenges. First, it is common that the positive samples of a disease are much rarer than the negative samples. This naturally creates an imbalance in the data, which is particularly challenging to induce unbiased computational models using that data. Second, for the geographically constrained facilities, both positive and negative samples are often too few to effectively train a computational model that can facilitate automated disease detection. Incidentally, global data sharing through public repositories also fails to fully resolve these issues due to the data privacy constraints. Whereas medical images are easily searchable content on the internet, their annotations related to a specific diagnostic task are seldom available.

It is well-established that deep learning [12] can induce computational models that can achieve expert-level accuracy for many disease detection tasks using medical images [8]. This fact has led to a wave of deploying deep learning solutions in medical image analysis [5]. However, this technology can only perform effective computational modelling if it is provided with a large amount of training data (e.g., a million samples). For the medical tasks, these samples need to be appropriately annotated by the experts. Thus, the challenges noted in the preceding paragraph present a bottleneck for fully exploiting deep learning in medical image analysis. Currently, Transfer Learning (TL) [22] is a common strategy to side-step this bottleneck [3,6,17].

Transfer Learning takes a deep learning model pre-trained for a *source* domain, and fine-tunes it with a *target* domain data. For the medical tasks, natural images usually form the source domain [5] due to their convenient annotations. The central idea behind TL is that by using a large amount of training images, the pre-trained model (a.k.a. source model) learns a detailed representation of the source domain. This representation also encodes the primitive patterns that form the fundamental image ingredients. Since the target medical domain also comprises images, it is likely that a slight modification to this encoding can already be sufficient to represent the target domain samples reasonably well. Transfer learning seeks to induce the desired modification with the scarcely available data for the medical task at hand.

Altaf et al. [6] recently noted that the large domain gap between the natural and medical images compromises the performance of TL for the medical tasks. They argued that this large gap requires proportionally large data of the target domain for an effective model transfer. Hence, they proposed to first transfer the source model to the target domain with a large-scale annotated medical data, albeit under a different auxiliary imaging modality. Their assumption is that, for a target data modality (e.g., CT scans), large-scale annotated samples are available for a related auxiliary modality (e.g., radiographs) in the medical domain. First transferring the model to the medical domain with the auxiliary modality, and then transferring it further to the target modality, is shown to improve the TL performance [6]. Though effective, availability of annotated auxiliary large-scale data is still a strong assumption for the medical imaging domain. Moreover, the imaging modality disparity within the target domain (e.g., CT scans vs radiographs) can still be problematic.

This paper introduces a novel concept of Pre-text Representation Transfer (PRT), which enables effectively transferring the source domain representation to the target domain without any data modality disparity, or assuming additional data annotations. It formulates an unsupervised learning task, termed *pre-text* task inspired by the self-supervised learning literature [14], which enables the use of a large amount of original target domain data. This data is un-annotated or has irrelevant annotations w.r.t. the target downstream task. We meticulously transfer the source domain 'representation' to the medical domain using this task. This allows us to use the transferred representation both as an effective feature extractor and as a source model to perform further transfer learning. We leverage both options, and fuse their predictions to compute the final output. In the process, we also adapt a collaborative representation scheme to serve as a classification layer for our feature extractor, such that its predictions can be intelligibly fused with the predictions of the transferred model for a performance boost.

Owing to the sensitivity of Computed Tomography (CT) to COVID-19 [20], and available benchmark studies on the exploration of transfer learning for the CT-based COVID-19 detection [4], we showcase the efficacy of our approach for this problem with an extensive evaluation that performs over 75 deep learning model training sessions. With a five-fold validation for three different models, using five different imbalance ratios of the limited CT training images, we demonstrate a consistent improvement over the conventional transfer learning performance with our technique.

The main contributions of this paper can be summarized as follows.

- We introduce a novel concept of Pre-text Representation Transfer (PRT) that allows effective transfer of the representation component of a model to the target domain on (large) unlabelled data.
- We develop a method that leverages the model resulting from PRT as a feature extractor, and fuses its predictions with subsequent transfer learning over the resulting model.
- With extensive five-fold experiments for CT-based COVID-19 detection, we establish the effectiveness of our method for limited and imbalanced data.

2 Proposed Method

We illustrate the central concept of our Pre-text Representation Transfer (PRT) based method in Fig. 1. To concisely present our contribution, we first formalize the model transfer mechanism under the traditional transfer learning (Sect. 2.4). Then, the components of our methods are discussed in detail (Sect. 2.2–2.5).

2.1 Traditional Model Transfer

From a computational perspective, we typically see medical image based disease detection as an image classification task. Let $\mathcal{M}(\mathbf{x} \sim \mathcal{X}, \boldsymbol{\Theta}) : \mathbf{x} \rightarrow \boldsymbol{\ell}_x$ denote an image classifier, where $\mathbf{x}$ is sampled from the image distribution $\mathcal{X}$, $\boldsymbol{\Theta}$ denotes

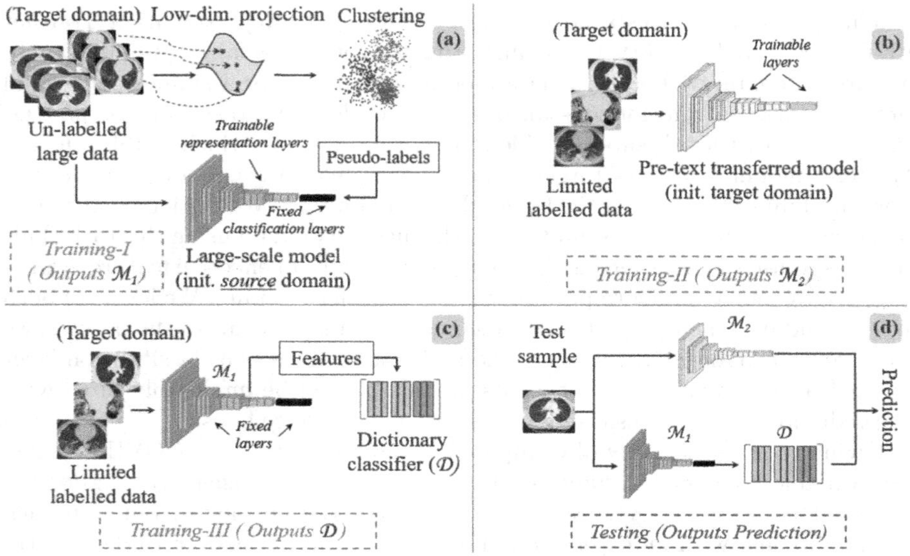

Fig. 1. (a) A large amount of un-annotated data of the target domain is clustered in a lower dimensional space to compute pseudo-labels. Using these labels, pre-text representation transfer is performed using a source domain model. **(b)** The pre-text transferred model $\mathcal{M}_1$ is further modified with the conventional Transfer Learning to induce model $\mathcal{M}_2$. **(c)** The $\mathcal{M}_1$ is used as a feature extractor to construct a feature dictionary $\mathcal{D}$ of the labelled training data. **(d)** A classifier defined by $\mathcal{D}$ fuses its predictions with the output of $\mathcal{M}_2$ to provide the final prediction.

the model parameters, and $\boldsymbol{\ell}_x \in [0,1]^L$ is an L-dimensional binary class vector, encoding the labels of L classes. For a deep neural model, a large amount of annotated samples is normally required to induce an effective $\mathcal{M}(.,.)$. Owing to the unavailability of adequate samples, medical image classification often seeks a transform $\boldsymbol{\Psi} : \widetilde{\mathcal{M}}(\mathbf{z} \sim \mathcal{Z}, \widetilde{\boldsymbol{\Theta}}) \rightarrow \mathcal{M}(\mathbf{x} \sim \mathcal{X}, \boldsymbol{\Theta})$, where $\mathcal{Z}$ is the distribution over natural images, and $\widetilde{\mathcal{M}}(.)$ is a natural image classifier with parameters $\widetilde{\boldsymbol{\Theta}}$. For $\mathbf{z}$, normally $\boldsymbol{\ell}_z \in [0,1]^{\widetilde{L}}$, s.t. $\widetilde{L} >> L$. The function $\boldsymbol{\Psi}$ can be understood as the core model transfer function in the traditional transfer learning. Due to the hierarchical nature of the neural models, we can write $\widetilde{\mathcal{M}}(.,.) = \widetilde{\mathcal{C}}(\widetilde{\boldsymbol{\Theta}_C}, \widetilde{\mathcal{R}}(\widetilde{\boldsymbol{\Theta}_R}, .))$. Here, $\widetilde{\mathcal{C}}(.,.)$ is composed of the deeper 'classification' layers of the model, having their respective parameters $\widetilde{\boldsymbol{\Theta}_C}$. The remaining earlier layers with parameters $\widetilde{\boldsymbol{\Theta}_R}$ are given by $\widetilde{\mathcal{R}}(.,.)$, which encode a 'representation' of the distribution $\mathcal{Z}$ of the source domain.

In the traditional transfer learning, the function $\boldsymbol{\Psi}$ replaces $\widetilde{\boldsymbol{\Theta}_C}$ with a new set of parameters $\boldsymbol{\Theta}_C$, and provides a slightly modified version of $\widetilde{\boldsymbol{\Theta}_R}$, say $\widehat{\boldsymbol{\Theta}_R}$ for the representation of the target image distribution. The eventual transferred model for the medical domain then becomes $\mathcal{M}(.,.) = \mathcal{C}(\boldsymbol{\Theta}_C, (\widehat{\mathcal{R}}(\widehat{\boldsymbol{\Theta}_R}, .))$. Both

$\boldsymbol{\Theta}_C$ and $\widehat{\boldsymbol{\Theta}_R}$ are learned with the samples of $\mathcal{X}$. However, due to the limited data availability, the latter is only a slight modification of $\widetilde{\boldsymbol{\Theta}_R}$ under an extremely small learning rate. It is known that if the domain gap $||\mathcal{Z} - \mathcal{X}||$ is large, only a slight perturbation to the source domain representation may not be sufficient to implement $\boldsymbol{\Psi}$ [6]. This is a major limitation of the traditional transfer learning, which renders the model transfer between the domains with a large covariate shift ineffective.

2.2 Pre-text Representation Transfer (Training-I)

Within the mainstream Machine Learning literature, self-supervised representation learning [7] is an established paradigm to address the limited training data problem. A stream of methods following this paradigm, uses the notion of a *pre-text* task [2], which does not require human labelling of the data. For instance, differentiating between known transformations of a given image and other images, is a pre-text task used by the contrastive learning methods [23]. The hope here is that by solving the pre-text task with a very large amount of unlabelled data, the model can still learn the general representation of the domain. This representation can then be fined-tuned to a given task using a limited amount of labelled data under the optimization objective of that downstream task.

We highlight two challenges in directly using the pre-text task based self-supervised learning for the medical domain. Firstly, it needs to train the original model $\widetilde{\mathcal{M}}(.,.)$ on an enormous amount of data for a very long duration. This extra-ordinary computational requirement is a major limitation of the self-supervised learning paradigm in general. This issue becomes even more pronounced in the sub-domain of medical imaging due to resource limitations in geographically constrained facilities. Secondly, this type of learning does not consider domain transfer for the downstream task. Notice in the preceding paragraph that the pre-text and the downstream tasks are both defined in the same domain. This preempts the possibility of transferring a self-supervised model learned with a natural image pre-text task to the medical domain. Addressing the issues, we introduce the notion of Pre-text Representation Transfer (PRT).

We let our source model $\widetilde{\mathcal{M}}(\mathbf{z} \sim \mathcal{Z}, .) : \mathbf{z} \rightarrow \boldsymbol{\ell}_z$ to be a large-scale pre-trained model of the natural images. We then form a set $\widehat{X}$, such that each element of this set $\widehat{x} \in \widehat{X}$ is still taken from the medical image distribution $\mathcal{X}$. We further impose that the imaging modality of $\widehat{x}$ is also the same as the modality of our eventual downstream task. That is, if the downstream task uses chest CT scans, $\widehat{X}$ only contains chest CT images. We are not concerned with the labels of $\widehat{X}$. Hence, any available image that satisfies our constraint of having the same imaging modality, belongs to $\widehat{X}$. In this work, we scrap the data from public repositories over the internet to create this set. This set is subsequently transformed into $\widetilde{L}$ clusters. It is emphasized that in our settings, $\widetilde{L}$ is also the dimensionality of $\boldsymbol{\ell}_z$. This is intentional, as it helps us in systematically leveraging clustering, which is an unsupervised process. Our aim here is to transform the representation component $\widetilde{\mathcal{R}}(\widetilde{\boldsymbol{\Theta}_R}, .)$ of $\widetilde{\mathcal{M}}(.,.)$, such that it can perform a conservative clustering of $\widehat{X}$ as a pre-text task. To make the process computationally efficient, we perform clustering over low-dimensional projections of $\widehat{x}$. In this work, we use ResNet50 [10] features of $\widehat{x}$ for the projection purpose. However, using any other projection method is also viable under the proposed pipeline.

Using the indices of the resulting clusters as pseudo-labels for $\widehat{X}$, we perform model training for our pre-text representation transfer. To that end, we initialize our network with the natural image model $\widetilde{\mathcal{M}}(.,.)$, and keep the classifier component $\widetilde{\mathcal{C}}(\widetilde{\boldsymbol{\Theta}_C},.)$ fixed during the training. This induces maximal modifications in the representation component $\widetilde{\mathcal{R}}(\widetilde{\boldsymbol{\Theta}_R},.)$ according to $\widehat{X}$. This is in sharp contrast to the traditional transfer learning that enforces minimal change on the representation component. The model resulting from this training is $\mathcal{M}_1(\widehat{x} \sim \widehat{X}, \widehat{\boldsymbol{\Theta}}) : \widehat{x} \rightarrow \boldsymbol{\ell}_{\widehat{x}} \in [0,1]^{\widetilde{L}}$. In the text to follow, we denote this model as $\mathcal{M}_1(.,.)$ for brevity.

2.3 Subsequent Transfer Learning (Training-II)

Following our notational convention, $\mathcal{M}(\mathbf{x} \sim \mathcal{X}, \boldsymbol{\Theta}) : \mathbf{x} \rightarrow \boldsymbol{\ell}_x$ is a model that results from the traditional transfer learning under $\boldsymbol{\Psi} : \widetilde{\mathcal{M}}(.,.) \rightarrow \mathcal{M}(.,.)$. It is possible to simply substitute $\widetilde{\mathcal{M}}(.,.)$ with $\mathcal{M}_1(.,.)$ in this transformation, where $\mathcal{M}_1(.,.)$ is the model resulting from Sect. 2.2. This can be done because we have ensured that $\boldsymbol{\ell}_z, \boldsymbol{\ell}_{\widehat{x}} \in [0,1]^{\widetilde{L}}$. In other words, we have not changed the number of labels for the underlying classification task. Hence, $\mathcal{M}_1(.,.)$ is directly substitutable in the training process. It can be expected that this substitution can benefit $\boldsymbol{\Psi}$ because the representation of $\mathcal{M}_1(.,.)$ is not only in the target domain $\mathcal{X}$, but it is also strictly restricted to the imaging modality of the samples in $\mathcal{X}$. Hence, we can subsequently perform the traditional transfer learning on $\mathcal{M}_1(.,.)$ to obtain a model $\mathcal{M}_2(.,.)$. This model modifies the representation component of $\mathcal{M}_1(.,.)$ only slightly and its classifier component more aggressively, with $10\times$ learning rate. This is in-line with the conventional transfer learning paradigm.

2.4 Feature Extraction and Dictionary Classifier (Training-III)

The traditional transfer learning updates the whole model, including its representation component. Since the proposed PRT already brings that component of $\mathcal{M}_1(.,.)$ in the target domain, we can also exploit $\mathcal{M}_1(.,.)$ separately as an effective feature extractor for the samples in $\mathcal{X}$. We leverage this fact in our method. Following [6], we use a collaborative representation [1] based classifier to predict the labels of the extracted features. The classifier constructs a dictionary $\mathcal{D}$ with the $\mathcal{M}_1(.,.)$ features of the training data. This dictionary is used to compute a class probability of a test feature extracted from $\mathcal{M}_1(.,.)$. The construction of the dictionary is relatively simple. Say, we have 'n' training samples available for a given class. We first construct a sub-matrix $\boldsymbol{d}_c = [\mathcal{R}_1(\hat{x}_1), \mathcal{R}_1(\hat{x}_2), ..., \mathcal{R}_1(\hat{x}_n)]$ for each class, where $\mathcal{R}_1(.)$ is the representation component of $\mathcal{M}_1(.)$. Then, we concatenate the sub-matrices for all the C classes to form $\mathcal{D} = [\boldsymbol{d}_1, \boldsymbol{d}_2, ...\boldsymbol{d}_C]$. This allows us to form a structured dictionary without requiring balanced training data. That is, we allow $\text{col}(\boldsymbol{d}_i) \neq \text{col}(\boldsymbol{d}_j)$, s.t. $i \neq j$ and col(.) computes the number of columns of the matrix in its argument. This construction of a structured dictionary is largely inspired by [6]. However, we extend the prediction mechanism of [6] to better adapt to our methodology. This is discussed in Sect. 2.5.

2.5 Output Predictions (Testing)

From the implementation perspective, the output of an L-class classifier is a probability vector $\boldsymbol{\rho} \in \mathbb{R}^L$, s.t. $\sum_{i=1}^{L} \rho_i = 1$, where ρ_i is the ith coefficient of $\boldsymbol{\rho}$. The label 'ℓ' of a sample is predicted as $\ell = \underset{i}{\text{argmax}} < \rho_i >$. In an analogous manner, the existing dictionary based classifier [6] also chooses to maximize the coefficients of a vector $\boldsymbol{q} \in \mathbb{R}^L$ to predict the class label. However, there is no external constraint over the coefficients q_i of $\boldsymbol{q}$. This is problematic because we eventually want to combine the predictions of dictionary based classification with the predictions of a deep model. Hence, we introduce a constraint in the mechanism, i.e. $\sum_{i=1}^{L} q_i = 1$ to render $\boldsymbol{q}$ into a probability vector. This allows a meaningful fusion of $\boldsymbol{\rho}$ and $\boldsymbol{q}$ as two probability vectors. We eventually predict the class label for a test image as $\ell = \underset{i}{\text{argmax}} < \mathbb{E}[\rho_i, q_i] >$.

It is noteworthy that besides being tailored for limited training data, our overall technique is intrinsically amenable to data imbalance. This is because, firstly, the central idea of PRT does not assume uniform data clustering. Thereby, it learns a representation without asserting a uniform class distribution in the target domain. Secondly, the used dictionary based classifier [6] performs linear operations using an over-complete basis representation [19]. Provided the availability of relevant basis vectors of the desired class in the dictionary, this scheme does not favor a class because of the relative number of the training samples. In our overall framework, transfer learning (*Training II* in Fig. 1) is the only process that conventionally operates under balanced data assumption. However, we perform transfer learning on a model resulting from PRT. Moreover, the output of this model is further fused with the dictionary based classifier's prediction. This compensates for the implicit data balance assumption.

3 Experiments

The problem of dealing with limited and imbalanced data is particularly relevant to the medical imaging domain. Considering the currently prevalent COVID-19 pandemic, we use the task of COVID-19 detection as a test bed for our technique. It is now established that Computed Tomography (CT) demonstrates even higher sensitivity to COVID-19 than Reverse Transcription Polymerase Chain Reaction (RT- PCR) [9,13,20]. Hence, advancing our understanding of CT-based COVID-19 detection is particularly important. Thus, in our experiments, we focus on COVID-19 detection using CT images.

Datasets and Settings: Our experiments utilize two sets of data. (1) Unlabelled large image set $\widehat{X}$. (2) Labelled limited data. The former is required for *Training I* in Fig. 1. We scrap 10K samples from the internet repositories [16,21] for form the set $\widehat{X}$. It is emphasized that we do not require any annotations for $\widehat{X}$, hence the labels provided by the repositories are not used. This is a major advantage of our technique, as it allows us to use any CT image (annotated or not) to improve performance over the downstream task of COVID-19 detection. For (2), we choose the Covid CT Dataset (CCD) [24] that contains 349 images of infected and 397 images of non-infected patients. A further 48 images from the negative samples were dropped to emulate balanced data. Following [4], we employ a five-fold evaluation protocol that sequentially splits the data into training and testing sets. To emulated imbalanced scenarios, we keep 10, 25, 50 and 75% data points from the positive samples in the training folds (discarding the rest), while the negative samples are always fixed to 349. It is worth emphasizing that these settings are particularly challenging because of the high level of data scarcity and imbalance. We transfer ImageNet models ResNet101 [10], VGG16 [18] and DenseNet201 [11] in our experiments. These are commonly used standard large-scale models trained over 1 million natural images of 1,000 categories of daily-life objects.

Implementation Details: We use the ImageNet models [10,11,18] provided by Mathworks© and conduct experiments with Matlab® on an NVIDIA GeForce GTX 1070 GPU with 8GB RAM. Based on our models and setup, $\widetilde{L} = 1,000$ and $L = 2$. In the *Training I* session - see Fig. 1, K-Means clustering is performed with $K = \widetilde{L} = 1,000$. The PRT is conducted with 15 epochs of training with a learning rate of 3e-4, using a batch size 16. The *Training II* session followed a similar settings, except that we reduced the number of epochs to 7. This is inline with [15]. The hyper-parameter settings for *Training III* session followed [6] for the dictionary based classifier.

Results: We report the results of our experiments in Fig. 2 and Tables 1, 2, and 3. In the figure, we plot the Accuracies (ACC) of the three models, whereas the tables summarise the Specificity (SPE), Sensitivity (SEN) and F1-Score (F1) values of our five-fold experiments. Let us denote the true positive outputs as TP, true negatives as TN, false positives as FP and false negatives as FN. The definitions of these metrics can then be given as

$$\mathrm{SPE} = \frac{\mathrm{TN}}{\mathrm{TN}+\mathrm{FP}} \times 100\% \ , \ \mathrm{SEN} = \frac{\mathrm{TP}}{\mathrm{TP}+\mathrm{FN}} \times 100\%,$$

$$\mathrm{F1} = 2 \times \frac{\mathrm{PPV} \times \mathrm{TPR}}{\mathrm{PPV}+\mathrm{TPR}},$$

where we compute PPV as TP/(TP+FP) and TPR as TP/(TP+FN). We compute Accuracy (ACC) as

$$\mathrm{ACC} = \frac{\mathrm{TP}+\mathrm{TN}}{\mathrm{TP}+\mathrm{TN}+\mathrm{FP}+\mathrm{FN}} \times 100\%.$$

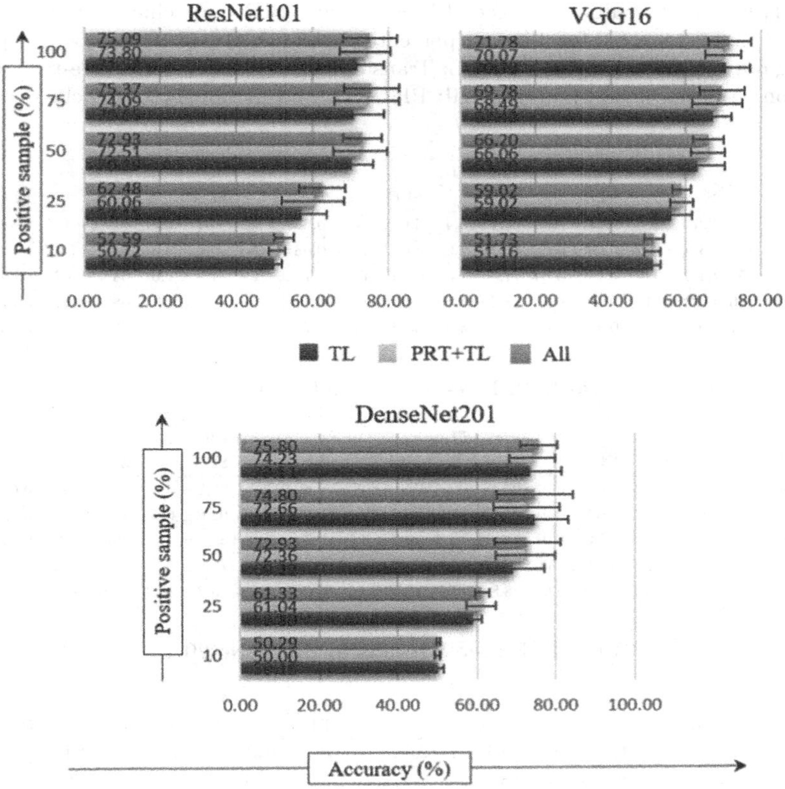

Fig. 2. Average accuracies for the five-fold experiments with three models. Results for **TL:** Transfer Learning, **PRT+TL:** Pre-text Representation Transfer (PRT) and TL and **All:** PRT+TL and dictionary based classification. Percentage of COVID-19 positive samples in the training data varies along the y-axis.

The other abbreviations used in reporting the results have the following meanings. **TL:** only conventional Transfer Learning (TL) is used. **PRT+TL:** Conventional TL is performed over the proposed Pre-text Representation Transfer (PRT). **All:** The PRT+TL predictions are also fused with the dictionary based classifier predictions.

From the plots in Fig. 2, we can make a few interesting observations. First, there is almost a consistent gain in the accuracy over the conventional TL with our eventual method (i.e., All). Second, in general, both PRT and dictionary based components are able to contribute to the final performance. This verifies our intuition that whereas a subsequent TL on a pre-text transferred model is useful, that model already learns an effective representation of the target domain that can be used for feature extraction. With respect to the TL, the average absolute gains in the accuracy of ResNet101, VGG16 and DenseNet201

Table 1. Five-fold results for ResNet101. Sensitivity (SEN), Specificity (SPE) and F1-score (F1) score reported for different percentage (%) of COVID-19 positive samples in training data. Results provided for **TL:** Transfer learning, **PRT+TL:** Pre-text Representation Transfer (PRT) and TL, **All:** PRT+TL and dictionary based classification.

%	TL			PRT+TL			All		
	SEN	SPE	F1	SEN	SPE	F1	SEN	SPE	F1
10	97.1 ± 4.1	2.6 ± 1.8	0.66 ± 0.02	97.4 ± 3.4	4.0 ± 2.8	0.66 ± 0.02	99.1 ± 1.2	6.1 ± 6.1	0.68 ± 0.01
25	90.7 ± 10.8	23.5 ± 13.6	0.68 ± 0.05	85.7 ± 7.9	34.4 ± 17.4	0.68 ± 0.05	90.3 ± 3.5	34.7 ± 13.7	0.71 ± 0.03
50	76.2 ± 4.5	64.3 ± 9.8	0.72 ± 0.04	77.3 ± 14.1	67.6 ± 7.1	0.73 ± 0.08	81.3 ± 7.3	64.4 ± 5.7	0.75 ± 0.05
75	75.1 ± 16.6	66.2 ± 14.9	0.71 ± 0.09	78.2 ± 15.3	69.9 ± 17.2	0.75 ± 0.09	82.3 ± 13.7	68.5 ± 13.2	0.77 ± 0.08
100	62.4 ± 14.1	80.6 ± 16.1	0.68 ± 0.09	81.3 ± 4.2	66.2 ± 11.4	0.76 ± 0.05	79.6 ± 2.6	70.5 ± 13.8	0.76 ± 0.05

Table 2. Five-fold results for VGG16.

%	TL			PRT+TL			All		
	SEN	SPE	F1	SEN	SPE	F1	SEN	SPE	F1
10	99.4 ± 1.3	3.5 ± 4.8	0.67 ± 0.01	95.1 ± 6.3	7.2 ± 6.2	0.66 ± 0.02	97.7 ± 4.3	5.7 ± 7.9	0.67 ± 0.01
25	94.8 ± 8.4	18.0 ± 13.1	0.69 ± 0.04	87.4 ± 11.3	30.6 ± 8.2	0.68 ± 0.04	92.8 ± 6.2	25.2 ± 4.2	0.69 ± 0.03
50	89.4 ± 13.9	37.2 ± 23.4	0.71 ± 0.04	82.8 ± 12.3	49.3 ± 6.5	0.71 ± 0.06	83.1 ± 13.1	49.3 ± 6.3	0.71 ± 0.06
75	84.8 ± 19.3	54.7 ± 22.3	0.74 ± 0.06	67.3 ± 7.1	65.9 ± 16.2	0.67 ± 0.04	73.6 ± 5.3	65.3 ± 10.7	0.71 ± 0.05
100	71.1 ± 13.1	70.5 ± 19.7	0.71 ± 0.05	68.8 ± 14.7	71.3 ± 6.9	0.69 ± 0.08	73.7 ± 13.1	69.9 ± 4.3	0.72 ± 0.08

Table 3. Five-fold results for DenseNet201.

%	TL			PRT+TL			All		
	SEN	SPE	F1	SEN	SPE	F1	SEN	SPE	F1
10	98.8 ± 1.2	1.4 ± 1.8	0.66 ± 0.01	98.0 ± 2.4	2.0 ± 2.2	0.66 ± 0.01	99.1 ± 0.8	1.4 ± 1.4	0.67 ± 0.01
25	93.7 ± 1.6	24.1 ± 5.5	0.70 ± 0.01	90.6 ± 8.4	31.5 ± 6.9	0.70 ± 0.04	91.1 ± 10.5	31.5 ± 9.9	0.70 ± 0.03
50	86.3 ± 6.2	52.2 ± 10.0	0.74 ± 0.06	83.9 ± 7.1	60.7 ± 10.5	0.75 ± 0.07	85.4 ± 6.3	60.5 ± 11.7	0.76 ± 0.07
75	77.4 ± 7.3	71.9 ± 14.1	0.75 ± 0.07	75.9 ± 11.0	69.3 ± 6.9	0.73 ± 0.09	79.1 ± 9.6	70.5 ± 10.9	0.76 ± 0.09
100	71.8 ± 8.6	75.4 ± 9.6	0.73 ± 0.08	73.9 ± 5.2	74.5 ± 9.4	0.74 ± 0.05	76.2 ± 1.8	75.4 ± 8.2	0.76 ± 0.04

for PRT+TL are 2.35, 1.07 and 0.77 respectively. Whereas the overall gains are 3.8, 1.81 and 1.75. We can also observe analogous trends in the sensitivity (SEN), specificity (SPE) and F1-score (F1) values for our five-fold experiments in Tables 1, 2 and 3. In general, results corresponding to 'All' are the best, followed by PRT+TL, followed by the conventional TL.

4 Conclusion

Data scarcity and class imbalance are common problems faced in medical imaging and other practical domains. We proposed the concept of Pre-text Representation Transfer (PRT) that can mitigate the adverse effects of these problems. The PRT allows us to tap into the cheaply available unlabelled data of the domain. This unlabelled data is used to systematically transfer the representation component of the deep model to the target domain, without changing the classification component. This allows us to use potentially unlimited data in the transfer. This is in sharp contrast to the conventional Transfer Learning that can only use limited annotated data for model transfer. By applying this concept to CT-based COVID-19 detection task, we demonstrated that PRT can not only be used to construct a more effective feature extractor of the target domain, but it can also be used to boost the performance of the conventional Transfer Learning. Moreover, we also devised a mechanism to fuse the outputs of a PRT-enhanced model with a PRT-based feature extractor to further enhance the final performance. Our five-fold experiments with three large-scale visual models, using five data imbalance settings, thoroughly established the effectiveness of the proposed technique.

Acknowledgment. This work was supported by Australian Government Research Training Program Scholarship. Dr. Akhtar is a recipient of the Office of National Intelligence National Intelligence Postdoctoral Grant # NIPG-2021-001 funded by the Australian Government.

References

1. Akhtar, N., Shafait, F., Mian, A.: Efficient classification with sparsity augmented collaborative representation. Pattern Recogn. **65**, 136–145 (2017)
2. Albelwi, S.: Survey on self-supervised learning: auxiliary pretext tasks and contrastive learning methods in imaging. Entropy **24**(4), 551 (2022)
3. Altaf, F., Islam, S., Janjua, N.K.: A novel augmented deep transfer learning for classification of COVID-19 and other thoracic diseases from X-rays. Neural Comput. Appl. **33**(20), 14037–14048 (2021)
4. Altaf, F., Islam, S.M., Akhtar, N.: Resetting the baseline: CT-based COVID-19 diagnosis with deep transfer learning is not as accurate as widely thought. In: 2021 Digital Image Computing: Techniques and Applications (DICTA), pp. 1–8. IEEE (2021)
5. Altaf, F., Islam, S.M., Akhtar, N., Janjua, N.K.: Going deep in medical image analysis: concepts, methods, challenges, and future directions. IEEE Access **7**, 99540–99572 (2019)
6. Altaf, F., Islam, S.M., Janjua, N.K., Akhtar, N.: Boosting deep transfer learning for COVID-19 classification. In: 2021 IEEE International Conference on Image Processing (ICIP), pp. 210–214. IEEE (2021)
7. Ericsson, L., Gouk, H., Loy, C.C., Hospedales, T.M.: Self-supervised representation learning: introduction, advances and challenges. arXiv preprint arXiv:2110.09327 (2021)
8. Esteva, A., et al.: Deep learning-enabled medical computer vision. NPJ Digit. Med. **4**(1), 1–9 (2021)
9. Fang, Y., et al.: Sensitivity of chest CT for COVID-19: comparison to RT-PCR. Radiology **296**(2), E115–E117 (2020)
10. He, K., Zhang, X., Ren, S., Sun, J.: Deep residual learning for image recognition. In: Proceedings of the IEEE Conference on Computer Vision and Pattern Recognition, pp. 770–778 (2016)
11. Huang, G., Liu, Z., Van Der Maaten, L., Weinberger, K.Q.: Densely connected convolutional networks. In: Proceedings of the IEEE Conference on Computer Vision and Pattern Recognition, pp. 4700–4708 (2017)

12. LeCun, Y., Bengio, Y., Hinton, G.: Deep learning. Nature **521**(7553), 436–444 (2015)
13. Long, C., et al.: Diagnosis of the coronavirus disease (COVID-19): RRT-PCR or CT? Eur. J. Radiol. **126**, 108961 (2020)
14. Misra, I., Maaten, L.V.D.: Self-supervised learning of pretext-invariant representations. In: Proceedings of the IEEE/CVF Conference on Computer Vision and Pattern Recognition, pp. 6707–6717 (2020)
15. Pham, T.D.: A comprehensive study on classification of COVID-19 on computed tomography with pretrained convolutional neural networks. Sci. Rep. **10**(1), 1–8 (2020)
16. Reeves, A.P., et al.: A public image database to support research in computer aided diagnosis. In: 2009 Annual International Conference of the IEEE Engineering in Medicine and Biology Society, pp. 3715–3718. IEEE (2009)
17. Roberts, M., et al.: Common pitfalls and recommendations for using machine learning to detect and prognosticate for COVID-19 using chest radiographs and CT scans. Nat. Mach. Intell. **3**(3), 199–217 (2021)
18. Simonyan, K., Zisserman, A.: Very deep convolutional networks for large-scale image recognition. arXiv preprint arXiv:1409.1556 (2014)
19. Tošić, I., Frossard, P.: Dictionary learning. IEEE Signal Process. Mag. **28**(2), 27–38 (2011)
20. Wynants, L., et al.: Prediction models for diagnosis and prognosis of COVID-19: systematic review and critical appraisal. BMJ **369** (2020)
21. Yan, K., Wang, X., Lu, L., Summers, R.M.: Deeplesion: automated mining of large-scale lesion annotations and universal lesion detection with deep learning. J. Med. Imaging **5**(3), 036501 (2018)
22. Yu, X., Wang, J., Hong, Q.Q., Teku, R., Wang, S.H., Zhang, Y.D.: Transfer learning for medical images analyses: a survey. Neurocomputing **489**, 230–254 (2022)
23. Zhang, Y., Jiang, H., Miura, Y., Manning, C.D., Langlotz, C.P.: Contrastive learning of medical visual representations from paired images and text. arXiv preprint arXiv:2010.00747 (2020)
24. Zhao, J., Zhang, Y., He, X., Xie, P.: COVID-CT-dataset: a CT scan dataset about COVID-19. arXiv preprint arXiv:2003.13865 (2020)

FNR-GAN: Face Normalization and Recognition with Generative Adversarial Networks

Amina Kammoun[1(✉)], Rim Slama[2], Hedi Tabia[3], Tarek Ouni[1], and Mohamed Abid[1]

[1] CES-Lab, ENIS Engineering School, Sfax University, Sfax, Tunisia
amina.kammoun@enis.tn
[2] LINEACT-Lab, CESI Lyon, Lyon, France
rsalmi@cesi.fr
[3] IBISC, Univ. Evry, Paris-Saclay University, Paris, France
hedi.tabia@univ-evry.fr

Abstract. The normalization of faces in the wild is an interesting process which can improve face recognition performances and avoid the complex computation of face generation in cross different variations. In this paper, we propose a multi-objective approach that generates and recognizes normalized faces while preserving their identities. An unsupervised Face Normalization and Recognition framework using discriminant normalized features is presented. This latter is based on an optimized combination of Generative Adversarial Network (GAN) generators and Convolutional Neural Network (CNN) classifiers. The main power of our approach is to generate optimized features representing normalized faces finding a trade-off between improving identity preservation and minimizing the architecture complexity. Additionally, it can be adapted to impaired and unlabeled datasets which can respond to real-world face variations and available data. Experimental results show that the proposed method outperforms other models on face normalization and achieves state-of-the-art frontal-frontal face verification in CFP protocol and face recognition in LFW. The code and results are available at github/FNR-GAN.

Keywords: Generative Adversarial Networks (GANs) · Unsupervised face normalization · Face recognition in the wild · Identity preservation

1 Introduction

Face recognition in the wild is still a topical issue in computer vision because its important need of large training datasets. Deep face recognition networks consist in learning face features cross real world variations in order to estimate the face identity. Therefore, most of existent models are trained on large datasets for better efficiency. The important need of big data inspired research committees to

W. Q. Yan et al. (Eds.): IVCNZ 2022, LNCS 13836, pp. 131–143, 2023.
https://doi.org/10.1007/978-3-031-25825-1_10

focus lately on face generation for augmenting datasets of different face variations in order to obtain better face recognition performances.

Recently [1], Generative Adversarial Networks (GANs) are overly used for face generation and augmentation under different face variations. Supervised GANs learn to generate face variations through the variation code. The main problem of these models is that they need labelled and paired datasets which are hard and expensive to collect. However, unsupervised GANs are able to learn different data distributions from unconstrained datasets. Most of existent unsupervised models focus on data augmentation. Nevertheless, the many-to-one face generation or face normalization is explored only by few researches [2,3].

Face normalization consists in recovering the canonical view of face images from one or many images of non-frontal view, in addition to neutralizing their emotions. Actually, many-to-one face normalization reduces the appearance variability of data to make faces align and compare them easily. Therefore, face recognition can be performed as if it were under controlled conditions.

Despite the recent advancements, the existent GAN based models still suffer from identity preservation, and stability. Furthermore, they have additional architecture complexity and computation issues. In addition to the normalization, face recognition requires further computational demands, whereas only few models [24] are designed, recently, to both generate and recognize faces. In this work, we aim to improve the identity preservation and reduce the architecture complexity of unsupervised face normalization GANs. For this end, we present a multi-objective framework for face normalization and recognition. Several researches have been made to optimize the different architecture blocks to obtain more stable generator and efficient classifier.

The main contributions of this paper can be summarized as follow:

- proposing a multi-objective approach for face normalization and recognition
- improving identity preservation of GAN based models
- minimizing the architecture complexity of face generators and classifiers
- achieving state-of-the art performances on face recognition and verification

Evaluating our framework on CFP [4] and LFW [5,6] benchmarks, we achieve state-of-the art performances on frontal-frontal face verification and face recognition in the wild. The rest of this paper is presented as follow: In Sect. 2, we review some related works in face generation and recognition. Section 3 presents the proposed approach. In Sect. 4, we detail the implementation steps and the evaluation of our method on different face recognition and verification benchmarks. Section 5 concludes the paper with interpretations and open issues.

2 Related Works

In this section, we present an overview of GAN based models proposed in history for face generation and recognition. Face generation includes face normalization (Many-to-one face generation) and facial data augmentation (One-to-many face generation). While, some recent researches on GANs involve the face identity preservation and recognition.

2.1 Face Augmentation: One-to-Many Face Generation

The related works focused on the GAN models which generate faces for data augmentation with respect to the pose and expression variations. Zhang et al. [7] presented an end-to-end supervised model which explicitly disentangles pose and expression through their corresponding codes. Some other works [8,9] proposed GAN models for 3D face generation cross expression variation. Authors in [10, 11] presented semi-supervised adversarial training frameworks to generate faces with wide ranges of poses and expressions. Many works [12–16] proposed GAN based models for dynamic face generation in which the animation of faces is controlled by human interpretable attributes.

2.2 Face Normalization: Many-to-One Face Generation

Many-to-one face generation models aim to normalize faces by dispelling the pose and expression variations. Qian et al. [2] proposed an unsupervised Face Normalization Model (FNM), which is based on face attention mechanism of 5 discriminators to refine the local texture of face images. Yin et al. [17] presented a Feature Adaptation Network (FAN) for surveillance face recognition and normalization, which consists of a feature disentanglement learning stage, and an adaptation network to impose feature-level and image-level similarity regularization. Hsu et al. [18] introduced DVN as Dual-View Normalization with a double-layer architecture which consists of identity preservation, normalized pose transformation and domain transformation.

2.3 Identity Preservation

Some works [19, 20] proposed multi-stages GANs for disentangling the identity and facial attributes which enable the identity preservation of synthesized faces. Other works [21–23] incorporated additional losses for better identity preservation while generating faces.

2.4 Face Recognition

Face generation and recognition based GANs [24] consist of three-player models, in which an identity classifier is introduced as an additional competitor which cooperates with the discriminator to compete with the generator.

3 Proposed Method

FNR-GAN is a multi-objective approach which both normalizes and recognizes faces. The main focus of our model is to improve the identity preservation of GANs while reducing the computation complexity of their architectures. We notice that most of existent GANs focused on generating faces. However, only, few works studied face recognition via GANs. Recently, CNNs achieved state-of-the art results in face recognition which outperform the human performance. Nevertheless, more researches should be made to improve the performances of GANs to preserve better the identity from real to generated faces. In this work, we focus on many-to-one face normalization rather than augmentation. Moreover, data augmentation using GANs requires huge computational resources and time. Nevertheless, face normalization dispels most of face variation, especially pose and expression, with less computation complexity.

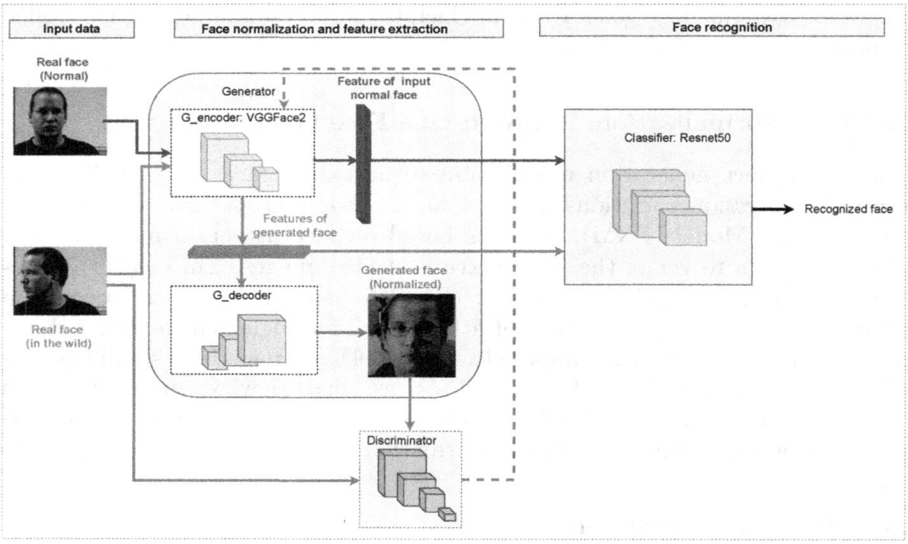

Fig. 1. Overview of the proposed FNR-GAN for face normalization and recognition. It consists of 2 blocks: a feature extraction and normalization block composed of a generator and a discriminator; and a face recognition block which consists in an identity classifier.

Hence, our goal is to synthesise normalized face features that are realistic and identity-preserved. Figure 1 illustrates the proposed network consisting of: (i) face generation block composed of a generator and a discriminator (ii) face recognition block consisting in identity classifier. First, we normalize posed and expressed faces. We notice that the use of an auto-encoder generator with a series of attention discriminators generates neutral faces with good identity preservation. However, due to face occlusion, the frontal face generation suffers from symmetry loss, as well as artifacts especially when frontalizing large poses.

Our approach can be applied for both paired/impaired and labelled/unlabelled face datasets thanks to its unsupervised architecture.

3.1 Face Normalization

Face normalization is the process of dispelling variations from face, mainly pose and facial expression, which can improve face recognition and verification in the wild. Nevertheless, it is less explored by research committees compared to face augmentation cross real variations. Actually, the many-to-one face generation is a delicate process because of the difficulty of preserving identity features in face rotation or reshape. We propose a GAN based face normalization model with less identity preservation loss compared to the existent models. The proposed normalization process consists of a generator encoder and a discriminator.

The Generator: VGGFace2 [25] is adapted as the generator of our model which outputs normalized face features. Our generator design is inspired from FNM [2] model's generator. The latter is an auto-encoder which consists of a VGGFace2 encoder associated to a decoder. The proposed study focus on the optimization of the generator. Therefore, from the researches done, it is obvious that the features from the generator encoder preserve better identity than the features from the decoder. Therefore, the ablation of the decoder reduces the additional noise generated which affects the identity preservation in face normalization. Consequently, we adopt an encoder based generator without a decoder, which outputs 2048 features of normalized face.

The Discriminator: The proposed discriminator [2] consists of 5 attention discriminators of fixed regions of output normalized face. The series of discriminators aim to distinguish between real and generated face features.

3.2 Face Recognition

In order to improve the identity preservation in normalized face generation and recognition, we incorporated a Resnet50 based classifier in addition to the generator block. The bottleneck of the generator encoder outputs feature vectors which are inputted to the classifier. The latter is trained on real normal features and tested on normalized ones. Face recognition is evaluated using both paired and in-the-wild datasets.

4 Experiments and Results

For the evaluation and the validation of our approach, several experiments have been carried out in this research. First, the presented protocol evaluates the model in the case of controlled conditions of pose and expression. Then, the assessment has been done on unconstrained in-the-wild environment. The constrained evaluation consists of two experiments using two paired datasets of pose and expression. On one hand, these experiments aim to examine in-depth the main limits of existent face normalization based GANs for the various poses and expressions. On the other hand, architecture improvements have been proposed for optimized face normalization with identity preserved. While the unconstrained evaluation involves two experiments of in-the-wild face recognition and verification.

4.1 Expression Neutralization

In this experiment, we define an expression neutralization protocol and evaluate our framework on JAFFE dataset.

JAFFE [26,27] dataset consists of 213 images of different facial expressions from 10 different Japanese female subjects. Each subject was asked to do 7 facial expressions (6 basic facial expressions and neutral).

Expression Neutralization Protocol: We define in this experiment an expression neutralization protocol which aims to generate optimized neutral facial features. First, we chose JAFFE as a paired dataset in order to visualize the results of neutralization for each expression (Fig. 2).

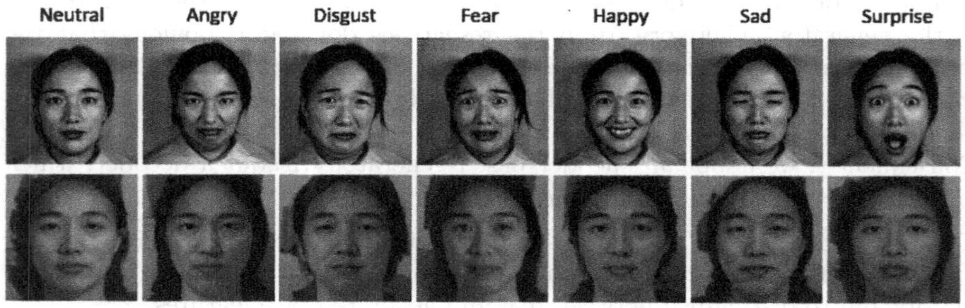

Fig. 2. Example of face normalization: expression neutralization. The first row represents real emotions and the second row the generated neutral faces using the FNM model.

Our evaluation protocol consists in training the model on real neutral faces while testing the generated neutralized ones. The face recognition system estimate the face identity by comparing between the neutral and neutralized faces. We examine from the recognition accuracy, the capacity of the model to preserve the identity of generated faces. For this end, three tests have been proposed. In the first test, we evaluated the FNM model using a pretrained Resnet50 classifier. In the second test, we notice that the retrain of three final layers of Resnet50 on the defined dataset improved the accuracy from 70 to 85.49%. Then, in the thirst test, we evaluate our model FNR and we see that the performance rise to 96.37%. The obtained results in Table 1 approve that our approach improves the identity preservation and face recognition accuracy of normalization GANs.

Actually, the train of Resnet50 on normalized features of the generator encoder gives better results compared to features outputted from the decoder. From these results, we notice that the decoder generates additional noise which affects the identity preservation in face generation. Therefore, the ablation of decoder improves the recognition rate by reducing the identity preservation loss.

Table 1. Expression neutralization and recognition on JAFFE dataset.

Method	Face verification accuracy (%)
FNM+Pretrained Resnet50	70
FNM+Resnet50	85.49
Our approach (FNR)	**96.37**

4.2 Head Pose Frontalization

In this experiment, we represent a face frontalization protocol, which we evaluated on Head Pose dataset.

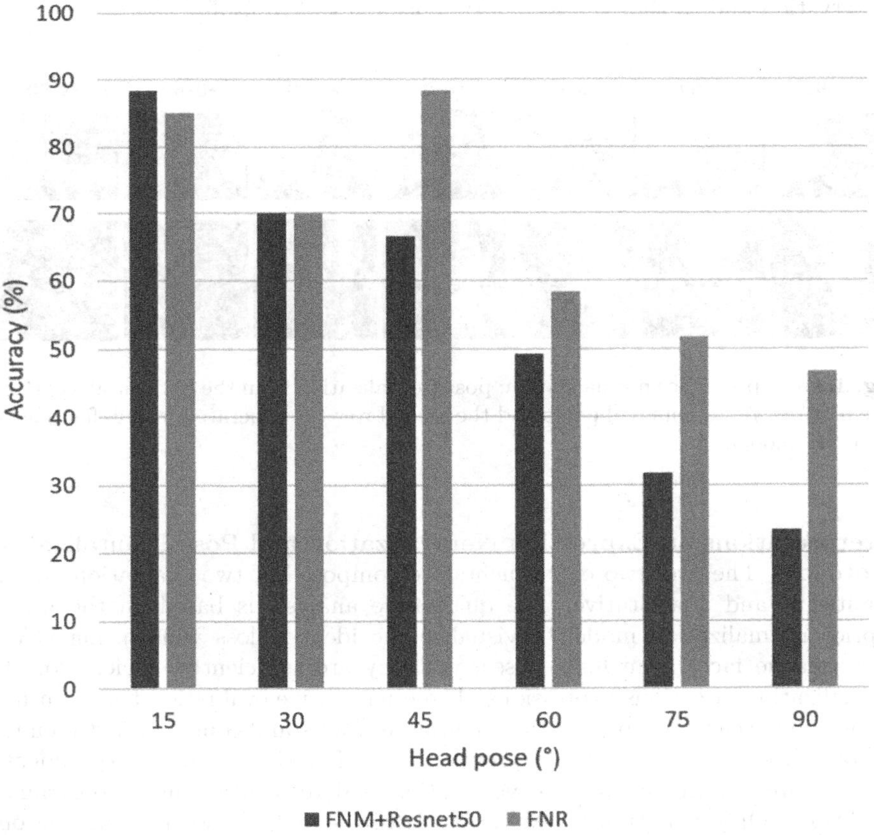

Fig. 3. Face recognition performances of FNM and FNR models on Head Pose dataset

Head Pose [28] database is a benchmark of 2790 monocular face images of 15 persons with variations of pan and tilt angles from −90 to +90°. For every person, 2 series of 93 different poses are available. People in the database wear glasses or not and have various skin color.

Pose Frontalization Protocol: This protocol is inspired from the previous protocol that we defined for expression neutralization. We trained Resnet50 classifier on real frontal faces and tested the generated frontalized ones. We evaluate the results for each pose in the horizontal ax from 0° to 90° (Fig. 3).

It is clear from the graph that especially for large pose frontalization, our model improves the face recognition accuracy compared to previous GANs. Additionally, as we visualize in Fig. 4, the obtained generated faces using a previous model preserved less identity when the pose is larger. Therefore, the identity loss rises with the rise of the pose to be frontalized. From the ablation of the decoder block and the train of an optimised classifier, we improve the identity preservation and so that the recognition accuracy of frontalization GANs.

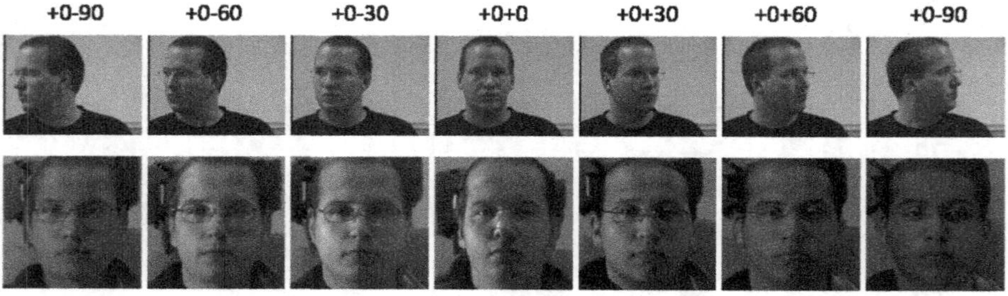

Fig. 4. Example of face normalization: pose frontalization from the horizontal ax (Pan). The first row represents real poses and the second row the generated frontal faces using the FNM model.

Interpretations on Expression Neutralization and Pose Frontalization Protocols: The first two experiments are composed of two evaluation stages: qualitative and quantitative. The qualitative analysis is based on the use of a prior normalization model to visualize the identity loss while using GANs to normalize faces from huge poses, yet they are sufficiently efficient for the normalization of various expressions. The quantitative evaluation based on face recognition accuracy approved the qualitative results and compared between the performances of our model and previous ones. For the following experiments, we evaluate our model for face verification and recognition in unconstrained conditions. Only quantitative evaluation is considered for two reasons. On one hand, a qualitative analysis is already done in the first experiments. On the other hand, the major aim of the proposed framework is to improve the identity preservation of GANs for the purpose of optimized face recognition considering accuracy and computation requirements.

4.3 Face Verification

After the validation of our approach separately in cases of expression neutralization and pose frontalization, we define the third protocol to validate our framework in the case of in-the-wild face verification based on the Celebrities in the Frontal Profile dataset (CFP).

CFP [4] consists of 7000 images of 500 subjects. Each person has 10 frontal and 4 profile images. The evaluation protocol includes Frontal-Frontal (FF) and Frontal-Profile (FP) face verification. The dataset is divided in 10 splits with 350 same and 350 different pairs each.

Our protocol is inspired from the evaluation protocol defined by Sengupta et al. [4] which is used for the evaluation of previous models [29–31]. This evaluation is based on the computation of the mean accuracy of the 10 splits of the database with standard variation. Various models are defined previously to handle the unconstrained face verification problem. For instance, Sankarana et al. [29] suggested a method that combines a deep CNN-based technique with a low-dimensional discriminative embedding step. Chen et al. [30] defined (FV-DCNN) representation, which is created from a combination of Fisher vector representation and Deep Convolutional Neural Network (DCNN) features. Tran et al. introduced a Disentangled Representation learning-Generative Adversarial Network (DR-GAN) [31], which is made of an encoder-decoder structured generator, pose coding, pose classification for integrated multi-image fusion.

Table 2. Face verification performances on CFP protocol (Mean Accuracy and standard deviation over 10 folds)

Method	FF face verification accuracy (%)
Sengupta et al. [4]	96.40 ± 0.69
Sankarana et al. [29]	96.93 ± 0.61
Chen et al. [30]	98.67 ± 0.36
Human [31]	96.24 ± 0.67
DR-GAN [31]	97.08 ± 0.62
FNM+Resnet50	**99.17 ± 0.56**
Our approach (FNR)	**99.4 ± 0.38**

For performing face verification, we take the same pairs and splits of faces defined in the protocol adopted previously. We first input the pairs in the generator encoder for obtaining the normalized features. Then, we compared between the feature vectors for each pair. Resnet50 is trained on 30 epochs in 64 batch size for every split. The comparison with previous models is presented in Table 2. The average face verification accuracy with a standard deviation over 10 folds is presented as the outcome. The last two rows represent the experiments made in this research. First, the features encoded from FNM generated images are classified using our Resnet based classifier. Finally, the results of face recognition using the proposed FNR optimized framework are presented.

From the results on Table 2, it is noticeable that we achieve state of the art frontal-frontal (FF) face verification on CFP protocol compared to previous models, and which exceeds the human performance. The obtained results confirm the first hypothesis that our model is well optimized and better preserves the face identity for unsupervised in-the-wild expression neutralization and pose frontalization.

4.4 Face Recognition

The fourth protocol is defined to validate our framework in the case of in-the-wild face recognition based on the LFW benchmark.

LFW dataset [5,6] contains more than 13k images of faces collected from the web. Each face has been labeled with the name of the person pictured. 1680 of the people pictured have two or more distinct photos in the dataset.

This experiment aims to optimize the blocks of our framework for both face normalization and recognition. Therefore, we defined 3 tests for face recognition in the wild. In our evaluation, we adopt the same train and test splits defined in the LFW benchmark. In the first test, we evaluated the FNM model. In fact, FNM is composed of a VGGFace encoder, a decoder and a discriminator. First, we encoded the normalized face images outputted from the FNM model. Then, we trained our Resnet50 based classifier which estimates the identity of test samples, by comparing them with the normalized features of train samples. In the second test, we evaluated VGGFace model with the proposed protocol after the ablation of both the discriminator and the decoder. In the third test, we evaluated the proposed FNR model after only the ablation of the decoder and the incorporation of our Resnet based classifier.

In Table 3, we compare the results of the proposed tests in bold, with those of previous models [24,32,33]. Hassner et al. [33] proposed a straightforward method for producing frontalized views by employing a single, unaltered 3D reference. Yin et al. introduced a deep 3D Morphable Model (3DMM) conditioned Face Frontalization Generative Adversarial Network (FF-GAN) [32] to generate neutral head pose face images. Shen et al. proposed FaceID-GAN [24], which generates face images from arbitrary poses and emotions while maintaining face identity, by incorporating an identity classifier as an additional competitor to the generator in tandem with the discriminator for optimizing image quality and facial identity preservation. We notice that the FNR framework achieved state-of-the art face recognition accuracy in LFW benchmark compared to the previous GAN models. The main advantages of the proposed architecture is to improve in-the-wild face recognition accuracy of facial GANs and reduce the computation complexity and resources. Actually, the researches made to optimize the architecture blocks as well as the number of layers allow to reduce the computational requirements and time. Moreover, we demonstrate that the generation and the training based on optimized normalized features reduce both the training requirements of GANs and classifiers, in addition to the face recognition performances. From the obtained results, we notice that the inclusion of the discriminator block improves the performance of VGGFace from 91.12% to 98.5%. In addition, the ablation of the decoder improves the identity preservation, as well as the recognition accuracy from 97.9% to 98.5%.

Table 3. Face recognition performances on LFW

Method	Face recognition accuracy (%)
Hassner et al. [33]	93.62
FF-GAN [32]	96.42
FaceID-GAN [24]	97.01
VGGFace	**91.12**
FNM+Resnet	**97.9**
Our approach (FNR)	**98.50**

5 Conclusions

This paper presented a multi-objective framework for face feature extraction, normalization and recognition. The proposed framework improves the identity preservation in normalized face generation for better face recognition in-the-wild. Experimental results on four datasets show that the proposed FNR-GAN model achieves state-of-art face verification and recognition in both constrained and unconstrained datasets. In the future researches, we plan to evaluate our framework on more challenging datasets. Furthermore, more qualitative and quantitative evaluation metrics will be used for the evaluation of generated faces.

References

1. Kammoun, A., Slama, R., Tabia, H., Ouni, T., Abid, M.: Generative adversarial networks for face generation: a survey. ACM Comput. Surv. **55**(5), 1–37 (2022)
2. Qian, Y., Deng, W., Hu, J.: Unsupervised face normalization with extreme pose and expression in the wild. In: 2019 IEEE/CVF Conference on Computer Vision and Pattern Recognition (CVPR) (2019)
3. Wang, M., Deng, W.: Deep face recognition: a survey. Neurocomputing **429**, 215–244 (2021)
4. Sengupta, S., Chen, J.-C., Castillo, C., Patel, V.M., Chellappa, R., Jacobs, D.W.: Frontal to profile face verification in the wild. In: 2016 IEEE Winter Conference on Applications of Computer Vision (WACV) (2016)
5. Huang, G.B., Learned-Miller, E.: Labeled faces in the wild: updates and new reporting procedures. Department Computer Science, University Massachusetts Amherst, Amherst, MA, USA, Technical report, 14-003 (2014)
6. Learned-Miller, E., Huang, G.B., RoyChowdhury, A., Li, H., Hua, G.: Labeled faces in the wild: a survey. In: Advances in Face Detection and Facial Image Analysis, pp. 189–248 (2016)

7. Zhang, F., Zhang, T., Mao, Q., Xu, C.: Joint pose and expression modeling for facial expression recognition. In: 2018 IEEE/CVF Conference on Computer Vision and Pattern Recognition (2018)
8. Deng, Y., Yang, J., Chen, D., Wen, F., Tong, X.: Disentangled and controllable face image generation via 3D imitative-contrastive learning. In: 2020 IEEE/CVF Conference on Computer Vision and Pattern Recognition (CVPR) (2020)
9. Cheng, S., Bronstein, M., Zhou, Y., Kotsia, I., Pantic, M., Zafeiriou, S.: MeshGAN: non-linear 3D morphable models of faces. arXiv (2019)
10. Gecer, B., Bhattarai, B., Kittler, J., Kim, T.-K.: Semi-supervised adversarial learning to generate photorealistic face images of new identities from 3D morphable model. In: Computer Vision - ECCV 2018, pp. 230–248 (2018)
11. Gao, Z., Zhang, J., Guo, Y., Ma, C., Zhai, G., Yang, X.: Semi-supervised 3D face representation learning from unconstrained photo collections. In: 2020 IEEE/CVF Conference on Computer Vision and Pattern Recognition Workshops (CVPRW) (2020)
12. Tripathy, S., Kannala, J., Rahtu, E.: FACEGAN: facial attribute controllable reenactment GAN. In: 2021 IEEE Winter Conference on Applications of Computer Vision (WACV) (2021)
13. Tripathy, S., Kannala, J., Rahtu, E.: ICface: interpretable and controllable face reenactment using GANs. In: 2020 IEEE Winter Conference on Applications of Computer Vision (WACV) (2020)
14. Zakharov, E., Shysheya, A., Burkov, E., Lempitsky, V.: Few-shot adversarial learning of realistic neural talking head models. In: 2019 IEEE/CVF International Conference on Computer Vision (ICCV) (2019)
15. Wiles, O., Koepke, A.S., Zisserman, A.: X2Face: a network for controlling face generation using images, audio, and pose codes. In: Computer Vision - ECCV 2018, pp. 690–706 (2018)
16. Bansal, A., Ma, S., Ramanan, D., Sheikh, Y.: Recycle-GAN: unsupervised video retargeting. In: Computer Vision - ECCV 2018, pp. 122–138 (2018)
17. Yin, X., Tai, Y., Huang, Y., Liu, X.: Fan: feature adaptation network for surveillance face recognition and normalization. In: Computer Vision - ACCV 2020, pp. 301–319 (2021)
18. Hsu, G.-S., Tang, C.-H.: Dual-view normalization for face recognition. IEEE Access **8**, 147765–147775 (2020)
19. Bao, J., Chen, D., Wen, F., Li, H., Hua, G.: Towards open-set identity preserving face synthesis. In: 2018 IEEE/CVF Conference on Computer Vision and Pattern Recognition (2018)
20. Shen, Y., Zhou, B., Luo, P., Tang, X.: FaceFeat-GAN: a two-stage approach for identity-preserving face synthesis. arXiv (2018)
21. Cheng, J., Chen, Y.-P.P., Li, M., Jiang, Y.-G.: TC-GAN. In: Proceedings of the 27th ACM International Conference on Multimedia (2019)
22. Yang, H., Huang, D., Wang, Y., Jain, A.K.: Learning face age progression: a pyramid architecture of GANs. In: 2018 IEEE/CVF Conference on Computer Vision and Pattern Recognition (2018)
23. Zhao, J., et al.: Dual-agent GANs for photorealistic and identity preserving profile face synthesis. In: Advances in Neural Information Processing Systems, vol. 30 (2017)
24. Shen, Y., Luo, P., Luo, P., Yan, J., Wang, X., Tang, X.: FaceID-GAN: learning a symmetry three-player GAN for identity-preserving face synthesis. In: 2018 IEEE/CVF Conference on Computer Vision and Pattern Recognition (2018)

25. Cao, Q., Shen, L., Xie, W., Parkhi, O.M., Zisserman, A.: VGGFACE2: a dataset for recognising faces across pose and age. In: 2018 13th IEEE International Conference on Automatic Face & Gesture Recognition (FG 2018) (2018)
26. Lyons, M.J., Kamachi, M., Gyoba, J.: The Japanese Female Facial Expression (JAFFE) dataset. Semantic scholar (1998)
27. Shih, F.Y., Chuang, C.F., Wang, P.S.: Performance comparisons of facial expression recognition in JAFFE database. Int. J. Pattern Recognit. Artif. Intell. **22**(03), 445–459 (2008)
28. Kara, Y.E., Genc, G., Aran, O., Akarun, L.: Actively estimating crowd annotation consensus. J. Artif. Intell. Res. **61**, 363–405 (2018)
29. Sankaranarayanan, S., Alavi, A., Castillo, C.D., Chellappa, R.: Triplet probabilistic embedding for face verification and clustering. In: 2016 IEEE 8th International Conference on Biometrics Theory, Applications and Systems (BTAS) (2016)
30. Chen, J.-C., Zheng, J., Patel, V.M., Chellappa, R.: Fisher vector encoded deep convolutional features for unconstrained face verification. In: 2016 IEEE International Conference on Image Processing (ICIP) (2016)
31. Tran, L., Yin, X., Liu, X.: Disentangled representation learning GAN for pose-invariant face recognition. In: 2017 IEEE Conference on Computer Vision and Pattern Recognition (CVPR) (2017)
32. Yin, X., Yu, X., Sohn, K., Liu, X., Chandraker, M.: Towards large-pose face frontalization in the wild. In: 2017 IEEE International Conference on Computer Vision (ICCV) (2017)
33. Hassner, T., Harel, S., Paz, E., Enbar, R.: Effective face frontalization in unconstrained images. In: 2015 IEEE Conference on Computer Vision and Pattern Recognition (CVPR) (2015)

Object Tracking with Multiple Dynamic Templates Updating

Mingyang Zhang(✉), Kristof Van Beeck, and Toon Goedemé

PSI-EAVISE Research Group, Department of Electrical Engineering, KU Leuven, Leuven, Belgium
{mingyang.zhang,kristof.vanbeeck,toon.goedeme}@kuleuven.be
http://iiw.kuleuven.be/onderzoek/eavise

Abstract. Most existing Siamese visual trackers see the object tracking task as similarity learning between a search image and a single template image. Utilizing only one template leads to the negligence of the rich semantic information in other frames. Meanwhile, those Siamese trackers with temporal context exploitation either incorporate specially designed non-generic modules or include online-learning parts which compromise real-time performance. In this paper, we propose a novel model architecture incorporating multiple dynamic templates in a Siamese visual tracker to maximize temporal information utilization. To attain a favorable appearance representation from these templates, we propose an online dynamic template pool updater that leverages the frames with dissimilar appearances. Furthermore, we design a new hard positive sampling strategy to train the tracker with dissimilar templates. With the proposed methods, a Siamese tracker can be straightforwardly transformed and trained to benefit from the temporal correlations among frames. Comprehensive experiments on various tracking datasets show positive results and prove the effectiveness of the proposed methods.

Keywords: Object tracking · Deep learning · Siamese networks · Dynamic templates · Hard positive sampling

1 Introduction

Visual object tracking has gained increasing attention in the past decade as it is a fundamental and challenging task in computer vision. It has various applications in self-driving vehicles [24], augmented reality [4], sport [14], etc. A generic visual object tracker is asked to predict the location of the tracked target in a video sequence, often only based on the annotation in the first frame. Significant progress has been made on the visual tracking task in recent years, mainly due to the development of deep-learning-based visual trackers.

Among the deep-learning-based methods, a prominent type is the Siamese trackers which have demonstrated notable improvement in visual tracking performance in recent years. Siamese trackers formulate the visual object tracking task as a similarity learning problem [1,11]. SA-Siam [9] separately trains two

W. Q. Yan et al. (Eds.): IVCNZ 2022, LNCS 13836, pp. 144–158, 2023.
https://doi.org/10.1007/978-3-031-25825-1_11

branches to keep the heterogeneity of the semantic and appearance features. SiamRPN++ [19] introduces a simple yet effective method to train the Siamese tracker with a deep backbone. SiamFC++ [33] builds an anchor-free structure to improve localization precision. However, these Siamese trackers usually lack the utilization of temporal information in the video because of the sole and static template.

To fill this gap, some trackers were proposed to exploit the temporal information in other frames better. STARK [34] integrates historical template appearance through concatenation in a Transformer-based architecture. UpdateNet [35] estimates the optimal template for the next frame with a dedicated convolutional neural network. TCTrack [2] utilizes a CNN backbone with online calibrated weights, and an adaptive temporal transformer, to incorporate temporal information in the extraction of features stage and the refinement of similarity maps stage, respectively. Despite the improvement made on the visual tracking task, these trackers often incorporate additional dedicated modules for their temporal context exploitation strategy.

The contributions of this work are threefold. First, we propose a method to leverage the temporal information in the video by utilizing extra templates. The presented method updates extra dynamic templates in a heuristic way to compose a template pool with dissimilar templates that include diverse appearance information. Second, to update such templates and to maintain a diverse template pool, we propose a template pool updater that dynamically selects suitable frames as new templates. Finally, we design a novel hard positive sampling strategy to enhance the training of our model with dissimilar examples.

As a result of applying our methods, we successfully trained a multi-template Siamese tracker named SiamDEPU. We validate our trained model SiamDEPU on challenging object tracking datasets, namely VOT2018 [15], GOT-10k [12] and LaSOT [5]. The extensive experimental results show that our methods can effectively enhance updating capabilities and improve the performance of a Siamese tracker.

2 Related Work

Visual tracking is an active topic in the computer vision field. Delivering a comprehensive literature review is beyond the scope of this work. We will only focus on trackers based on Siamese networks and those with temporal information exploitation.

2.1 Siamese Trackers

The pioneering work SiamFC [1] adopts the Siamese network and cross-correlation to construct an offline trained similarity metric. SA-Siam [9] composes two similarity-learning Siamese networks for semantic features in the classification task and appearance features in the similarity matching task, respectively. It keeps the heterogeneity of the two types of features by training these two branches separately. SiamRPN [20] improves the Siamese tracker with a region proposal network and captures scale changes of the tracked object by predefined anchor boxes. DaSiamRPN [36] implements explicit generation of diverse semantic negative pairs in the training process. With such data augmentation, the tracker learns discriminative features that lead to improved robustness. SiamRPN++ [19] successfully replaces prevailingly used AlexNet [17] with ResNet [10]. It removes padding and shifts the object location randomly during training to eliminate the center bias in prevenient Siamese trackers. SiamMask [30] improves the off-line training process by augmenting the loss with a binary segmentation task. By unifying the object tracking and segmentation tasks with one approach, it achieves enhanced performance on both accuracy and robustness metrics. SiamBAN [3] and SiamCAR [7] exploit the fully convolutional network to avoid the use of pre-defined anchor boxes which call for heuristic configurations.

2.2 Temporal Exploitation for Tracking

DSiam [8] introduces online learning to construct a dynamic Siamese network via a fast transformation model. With such a model, DSiam [8] learns the object appearance variation from previous frames. ROLO [23] feeds a Long Short-Term Memory(LSTM) network the concatenated high-level features from an object detector. The performance of such detection-based trackers heavily relies on the detectors' quality, and these trackers often run at a low speed because of the redundancy in the simple concatenation structure. Siam R-CNN [28] models the history of both the tracked object and the distractors with the re-detection architecture. By re-detecting both the first-frame template and the previous-frame predictions, the tracker performs better on hard scenarios, e.g., after long occlusions. STMTrack [6] builds their tracking framework with a space-time memory network that saves the historical information about the tracked object to tackle appearance variations. UpdateNet [35] predicts an accumulated template for the next frame with the initial template, the last accumulated template, and the predicted template as inputs. The performance of a Siamese tracker can be improved by replacing the template updating strategy in the tracker with the UpdateNet module. STARK [34] builds a spatio-temporal tracking framework by implementing a template updating module that takes the results from a Transformer-based encoder-decoder architecture as inputs to make template update decisions. TrDiMP/TrSiam [29] propagates temporal contexts in its Transformer-based decoder by doing cross-attention operations on both the template and the search backbone outputs.

3 Method

In this work, we aim to enhance the temporal information exploitation capability in Siamese trackers. To achieve this objective, we propose our Dynamic Template

Pool Updating Siamese Tracker called SiamDEPU. To efficiently update the templates, we propose a novel dissimilarity-based template updater. For the training of a multi-template tracker, we further propose a hard positive example sampling approach based on group dissimilarity.

3.1 Dynamic Template Pool Updating Siamese Tracker

Overview. Figure 1 shows the architecture of SiamDEPU in the tracking phase. Instead of a single template, we propose to compose a template pool as the appearance representation of tracked objects. Together with the search frame, the template pool is fed to the backbone network. In this paper, we mainly implement our experiments with ResNet-50 [10] as the feature extractor. The output of the backbone network is a feature group including features coming from different layers in the backbone. These features with different numbers of channels are passed through a 1×1 convolutional layer, which reduces the number of channels to 256. The resulting features of templates are then compared with the search image features by performing cross-correlation on them. The cross-correlation response maps are inputs of the bounding box regression head and classification head. The tracker predicts the tracked object's new location by fusing the outputs from the heads. Based on the new predictions, the Template Pool Updater updates the template pool dynamically during inference with the latest predictions.

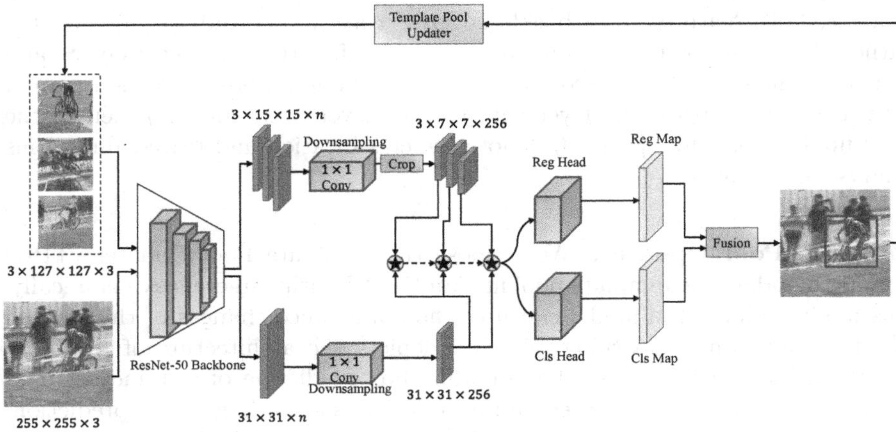

Fig. 1. Overview of the proposed method in the tracking phase. The network takes a template pool and a search image as inputs. ⊛ denotes depth-wise cross-correlation.

Temporal Information Exploitation with Multiple Templates. Most previous works only utilize one template as the appearance representation of the tracked object. However, features from different frames have rich semantic information which is neglected in one-template trackers. The use of features from dissimilar templates is hypothesized to improve the performance of a Siamese tracker. To test our hypothesis, we propose to exploit temporal information with a template pool composed of several dissimilar templates. We take SiamRPN++ [19] as our initial architecture to apply our method of utilizing multiple templates. SiamRPN++ [19] feeds the cross-correlation maps of features from different layers in ResNet-50 [10], namely *conv*3, *conv*4 and *conv*5, to three individual heads. In our SiamDEPU, three independent modules are placed in a regression/classification head. These modules process the cross-correlation maps of features from the three convolutional layers respectively. Each module aggregates information in the features of all the templates in the pool. The final regression/classification map is the weighted sum of the outputs of the three individual modules in a head. This overall process of the aggregation of n templates can be expressed as follows:

$$S_j = \phi(F_j(x), F_j(z_0), F_j(z_1) \dots F_j(z_{n-1})), \tag{1}$$

$$B_j = \psi(F_j(x), F_j(z_0), F_j(z_1) \dots F_j(z_{n-1})) \tag{2}$$

and

$$S_{all} = \sum_{j=3}^{5} \alpha_j * S_j, B_{all} = \sum_{j=3}^{5} \beta_j * B_j \tag{3}$$

where $F_j(z_i)$ denotes the features of template z_i and $F_j(x)$ the features of search image x, both features from backbone layer *convj*; $\phi(\cdot)$ and $\psi(\cdot)$ denote the learned classification and bounding box regression functions, respectively; S_j and B_j denote the classification score map and the bounding box regression map of the features from backbone layer *convj*, respectively; S_{all} and B_{all} denote the fused final maps; and α_j and β_j denote the classification and regression fusions weights, respectively.

Template Pool Updater. We propose the Template Pool Updater (TPU) module to update the template pool in SiamDEPU during inference dynamically. This module allows the model captures the appearance changes of the tracked object without online learning. Figure 2 depicts the architecture of the TPU module in SiamDEPU. After the bounding box prediction of a frame becomes available, the TPU crops the search image features according to the prediction. The cropped features and the template features in the last template pool together form a template candidate pool. Then pairwise dissimilarities are calculated, based on which the template group with the highest group dissimilarity value is selected as the new template pool. In the dissimilarity module, for each feature group from one of the three convolutional layers in the backbone, a 7×7 cosine similarity map is calculated. The module takes the mean value of this map as the similarity value and uses one minus cosine similarity as the dissimilarity metric. The dissimilarity of two templates is defined as the mean of the dissimilarities of the features from the three convolutional layers. The sum of these cosine dissimilarities is defined as the group dissimilarity of a template pool candidate. Additionally, throughout the inference of a video, the initial template is always kept in the template pool without being replaced, because it represents the appearance ground truth of the target.

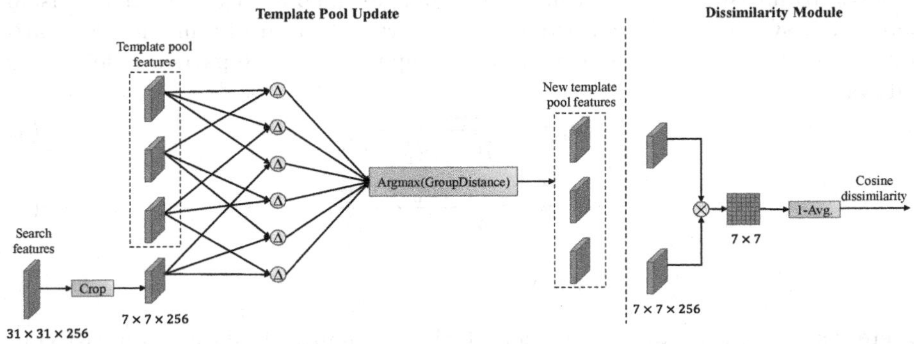

Fig. 2. Structure of the Template Pool Updater. Ⓐ and ⊗ denote the dissimilarity module and cosine similarity module, respectively.

3.2 Dissimilarity Based Hard Example Sampling

In a training dataset, hard positive examples are the examples from the same class but far from each other according to some distance metric, and hard negative examples are the examples from different classes but close to each other [13]. Selecting hard examples for more effective and efficient training, instead of randomly selecting examples, is called hard exampling sampling. It is now often used in the training of deep embeddings and results in better gradients with the selected hard examples.

One way of performing hard example sampling is to select hard examples from training data with the outputs of the model, known as hard example mining. For example, the online hard example mining (OHEM) algorithm [27] selects hard negative examples by sorting the input RoIs by loss, making the training of an object detector more effective and efficient. However, hard example sampling has not been fully exploited for the object tracking problem. SiamRPN++ [19] randomly selects positive pairs from the same videos and negative pairs from different videos without any manually designed hard example selection method. Siam R-CNN [28] finds negative examples, which are hard for re-detection, to guide the model learning.

By analyzing the training process of prevailing Siamese trackers, we find that the randomly sampled positive pairs become easy for the trackers during the training and hinder the improvement of tracking performance. To fill this gap, we propose a dissimilarity-indexing-based hard positive example sampling method to maximize the discriminative power of a Siamese tracker. In this paper, we utilize the Structural Similarity Index to find hard positive examples in the training.

Structural Dissimilarity Index. Structural Similarity Index (SSIM) [31] is a metric to measure the similarity of two images, inspired by the human visual perception system. SSIM consists of three parts which are luminance comparison, contrast comparison, and structural comparison. These parts are defined as follows:

$$l(\mathbf{x}, \mathbf{y}) = \frac{2\mu_x\mu_y + C_1}{\mu_x^2 + \mu_y^2 + C_1}, \tag{4}$$

$$c(\mathbf{x}, \mathbf{y}) = \frac{2\sigma_x\sigma_y + C_2}{\sigma_x^2 + \sigma_y^2 + C_2}, \tag{5}$$

$$s(\mathbf{x}, \mathbf{y}) = \frac{\sigma_{xy} + C_3}{\sigma_x\sigma_y + C_3}, \tag{6}$$

where $l(\mathbf{x}, \mathbf{y})$, $c(\mathbf{x}, \mathbf{y})$ and $s(\mathbf{x}, \mathbf{y})$ denote the luminance similarity, contrast similarity and structural similarity of two images, respectively. In Eq. 4, Eq. 5 and Eq. 6, μ and σ denote the mean and the standard deviation of an image, respectively, and σ_{xy} is defined as:

$$\sigma_{xy} = \frac{1}{N-1}\sum_{i=1}^{N}(x_i - \mu_x)(y_i - \mu_y). \tag{7}$$

SSIM is defined as a combination of the three comparisons:

$$\mathrm{SSIM}(\mathbf{x}, \mathbf{y}) = [l(\mathbf{x}, \mathbf{y})]^{\alpha}[c(\mathbf{x}, \mathbf{y})]^{\theta}[s(\mathbf{x}, \mathbf{y})]^{\gamma}, \tag{8}$$

where α, β and γ denote the weights of the three comparisons, and $\alpha = \beta = \gamma = 1$ in this work. $-1 \leq \mathrm{SSIM}(\mathbf{x}, \mathbf{y}) \leq 1$ and $\mathrm{SSIM}(\mathbf{x}, \mathbf{y}) = 1$ means the two image are identical to each other. As mentioned in [31], it is useful to apply the SSIM index locally rather than globally. In practice, it is more common to divide the image into regions, calculate the SSIM values on those values and take the mean value as the similarity. Such mean similarity is called Mean Structural Similarity Index(MSSIM) which defined as

$$\mathrm{MSSIM}(\mathbf{X}, \mathbf{Y}) = \frac{1}{M}\sum_{j=1}^{M}\mathrm{SSIM}(\mathbf{x_j}, \mathbf{y_j}) \tag{9}$$

where $\mathbf{X}$ and $\mathbf{Y}$ are the two images under comparison, and $\mathbf{x_j}$ and $\mathbf{y_j}$ denote the local image windows. Based on MSSIM, we define the dissimilarity metric used in this paper as

$$\mathrm{DSSIM}(\mathbf{X}, \mathbf{Y}) = \frac{C_4 - \mathrm{MSSIM}(\mathbf{X}, \mathbf{Y})}{C_5}. \tag{10}$$

Throughout this paper, to have $0 \leq \mathrm{DSSIM}(\mathbf{x}, \mathbf{y}) \leq 1$, the following parameters are used in DSSIM calculation: $C_4 = 1$ and $C_5 = 2$.

Group Dissimilarity. With the metric defined in Eq. 10, a dissimilarity value of two frames from the same video can be calculated. Based on such pairwise dissimilarities, a triplet dissimilarity can be defined as

$$d(\mathbf{x}, \mathbf{y}, \mathbf{z}) = \mathrm{DSSIM}(\mathbf{x}, \mathbf{y}) + \mathrm{DSSIM}(\mathbf{x}, \mathbf{z}) + \mathrm{DSSIM}(\mathbf{y}, \mathbf{z}) \tag{11}$$

by denoting the dissimilarity of three images $\mathbf{x}, \mathbf{y}, \mathbf{z}$ as $d(\mathbf{x}, \mathbf{y}, \mathbf{z})$. The group dissimilarity definition can be further extended to the n-tuple-dissimilarity as

$$d(\mathbf{x_0}, \mathbf{x_1}, ..., \mathbf{x_{N-1}}) = \sum_{i=0}^{N-1} \sum_{j=i+1}^{N} \mathrm{DSSIM}(\mathbf{x_i}, \mathbf{x_j}). \tag{12}$$

Such dissimilarity can also be obtained from the dissimilarity matrix $\mathbf{D} \in \mathbb{R}^{N \times N}$ defined as

$$\mathbf{D} = \begin{bmatrix} 0 & d_{0,1} & \cdots & d_{0,N-1} \\ d_{1,0} & 0 & \cdots & d_{1,N-1} \\ \vdots & \vdots & \ddots & \vdots \\ d_{N-1,0} & d_{N-1,1} & \cdots & 0 \end{bmatrix}. \tag{13}$$

In the dissimilarity matrix $\mathbf{D}$, the following equations are true:

$$d_{i,j} = d_{j,i} \tag{14}$$

$$d_{i,i} = 0 \tag{15}$$

The symmetry of $\mathbf{D}$ expressed by Eq. 14 can be derived from the symmetry of SSIM. As mentioned in [31], SSIM = 1 is true for the comparison of two same images, resulting in Eq. 15.

With Eq. 14 and Eq. 15 above, the group dissimilarity of images $\mathbf{x_0}, \mathbf{x_1}, ..., \mathbf{x_{N-1}}$ can be defined as the sum of the elements above the diagonal in the dissimilarity matrix $\mathbf{D}$, because those elements include all the possible pairwise combinations from $\mathbf{x_0}, \mathbf{x_1}, ..., \mathbf{x_{N-1}}$.

Hard Positive Example Sampling with Group Dissimilarity. Given a frame I_{t_0} from a video, the hard positive examples for I_{t_0} are the frames that yield the highest group dissimilarity together with I_{t_0} among all the possible frame combinations that include I_{t_0}. The triplet-DSSIM-based hard positive example sampling process is shown in Algorithm 1. The inner loop starts from $i + 1$ instead of 0 because of the symmetry of dissimilarity matrix $\mathbf{D}$ shown in Eq. 14.

A new training dataset with hard positive example groups is generated by executing Algorithm 1 on every frame in a video and repeating it on every video in the training dataset. Qualitative results of hard positive triplet example sampling on GOT-10k [12] are shown in Fig. 3.

Algorithm 1 : DSSIM Based Hard Positive Example Sampling

Notations: Video Sequence ($\mathcal{V}$), Total number of frames in $\mathcal{V}$ (T), Index of the frame to be included in the group dissimilarity (t_0), Dissimilarity matrix for t_0 frame ($\mathbf{D^{t_0}} \in \mathbb{R}^{T\times T}$), Element of $\mathbf{D^{t_0}}$ at the ith row and j^{th} column (d_{ij}), Function that returns indices of the minimum element in matrix $\mathbf{D}$ argmin($\mathbf{D}$), Hard positive examples indices (t_1, t_2)
Input: $\mathcal{V} = \{I_0, I_1, ..., I_{T-1}\}$, Current frame t_0 ($0 \leq t_0 \leq T-1$)
Output: t_1, t_2

```
for i = 0 : T − 1 do
    for j = i + 1 : T − 1 do
        d_ij = DSSIM(I_i, I_t0) + DSSIM(I_j, I_t0) + DSSIM(I_i, I_j)
    end
end
(t1, t2) = argmax(D^t0)
return t1, t2
```

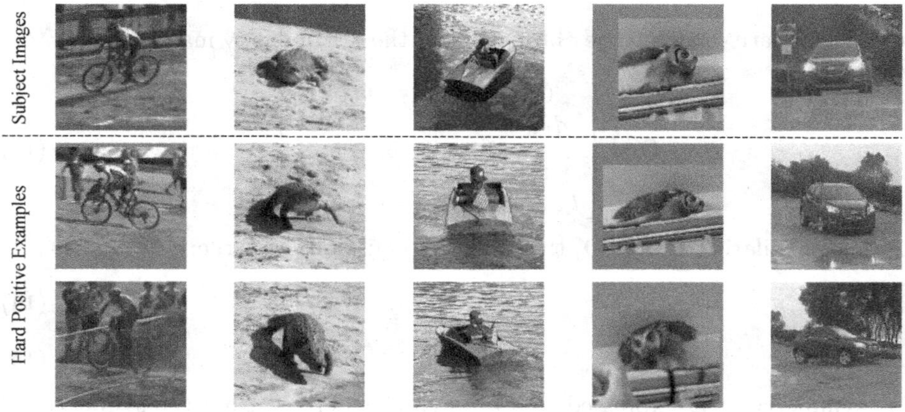

Fig. 3. Qualitative results of hard positive example sampling on GOT-10k dataset

4 Experiments

4.1 Implementation Details

The backbone of SiamDEPU is pre-trained on ImageNet [26]. We train the network on the training set of GOT-10k [12]. The size of exemplar images and search images are 127 × 127 and 255 × 255 pixels, respectively. To have a fair comparison with the baseline SiamRPN++, we also set our training loss as the sum of classification loss and the standard smooth L1 loss for regression. We use synchronized SGD [18] over 2 GPUs with a mini-batch size of 64 image groups, thus each GPU hosting 32 image groups. The network is trained for 20 epochs in total. For the first 5 epochs, a warmup learning rate exponentially grown from 0.001 to 0.005 is applied while keeping the parameters of the backbone frozen. In the remaining epochs, the model is trained end-to-end with a learning rate that exponentially decays from 0.005 to 0.0005. The weight decay and momentum in the optimizer are set to 0.0001 and 0.9, respectively. Our experiments are implemented in Python using Pytorch on a server with 8 T V100 GPUs and a Xeon E5-2698 2.20 GHz CPU.

4.2 Comparison with Baselines

The original SiamRPN++ model is trained on the training datasets of COCO [21], ImageNet DET [26], ImageNet VID, and YouTube-BoundingBoxes Dataset [25], but GOT-10k [12] requires trackers not to be trained with extra data. To have a fair comparison and to follow the policy of GOT10k, we trained a SiamRPN++ model with the training set of GOT-10k [12], following the same procedure and settings of training our model. The hyper-parameters in both trackers trained on GOT-10k [12] are fine-tuned on VOT2018. We compare the original SiamRPN++ (denoted as SiamRPNpp-plain), the SiamRPN++ we trained on GOT-10k [12] (denoted as SiamRPNpp-got10k), and our model SiamDEPU on prevailing datasets, namely VOT2018 [15], GOT10k [12] and LaSOT [5], UAV123 [22], OTB2015 [32], VOT2019 [16] and VOT2018LT [15]. The number of templates in our model is set to 3 in this section.

VOT2018. VOT2018 [15] contains 60 videos with varying challenging factors. We take Accuracy (A), Robustness (R) and Expected Average Overlap (EAO) into account for the performance evaluation, following the evaluation protocol of VOT2018. For Accuracy and EAO, higher values indicate better performance while for Robustness, a lower value indicates better performance. As shown in Table 1, the performance of SiamRPN++ drops when only trained on GOT-10k [12], but our model maintains a stronger generalization ability on unseen objects, improving the performance of the SiamRPN++ model trained on GOT-10k by 22% in terms of EAO.

Table 1. Performance comparisons on VOT-2018 [15] benchmark

	SiamRPNpp-plain [19]	SiamRPNpp-got10k	SiamDEPU
$EAO \uparrow$	0.414	0.327	**0.399**
$A \uparrow$	0.600	0.572	**0.582**
$R \downarrow$	0.234	0.314	**0.225**

GOT-10k. GOT-10k [12] is a large-scale generic object tracking benchmark with zero overlap between its train and test tests. As proposed in [12], we use mean average overlap (mAO) and success rate (SR) with two overlap thresholds ($SR_{0.5}$ and $SR_{0.75}$) as the performance metrics. As shown in Table 2, our method improves the performance of the SiamRPN++ trained on GOT-10k by 5%, 7.5% and 4% in terms of mAO, $SR_{0.5}$ and $SR_{0.75}$, respectively. SiamDEPU even slightly outperforms the original SiamRPN++ trained on a much larger training dataset compound. This shows our model's strong ability to adjust to appearance changes that are common in GOT-10k.

Table 2. Performance comparisons on GOT-10k [12] benchmark

	SiamRPNpp-plain [19]	SiamRPNpp-got10k	SiamDEPU
mAO	0.518	0.494	**0.519**
$SR_{0.5}$	0.616	0.577	**0.620**
$SR_{0.75}$	0.325	0.316	**0.329**

LaSOT. LaSOT [5] is a large-scale long-term object tracking dataset including 280 videos with an average length of 2448 frames in its test set. SiamDEPU achieves a gain of 4.2% and 2.2% over the SiamRPN++ trained on GOT-10k, in terms of success (AUC) and precision, respectively (Table 3).

Table 3. Performance comparisons on LaSOT [5] benchmark

	SiamRPNpp-plain [19]	SiamRPNpp-got10k	SiamDEPU
AUC	0.496	0.447	**0.466**
Precision	0.491	0.463	**0.473**

UAV123. The UAV123 [22] contains 123 videos captured from low-altitude UAVs. As shown in Table 4, our SiamDEPU outperforms the SiamRPN++ model trained on GOT-10k by 2.2% and 2.2% in terms of AUC (success) and precision, respectively.

Table 4. Performance comparisons on UAV123 [22] benchmark

	SiamRPNpp-plain [19]	SiamRPNpp-got10k	SiamDEPU
AUC	0.613	0.589	**0.602**
Precision	0.807	0.784	**0.801**

OTB2015. OTB-2015 [32] is a popular tracking benchmark including 100 videos. As shown in Table 5, our network increases the performance of SiamRPN++ trained on GOT-10k by 2.0% and 3.3% in terms of AUC (success) and precision, respectively.

Table 5. Performance comparisons on OTB2015 [32] benchmark

	SiamRPNpp-plain [19]	SiamRPNpp-got10k	SiamDEPU
AUC	0.696	0.645	**0.658**
Precision	0.914	0.846	**0.874**

VOT2019. VOT2019 [16] replaces the least challenging videos in VOT2018 [15]. The comparison of performance is shown in Table 6. Our model increases the baseline by a large margin of 15.5% on EAO and even outperforms the original SiamRPN++ [19] trained on a much larger training dataset compound. Such improvement proves the robustness of our model contributed by the dynamic template updating method.

Table 6. Performance comparisons on VOT-2019 [16] benchmark

	SiamRPNpp-plain [19]	SiamRPNpp-got10k	SiamDEPU
$EAO \uparrow$	0.287	0.252	**0.291**
$A \uparrow$	0.595	0.574	**0.580**
$R \downarrow$	0.467	0.547	**0.451**

VOT2018 Long-Term Dataset. VOT2018-LT [15] is a challenging long-term tracking benchmark composed of 32 long sequences. As shown in Table 7, our model gains 4.9% improvement in F-score on VOT2018-LT.

Table 7. Performance comparisons on VOT2018-LT [15] benchmark

	SiamRPNpp-plain [19]	SiamRPNpp-got10k	SiamDEPU
$F1$	0.629	0.534	**0.560**

4.3 Ablation Experiments

The ablation study results of our proposed methods on VOT2018 [15] are shown in Table 8. The proposed dynamic template updating architecture gains a 9.2% improvement on VOT2018, and the dynamic template updating model trained with the hard positive sampling strategy gains a 22% improvement. This proves the effectiveness of our dynamic template updating architecture and hard positive example sampling strategy.

Table 8. Ablation experiment results on VOT2018 [15]

Component	SiamRPNpp-got10k	SiamDEPU	
Dynamic templates update?		✓	✓
Hard positive example sampling in training?			✓
EAO	0.327	0.357	0.399

5 Conclusion

In this work, we present a method to adapt a Siamese tacker for temporal information exploitation with multiple dynamic templates. Our model SiamDEPU directly extracts appearance representations from multiple templates and aggregates features at the late stage of the network. Then, to update the templates, we propose a template pool updater that selects dissimilar templates to compose a template pool with diverse appearance representations. Moreover, we design a hard positive example sampling strategy to improve the training of a Siamese tracker. Extensive experiments demonstrate that our methods can effectively improve a Siamese tracker's performance.

Acknowledgements. This work is supported by the Chinese Scholarship Council (CSC), grant number 201907820021.

References

1. Bertinetto, L., Valmadre, J., Henriques, J.F., Vedaldi, A., Torr, P.H.S.: Fully-convolutional siamese networks for object tracking. In: Hua, G., Jégou, H. (eds.) ECCV 2016. LNCS, vol. 9914, pp. 850–865. Springer, Cham (2016). https://doi.org/10.1007/978-3-319-48881-3_56
2. Cao, Z., Huang, Z., Pan, L., Zhang, S., Liu, Z., Fu, C.: TCTrack: temporal contexts for aerial tracking. In: Proceedings of the IEEE/CVF Conference on Computer Vision and Pattern Recognition (CVPR), pp. 14798–14808 (2022)
3. Chen, Z., Zhong, B., Li, G., Zhang, S., Ji, R.: Siamese box adaptive network for visual tracking. In: Proceedings of the IEEE/CVF Conference on Computer Vision and Pattern Recognition (CVPR), pp. 6668–6677 (2020)
4. Deshmukh, S.S., Joshi, C.M., Patel, R.S., Gurav, Y.: 3D object tracking and manipulation in augmented reality. Int. J. Emerg. Technol. Innov. Res. **5**(1), 287–289 (2018)
5. Fan, H., et al.: LaSOT: a high-quality benchmark for large-scale single object tracking. In: Proceedings of the IEEE/CVF Conference on Computer Vision and Pattern Recognition (CVPR) (2019)
6. Fu, Z., Liu, Q., Fu, Z., Wang, Y.: STMTrack: template-free visual tracking with space-time memory networks. In: Proceedings of the IEEE/CVF Conference on Computer Vision and Pattern Recognition (CVPR), pp. 13774–13783 (2021)
7. Guo, D., Wang, J., Cui, Y., Wang, Z., Chen, S.: SiamCAR: siamese fully convolutional classification and regression for visual tracking. In: Proceedings of the IEEE/CVF Conference on Computer Vision and Pattern Recognition (CVPR), pp. 6269–6277 (2020)

8. Guo, Q., Feng, W., Zhou, C., Huang, R., Wan, L., Wang, S.: Learning dynamic siamese network for visual object tracking. In: Proceedings of the IEEE International Conference on Computer Vision (ICCV), pp. 1763–1771 (2017)
9. He, A., Luo, C., Tian, X., Zeng, W.: A twofold Siamese network for real-time object tracking. In: Proceedings of the IEEE Conference on Computer Vision and Pattern Recognition (CVPR), pp. 4834–4843 (2018)
10. He, K., Zhang, X., Ren, S., Sun, J.: Deep residual learning for image recognition. In: Proceedings of the IEEE Conference on Computer Vision and Pattern Recognition (CVPR), pp. 770–778 (2016)
11. Held, D., Thrun, S., Savarese, S.: Learning to track at 100 FPS with deep regression networks. In: Leibe, B., Matas, J., Sebe, N., Welling, M. (eds.) ECCV 2016. LNCS, vol. 9905, pp. 749–765. Springer, Cham (2016). https://doi.org/10.1007/978-3-319-46448-0_45
12. Huang, L., Zhao, X., Huang, K.: GOT-10k: a large high-diversity benchmark for generic object tracking in the wild. IEEE Trans. Pattern Anal. Mach. Intell. **43**(5), 1562–1577 (2021)
13. Jin, S., et al.: Unsupervised hard example mining from videos for improved object detection. In: Proceedings of the European Conference on Computer Vision (ECCV) (2018)
14. Kamble, P.R., Keskar, A.G., Bhurchandi, K.M.: A deep learning ball tracking system in soccer videos. Opto-Electron. Rev. **27**(1), 58–69 (2019)
15. Kristan, M., et al.: The sixth visual object tracking VOT2018 challenge results. In: Proceedings of the European Conference on Computer Vision (ECCV) Workshops (2018)
16. Kristan, M., et al.: The seventh visual object tracking VOT2019 challenge results. In: Proceedings of the IEEE/CVF International Conference on Computer Vision (ICCV) Workshops (2019)
17. Krizhevsky, A., Sutskever, I., Hinton, G.E.: Imagenet classification with deep convolutional neural networks. In: Pereira, F., Burges, C., Bottou, L., Weinberger, K. (eds.) Advances in Neural Information Processing Systems (NIPS), vol. 25. Curran Associates, Inc. (2012)
18. LeCun, Y., et al.: Backpropagation applied to handwritten zip code recognition. Neural Comput. **1**(4), 541–551 (1989)
19. Li, B., Wu, W., Wang, Q., Zhang, F., Xing, J., Yan, J.: SiamRPN++: evolution of siamese visual tracking with very deep networks. In: Proceedings of the IEEE/CVF Conference on Computer Vision and Pattern Recognition (CVPR), pp. 4282–4291 (2019)
20. Li, B., Yan, J., Wu, W., Zhu, Z., Hu, X.: High performance visual tracking with Siamese region proposal network. In: Proceedings of the IEEE Conference on Computer Vision and Pattern Recognition (CVPR), pp. 8971–8980 (2018)
21. Lin, T.-Y., et al.: Microsoft COCO: common objects in context. In: Fleet, D., Pajdla, T., Schiele, B., Tuytelaars, T. (eds.) ECCV 2014. LNCS, vol. 8693, pp. 740–755. Springer, Cham (2014). https://doi.org/10.1007/978-3-319-10602-1_48
22. Mueller, M., Smith, N., Ghanem, B.: A benchmark and simulator for UAV tracking. In: Leibe, B., Matas, J., Sebe, N., Welling, M. (eds.) ECCV 2016. LNCS, vol. 9905, pp. 445–461. Springer, Cham (2016). https://doi.org/10.1007/978-3-319-46448-0_27
23. Ning, G., et al.: Spatially supervised recurrent convolutional neural networks for visual object tracking. In: 2017 IEEE International Symposium on Circuits and Systems (ISCAS), pp. 1–4 (2017)

24. Rangesh, A., Trivedi, M.M.: No blind spots: full-surround multi-object tracking for autonomous vehicles using cameras and lidars. IEEE Trans. Intell. Veh. **4**(4), 588–599 (2019)
25. Real, E., Shlens, J., Mazzocchi, S., Pan, X., Vanhoucke, V.: YouTube-BoundingBoxes: a large high-precision human-annotated data set for object detection in video. In: Proceedings of the IEEE Conference on Computer Vision and Pattern Recognition (CVPR), pp. 5296–5305 (2017)
26. Russakovsky, O., et al.: ImageNet large scale visual recognition challenge. Int. J. Comput. Vision **115**(3), 211–252 (2015)
27. Shrivastava, A., Gupta, A., Girshick, R.: Training region-based object detectors with online hard example mining. In: Proceedings of the IEEE Conference on Computer Vision and Pattern Recognition (CVPR), pp. 761–769 (2016)
28. Voigtlaender, P., Luiten, J., Torr, P.H., Leibe, B.: Siam R-CNN: visual tracking by re-detection. In: Proceedings of the IEEE/CVF Conference on Computer Vision and Pattern Recognition (CVPR), pp. 6578–6588 (2020)
29. Wang, N., Zhou, W., Wang, J., Li, H.: Transformer meets tracker: exploiting temporal context for robust visual tracking. In: Proceedings of the IEEE/CVF Conference on Computer Vision and Pattern Recognition (CVPR), pp. 1571–1580 (2021)
30. Wang, Q., Zhang, L., Bertinetto, L., Hu, W., Torr, P.H.: Fast online object tracking and segmentation: a unifying approach. In: Proceedings of the IEEE/CVF Conference on Computer Vision and Pattern Recognition (CVPR), pp. 1328–1338 (2019)
31. Wang, Z., Bovik, A., Sheikh, H., Simoncelli, E.: Image quality assessment: from error visibility to structural similarity. IEEE Trans. Image Process. **13**(4), 600–612 (2004)
32. Wu, Y., Lim, J., Yang, M.H.: Object tracking benchmark. IEEE Trans. Pattern Anal. Mach. Intell. **37**(9), 1834–1848 (2015)
33. Xu, Y., Wang, Z., Li, Z., Yuan, Y., Yu, G.: SiamFC++: towards robust and accurate visual tracking with target estimation guidelines. In: Proceedings of the AAAI Conference on Artificial Intelligence, vol. 34, no. 07, pp. 12549–12556 (2020)
34. Yan, B., Peng, H., Fu, J., Wang, D., Lu, H.: Learning spatio-temporal transformer for visual tracking. In: Proceedings of the IEEE/CVF International Conference on Computer Vision (ICCV), pp. 10448–10457 (2021)
35. Zhang, L., Gonzalez-Garcia, A., Weijer, J.V.D., Danelljan, M., Khan, F.S.: Learning the model update for Siamese trackers. In: Proceedings of the IEEE/CVF International Conference on Computer Vision (ICCV) (2019)
36. Zhu, Z., Wang, Q., Li, B., Wu, W., Yan, J., Hu, W.: Distractor-aware Siamese networks for visual object tracking. In: Proceedings of the European Conference on Computer Vision (ECCV) (2018)

Detection and Tracking of Pinus Radiata Catkins

Eric Song, Sam Schofield(✉), and Richard Green

Computer Science and Software Engineering, University of Canterbury, Christchurch, New Zealand
sam.schofield@pg.canterbury.ac.nz

Abstract. Pinus Radiata trees form pollen-producing catkins that can be harvested for pharmaceutical uses. Unmanned Aerial Vehicles (UAVs) may be well suited to the task of autonomously harvesting these catkins. We propose a method to reliably detect and track P. Radiata catkins in three dimensions that can be used for real-time guidance of a UAV. The method applies the YOLOv5 deep learning algorithm to detect catkins in the X-Y plane. A novel optimisation of the MeanShift algorithm is utilised to assist existing contour detection algorithms in segmenting individual catkins in the Z plane. A Kanade-Lucas-Tomasi tracker was used with RANSAC for accurate frame-to-frame tracking. The method achieved a Mean Average Precision of 0.87 on images taken at a commercial pine pollen farm. The method detected the depth of catkins at distances of up to 1200 mm to an accuracy of 2 mm, or 8 mm for occluded catkins. Detected catkins can be reliably tracked at speeds of 1ms. An average frame rate of 22 frames per second was achieved on an Intel i5 CPU, with the Meanshift optimisation performing up to 41 times faster than existing implementations. These results indicate that the proposed method could be used to successfully assist in the automated harvesting of P. Radiata catkins.

Keywords: Object detection · Object tracking · Aerial manipulation

1 Introduction

Pinus Radiata trees form pollen-producing catkins that can be harvested for pharmaceutical uses (Fig. 1). Currently, catkins are harvested by hand. As P. Radiata can grow to over 20 m at an age of 20 years [1], these catkins are only manually harvested from younger, shorter P. Radiata trees. However, as P. Radiata is the most common tree species grown in New Zealand forestry [17], a cost-effective method for harvesting catkins from mature P. Radiata could greatly increase the scalability and commercial viability of this industry. Unmanned Aerial Vehicles (UAVs), or drones, may be well suited to the task of autonomously harvesting catkins at these heights. To achieve this task, a drone must first be able to reliably detect and track the catkins in three dimensions.

We propose a method to reliably detect and track P. Radiata catkins using a drone-mounted Intel® RealSense™ D435 depth camera. To achieve this objective, three sub-goals must be met. Firstly, bounding boxes of catkins must be

W. Q. Yan et al. (Eds.): IVCNZ 2022, LNCS 13836, pp. 159–174, 2023.
https://doi.org/10.1007/978-3-031-25825-1_12

Fig. 1. P. Radiata catkin on a branch.

detected in real time from the camera's video feed. Secondly, the points at which to cut each catkin must be determined in three dimensions. Finally, the movement of multiple catkins relative to the drone must be tracked to provide guidance for harvesting. We show that the tracking data provided by proposed method is sufficient to guide a harvesting drone to harvest catkins on mature P. Radiata trees.

2 Previous Research

While there is no existing research on automated harvesting of P. Radiata catkins, many methods have been proposed for the automated harvesting of other plants. Many of these methods use convolutional neural networks to detect fruit, all of which could be trained on catkin images and used for catkin detection. Williams et al. [21] presented a kiwifruit harvesting system that used the Faster R-CNN 2.0 network to detect kiwifruit at an accuracy of 89.6%. A Single Shot MultiBox Detector (SSD) was used in [13] to detect apples with a precision of 100% and recall of 92.31%. The authors in [12] developed a pine cone detection algorithm using a modification of the existing YOLOv4 object detection network. This achieved a high average precision (AP) of 95.8% - a 7.4% increase compared to the original YOLOv4 network. A newer iteration of the YOLO network - YOLOv5 - was used in [6] to detect apples, achieving low false positive and negative rates of 3.5% and 2.8% respectively.

The computational cost of the networks used in the above papers are shown in Table 1.

Table 1. Computation cost (in GFLOPS) for previous neural networks

	Faster R-CNN 2.0	SSD	Modified YOLOv4	YOLOv5
GFLOPS	850	31	75.3	16.5

Given the drone-mounted Intel NUC's limited processing power, at around 57.21 GFLOPS, it is unlikely that any of the networks will be sufficiently fast for real-time detection. Thus, the approaches may not be suitable for real-time feedback in a catkin harvesting system in which the camera is attached to a moving drone.

Some fruit harvesting approaches used other machine learning techniques to perform a segmentation of the image, labelling all pixels belonging to a fruit. In [19], a Support Vector Machine (SVM) was to detect aubergines. Watershed transforms were then used to separate the image into separate blobs for each aubergine. The SVM achieved a recall of 88.10% and a precision of 88.35%, with most errors occurring due to changing lighting conditions. As the testing occurred in an indoor laboratory, the algorithm may perform more poorly in outdoor settings with a larger range of lighting conditions, as is required for catkin harvesting.

A different segmentation method for spherical or cylindrical fruits was proposed in [7]. This applied a probabilistic model on the HSV values of each pixel to exclude background pixels, then used region growing to generate a set of point clouds. A shape detection algorithm and SVM classifier were then applied to identify which point clouds were true positives. This approach was however unable to distinguish fruit which resembled the color or shape of leaves of the same plant, and would in some cases identify a single fruit occluded by branches or leaves as several different fruit. Catkins will often be partially occluded by pine needles, and may have a similar green colour to the needles at harvesting time.

Most automated agricultural systems used stereo point matching to determine the 3D locations of fruit. The approach used in [21] relied on the networks' ability to detect the calyx (a distinctive protruding feature) of each kiwifruit. Each calyx was used as an accurate point on which stereo matching could be performed. However, this method was only successful if the calyx was visible. Additionally, as this method depended on each kiwifruit having at most one single, distinctive calyx for point matching, the method may not generalise to catkins, which do not have a single, unique, distinctive feature. A stereo camera was used in [13] to determine the 3D position of each apple by simply taking the distance of the center point of the detected apple's bounding box. For round apples, the center point is guaranteed to be within the apple - however, for catkins that are irregularly shaped or not perpendicular to the ground, this may not be guaranteed.

An advantage of the semantic segmentation approaches discussed previously are that they allow for simple calculation of the depth value - as they label all pixels belonging to a fruit, the corresponding pixels in the depth image could be taken as the depth of the fruit. However, labelling data for segmentation tasks require the individual pixels of each object to be masked, whereas object detection only require bounding boxes drawn around each object [14]. A deep-learning based segmentation approach would be significantly more labour-intensive than an object detection approach.

The automated harvesting systems discussed above all assume a stationary camera. However, for a drone-mounted video feed, frame-to-frame tracking of catkins must be performed to provide real-time guidance for the drone. A Kalman Filter corrected Kanade-Lucas-Tomasi tracker was used for real time tracking of fruit in [10]. A Lucas-Kanade based method was also used in [18] for apple tracking, but the detected contours of each apple was used to match between frames, rather than point features as used in [10]. To create a 3D reconstruction of the fruit and environment, a Structure-from-Motion technique was used. Structure-from-Motion approaches assume a static environment; they therefore may not be suitable for catkin harvesting as wind generated by the drone frequently causes pine branches and needles to move. [24] used a modification of the SORT tracking algorithm for orange tracking, which estimated the positions of missed detections by using the average displacement of the other detected oranges. This estimation was accurate as the oranges were all in approximately the same plane, an assumption that may not be true for catkins on different branches.

3 Methodology

3.1 Catkin Detection

The proposed method uses the YOLOv5-nano neural network [5] for the detection of catkins. This network was chosen due to its real-time performance on CPU hardware, requiring only 4.5 GFLOPS [5]. To reduce the training time required, the network weights were initialised to the weights of a model pretrained on the COCO dataset [8].

The training data set consisted of 205 images. 60 of these images were of catkins obtained from the internet. 85 of these images were taken during fields trips to a catkin farm. 60 images were null examples - forested scenes containing no catkins - to reduce the network's tendency to detect catkins when none were in frame.

Most non-null examples in the data set contained exclusively catkins, pine branches and needles, and a background such as the sky. All yellow objects in the training data were therefore catkins, potentially resulting in a network that incorrectly labels all yellow objects as catkins. This may result in the tracking of yellow non-catkin objects in the environment, such as a high-visibility vest or an operator's fingers. To prevent this, 45 images of other yellow objects were added to the training set. To ensure that the network would be able to distinguish between catkins and other yellow objects in the same image, the test data was augmented by superimposing a random selection of the yellow objects into each image. Each instance of each objects was superimposed at a random angle, and a Gaussian blur was applied with a random radius to increase the diversity of the augmentations. An example of this is shown in Fig. 2. The validation data set was also augmented in a similar way, using a different set of 20 yellow objects to detect any over-fitting to the objects used in the training data.

Training occurred on a single computer with a 3.20 GHz Intel® Core^{tm} i7-8700 CPU and an nVidia GeForce RTX^{TM} 2070 GPU. The network was trained with a batch size of 32 - the maximum possible size given the available hardware - for 150 epochs, taking around 20 min to complete. The epoch in which the network yielded the best performance on a validation set of 71 images taken at ProSeed's catkin farm was chosen for the final weights for the network. This set of images are likely to best reflect the environment that the network is expected to be used in.

Fig. 2. Training image augmented with superimposed objects

3.2 Depth Estimation

The YOLOv5-nano network only gives bounding boxes for each detected catkin. Thus, there will be background pixels within each bounding box that do not correspond to the catkin. To accurately determine the depth of each catkin, the pixels actually belonging to the catkin must be discriminated. Branches and pine needles immediately behind the catkin may have depth values that are closer to the catkin than the background. Therefore, straightforward binary thresholding techniques such as Otsu's method [15] may include these unwanted elements in the foreground. A segmentation approach which can distinguish multiple regions is needed.

Machine-learning based approaches were not feasible due to the lack of existing depth images. An unsupervised clustering approach was instead used. The MeanShift (MS) algorithm [2] was used to find clusters in the histogram of depth

Fig. 3. Catkin detection by YOLOv5 (bounding box in red) (Color figure online)

values of the image. MS uses a set of initial seeds and a single hyper-parameter: the bandwidth b [2]. In each iteration and for each seed, the mean of all data points within distance b of the seed is calculated. The seed is then set to that mean for the next iteration. The algorithm iterates until all seeds converge to a mode. A significant advantage of this algorithm is its ability to distinguish an arbitrary number of clusters; other clustering algorithms such as K-means requires a number of clusters to be determined beforehand.

While MS has been shown to be effective at segmenting both color images and depth images [2,22], the color values are not used in clustering. The color values of different pixels within the same catkin may vary significantly due to shadows, potentially resulting in an incorrect segmentation. Existing MS implementations use generalised data structures such as ball trees [9,16] to find the distance-b neighbours of each seed. While these data structures can operate on arbitrary n-dimensional data, they are inefficient at very low or high dimensions [23].

A novel implementation of MS is used to achieve significantly faster performance on the one-dimensional histogram of depth values. The proposed MS implementation sorts the data points and pre-computes the prefix sums of the sorted points. To find the mean of all points within distance b of a point p, a binary search is used to find the prefix sums corresponding to the points $p - b$ and $p + b$. The difference between the two prefix sums will be the sum of all b-distance points, which can be divided by the size of the interval to find the mean.

The algorithm is described in Algorithm 1. Note that lines 9 and 10 can be achieved with a $\mathcal{O}(\log(N))$ binary search. This gives a $\mathcal{O}(IM\log(N) + N)$ running time, where I is the number of iterations required for convergence. When applied to depth images of catkins, the algorithm typically converges within 20 iterations.

The seeds used in the proposed method were not randomly generated. To reduce the number of seeds needed, the depth values were grouped into discrete bins, each of size $b/2$. Bins containing more than 10 depth values were chosen as seeds. This reduces the number of seeds needed, as very small clusters will not be processed.

Algorithm 1. Fast 1D MeanShift

Inputs: bandwidth h, tolerance η, seeds S, values A

$N[0] \leftarrow 0$
for $i \leftarrow 1$ to $length(A)$ **do**
    $N[i] \leftarrow N[i-1] + A[i-1]$
end for
$S' \leftarrow S$
do
    $S \leftarrow S'$
    for $i \leftarrow 0$ to $|S| - 1$ **do**
        $l \leftarrow$ smallest j where $A[j] \geq S[i] - \eta$
        $r \leftarrow$ largest j where $A[j] \leq S[i] + \eta$
        $S'[i] \leftarrow \frac{N[r]-N[l]}{r-l}$
    end for
while $\sum_{i=0}^{|S|-1} |S'[i] - S[i]| \leq \eta$
return S'

The means obtained from MS were used to cluster the entire depth image - each pixel was assigned to the cluster containing the closest mean. The result of this is shown in Fig. 4. Each cluster obtained from MeanShift was then used to generate a new binarised image, with all pixels belonging to the cluster becoming fully white and all other pixels becoming fully black (Fig. 5b). This allowed a morphological smoothing to be performed on each binarised image, reducing noise and small, irrelevant features. This consisted of an opening followed by a closing operation, both performed for three iterations using a 3×3 square structuring element.

Suzuki's algorithm for contour detection [20] was then applied to each binarised image to detect the outlines of each segmented region. Some binarised images may have consisted of multiple disconnected components that happened to share similar depth values. The application of Suzuki's algorithm will extract separate contours for each of these components, distinguishing regions that can not be detected through the depth segmentation alone. This is shown in Fig. 5a.

Finally, one of the regions must be selected as the one corresponding to the catkin. This region is likely to fill up the majority of the bounding box, so selecting the largest region is correct in most cases. However, in some cases the background region is larger than the catkin's region. In these cases, the background region will be highly non-convex as it will contain a large concavity in place of the catkin. To eliminate these cases, an initial filtering step removes any regions whose areas are less than half of the area of its convex hull, after

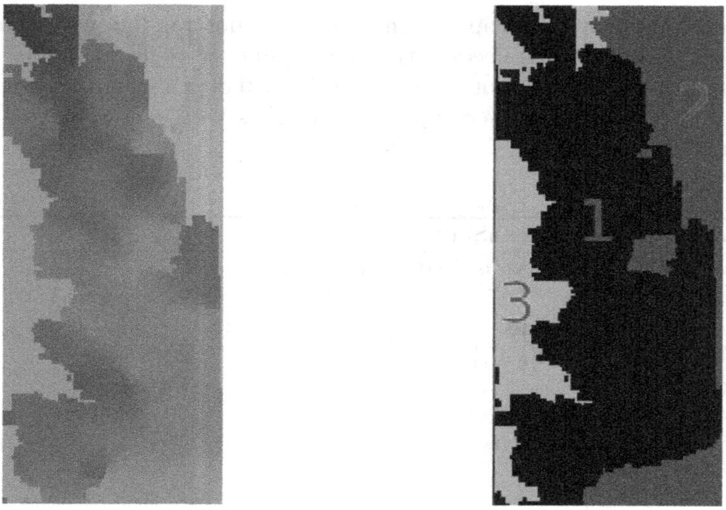

(a) Original depth image within bounding box

(b) Result of meanshift segmentation. Clusters are numbered from 1 to 3.

Fig. 4. Meanshift segmentation of the catkin in Fig. 3

which the largest region is chosen. The median of the depth values within this region is used as the depth of the catkin.

3.3 Catkin Tracking

Frame-to-frame tracking of catkins is achieved using a variant of a Kanade-Lucas-Tomasi tracker [11]. The implementation used detects corner points in each frame, and computes the movement of each corner point between consecutive frames. The relatively complex shape and surface of a catkin provides a large number of corner points to track - to reduce the required computational cost, the number of points used was limited to 30. This allows for a correspondence between corner points of consecutive frames to be maintained. A homography transform between the corresponding sets of corner points is then computed. This homography gives the transformation between the perspectives of the camera in the two different frames [4], which can be used to predict the likely bounding box of the catkin in the later frame. A prediction is associated with a new detection if it intersects significantly, allowing an association between detections of consecutive frames to be maintained.

Errors in the correspondence between corner points are reduced through the RANSAC framework [3] in calculating the homography between consecutive frames. This process excludes outliers in the correspondences from influencing the calculated homography. In this context, incorrect correspondences will imply a significantly different homography than the majority of correct corresponding points, and will be discarded.

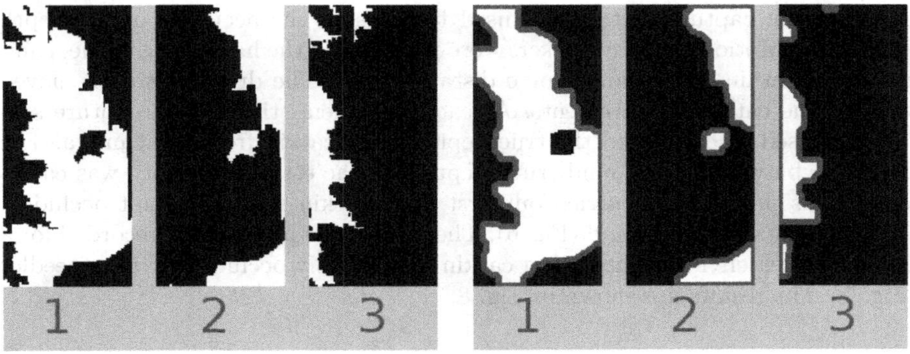

(a) Binarised images generated from the segmentation in Fig. 4.

(b) Contours (red) extracted from binarised and smoothed images.

Fig. 5. Binarisation and contour extraction. Numbering corresponds to clusters in Fig. 4.

4 Results

The method was implemented on an Intel® NUC with the specifications given in Table 2.

Table 2. Specifications of the computer used.

CPU	Intel® Core™ i5-6260U
RAM	8 GB DDR4
Camera	Intel® RealSense™ D435
Resolution	640 × 480 pixels
Frame Rate	60 frames per second
Operating System	Ubuntu 18.04
Inference Engine	ONNX with CPU Execution Provider
Python Version	3.7
OpenCV Version	4.5
NumPy Version	1.21

The trained YOLOv5 model achieved a Mean Average Precision (mAP) of 0.87 at an IoU threshold of 0.5 (mAP@0.5). The mean mAP at IoU thresholds of 0.5 to 0.95 (mAP@[0.5:0.95]) was 0.57. The precision and recall at an IoU threshold of 0.5 were 0.85 and 0.82 respectively, giving false positive and false negative rates of 0.15 and 0.18 respectively.

A motion capture system was used to measure the accuracy of the depth estimation. Motion capture markers were placed onto the harvesting drone, camera and a catkin on a branch some distance away. The drone was then moved towards the catkin in increments. At each increment, the motion capture system was used to calculated the true depth of the catkin from the camera. The deviation between this ground truth depth and the estimated depth was calculated. This process was carried out first for a catkin which was not occluded, against a distant background (Fig. 6). Then, the same process was recorded for a more complex environment with a catkin significantly occluded by pine needles (Fig. 7). The results are shown in Fig. 8.

Fig. 6. RGB image, depth image, and segmentation of unoccluded catkin.

Fig. 7. RGB image, depth image, and segmentation of partially occluded catkin.

Note that the YOLOv5 model could not reliably detect the catkin in the simple environment at distances greater than 3000 mm; thus, no results are included past that distance. Similarly, as the catkin could not be reliably detected past 1600 mm in the complex environment, no results are included for distances greater than 1600 mm. The harvesting drone also cannot approach closer than around 700 mm to the catkin due to the length of the cutting tool; thus distances significantly closer than this were not tested.

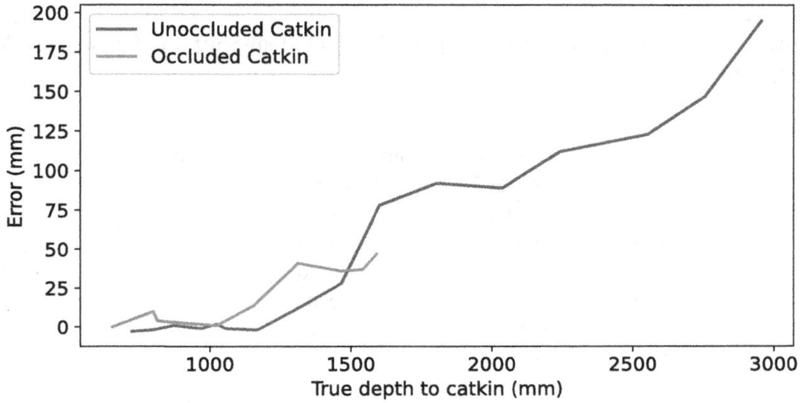

Fig. 8. Difference in mm between estimated and true depth for occluded and unoccluded catkins.

The calculated depths for the simple environment remain within 3 mm of the true depth up to a distance of 1200 mm, with a root-mean-square deviation (RMSD) of 1.9 mm. The calculated values then vary significantly for longer distances; the error appears to scale approximately proportionally with the distance to the catkin. The estimated depth is also always greater than the true depth at longer distances. A similar pattern is found for the unoccluded catkin, with a RMSD of 7.9 mm within 1200 mm and significantly larger deviations at longer distances. The higher error within 1200 mm of the occluded catkin (as compared to the unoccluded catkin) is a result of the less accurate segmentation shown in Fig. 6.

Manual inspection of the depth images at greater distances indicated that the estimated depths are similar to the depth values corresponding to the catkins. This indicates that the increasing error past 1200 mm is a result of inaccuracies in the depth values supplied by the depth camera.

The average computation times for both the proposed Fast 1-D MeanShift implementation and the default MeanShift implementation provided by the Scikit-learn library [16] are shown in Table 3. The Fast 1-D MeanShift was implemented in Python using the NumPy library. The computation times were measured over an average of 500 frames. In each frame, the input data to both implementations were random samples of the depth values of the catkin shown in Fig. 7.

The 1-D MeanShift implementation is 8 times faster than the Scikit-learn implementation at a sample size of 500. At a sample size of 100 000, the 1-D implementation is 41 times faster. These results show that the Fast 1-D MeanShift implementation is significantly faster, and more scalable, than previous implementations.

Table 3. Average computation times (s) for Fast 1-D MeanShift and Scikit-learn MeanShift implementations

Sample size	Average computation time (s)	
	Fast 1-D MeanShift	MeanShift (Scikit-learn)
500	0.00328	0.0265
1 000	0.00342	0.0325
5 000	0.00379	0.0634
10 000	0.00423	0.0846
50 000	0.00816	0.296
100 000	0.0128	0.537

The average computational times for each component of the proposed method are shown in Table 4. These computational times were measured with the drone facing a tree containing at least 20 catkins. The total computational time per frame is 0.0453 s. This equates to an average of 22 frames per second, which is sufficient for real-time guidance.

Table 4. Average computational times (s) per component.

Detection	Depth Estimation	Tracking	Total
0.0386	0.0042	0.0025	0.0453

Tracking was tested by moving the camera at a constant speed of 1m/s at a distance of approximately 1.5m from a branch containing eight catkins. Six of the catkins were tracked from first frame until the catkins moved out of frame. Tracking for two of the catkins were lost after the neural network failed to detect them for 15 frames. However, both catkins were re-acquired after they moved back into view and were successfully tracked until they moved out of view.

5 Limitations

Some of the false negative detections given by the neural network were failures to detect catkins in relatively low light conditions. During clear days with bright sunlight, catkins in shadow appeared with very low contrast. This is compounded by the severe glare in the images captured while facing the sun (see Fig. 9).

Currently, in sunny conditions, only catkins on the sun-facing side of a tree can be harvested at one time. The other side of each tree must be harvested later in the day. Due to the low cycle rate of the current harvesting drone, this is not currently a significant limitation, as only a small area can be harvested in half a day. However, if the cycle time improves, the drone may be capable of harvesting a larger area. This limitation would cause the drone to retrace its path across a large area each day.

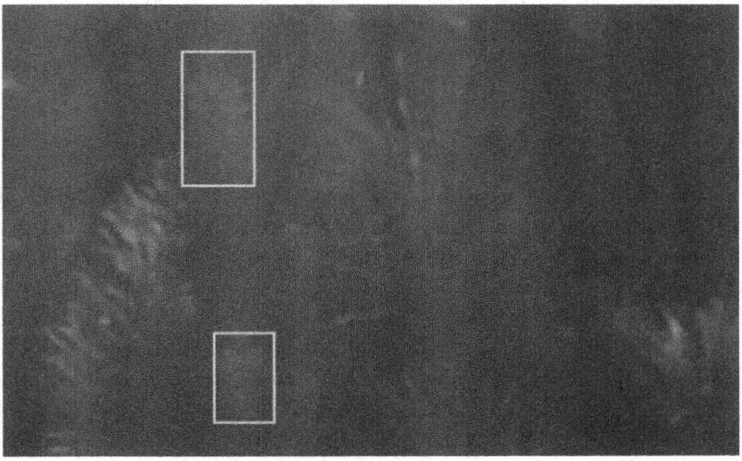

Fig. 9. Catkins in shadow in high-glare environment, indicated by green boxes. These catkins were not successfully detected by the trained network. (Color figure online)

The lower accuracy of the depth values at greater distances is likely due to failure of the D435 depth camera's stereo point matching to measure distances accurately. At distances further than 1200 mm the catkin appears only approximately five pixels in height and width. There are likely few corner points to match within this region - it is possible that the depth camera must extrapolate some depth values from neighbouring areas' depths. This may also make it difficult for the MeanShift segmentation to accurately detect clusters, as few distinct data points exist.

The decreased accuracy in complex environments indicates that the proposed method is limited in its ability to accurately judge distances to partially occluded catkins. The MeanShift segmentation step does not take into account spatial data when performing the segmentation. This leads to cases in which the segmentation result is incorrect due to clusters that have similar depth values, but are not spatially connected. One example is shown in Fig. 7. Additionally, as the segmentation only takes into account depth values, pine branches that are at a similar depth to the catkin will be considered to be part of the same cluster as the catkin. This may affect the calculated median distance in cases where the depth of the pine branches are sufficiently different from the actual catkin depth, but similar enough to be counted as the same region.

The limited range at which a catkin can be detected and reliably localised, as shown in Fig. 8, limits the proposed solution to harvesting catkins within two to three meters of the drone's initial position. While this may be partially due to the small size of the YOLOv5-n model used, it is also a limitation of the 640×480 pixel resolution mode of the D435. No model would be able to detect catkins at further than 4 m, as catkins at these distances appear as a single pixel. Higher resolution modes from the D435 are available, but would result in a lower throughput as the larger images would be more computationally intensive to process. Possible solutions to this issue are discussed in Sect. 6.1.

6 Conclusion

In this paper we proposed a method to detect and track P. Radiata catkins in 3D space. The method uses a trained YOLOv5n convolutional neural network to detect catkins. A novel optimisation of the MeanShift algorithm is utilised to assist existing contour detection algorithms in segmenting individual catkins in the depth image. A Kanade-Lucas-Tomasi tracker was used with RANSAC for accurate frame-to-frame tracking. The neural network achieved a mAP@0.5 of 0.87 and a mAP@[0.5:0.95] of 0.57 on images taken at a commercial pine pollen farm. The proposed method was able to detect the depth of an unoccluded catkin at distances of up to 1200 mm to an accuracy of 2 mm, or 8 mm for an occluded catkin. Detected catkins can be reliably tracked at speeds of 1ms. An average frame rate of 22 frames per second was achieved, with the Meanshift optimisation performing up to 41 times faster than existing implementations. These results indicate that the proposed method could be used to successfully guide a drone to harvest P. Radiata catkins. To the best of our knowledge, this paper proposes the first method in academic literature for automated detection and tracking of P. Radiata catkins.

6.1 Future Research

One problem experienced by the catkin harvesting drone is that pine needles occasionally get caught in the pruning tool when performing a cut. This problem could be overcome by developing a method to determine the optimal cut point for a catkin. Paired with a larger and more powerful cutting tool, this could significantly increase the harvesting success rate. Cut point detection could potentially be achieved by detecting any densely clustered pine needles around the catkin and extrapolating a point below the needles. A semantic segmentation approach which detects branches, needles and catkins, similar to that taken in [19], could also be used to find an exact cut point on the branch supporting the catkin. Alternatively, another neural network could be trained to detect the cut points, although this may impose a higher computation cost and may not be effective when the branches are obscured by pine needles.

Acknowledgements. The research reported in this article was conducted as part of "Enabling unmanned aerial vehicles (drones) to use tools in complex dynamic environments UOCX2104", which is funded by the New Zealand Ministry of Business, Innovation and Employment.

References

1. van der Colff, M., Kimberley, M.O.: A national height-age model for Pinus radiata in New Zealand. NZ J. Forest. Sci. **43**(1), 4 (2013). https://doi.org/10.1186/1179-5395-43-4
2. Comaniciu, D., Meer, P.: Mean shift: a robust approach toward feature space analysis. IEEE Trans. Pattern Anal. Mach. Intell. **24**(5), 603–619 (2002). https://doi.org/10.1109/34.1000236
3. Fischler, M.A., Bolles, R.C.: Random sample consensus: a paradigm for model fitting with applications to image analysis and automated cartography. Commun. ACM **24**(6), 381–395 (1981). https://doi.org/10.1145/358669.358692
4. Hartley, R.I., Zisserman, A.: Multiple View Geometry in Computer Vision, 2nd edn. Cambridge University Press, Cambridge (2004). ISBN 0521540518
5. Jocher, G., et al.: Marc, albinxavi, fatih, oleg, wanghaoyang0106: ultralytics/yolov5: v6.0 - YOLOv5n 'Nano' models, Roboflow integration, TensorFlow export, OpenCV DNN support (2021). https://doi.org/10.5281/zenodo.5563715
6. Kuznetsova, A., Maleva, T., Soloviev, V.: Detecting apples in orchards using YOLOv3 and YOLOv5 in general and close-up images. In: Han, M., Qin, S., Zhang, N. (eds.) ISNN 2020. LNCS, vol. 12557, pp. 233–243. Springer, Cham (2020). https://doi.org/10.1007/978-3-030-64221-1_20
7. Lin, G., Tang, Y., Zou, X., Xiong, J., Fang, Y.: Color-, depth-, and shape-based 3D fruit detection. Precis. Agric. **21**(1), 1–17 (2019). https://doi.org/10.1007/s11119-019-09654-w
8. Lin, T.-Y., et al.: Microsoft COCO: common objects in context. In: Fleet, D., Pajdla, T., Schiele, B., Tuytelaars, T. (eds.) ECCV 2014. LNCS, vol. 8693, pp. 740–755. Springer, Cham (2014). https://doi.org/10.1007/978-3-319-10602-1_48
9. Liu, T., Moore, A.W., Gray, A., Cardie, C.: New algorithms for efficient high-dimensional nonparametric classification. J. Mach. Learn. Res. **7**(6) (2006)
10. Liu, X., et al.: Robust fruit counting: Combining deep learning, tracking, and structure from motion. In: 2018 IEEE/RSJ International Conference on Intelligent Robots and Systems (IROS), pp. 1045–1052 (2018). https://doi.org/10.1109/IROS.2018.8594239
11. Lucas, B., Kanade, T.: An iterative image registration technique with an application to stereo vision (IJCAI), vol. 81 (1981)
12. Luo, Z., Zhang, Y., Wang, K., Sun, L.: Detection of pine cones in natural environment using improved YOLOv4 deep learning algorithm. Comput. Intell. Neurosci. **2021**, 1–12 (2021). https://doi.org/10.1155/2021/5601414
13. Onishi, Y., Yoshida, T., Kurita, H., Fukao, T., Arihara, H., Iwai, A.: An automated fruit harvesting robot by using deep learning. ROBOMECH J. **6**(1) (2019). https://doi.org/10.1186/s40648-019-0141-2
14. Osco, L.P., et al.: A review on deep learning in UAV remote sensing. Int. J. Appl. Earth Obs. Geoinform. **102**, 102456 (2021). https://doi.org/10.1016/j.jag.2021.102456. https://www.sciencedirect.com/science/article/pii/S030324342100163X

15. Otsu, N.: A threshold selection method from gray-level histograms. IEEE Trans. Syst. Man Cybern. **9**(1), 62–66 (1979). https://doi.org/10.1109/TSMC.1979.4310076
16. Pedregosa, F., et al.: Scikit-learn: machine learning in Python. J. Mach. Learn. Res. **12**, 2825–2830 (2011)
17. New Zealand's forests. https://www.mpi.govt.nz/forestry/new-zealand-forests-forest-industry/new-zealands-forests/
18. Roy, P., Isler, V.: Surveying apple orchards with a monocular vision system. In: 2016 IEEE International Conference on Automation Science and Engineering (CASE), pp. 916–921 (2016). https://doi.org/10.1109/COASE.2016.7743500
19. SepúLveda, D., Fernández, R., Navas, E., Armada, M., González-De-Santos, P.: Robotic aubergine harvesting using dual-arm manipulation. IEEE Access **8**, 121889–121904 (2020). https://doi.org/10.1109/ACCESS.2020.3006919
20. Suzuki, S., Be, K.: Topological structural analysis of digitized binary images by border following. Comput. Vis. Graph. Image Process. **30**(1), 32–46 (1985). https://doi.org/10.1016/0734-189X(85)90016-7. https://www.sciencedirect.com/science/article/pii/0734189X85900167
21. Williams, H., et al.: Improvements to and large-scale evaluation of a robotic kiwifruit harvester. J. Field Robot. **37**(2), 187–201 (2019). https://doi.org/10.1002/rob.21890
22. Xiao, W., Zaforemska, A., Smigaj, M., Wang, Y., Gaulton, R.: Mean shift segmentation assessment for individual forest tree delineation from airborne lidar data. Remote Sens. **11**(11) (2019). https://doi.org/10.3390/rs11111263. https://www.mdpi.com/2072-4292/11/11/1263
23. Yang, J., Rahardja, S., Fränti, P.: Mean-shift outlier detection and filtering. Pattern Recognit. **115**, 107874 (2021). https://doi.org/10.1016/j.patcog.2021.107874. https://www.sciencedirect.com/science/article/pii/S0031320321000613
24. Zhang, W., et al.: Deep-learning-based in-field citrus fruit detection and tracking. Hortic. Res. **9** (2022). https://doi.org/10.1093/hr/uhac003

Assessing the Condition of Copper Conductors Using Deep Learning

Zhicheng Pan[1], David Wilson[1(✉)], Martin Stommel[1], and Alex Castellanos[2]

[1] Department of Electrical and Electronic Engineering, Auckland University of Technology, Auckland 1010, New Zealand
{steven.pan,david.wilson,mstommel}@aut.ac.nz
[2] Unison Networks Limited, Hastings 4175, New Zealand
alex.castellanos@unison.co.nz

Abstract. The timely replacement of in-service overhead electrical conductors in a distribution network is crucial for the reliable operation of the grid. Visually inspecting the conductors is a non-destructive testing method that can quickly identify sign of damage and degradation on the conductor's surface and the result is used to inform decision in the risk management framework. In this study, a deep neural network is employed to classify a conductor's condition into multiple classes ranked by the type and severity of the degradation. The best test accuracy of 83.02% was reached. This model can be used to improve on the current manually-intensive and time consuming inspection practice commonly used in the electrical distribution industry.

Keywords: Power lines visual inspection · Deep learning · Convolutional Neural Network (CNN) · Transfer learning

1 Introduction

The reliable operation of an electricity distribution network is vital to a power company and this is ensured by regular asset maintenance and renewal. Condition-Based Risk Modelling (CBRM) is the tool adapted, to identify the assets with the highest risk of failure, hence to best allocate limited resources in the most beneficial way. The main objective of this study is to achieve automatic visual inspection of bare overhead copper conductors performed in asset replacement routine. Such a strategy allows the conductor's health condition to be assessed without de-energizing the lines and thus interfere with the normal operation of the grid.

The three major types of conductor degradation modes considered in this study are: (1) broken strands, (2) loss or discolouration of surface material, and

This project is funded by the R&D Fellowship Grants from New Zealand Callaghan Innovation, and is in partnership with and sponsored by Unison Networks Ltd. https://www.unison.co.nz/.

W. Q. Yan et al. (Eds.): IVCNZ 2022, LNCS 13836, pp. 175–187, 2023.
https://doi.org/10.1007/978-3-031-25825-1_13

(3) surface pitting. These degradation lead to the loss of tensile strength, and ultimately, conductor failure. Specific environmental factors such as elevated sulphur levels in geothermally active areas, [1] or marine climates exacerbate the conductor degradation. While manual inspection by suitably trained expert personnel can quickly assess the degradation level, this is time consuming, tedious and expensive. This work aims to address this.

The severity of degradation and hence remaining life expectancy is quantified from 1 to 5, (brand new to end of life), which is termed the *visual score*. It is one of the condition inputs to the CBRM model and is negatively correlated to the conductor's ultimate tensile strength [2]. Another similar scheme of correlating loss of tensile strength to a score is termed *health index* in [3]. We will use the literal form of the visual score interchangeably with the numeric form in this paper, to suit the presentation content.

This work is a component of a larger project where the conductor images are taken by a drone flying above the grid. Here we simplify the condition classification task by using images taken under a laboratory setting where the background and lighting are more closely controlled. The hypothesis we are testing is: Given a conductor image, an algorithm can automate the visual inspection process which evaluates its condition and outputs a visual score with respect to the descriptions listed in Table 1. This is a supervised learning problem and we have applied deep learning to approach it. The project team at the sponsoring lines company collected, curated, and labelled their own private dataset.

The remainder of the paper will be organized as follow: Sect. 2 reviews previous attempts of conductor condition recognition/classification using visual data. Section 3 introduces the image dataset we used in this work and the preprocessing steps applied before deep learning. Section 4 details the methods and experimental setup of the supervised learning task. Section 5 demonstrates the classification results with mis-classification analysis, and visualization including t-SNE [4] and Guided Grad-CAM [5], and finally, Sect. 6 summarizes the achievement we had so far and gives directions for future work.

2 Related Work

Given the economic importance of this asset management problem, it is no surprise that there has been some prior work on line inspection.

An early hand-crafted solution to detect arc marks and damaged conductors in [6] uses outlier detection where the mean surface pixel intensity/brightness and conductor border profile were the target features. The threshold α was chosen manually that represents the number of standard deviation away from mean statistics.

Huang et al. in [7] proposed a method to detect broken conductor and surface damage of the Aluminium-Conductor Steel-Reinforced (ACSR) cable, from images taken by a Line Suspended Robot (LSR). This method relies on the natural alternating brightness pattern of the strand winding. This pattern was binarized into square waves and by analysing the gap between two consecutive wave peaks, broken strands can be identified by assuming that the inner aluminium strands are darker than outer ones because of shading. The same authors later proposed another diagnosis model in [8], that focuses on the detection of loosen strands only. Their approach is model driven such that three strand features are extracted manually: the length, angle, and curvature. These features are then compared against that of a normal/ideal winding pattern to detect the anomaly.

Clearly a suspended robot that is physically close to the lines can extract a detailed image, but such a heavy and cumbersome suspended robot is unsuitable for the much lighter thinner distribution grid cables, so is not proposed for this application.

Song et al. in [9] also utilized the strand winding pattern to detect broken strands. The pattern is characterized by the Histogram of oriented Gradient (HoG) feature, then the HoG feature is classified into three classes: *normal wire*, *broken wire*, and *obstacles*, by two separate binary Support Vector Machine (SVM).

These work focus on a single type of conductor damages only while we are examine multiple types to evaluate a final score. Therefore, it could be difficult to design separate pipelines for each type alone. We can resort to the power of deep learning to automatically learn the mapping, given the images and their labels.

3 Dataset and Data Preparation

There are currently two major sets of image data, one from the field taken by a drone and another from a laboratory, the latter which is the dataset used in this study. It is worth noting that the sponsoring lines company is actively collecting new field data.

3.1 Laboratory Dataset

This project uses images of out-of-service conductors recovered from the field which were subsequently analysed for their mechanical properties in [2]. Approximately 2700 photos of conductor sections, taken at a length of approximately 7 cm, are used for the subsequent deep learning. These RGB photos were deliberately taken against a homogeneous background with consistent lighting conditions, (warm light and cold light), to reduce the influence of background on condition classification. The original photo size is 4608×2592 in pixels.

Table 1 illustrates some sample images with the corresponding visual score. The labelling of the conductor's visual image score was done by the maintenance experts employed by the sponsoring company. Within the industry it is well known, and evident from the subtle differences in Table 1, that this labelling is very subjective, and one of the unique characteristics of this project is that there is considerable variance even in the human expert assessors. Of the 5 listed classes, only 4, (classes 2 through 5), are of interest because the brand-new class 1 never appears in the geographic areas of interest. Consequently, this class can be removed from the data set.

Table 1. The visual score class descriptions and sample images

Class	Guidelines	Num. Images	Sample Images
1	**New** appearance	0	*Not considered in this study*
2	Partial, **light** discolouration	763	
3	**Major** discolouration **Light** dust type corrosion	684	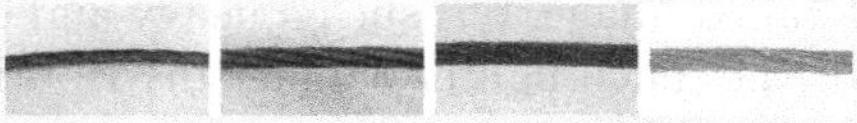
4	**Major** discolouration **Light** flaking type corrosion and/or **Minor** pitting and/or **One** localized broken strand	757	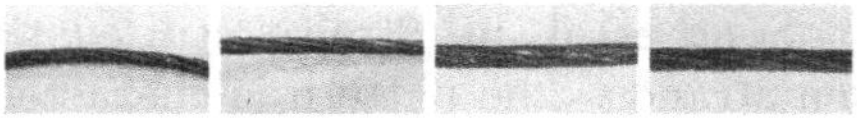
5	**Major** discolouration **Heavy** flaking type corrosion and/or **Heavy** pitting and/or **Multiple** broken strands	495	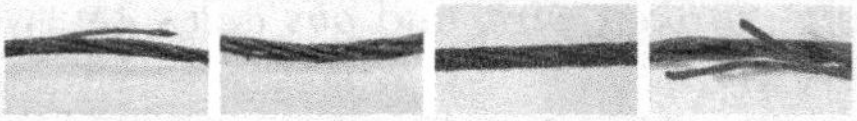

3.2 Data Preparation

The image is first segmented into the conductor foreground and the background, then the image is cropped using the foreground's bounding box, the background pixels are set to 0, this operation can be seen in Fig. 1. By doing this, we first filter out irrelevant information – the background, as it is not relating to conductor condition. Secondly, by cropping using the bounding box we effectively zoom in to the image and focus on the foreground only, this helps increase the information density as the image is now optimized to contain only the necessary information for condition recognition/classification. When the image is later down-sampled to a smaller size, information loss is reduced. However, as a prerequisite, it is critical to ensure a consistently successful foreground identification.

The training, validation, and test set were split in a 70%, 20% and 10% ratio. All images were shuffled before the split so to preserve the class distribution. The training and validation set were lastly combined after all the hyper-parameters were properly tuned, to train the network one last time.

4 Experimental Setup

As is common for these types of approaches, we use transfer learning to leverage the benefit of previously learned features and to speed up convergence, instead of designing a totally new CNN architecture and regress all the parameters from scratch. This allows us to train the network with a relatively small dataset while still achieving a good performance. Notwithstanding, we still need to modify the original network first by replacing the last fully-connected layer with a new one matching the class output size, 4 in this case. The standard multi-class cross-entropy loss function was used. The learn rate of this FC layer is also scaled up

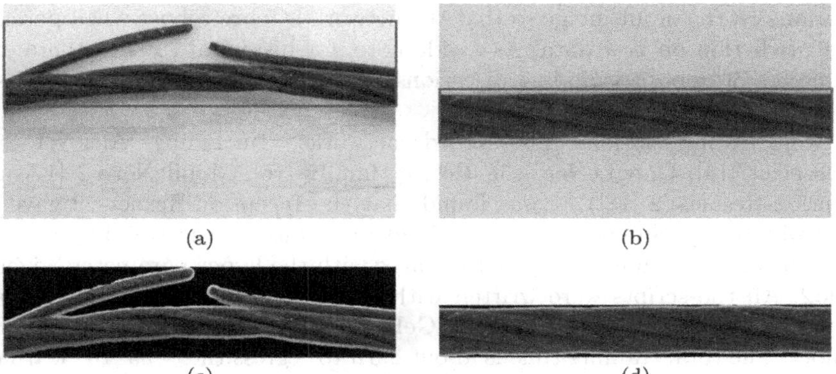

Fig. 1. Masking and cropping operation. (a–b) Original images in larger size and with background. (c–d) Masked and cropped images in smaller size and background set to 0.

by a factor of 5 to speed up the training, as it is initialized randomly with the Glorot initializer [10]. This modification can be seen in Fig. 2. We leave the CNN feature extractor unchanged and load all its layers with pre-trained parameters, the learn rate factor for these layers remains 1, so we are performing transfer learning with *fine tuning* through the entire network. The input image will be resized to 299(W) × 299(H) × 3(C) to match the original input size.

We did not use explicit l_2 regularization, but label smoothing instead with $\alpha = 0.05$. Label smoothing regularizes the model by injecting noise at the output target [11], thus to force the network to be less confident about the true labels to prevent overfitting as noted by [12]. It also models the mistakes in the ground truth labels, since in reality there is always some numbers of them in a dataset, and we will present later in Sect. 5. The original one-hot encoding target $y_k = \{1, 0, 0, 0\}$ is now transformed to

$$y_k^{\text{LS}} = y_k(1-\alpha) + \frac{\alpha}{K} = \{0.9625, 0.0125, 0.0125, 0.0125\}, \tag{1}$$

where K is the number of output classes [12]. The use of label smoothing is intuitive in our context, given that the class labels are not strictly mutually exclusive. Since the conductor naturally continuously degrades slowly, any boundaries are artificial, so the final decision on the ground truth is up to the subjective judgement of the human experts, and they might disagree with each other due to the difference in discernment of the qualitative descriptors *light*, *heavy*, *minor* and *major* from Table 1.

Image data augmentation was also applied to the combined training data set by introducing −15° to 15° of random rotation, 0.8 to 1.2 of random scaling, from a continuous uniform distribution within the intervals, and random x-axis reflection with 50% probability. This technique helps the network better generalize to unseen data and prevents overfitting. It simulates some of the possible variations on the input image so that the network is more robust when performing a prediction on new data. As a side note, for historical reasons there exist conductors of opposite winding directions so the random x-axis reflection is to simulate this variation.

After testing various pre-trained networks, (including AlexNet [13], SqueezeNet [14], GoogLeNet [15], ResNet family [16], MobileNet-v2 [17], and Inception-Resnet-v2 [18]), it was found that the Inception-Resnet-v2 gave the best validation accuracy after about 1520 iterations on the training set, each iteration contains a mini-batch of 16 images with the hyper-parameters given in Table 2. All the scripts were written with MATLAB® software packages, and the model was trained on a NVIDIA GeForce RTX 3080 GPU with 10 GB of memory. The total training time is about 8.5 h to regress the $\sim 54 \cdot 10^6$ learnable parameters.

Table 2. The hyper-parameters used for training the Inception-Resnet-v2 network.

Hyper-parameters	Value
Optimizer	Adam Gradient decay factor: 0.9 Square gradient decay factor: 0.999 $\epsilon = 10^{-8}$
Mini-batch Size	16
Number of Epochs	10
Image Augmentation	Random rotation: $[-15°, 15°]$ Random scale: $[0.8, 1.2]$ Random x-reflection
Initial Learn Rate	$lr_0 = 3 \cdot 10^{-5}$
Learn Rate Decay	$lr_t = lr_0/\sqrt{\text{epoch}}$
Data Shuffle	Every epoch
Label Smoothing, [12]	$\alpha = 0.05$

5 Results and Discussion

The result of the network training is shown in Fig. 3, giving both training and validation combined datasets. The faint orange line is the raw training accuracy data while the blue line is the smoothed moving average of the raw data across the last 60 points. The final smoothed training accuracy is 87.7%, the final training loss is 0.53. The raw data exhibits a large fluctuation due to the small mini-batch size.

The model's performance on test set is shown in the confusion chart in Fig. 4. The overall recognition accuracy is 83.02% (220 out of 265). In terms of F_1-score,

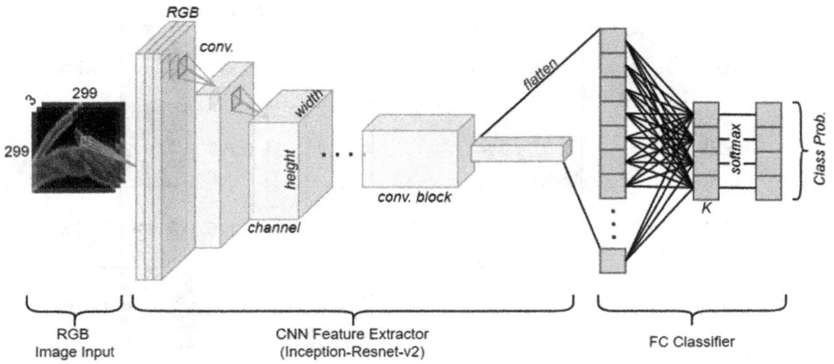

Fig. 2. General components of a deep CNN. It contains the image input, the feature extractor, and a classifier. In transfer learning, we replace the last fully-connected layer in the classifier with one matching the class output size which in this case is four visual score classes from 2 to 5, so $K = 4$.

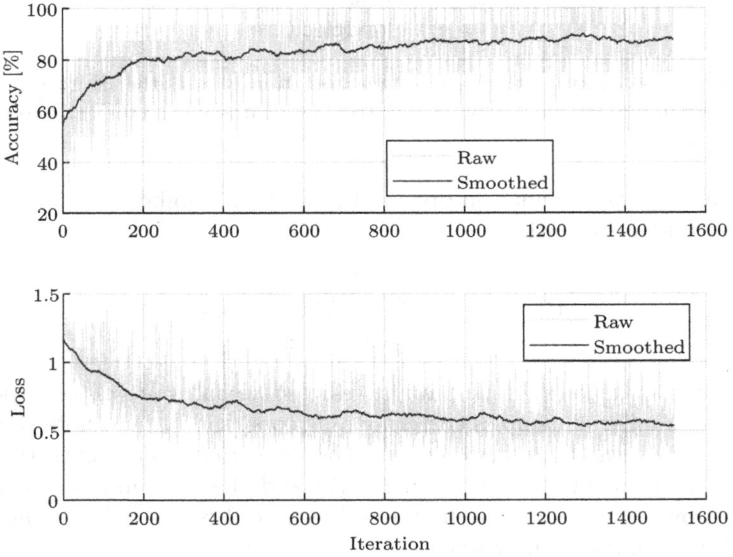

Fig. 3. Network training graph on the final combined training and validation set. The orange line is the raw data and the blue line is the smoothed moving average across last 60 data points. (Color figure online)

it performs the best on class 2 at 0.889, the highest recall is with class 3 at 94%, and the highest precision is with class 2 at 92.8%. One special metric of interest in this application is the Under-Estimation Rate (UER) since if the model were to mis-classify, we would rather prefer over-estimation to ensure that more of the severe/fatal conductor deterioration (class 5) are detected. The UER is defined as

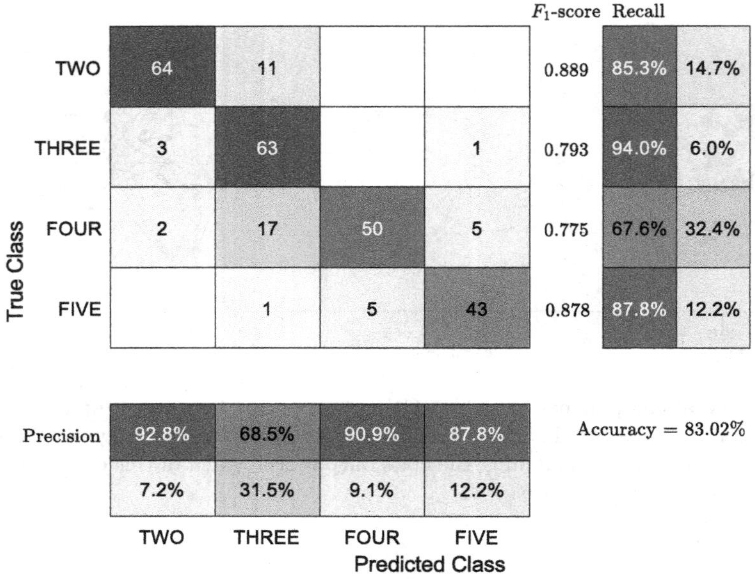

Fig. 4. Confusion chart of the test set classification result, the chart also displays precision, recall, and F_1-score. The overall test set accuracy is 83.02% (220 out of 265).

the number of under-estimation divided by the total number of samples subject to possible under-estimation

$$\text{UER} = \frac{\sum_{s_t=3}^{5} \sum_{s_p=2, s_p<s_t}^{5} n_{s_t,s_p}}{\sum_{s_t=3}^{5} \sum_{s_p=2}^{5} n_{s_t,s_p}}, \tag{2}$$

where s_t and s_p denote the true and predicted visual score class respectively, n_{s_t,s_p} is the number of samples corresponding to s_t and s_p, and note the $s_p < s_t$ condition indicating *under-estimation*. The numerator represents the entries below the diagonal in the confusion matrix and the denominator is the summation of rows apart from the first row, because 2 is the lightest class in the degradation/damage severity scale.

The UER for the test set is $28/190 = 14.74\%$ and ideally we want this measure to be as small as possible, alongside the standard accuracy measure. This rate can be reduced through biasing the network towards a higher class and one way to do it is to assign a larger weight to the higher class in the logit output, before the *softmax* operation, in those less confident predictions.

5.1 Assessing the Assessors

It is also interesting from a theoretical point of view, and helpful to our industrial partners, to analyze the mis-classified samples to better understand the

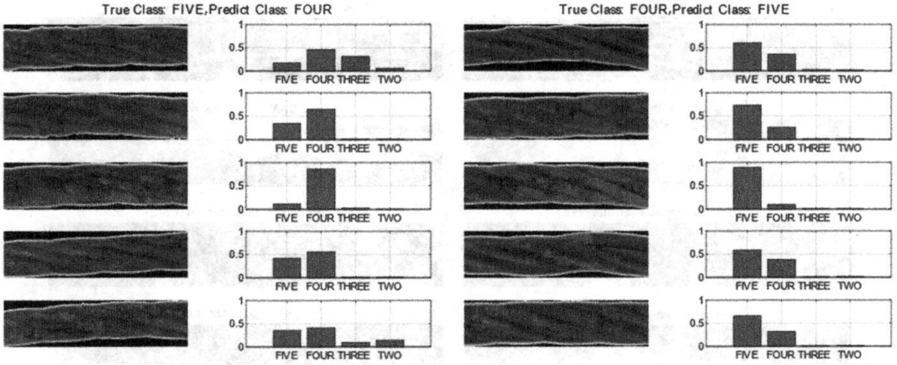

Fig. 5. Examples of mis-classification from the test set. Left: true class 5 mis-classified as 4, right: true class 4 mis-classified as 5, the bar graphs show the corresponding probability for each class. These are the hard/ambiguous examples the network mis-classifies. It may signals human error in ground truth as well.

network's failure modes or potentially highlight the human grading inconsistencies. Figure 5 shows 10 examples of mis-classified 4 and 5 classes as indicated by the confusion chart. It turns out, that these are the more difficult and ambiguous samples in the dataset, and if we carefully manually inspect at these images it is not hard to find evidence of suspected inevitable human error in the labelling process. Consider the pair of images at row 4, in both columns: the left one (labelled class 5) has less surface discolouration compare to the right one (labelled class 4) and its strand winding pattern is more regular too. This error, or if not, the ambiguity is evident from the similar class probabilities. Despite the left image is visually in compliance with class 4 but the possibly wrong label 5 drives the network much less confident about the prediction.

5.2 Inspecting the Network

Figure 6 illustrates two visualization methods which attempt to explain the network's prediction with concrete examples. Such diagnostics are particularly valuable and reinforce the opinions of the domain-experts currently used by the sponsoring industry.

Grad-CAM proposed by [5] produces an activation heatmap the same size as the spatial dimensions of the feature map in the last convolutional layer in the CNN feature extractor. In this case, it is the `conv_7b` from Inception-Resnet-v2 of size 8(W) $\times$ 8(H). The final heatmap is a weighted linear combination of all feature maps from all channels in this layer, the weight is the global-average-pooling of the gradient of the logit with respect to a feature map. Since it uses the last convolution layer, the heatmap is very coarse. To compensate this, we can produce another fine gradient attribution map with guided backpropagation [19] that only highlights pixels in the image that positively contribute to the gradient of the score with respect to the image. Now we have two activation

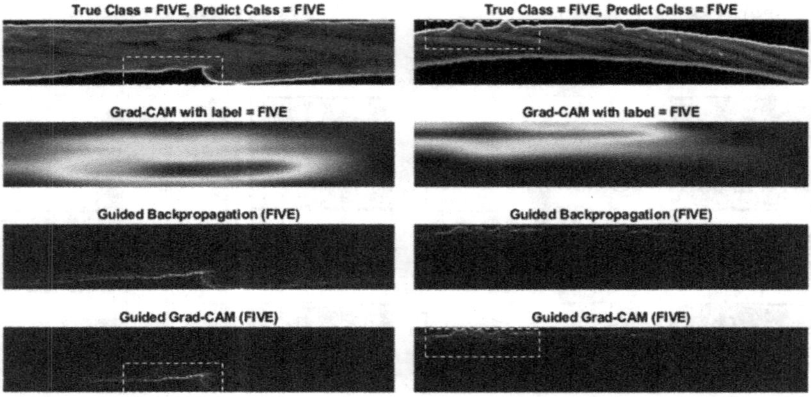

Fig. 6. Network prediction visualization. First row: original image, second row: Gradient-weighted Class Activation Mapping (Grad-CAM) [5], the dark red region of the map indicates high level of activation for the class 5, third row: Guided backpropagation gradient attribution map [19], fourth row: Guided Grad-CAM [5], the yellow dash box highlights the region that attribute the most to the prediction of class 5 for a particular image.

heatmaps from both the beginning in fine pixel level and the end in coarse feature level. These two heatmaps are then resized to match the original image size and element-wise multiplied to produce the final guided Grad-CAM image as how the authors did in [5].

Figure 6, shows not unsurprisingly, that the regions with a broken strand and surface pitting contribute the most to the prediction of class 5, which is in agreement with the class descriptions in Table 1. These two defects are the most visually appealing as well with a strong edge to the background.

5.3 Visualization Using t-SNE

The t-Distributed Stochastic Neighbor Embedding (t-SNE) from [4] is an alternative visualization method to better understand how the network transforms the input to make the prediction. This is a dimensionality reduction technique that is best suited for high-dimensional datasets which aim to preserve local structure of data distribution. Consequently, data points that are close together in the original high-dimensional space will stay close together in their lower dimensional embedding.

The top plot of Fig. 7 is from the input layer where we used the pixels directly on the gray-transformed resized image of size 299(W) $\times$ 299(H) $\times$ 1(C) giving a $299^2 = 89401$ dimensional data point. This plot displays no clear clusters of the visual score classes, despite the slight preponderance of purple points at the centre, orange points on the left, and blue points on the right. Clearly the defect characteristics that define the visual score are not strong enough in the 89401-dimensional space.

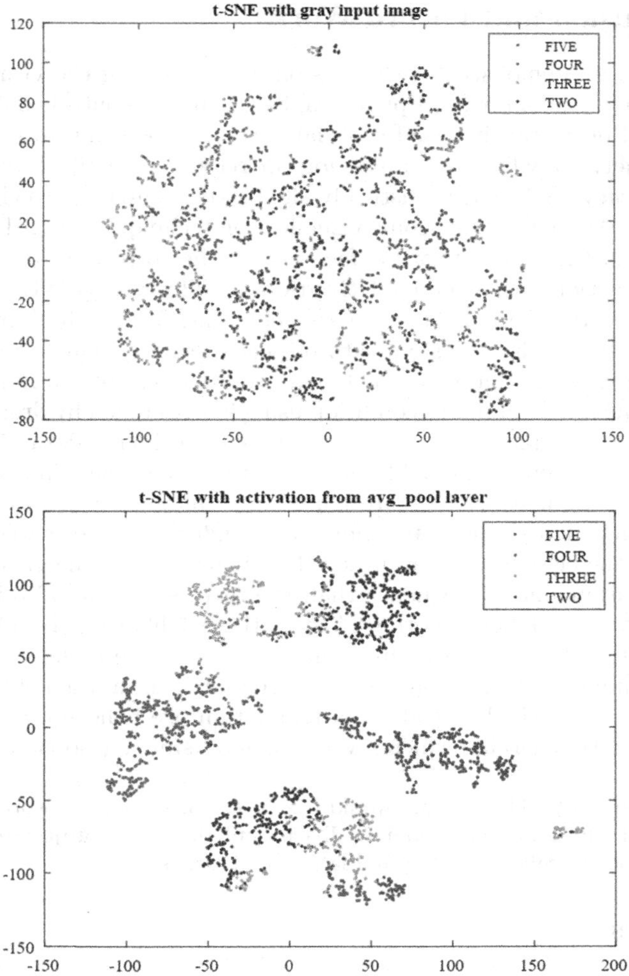

Fig. 7. 2D t-SNE visualization on training data, color labels are from the ground truth, each dot represents an image. (Top) result from the input level showing no meaningful clusters about the visual score classes. (Bottom) result from the `avg_pool` layer in Inception-Resnet-v2 showing clear clusters of different visual scores. The parameters used for generating these two plots are the same.

However, the bottom plot displays apparent clusters which are clearly highly correlated to each class. It is obtained from the `avg_pool` layer with feature vector of size $1(W) \times 1(H) \times 1536(C)$ giving a 1536-dimensional point. We can say that the network has learned the defect features that best separate the images into clusters according to the notion of visual score. The 1536-dimensional latent space representation of the image data is much more condensed, precise, and effective for predicting the visual score.

6 Conclusion and Future Work

In this paper, we demonstrated the feasibility of assessing the visual score of a piece of copper conductor from photographs taken in a controlled environment. We employed deep learning, and especially, transfer learning to benefit from a pre-trained network which is suitable for our relatively small dataset. The final test set accuracy 83.02% is statistically significant comparing to chance probability of 25%. We then showed some visualization techniques towards explainable AI and verified that the network is focusing on the correct region of the image to make predictions using Guided Grad-CAM. Through t-SNE, we saw that the network extracts useful latent space representation of the input that best separates the images into clusters with respect to their visual score labels.

This work was originally planned as a proof-of-concept study for the automated assessment of conductor condition using images taken by drones. In recent months, our industrial partners have expanded the image data set to include labelled images taken in the field and we will use these new images to continuously update the network. We are particularly interested in ways to bias the network so that it over-estimates more than under-estimate, without affecting too much on the classification accuracy. This is because an underestimate has a much higher economic consequence than the reverse. We are also interested in other ways to apply label smoothing just to the neighboring classes.

The result of this study is a milestone towards the overarching goal of recognizing the conductor's condition from its aerial image from the field. It is a more difficult task, but with this study as a foundation, we believe the experimental procedures are transferable which gives us a good starting point.

Acknowledgement. The authors would like to thank Unison Networks Limited for their technical supports and regular feedback, and the entire project team for their efforts and times to collect, curate, and label the datasets.

References

1. EPRI. Parameters that influence the aging and degradation of overhead conductors. Technical report, EPRI (2003)
2. Graham, M., et al.: Distribution overhead copper conductors, their condition and risk-based replacement. Electricity Engineers' Association (2021)
3. Naranpanawe, L., et al.: A practical health index for overhead conductors: experience from Australian distribution networks. IEEE Access **8**, 218863–218873 (2020). https://doi.org/10.1109/ACCESS.2020.3042486. ISSN 2169-3536
4. Van der Maaten, L., Hinton, G.: Visualizing data using t-SNE. J. Mach. Learn. Res. **9**(11) (2008)
5. Selvaraju, R.R., et al.: Grad-CAM: visual explanations from deep networks via gradient-based localization. In: Proceedings of the IEEE International Conference on Computer Vision, pp. 618–626 (2017)
6. Ishino, R., Tsutsumi, F.: Detection system of damaged cables using video obtained from an aerial inspection of transmission lines. In: IEEE Power Engineering Society General Meeting, vol. 2, pp. 1857–1862 (2004). https://doi.org/10.1109/PES.2004.1373201

7. Zhang, Y., et al.: A recognition technology of transmission lines conductor break and surface damage based on aerial image. IEEE Access **7**, 59022–59036 (2019). https://doi.org/10.1109/ACCESS.2019.2914766. ISSN 2169-3536
8. Huang, X., et al.: A method of transmission conductor-loosened detect based on image sensors. IEEE Trans. Instrum. Meas. **69**(11), 8783–8796 (2020). https://doi.org/10.1109/TIM.2020.2994475
9. Song, Y., Wang, H., Zhang, J.: A vision-based broken strand detection method for a power-line maintenance robot. IEEE Trans. Power Deliv. **29**(5), 2154–2161 (2014). https://doi.org/10.1109/TPWRD.2014.2328572. ISSN 1937-4208
10. Glorot, X., Bengio, Y.: Understanding the difficulty of training deep feedforward neural networks. In: AISTATS. JMLR Proceedings, vol. 9, pp. 249–256. JMLR.org (2010)
11. Goodfellow, I., Bengio, Y., Courville, A.: Deep Learning. MIT Press, Cambridge (2016). https://www.deeplearningbook.org
12. Müller, R., Kornblith, S., Hinton, G.E.: When does label smoothing help? In: Advances in Neural Information Processing Systems, vol. 32 (2019)
13. Krizhevsky, A., Sutskever, I., Hinton, G.E.: Imagenet classification with deep convolutional neural networks. In: Proceedings of the 25th International Conference on Neural Information Processing Systems - Volume 1, NIPS 2012, Lake Tahoe, Nevada, pp. 1097–1105. Curran Associates Inc. (2012)
14. Iandola, F.N., et al.: Squeezenet: Alexnet-level accuracy with 50x fewer parameters and <0.5 MB model size (2016). https://doi.org/10.48550/ARXIV.1602.07360
15. Szegedy, C., et al.: Going deeper with convolutions. In: 2015 IEEE Conference on Computer Vision and Pattern Recognition (CVPR), pp. 1–9 (2015). https://doi.org/10.1109/CVPR.2015.7298594
16. He, K., et al.: Deep residual learning for image recognition. In: 2016 IEEE Conference on Computer Vision and Pattern Recognition (CVPR), pp. 770–778 (2016). https://doi.org/10.1109/CVPR.2016.90
17. Sandler, M., et al.: Mobilenetv 2: inverted residuals and linear bottlenecks. In: Proceedings of the IEEE Conference on Computer Vision and Pattern Recognition, pp. 4510–4520 (2018)
18. Szegedy, C., et al.: Inception-v4, inception-resnet and the impact of residual connections on learning. In: Thirty-First AAAI Conference on Artificial Intelligence (2017)
19. Springenberg, J.T., et al.: Striving for simplicity: the all convolutional net. In: Bengio, Y., LeCun, Y. (eds.) 3rd International Conference on Learning Representations, ICLR 2015, San Diego, CA, USA, 7–9 May 2015, Workshop Track Proceedings (2015). https://arxiv.org/abs/1412.6806

Evolving U-Nets Using Genetic Programming for Tree Crown Segmentation

Wenlong Fu[1](✉), Bing Xue[1], Mengjie Zhang[1], and Jan Schindler[2]

[1] Victoria University of Wellington, PO Box 600, Wellington 6140, New Zealand
wenlong.fu@gmail.com

[2] Manaaki Whenua-Landcare Research, Wellington 6011, New Zealand

Abstract. The U-Net deep learning algorithm and its variants have been developed for biomedical image segmentation, and due to their success gained popularity in other science domains including remote sensing. So far no U-Net structure has been specifically designed to segment complex tree canopies from aerial imagery. In this paper, a handcrafted convolutional block is introduced to replace the raw convolutional block used in the standard U-Net structure. Furthermore, we proposed a Genetic Programming (GP) approach to evolving convolutional blocks used in the U-Net structure. The experimental results on a tree crown dataset show that both the handcrafted block and the GP evolved blocks have better segmentation results than the standard U-Net. Additionally, the U-Net using the proposed handcrafted blocks has fewer numbers of the learning parameters than the standard U-Net. Also, the proposed GP approach can evolve convolutional blocks used in U-Nets that perform better than the handcrafted U-Net and the standard U-Net, and can also achieve automation.

Keywords: Genetic programming · U-Net · Image segmentation · Convolutional neural network

1 Introduction

Image segmentation is a typical process to extract objects in images. Being different from classical image segmentation methods, such as clustering, deep convolutional neural networks have been developed successfully for image segmentation in recent years [2,9]. The U-Net, as a convolutional neural network (CNN), was proposed for biomedical image segmentation [15]. Different from other deep learning algorithms with a larger number of learning parameters and requiring more and more training images, the U-Net can achieve good performance on the biomedical image segmentation only using a limited number of

Funded by the New Zealand Ministry of Business, Innovation and Employment under contract C09X1923 (Catalyst: Strategic Fund).

W. Q. Yan et al. (Eds.): IVCNZ 2022, LNCS 13836, pp. 188–201, 2023.
https://doi.org/10.1007/978-3-031-25825-1_14

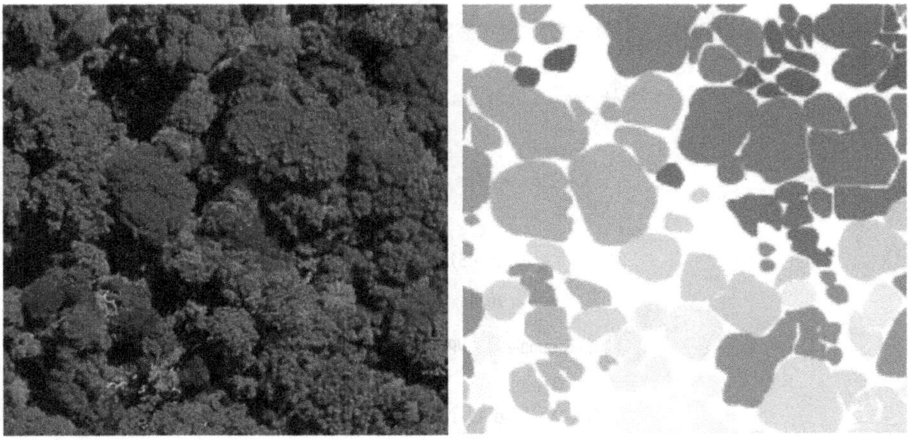

Fig. 1. An example of a tree crown image and its ground truth. (Color figure online)

training images [15]. Also, the U-Net variants have been recommended for further improving medical image segmentation performance [4,13]. However, when the U-Net technique is applied to other images with more complicated objects and background, improvements are needed to achieve promising segmentation results [12].

Tree crown segmentation from aerial imagery is helpful for informing a wide range of applications, including nature conservation efforts, urban planning, and forest management. These images include trees on land from different places, such as a city center, a park, etc. This paper uses a labelled tree crown dataset that includes images from Wellington, New Zealand. Figure 1 shows one example of tree image and its ground truth (tree segmentation using different gray-scale levels). The tree crown image has more complicated backgrounds than biomedical images. For instance, it is not easy to distinguish small green plants (as background) and green trees. In order to apply U-Nets more effectively to tree crown image segmentation, further improvement on the U-Net method are needed. In this paper, a handcrafted convolutional block is proposed to replace the raw convolutional block used in the U-Net for more effectively extracting features to obtain better segmentation results than the U-Net.

However, handcrafting neural network architectures is expensive and requires robust domain knowledge. It will be very helpful to design an automatic approach to search for a convolutional block without the need of (strong) domain knowledge. Evolutionary deep learning techniques, which require no or very little specific domain knowledge, have been proposed for neural network design [19,20]. Genetic Programming (GP) was successfully used to evolve CNN architectures for image classification tasks [16]. GP has first-class potential to evolve CNN blocks in the U-Net structure for the tree crown segmentation task.

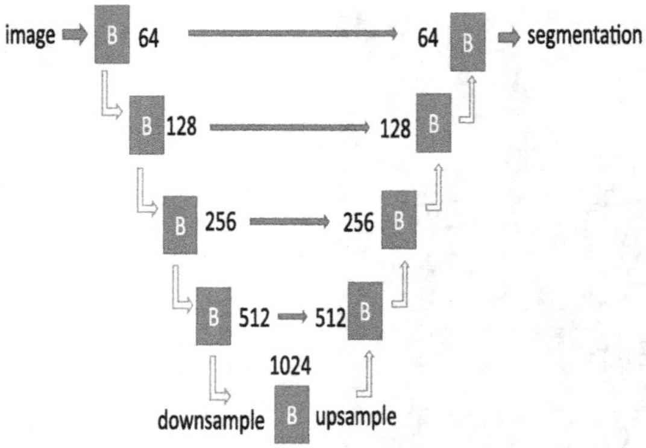

Fig. 2. The U-Net structure.

The goal of this paper is to propose a GP technique to evolve CNN blocks used in the U-Net structure for the tree crown segmentation task. Specifically, a new CNN block is firstly manually designed to see if it can achieve better performance than standard U-Net method. Then a new GP method is designed to automatically explore CNN blocks without any manual effort. The tree crown segmentation task is utilized to check whether the proposed handcrafted CNN block and the GP evolved CNN blocks can be used to obtain better segmentation results than the standard U-Net. Furthermore, to reasonably explain the GP evolved results, the GP evolved CNN blocks are analyzed.

2 Background

2.1 U-Net for Image Segmentation

The U-Net technique was proposed for biomedical images segmentation [15]. The "U" shape structure, shown in Fig. 2, has two parts: the left part employs maximum pooling operations to perform down-sampling, the right part performs up-sampling operations to get a predicted segmentation result that has the same size as the input image. Each layer combines the previous outputs from the up-sampling and the current outputs from the left part. Each block "B" in Fig. 2 has two convolution operators and a bath normalization operation to extract features. More details can be found in [15].

To effectively suppress irrelevant regions in an image, an attention gate block is proposed [13]. From the results on the two large CT abdominal datasets used in [13], the proposed attention U-Net outperforms the standard U-Net. Figure 3 shows the attention U-Net structure. The attention gate block "A" combines the outputs from the down-sampling and the up-sampling by the learned weights of the attention gate.

To utilize the outputs from the down-sampling and up-sampling effectively, more complicated network structures have been developed, such as UNet++ [21].

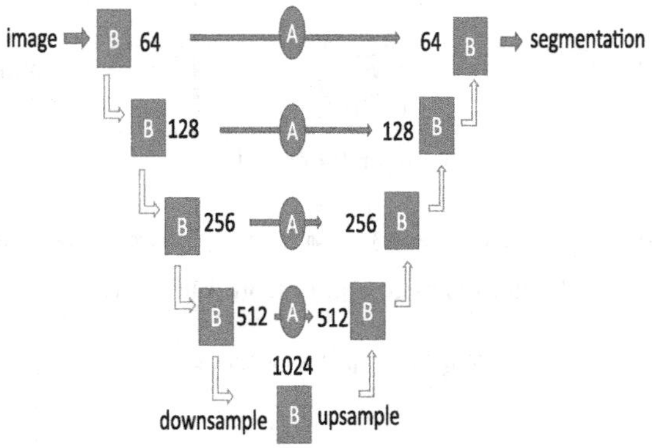

Fig. 3. The attention U-Net structure.

UNet++ borrowed the idea that convolution layers were used to bridge the outputs between the down-sampling and up-sampling operations in the Dense block [7]. Also, more complicated block structures are proposed to effectively extract features, such as the broad U-Net [4]. The broad U-Net utilized a multi-scale feature convolution block to replace the block used in the raw U-Net. The attention technique is also used to combine the outputs from the down-sampling and the up-sampling. The results from the broad U-Net shows that it is effective to use multi-scale-based CNN blocks in the U-Net structure to improve segmentation performance.

2.2 Evolutionary Automated Design of Deep Neural Networks

Besides manually designed novel CNN blocks, network architecture search algorithms [14] can automatically search for promising neural network architectures. Evolutionary deep learning techniques [19,20] have the ability to search for effective networks without the need for specific domain knowledge. Particle Swarm Optimization and Genetic Algorithm have been applied to search for CNN blocks for image classification [3,17]. A Cartesian GP [16] evolved whole network architectures for image classification tasks. More functions can be found in [10,19,20]. These existing functions motivated us to design an evolutionary deep learning method to automatically search for neural network architectures for image segmentation tasks.

3 The Proposed Approaches

This section describes the proposed handcrafted CNN block, called Layer block, used in the U-Net structure. The proposed GP system based on the standard U-Net structure is also presented. It is noted that the handcrafted CNN block and the GP proposed system are independent.

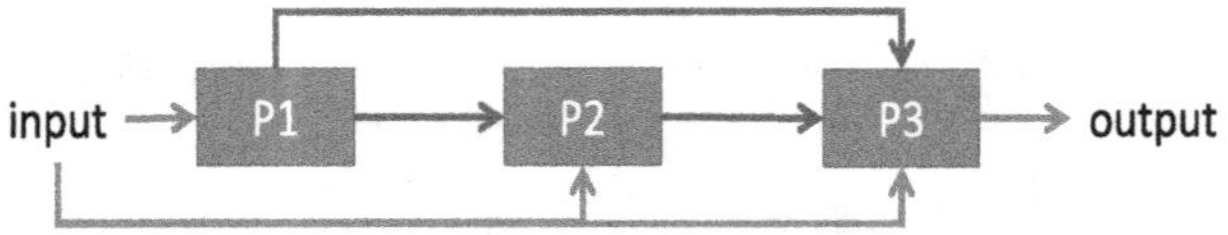

(a) layer block ("L").

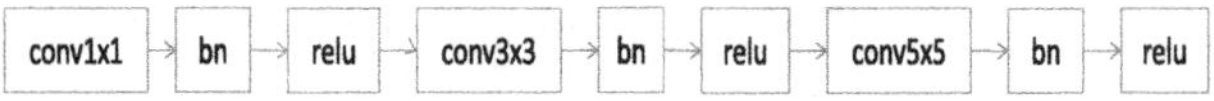

(b) details of one part from block "L".

Fig. 4. The layer block.

3.1 Layer U-Net

To obtain rich features from images, the broad U-Net [4] includes different convolution operators. The deep layer aggregation [18], which is used for image classification and is not one of the U-Net methods, also extracts features from different convolution operators. To extract features effectively from images, a layer CNN block ("L") is proposed (see Fig. 4(a)) that is inspired by the deep layer aggregation [18] and the Dense block [7]. The layer block includes three parts that have the same network structure. The second part "P2" takes the output of the first part "P1" and the input as its ("P2") input, The third part "P3" takes the output of the first part ("P1"), the output of the second part ("P2"), and the input (outside of the whole block) as its ("P3") input to get a final output. In this structure, different level outputs are considered to enrich features. The structure of each part is shown in Fig. 4(b). Here, "bn" is the batch normalization, and "relu" is the Rectified Linear Units [1]. Three different kernel sizes ($conv1 \times 1$, $conv3 \times 3$, and $conv5 \times 5$) are used in the block to extract rich features based on different scales.

Figure 5 shows the proposed layer U-Net structure using the layer block ("L") from Fig. 4(a). Different from the U-Net structure in Fig. 2, the layer U-Net has one fewer layer than the U-Net structure in order to reduce the computational cost. The number of starting input channels in the layer U-Net are less than the 64 used in the standard U-Net. It is noted that the part "P3" from the block "L" is connected to the input of the block "L", the output of the part "P1", and the output of the part "P2" as its ("P3") input. Therefore, the internal computational cost is higher than a block only having a single connection among its convolution operators. The attention gate is also used effectively to combine outputs from the down-sampling and the up-sampling in the proposed layer U-Net.

3.2 GP U-Net

An existing GP approach [16] evolved the whole network architectures for image classification tasks. However, the computational cost is very high. Rather than using a Cartesian GP [16] to evolve a whole network architecture, this paper

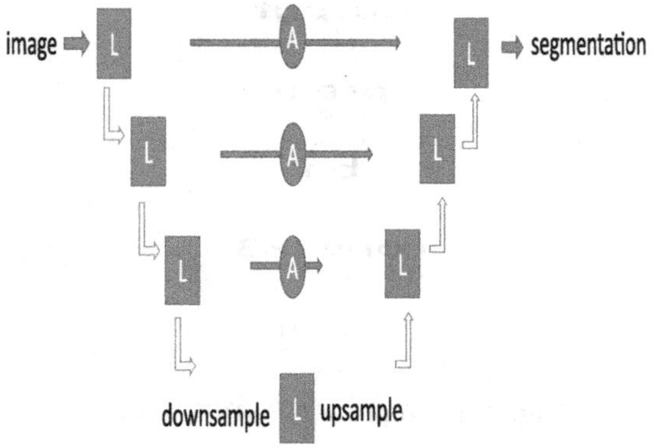

Fig. 5. The layer U-Net structure.

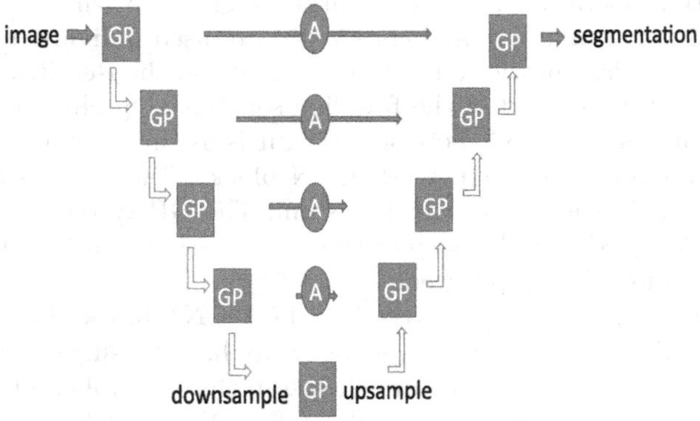

Fig. 6. The GP U-Net structure.

proposes a tree-based GP to evolve blocks used in the fixed U-Net structure. Figure 6 shows the GP U-Net structure. A block (GP tree) will be evolved by the tree-based GP approach and will be used in each layer. It is noted that the same architecture of the GP evolved CNN block will be used in the proposed GP U-Net from Fig. 6.

The proposed GP system includes the terminal set and the function set. The terminal set only includes the input of each block used in Fig. 6. The function set includes different functions and a typical CNN block, namely a basic block from the residual network [5] that has been widely applied to computer vision tasks [19]. The basic block ("basic") from the residual network is employed in the function set. The basic block includes a convolution with kernel size 3, a batch normalization function, another convolution operator with kernel size 3, another batch normalization function, and a short cut that includes a convolution operator with kernel size 1 and a batch normalization function. For the normalization process, the batch normalization function ("bn") and the local

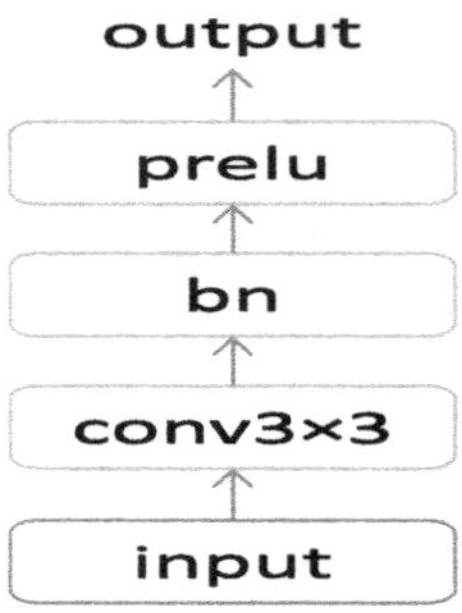

Fig. 7. Example of GP CNN block.

response normalization function ("lrn") from the ImageNet [8] are chosen. For basic convolution operations, the convolution operators with the kernel sizes 1, 3 and 5 ($conv1 \times 1$, $conv3 \times 3$, and $conv5 \times 5$) are used individually. For activation functions, the Parametric ReLU ("prelu") [6] and the Rectified Linear Units ("relu") [1] are also included in the function set. Since "prelu" and "relu" have been successfully used in CNN networks [19], it is assumed that both activation functions are reliable to construct GP CNN blocks. Table 1 lists the terminals and the functions in the proposed GP system. The GP system does not choose the manually designed layer bock since the computational cost of the layer block is high and the block is complicated.

Figure 7 shows an example of a tree-based GP CNN block. Since the function set only has single inputs, each evolved tree only has one single node. The functions $conv3 \times 3$, "bn", and "prelu" are chosen in this example and are combined as a block that will be used to construct a GP U-Net model.

When a single CNN block is created in the GP system as an individual, the single CNN block is used to construct a GP U-Net. The constructed GP U-Net will be trained with limited epochs, a validation data will be then used to measure the trained GP U-Net, and the measured value will be used as the fitness value of the GP individual (the single CNN block). Here, Intersection over Union (IoU) is chosen for segmentation performance measurement, since IoU is good for overlap measure [11]. IoU is defined in Eq. 1, where P is the predicted segmentation result, G is the ground truth, $|P \cap G|$ is the overlap area between P and G, and $|P \cup G|$ is the total area including P and G.

$$IoU = \frac{|P \cap G|}{|P \cup G|} \tag{1}$$

It is noted that the proposed GP system evolves CNN blocks with restricted inputs and outputs so that the GP evolved CNN block can be easily fitted into the U-Net structure. The existing GP approach [16] may have an invalid neural network, such as non-matched inputs for a concatenating function that requires the same shape or empty of its inputs. The proposed GP system have more concise structures and they will be always validated in the U-Net structure. The

Table 1. The terminal set and the function set used in the GP system.

Set	Name
Terminal	Input
Function (normalization)	*bn*, *lrn*
Function (activation)	*prelu*, *relu*
Function (convolution)	*basic*, *conv*1 × 1, *conv*3 × 3, *conv*5 × 5

Algorithm 1. GP U-Net.

Require: maximum generation g_m, training data d_{train}, validation data d_{val}, maximum training epochs e_m

1: initial population, set the current generation $g_i = 0$;
2: increase the current generation $g_i = g_i + 1$;
3: use each individual to train a model on d_{train} with e_m epochs, and evaluate each trained model on d_{val};
4: if $g_i > g_m$, output the best individual as the solution;
5: apply the mutation and crossover operations to generate new population;
6: replace the current population by the new population and the current elitism, then go to step 2.

GP algorithm is defined in Algorithm 1. A training dataset and a validation dataset are required.

4 Design of the Experiment

The task is to extract trees from different images. These trees come from different places in Wellington, New Zealand, such as the city center, and parks. Figure 8(a) shows two examples of training images from a residential area and a non-residential area. Figure 8(b) show their ground truth of the segmented trees, which are represented by different gray-scales. In general, trees from residential areas are easier than non-residential areas to be extracted since residential areas include buildings that are obviously different from trees. However, it is not easy to distinguish between green grasses or small green plants from both areas and trees. The tree crown dataset includes 983 training images, 164 validation images, and 492 test images, respectively. Each image includes 512 × 512 pixels. In this paper, 256 × 256 is used for computational cost and there are no obviously different segmentation results between using 256 × 256 and 512 × 512. Different from the standard U-Net using 64 channels (see Fig. 6), the proposed layer U-Net starts the block with 16 channels or 32 channels. Since the third block in Fig. 4 has large numbers of input channels when a large number of start input channels (as input) are used, it is expected that 32 channels would be enough to extract tree features. Therefore, the number of channels is smaller than 64. All models are trained on a NVIDIA GeForce RTX 3070 Laptop GPU (8.0GB).

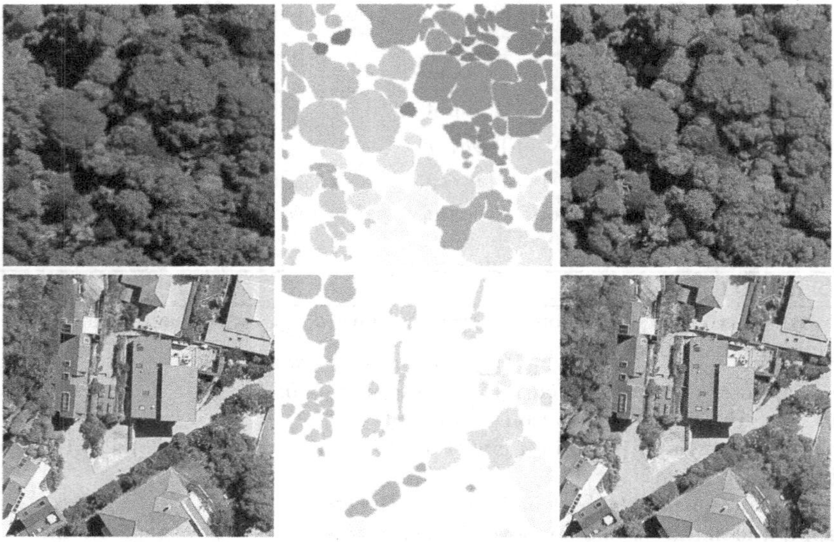

(a) training images (b) ground truth (c) highlight (Blue)

Fig. 8. Examples of training images and their ground truth. (Color figure online)

When the standard U-Net or one of its variants is trained, the loss function is the cross-entropy function [15]. The maximum epoch is 200. For GP, the population size is 15 and the maximum generations are 15. The probabilities for elitism (reproduction), mutation, and crossover are 0.05, 0.35, and 0.6, respectively. Since the population size is small, the high mutation probability 0.35 is chosen empirically. Because the population size is 15 and the selected number of the top best individuals is less than one ($0.05 \times 15 = 0.75$), Only the best individual (for reproduction) is copied to the next population. Each GP individual is used to construct a GP U-Net that will be trained by 10 epochs; the trained model will be evaluated by IoU as its fitness value.

5 Results and Discussions

Table 2 shows the test results from the standard U-Net model, the attention U-Net model, the proposed layer U-Net model with two parameters (16 and 32), and three GP U-Net models. The number of learning parameters and the training hours from these models are also given. The average time of a GP process is around 19 h. The three GP trees (CNN blocks) shown in Fig. 9 are used to construct the GP U-Net models independently.

From the table, four interesting observations can be made. First, the proposed layer U-Net model (16) has the smallest number of the learning parameters (4.3 million), the test performance IoU is similar to the standard U-Net model. It is noted that the training time difference between the layer U-Net model (16)

Table 2. Test results (*IoU*), number of parameters (million), and the training hours from different methods.

Method	*IoU*	Parameters (M)	Training time (hours)
U-Net	0.80	34.5	3.8
Attention U-Net	0.80	34.8	4.0
Layer U-Net (16)	0.80	**4.3**	**3.2**
Layer U-Net (32)	0.82	17.1	6.4
GP U-Net 1	0.82	19.2	4.9
GP U-Net 2	**0.83**	36.2	5.5
GP U-Net 3	**0.83**	36.2	5.7

and the U-Net model is not large. This is probably because the structure in the proposed layer U-Net is more complicated than the standard U-Net, although the layer U-Net model (16) has a smaller number of the learning parameters. Second, the layer U-Net model (32) improves the test performance via increasing the input channels. It is noted that the number of the parameters of the layer U-Net model (32) is two times larger than that of the layer U-Net model (16). Third, the GP U-Net (models 1, 2, and 3) have better test results (*IoU*) than the standard U-Net. The numbers of the learning parameters of the GP U-Net models are in a large range. Although the GP U-Net models 2 and 3 have the highest *IoU* values among all the models, the numbers of the learning parameters of the GP U-Net models 2 and 3 are larger than the other models. Last, the test performances of the standard U-Net model and the attention U-Net model are very close. From their comparison, it seems that the attention gate does not always improve the segmentation results, such as the tree crown dataset.

Figure 9 provides the details of the GP evolved CNN blocks. It is surprising that the GP evolved CNN block 1 ("GP1") only has one convolution operator with the kernel size 3. Although the GP evolved CNN block 1 is simple, the relevantly trained GP U-Net has a similar test performance to the standard U-Net model. For GP evolved CNN blocks 2 and 3 ("GP2" and "GP3"), the basic block is chosen. The only difference is that the "GP2" has an extra "prelu" function compared with "GP3". Since the batch normalization ("bn") is used after the basic block function, the "prelu" function may not have a strong influence on the trained model. The results of Table 2 show that the test performances of both models are very close. Both models (GP U-Net 2 and 3) have better results than the GP U-Net 1.

It is noted that the basic block is easily chosen by GP. The GP evolved CNN blocks are evaluated based on their trained models after 10 epochs. How to find potential better models but not using large numbers of epochs is a future research direction.

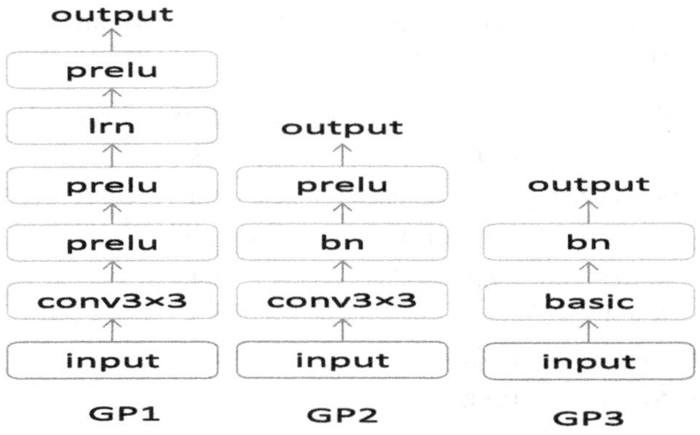

Fig. 9. The GP evolved trees (CNN blocks).

Figure 10 presents three examples of test images and the segmentation results from the ground truth, (the standard) U-Net, and the GP U-Net model 3. Two images come from the residential areas, and one from a non-residential area with many trees. There are three interesting results. First, in most cases the GP U-Net has better ability to extract trees than U-Net on small areas. The first segmentation result on the first (left) image from U-Net misses smaller trees at the middle part of the test image, which is marked by a circle. For the second (middle) image, on the middle area marked by a circle, U-Net only extracts around half of the tree segmentation. The GP U-Net segments most of the tree areas. Also, on the left bottom area, marked by a rectangle, U-Net misses one tree segmentation result, and provides only a partial result for another; however, the GP U-Net has both tree segmentation results. Second, when these trees are very close the GP U-Net and U-Net have issues marking tree segmentation as a whole. For instance, the second (middle) tree image has crowded trees on the middle top area marked by a circle. However, both models have connected segmentation results. It is possible that only the image of 256×256 pixels is used. Future work could include how to make a separated segmentation result is another. Last, the segmentation results from the GP U-Net are still required for improvement. In the top area of the first image, marked by a rectangle, the U-Net finds the right small tree area, but GP U-Net misses this area. Also, the GP U-Net incorrectly predicts small plants in the middle of the second image. In summary, the visual comparisons show that the GP U-Net has better segmentation results than U-Net.

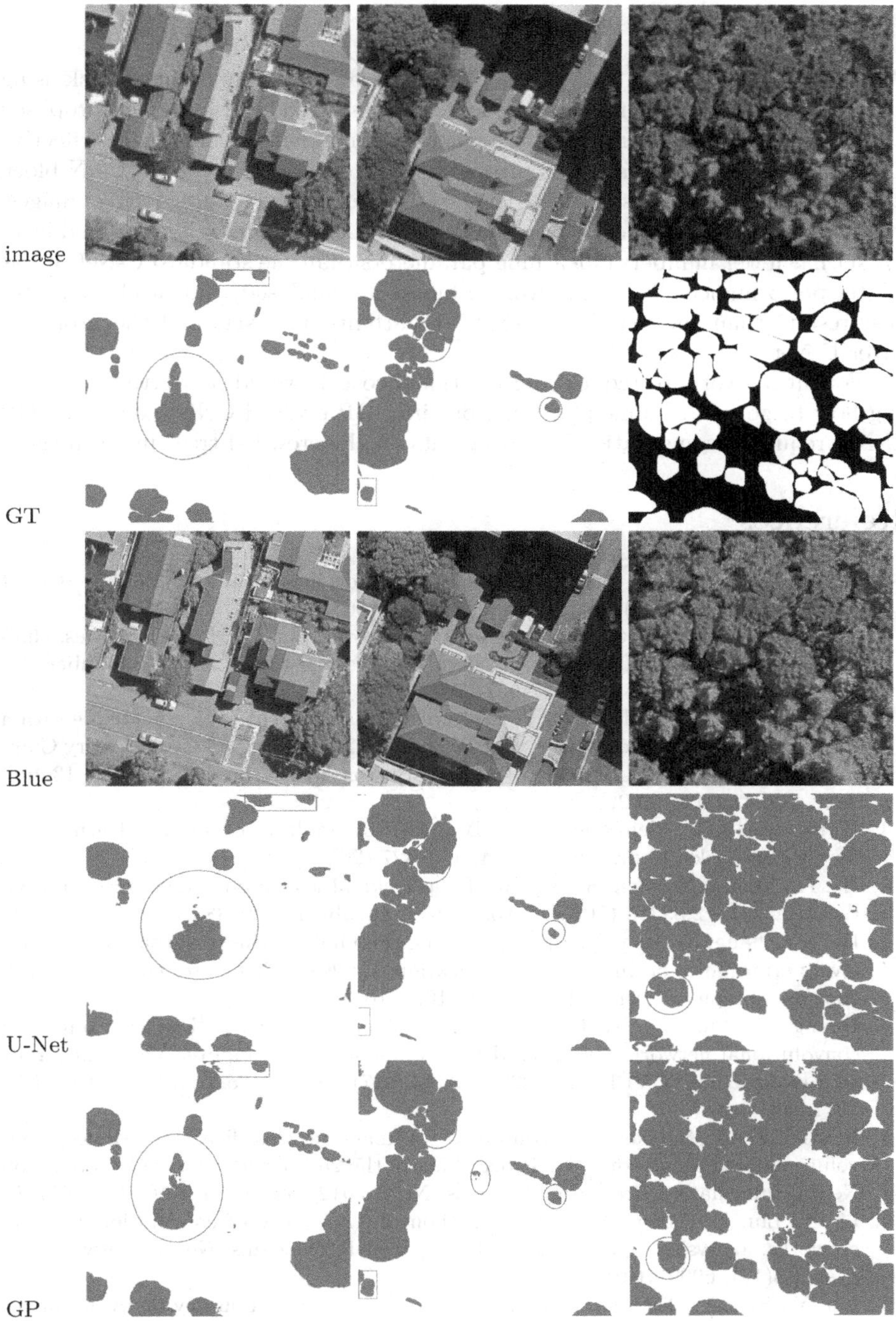

Fig. 10. Examples of test images, their ground truth (Gt) and the results from the U-Net and GP U-Net 3. Here, "Blue" highlights the ground truth on the test images. (Color figure online)

6 Conclusions

This paper conducted an investigation on the tree crown segmentation task using the U-Net structure. To improve segmentation results, a layer block was proposed to be used in U-Net. Furthermore, rather than manually designing an effective CNN block, a GP approach was proposed to automatically evolve a CNN block to be used in the U-Net structure. From the experimental results on tree images, the GP evolved blocks outperform the standard U-Net. While the proposed layer U-Net has fewer numbers of learning parameters than the standard U-Net, it has better performance. Three GP evolved blocks are analysed, and two have better test results than the standard U-Net, the attention U-Net, and the proposed layer U-Net.

For future work, to reduce computational cost, it would be worth investigating how to get rid of unhelpful functions in a GP evolved CNN block. The GP U-Net requires segmentation improvements on the crowded trees in an image.

References

1. Agarap, A.F.: Deep learning using rectified linear units (ReLU). CoRR abs/1803.08375 (2018). http://arxiv.org/abs/1803.08375
2. Alzubaidi, L., et al.: Review of deep learning: concepts, CNN architectures, challenges, applications, future directions. J. Big Data **8** (2021). Article number: 53. https://doi.org/10.1186/s40537-021-00444-8
3. Deng, S., Sun, Y., Galván, E.: Neural architecture search using genetic algorithm for facial expression recognition. In: GECCO 2022: Genetic and Evolutionary Computation Conference, Companion Volume, Boston, Massachusetts, USA, 9–13 July 2022, pp. 423–426 (2022)
4. Fernández, J.G., Mehrkanoon, S.: Broad-UNet: multi-scale feature learning for nowcasting tasks. Neural Netw. **144**, 419–427 (2021)
5. He, K., Zhang, X., Ren, S., Sun, J.: Deep residual learning for image recognition. CoRR abs/1512.03385 (2015). http://arxiv.org/abs/1512.03385
6. He, K., Zhang, X., Ren, S., Sun, J.: Delving deep into rectifiers: surpassing human-level performance on ImageNet classification. In: 2015 IEEE International Conference on Computer Vision (ICCV), pp. 1026–1034 (2015)
7. Huang, G., Liu, Z., Van Der Maaten, L., Weinberger, K.Q.: Densely connected convolutional networks. In: 2017 IEEE Conference on Computer Vision and Pattern Recognition (CVPR), pp. 2261–2269 (2017). https://doi.org/10.1109/CVPR.2017.243
8. Krizhevsky, A., Sutskever, I., Hinton, G.E.: ImageNet classification with deep convolutional neural networks. In: Proceedings of the 25th International Conference on Neural Information Processing Systems, NIPS 2012, vol. 1, pp. 1097–1105 (2012)
9. Li, Z., Liu, F., Yang, W., Peng, S., Zhou, J.: A survey of convolutional neural networks: analysis, applications, and prospects. IEEE Trans. Neural Netw. Learn. Syst. **33**(12), 6999–7019 (2022)
10. Liu, Y., Sun, Y., Xue, B., Zhang, M., Yen, G., Tan, K.: A survey on evolutionary neural architecture search. IEEE Trans. Neural Netw. Learn. Syst. (2021). https://doi.org/10.1109/TNNLS.2021.3100554

11. Minaee, S., Boykov, Y., Porikli, F., Plaza, A., Kehtarnavaz, N., Terzopoulos, D.: Image segmentation using deep learning: a survey. IEEE Trans. Pattern Anal. Mach. Intell. **44**(7), 3523–3542 (2022)
12. Mo, Y., Wu, Y., Yang, X., Liu, F., Liao, Y.: Review the state-of-the-art technologies of semantic segmentation based on deep learning. Neurocomputing **493**, 626–646 (2022)
13. Oktay, O., et al.: Attention U-Net: learning where to look for the pancreas. In: Medical Imaging with Deep Learning (2018). https://openreview.net/forum?id=Skft7cijM
14. Ren, P., et al.: A comprehensive survey of neural architecture search: challenges and solutions. ACM Comput. Surv. **54**(4), 1–34 (2021)
15. Ronneberger, O., Fischer, P., Brox, T.: U-Net: convolutional networks for biomedical image segmentation. In: Navab, N., Hornegger, J., Wells, W.M., Frangi, A.F. (eds.) MICCAI 2015. LNCS, vol. 9351, pp. 234–241. Springer, Cham (2015). https://doi.org/10.1007/978-3-319-24574-4_28
16. Suganuma, M., Shirakawa, S., Nagao, T.: A genetic programming approach to designing convolutional neural network architectures. In: Proceedings of the Genetic and Evolutionary Computation Conference, pp. 497–504 (2017)
17. Wang, B., Xue, B., Zhang, M.: Particle swarm optimisation for evolving deep neural networks for image classification by evolving and stacking transferable blocks. In: 2020 IEEE Congress on Evolutionary Computation (CEC), pp. 1–8 (2020). https://doi.org/10.1109/CEC48606.2020.9185541
18. Yu, F., Wang, D., Shelhamer, E., Darrell, T.: Deep layer aggregation. In: 2018 IEEE/CVF Conference on Computer Vision and Pattern Recognition, pp. 2403–2412 (2018). https://doi.org/10.1109/CVPR.2018.00255
19. Zhan, Z.H., Li, J.Y., Zhang, J.: Evolutionary deep learning: a survey. Neurocomputing **483**, 42–58 (2022)
20. Zhou, X., Qin, A.K., Gong, M., Tan, K.C.: A survey on evolutionary construction of deep neural networks. IEEE Trans. Evol. Comput. **25**(5), 894–912 (2021)
21. Zhou, Z., Rahman Siddiquee, M.M., Tajbakhsh, N., Liang, J.: UNet++: a nested U-Net architecture for medical image segmentation. In: Stoyanov, D., et al. (eds.) DLMIA/ML-CDS -2018. LNCS, vol. 11045, pp. 3–11. Springer, Cham (2018). https://doi.org/10.1007/978-3-030-00889-5_1

Probability Mapping of Spectral CT Material Decomposition to Aid in Determining Material Identification and Quantification Likelihood

Theodorus Dapamede[1,3](✉), Krishna M. Chapagain[1,3], Mahdieh Moghiseh[1,3], James Atlas[2], Philip H. Butler[1,2,3,4], Anthony P. H. Butler[1,2,3,4], and MARS Collaboration

[1] University of Otago Christchurch, Christchurch, New Zealand
{theo.dapamede,krishna.chapagain}@postgrad.otago.ac.nz
[2] University of Canterbury, Christchurch, New Zealand
james.atlas@canterbury.ac.nz
[3] MARS Bioimaging Limited, Christchruch, New Zealand
{mahdieh.moghiseh,phil.butler,anthony.butler}@marsbioimaging.com
[4] European Organisation for Nuclear Research (CERN), Geneva, Switzerland

Abstract. In this study we propose a method to generate probability maps for material identification and quantification in spectral computed tomography (CT). Our goal in producing these maps is to provide spectral CT users, such as clinicians and researchers, per-voxel confidence levels on the materials and concentrations identified, especially for complex biological specimens with unknown materials and unknown concentrations. Our method is based on a likelihood ratio approach that modifies prior (pre-test) probabilities to produce posterior (post-test) probabilities. Prior probabilities are calculated from the per-voxel material decomposition error conditioned on non-negative quantifications. To evaluate our method, we scanned a calibration phantom containing vials of water, lipid, hydroxyapatite (HA) and gadolinium (Gd) in various concentrations using a MARS Spectral Photon Counting CT scanner. In addition, for a validation dataset we also scanned a bovine patella sample incubated in Gd contrast agent solution with the same scanner and scan protocols. Material decomposition was performed on both datasets and likelihood ratios were calculated from the calibration phantom for each material at their respective concentrations. Posterior probability maps for HA and Gd were generated for both datasets using the calculated likelihood ratios on the prior probabilities. The results show that these maps provide additional information on the degrees of confidence for each voxel's material identification and quantification. We also demonstrate that our method can be applied to a complex biological sample.

Keywords: Material decomposition · Probability map · Photon counting · Spectral CT

W. Q. Yan et al. (Eds.): IVCNZ 2022, LNCS 13836, pp. 202–213, 2023.
https://doi.org/10.1007/978-3-031-25825-1_15

1 Introduction

Spectral computed tomography (CT) has advanced significantly in recent years. This next generation CT technology has energy resolving detectors that compartmentalize energies into various bins, which allow for materials to be identified and also quantified, also known as material decomposition (MD). Enhanced spatial and contrast resolutions, along with improved material differentiation, are likely to enable or enhance a variety of clinical applications. These include atherosclerotic plaque characterization [1], dental imaging [2], infectious disease imaging [3], detection of arthroplasty implant failure [4], nanoparticle imaging [5], as well as simultaneous imaging and measurement of multiple contrast agents [6–8].

Advances in material decomposition algorithms have also been seen, such as optimisation or variation methods [9] and deep learning approaches [10–12]. Most of these studies perform their MDs on simulated data or phantoms of known materials. Some studies also used datasets from scanned biological samples, however they only assessed the identification performance of the post-MD images against these scans, either by manually segmenting a ground truth map [11] or by generating a virtual monochromatic image to compare with a conventional CT image [10]. In contrast, the quantification part of the MD is usually only evaluated based on the degree of correlation against the measured concentrations of the compound in the phantoms. However, despite having good quantification correlations with phantoms, they do not take into account misidenfication errors that may go unrecognized. Furthermore, there is often no ground truth for the identity and concentration of materials in complex biological specimens. It is therefore important to know the degree of confidence that can be placed on the identity and quantity of any material provided by spectral CT MD algorithms [13].

Surprisingly, despite the advances in MD mentioned above, there is a very limited number of studies that have actually tried to show per-voxel confidence levels of post-MD or material images. The study by Raja, et al. [13] introduced a method to measure the net effect of errors on post-MD identification and quantification. They used sensitivity and specificity metrics to measure the MD performance at different material concentrations, thus providing a measure of confidence to how well the MD had correctly identified a target material at a specific concentration. However, the use of sensitivity or specificity only evaluates the performance of the MD against the ground truth of known material concentrations [14]. On the other hand, if we take a random voxel from a complex biological specimen with unknown materials at unknown concentrations, sensitivity or specificity values cannot be used to show how likely that voxel truly contains a material at that measured concentration.

Compared to sensitivity and specificity, positive predictive values (PPV) and negative predictive values (NPV) are more appropriate to show the probability of a given voxel being correctly identified [14]. This is because the PPV takes into account both the true positives and false positives, while NPV takes into account the true negatives and false negatives, which results in probability scores. The downside to this, however, is that PPV and NPV are correlated with the prevalence or prior probabilities of the materials of interest [15], which in the case of complex biological samples is unknown. In this regard, likelihood ratios are more useful since they can be used in order to calculate the likelihood of a voxel, while also adjusting for prior probabilities of the probability of a material based on a variety of contexts [16].

In this study, we build upon the prior works of Raja, et al. [13] and propose a new method to generate per-voxel probability maps for post-MD images using likelihood ratios.

2 Materials and Methods

2.1 Datasets

Calibration Phantom. Vials containing lipid, water, and gadolinium (Gd) as well as hydroxyapatite (HA) rods at different concentrations were arranged in a perspex holder as shown in Fig. 1A to serve as a calibration phantom. The phantom was scanned using the MARS 10 Spectral Photon Counting CT (SPCCT) (MARS Bioimaging Ltd, NZ) using the protocol summarised in Table 1.

Gadolinium in Cartilage Sample. A sectioned healthy bovine patella (Fig. 1B) was prepared by dipping into 50% diluted Dotarem (macrocyclic gadolinium contrast agent), incubated for 24 h at 37° C and washed in PBS. The sample was then scanned in the MARS 10 SPCCT with the same scanning protocol as the calibration phantom above.

Table 1. Spectral photon counting CT scanning parameters

Parameter	Value
MARS SPCCT scanner ID	10
Tube voltage (kVp)	120
Tube current (μA)	32
Exposure time (ms)	220
Energy thresholds (keV)	35, 50, 60, and 75
SDD	271.95 mm
SOD	211.95 mm
Voxel size	$0.09 \times 0.09 \times 0.09$ mm^3
Scan length	Phantom: 34.02 mm
	Bovine: 91.96 mm
Scan radius	Phantom: 37.08 mm
	Bovine: 37.08 mm

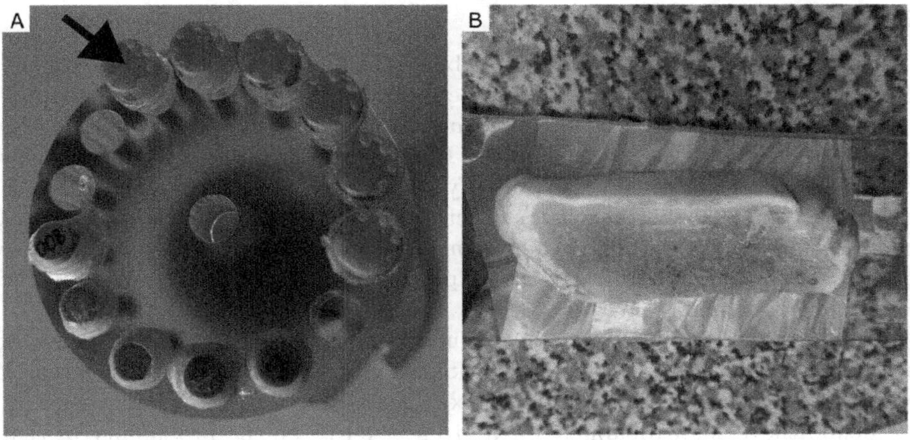

Fig. 1. (**A**) Calibration phantom configuration. Clockwise from arrow: lipid, water, Gd 0.5 mg/mL, Gd 2 mg/mL, Gd 4 mg/mL, Gd 8 mg/mL, HA 54.3 mg/mL, HA 104.3 mg/mL, HA 211.7 mg/mL, HA 402.3 mg/mL, HA 808.5 mg/mL; (**B**) Bovine patella sample preparation.

2.2 Material Decomposition (MD)

Material decomposition is the process of extracting material information from the linear attenuation coefficients (energy information) [17]. The linear attenuation (μ_E) of a composite material at a specific energy range (E) is defined as the linear combination of its constituents, and is given by [13]:

$$\mu_E = \sum_{i=1}^{m} \rho_m \left(\frac{\mu}{\rho}\right)_{mE} \tag{1}$$

where m is the index of the constituent materials, ρ_m is the concentration or density of the constituent material, and $(\mu/\rho)_{mE}$ is the mass attenuation of the respective material at energy range E.

If all voxels in a reconstructed multi-energy CT scan image containing all the μ_E are represented by $\mathbf{b}$, then we can use the system of linear equation $\mathbf{Ax} = \mathbf{b}$ to solve for material decomposition. Here, $\mathbf{A}$ is the material matrix containing K materials while vector $\mathbf{x}$ describes the composition of $\mathbf{b}$ [17]. For this study, we performed MD on all the datasets using an in-house MD software with a least squares based error function MD algorithm (MDLSQ) that minimises the $\mathcal{L}_2$ norm (Eq. 2).

$$\text{MDLSQ}(\mathbf{x}) = \min_{\mathbf{x}} \|\mathbf{Ax} - \mathbf{b}\|_2^2 \tag{2}$$

2.3 Probability Mapping

We propose to use the Posterior Probability calculated from the Positive Likelihood Ratio (LR^+) as the values for the probability map (p-map). We adapted the LR^+, commonly used to interpret medical diagnostic tests, to evaluate MD algorithms. In this case, we define the LR^+ to be the ratio of the probability of an MD algorithm to correctly identify a material or material combination in a voxel, to the probability of it incorrectly predicting that material or material combination. The LR^+ indicates how much the MD algorithm result will raise or lower the Prior Probability of the suspected materials, which is output as the Posterior Probability.

While the Prior Probability in diagnostic tests relates to the prevalence of a disease in a population and is often known through epidemiological research [15], the prevalence of a material in voxels, may not be known. For our study, we define the Prior Probability for a given voxel to be the probability of that voxel to be identified as a target material by the MD algorithm while taking into account the error and the quantification results for each solution.

To calculate the Prior Probability per-voxel from K material solutions, a probability is generated on the MD error value, in this case the least squares value of that voxel (ℓ_k), in each material solution (k) using the logistic function (Eq. 3). Then these probabilities are conditioned based on the quantification for each material solution. A quantification is only deemed valid if it returns a positive value (Eq. 4). Finally, all conditioned-probabilities (y'_k) are then passed through a soft-max function (Eq. 5) to produce a probability distribution across the different materials.

$$y_k = \frac{1}{1 + e^{\ell_k}} \times 100\% \tag{3}$$

$$y'_k = \begin{cases} y_k, & \text{if quantification} > 0 \\ 0, & \text{otherwise} \end{cases} \tag{4}$$

$$\text{Prior Probability} = \text{Softmax}(y'_k) = \frac{e^{y'_k}}{\sum_{i=1}^{K} e^{y'_i}} \tag{5}$$

The Posterior Probability, on the other hand, is the probability of that voxel to contain the target material after the MD is performed and evaluated against a calibration phantom. To calculate this, we first calculate the sensitivity and specificity of the MD result for each material and their corresponding concentrations from the calibration phantom. For each concentration, a 95% confidence interval is set as the voxel inclusion criteria. The sensitivity (Eq. 6), or the True Positive Rate, is the ability of the MD algorithm to correctly identify voxels with the target materials, while the specificity (Eq. 7), or the True Negative Rate, is the ability of the MD algorithm to correctly identify voxels without the target material.

$$\text{Sensitivity} = \frac{\text{True Positive}}{\text{True Positive} + \text{False Negative}} \tag{6}$$

$$\text{Specificity} = \frac{\text{True Negative}}{\text{True Negative} + \text{False Positive}} \tag{7}$$

The LR^+ is then calculated as the ratio of the True Positive Rate (sensitivity) and the False Positive Rate (1 – Specificity) (Eq. 8) for each material concentration. A Likelihood Ratio map is produced over the quantification image by interpolation of the fitted exponential curve. The LR^+ map is then multiplied with the Prior Odds (Eq. 9), which results in the Posterior Odds (Eq. 10). Finally, the Posterior Odds is then transformed to Posterior Probability values according to Eq. 11 [16].

$$LR^+ = \frac{\text{Sensitivity}}{1 - \text{Specificity}} \tag{8}$$

$$\text{Odds}_{(\text{Prior})} = \frac{\text{Prior Probability}}{1 - \text{Prior Probability}} \tag{9}$$

$$\text{Odds}_{(\text{Posterior})} = \text{Odds}_{(\text{Prior})} \times LR^+ \tag{10}$$

$$\text{Posterior Probability} = \frac{\text{Odds}_{(\text{Posterior})}}{1 + \text{Odds}_{(\text{Posterior})}} \tag{11}$$

3 Results and Discussion

Regions of Interests (ROIs) were manually selected on the calibration phantom to mark ground truth data as shown in Fig. 2. The ROIs had a radius of 2.7 mm and span across 10 continuous slices. This resulted in a voxel count of 28,210 for each material concentration with a total of 310,310 voxels. Table 2 summarises the material identification and quantification results of the calibration phantom using the MDLSQ algorithm.

Linear regression was performed to evaluate the correlation between measured concentrations and the expected concentrations, and the plots for each material are shown in the top row of Fig. 3. The results show significantly strong correlations between the measured and expected concentrations for HA ($R^2 = 0.99$, $p < 0.01$, $SE = 9.05$ mg/mL) and Gd ($R^2 = 0.99$, $p < 0.01$, $SE = 0.27$ mg/mL).

As the likelihood ratio has a skewed exponential distribution [18], an exponential function was then fitted to the positive likelihood ratio values in Table 2 to enable interpolation and generate the material likelihood ratio maps. The fitted functions are shown in the bottom row of Fig. 3. The positive likelihood ratio for HA ranged from 5.28 at the lowest concentration to undefined value for HA 800 mg/mL as the specificity reached 1. Therefore, HA 800 mg/mL was omitted from the plot and calculations.

The material images of HA and gadolinium in the calibration phantom, along with their respective probability maps, are shown in Fig. 4. In the prior probability maps (B and E), we see voxels of the other materials also having some

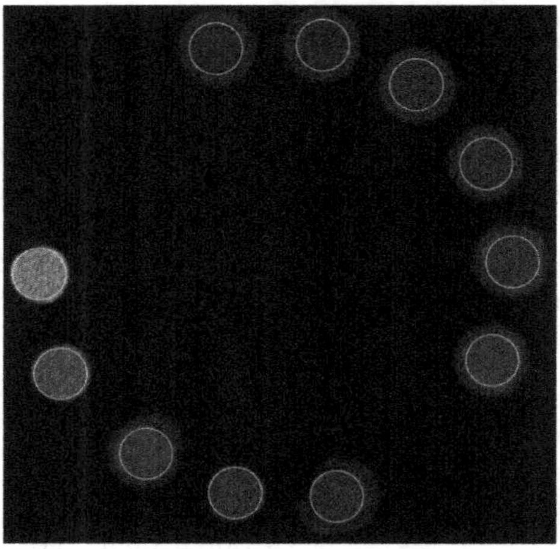

Fig. 2. ROI selection

Table 2. HA and Gd material identification and quantification (n = 28210 per material concentration)

Material	Concentration (mg/mL)	Measured (mg/mL)	SD (mg/mL)	SE×1.96 (mg/mL)*	Sens.	Spec.	LR^+
Gd	0.5	0.7	1.4	0.02	0.27	0.94	4.81
Gd	2	1.8	2.2	0.03	0.53	0.93	7.22
Gd	4	3.9	2.7	0.03	0.83	0.92	10.34
Gd	8	8.1	3.0	0.04	0.95	0.96	22.73
HA	54.3	41.9	50.3	0.59	0.53	0.90	5.28
HA	104.3	89.3	62.5	0.73	0.83	0.90	8.04
HA	211.7	203.0	72.5	0.85	0.95	0.96	26.24
HA	402.3	377.9	87.1	1.02	0.95	0.99	94.92
HA	808.5	808.7	114.0	1.33	0.95	1	Undef

*SE: Standard Error (SD $/\sqrt{n}$)

probability of being selected as the target material. For example in the HA maps we see vials of Gd and water showing almost equal probabilities to HA. However, in contrast, after calculating the posterior probability (C and F), we see significant increase in the probabilities for the true target materials. If the likelihood ratio equals 1, then the prior probability is not changed, which is also shown in C and F where the non target materials have the same probabilities as their prior.

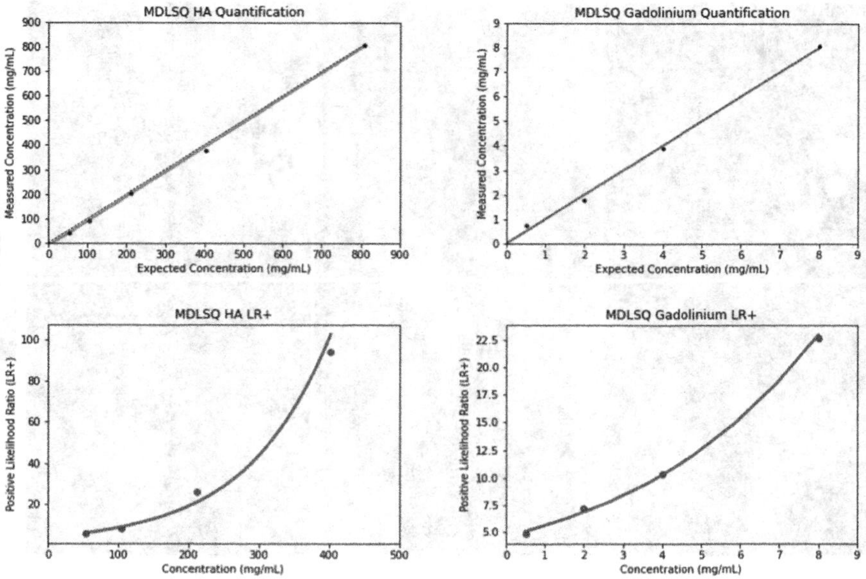

Fig. 3. Quantification and LR+ Plots for HA and gadolinium

We applied the calculated likelihood functions from the calibrated phantom above to the bovine patella with contrast agent sample to produce the probability maps. Figure 5 displays the output images from this sample. A slice from the spectral image at energy range 7–35 keV is shown in A, while the material decomposition images for HA and gadolinium are shown in B and C, respectively. The highest measured HA concentration is 1957 mg/mL while the highest measured Gd is 122 mg/mL. Both of these are significantly higher than the concentrations available in the calibration phantom. This resulted in very high likelihood ratios which significantly increased the posterior probabilities to values nearing 1 (as seen in sub-figures E and G). The final sub-figure H is a 3D rendering of both HA and Gd material maps combined using our in-house software MARS Vision (MARS Bioimaging Ltd, NZ).

In this study we generated prior probabilities from the least square error of each voxel being identified as a material solution. With the 4 materials present, i.e. Lipid, Water, HA and Gd, there were 7 possible material solutions that were tested (Water, Lipid, Lipid + Water, HA alone, HA + Water, Gd alone, and Gd + Water). The lower the least squares error for a particular solution, the

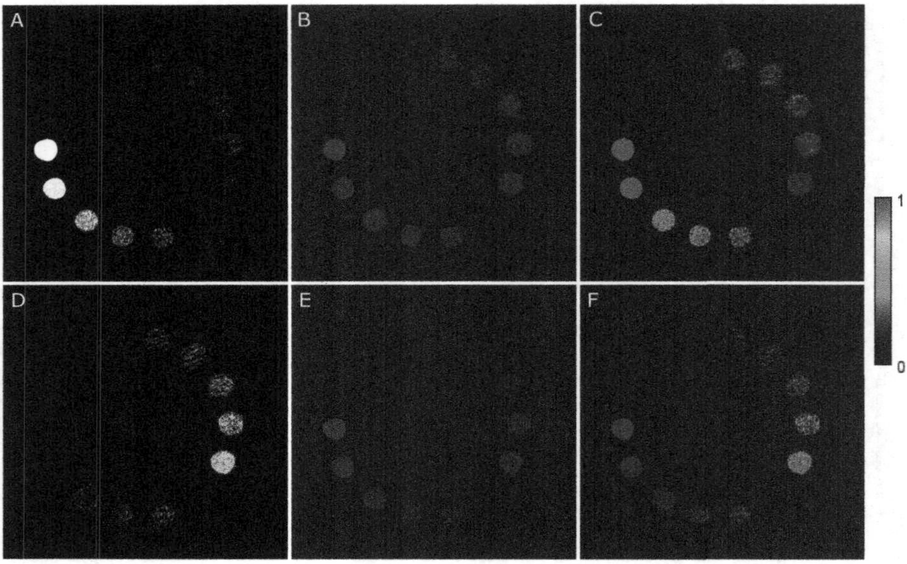

Fig. 4. HA (**A**) and Gd (**D**) material images along with the prior (**B, E**) and posterior probability maps (**C, F**), respectively. Right colour bar indicating the probability values for the p-maps.

higher the probability of that voxel being classified into that material solution. Other factors that may be taken into account when determining the prior probability are the angle between bases and information regarding the interaction between materials in certain spaces or conditions. This may generate an even more accurate estimation of a voxel's prior probability. Nonetheless, the advantage of using likelihood ratio to generate posterior probabilities is that it can be applied to any prior probability map. Therefore, if a more advanced prior probability method is generated, our method to generate a posterior probability will still be applicable.

The prior and posterior probability maps are additional information for end users, such as clinicians and researchers, as an estimate of the degree of confidence for each voxel in the post-MD images. A clinician, for example, might only be interested in areas where the probability of finding a certain uptake of a contrast agent to be above a certain threshold, e.g. 50%. Or a researcher who is interested in certain regions may like to know the probability of that region actually containing the material of interest. To evaluate these assumptions, however, further use case studies are required.

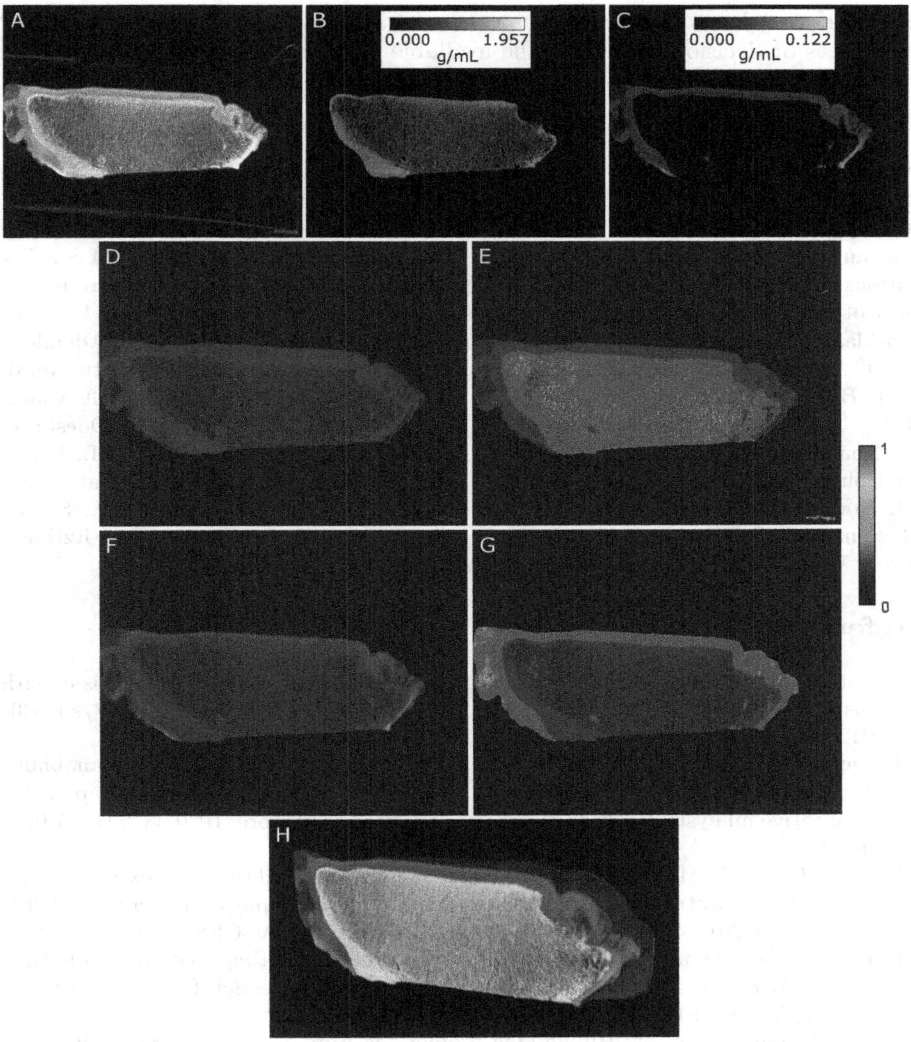

Fig. 5. Bovine cartilage with Gd contrast sample: (**A**) Energy image [7–35 keV], (**B**) HA material image, (**C**) Gd material image, (**D**) HA prior probability map, (**E**) HA posterior probability map, (**F**) Gd prior probability map, (**G**) Gd posterior probability map, and (**H**) 3D rendering of HA and Gd material maps.

4 Conclusion

In this study we have introduced a method to generate probability maps to be used alongside material decomposition images from spectral CT scanners. We have demonstrated a proof of concept for its utility using a bovine patella sample incubated in gadolinium contrast agent solution. Probability maps provide extra information to users, such as researchers and clinicians, to aid in further inferences or decisions based on the available data. Future studies to show the probability maps on different MD algorithms as well as on clinical patient data are needed.

Acknowledgements. This project was funded by the Ministry of Business, Innovation and Employment (MBIE), New Zealand under contract number UOCX1404, by MARS Bioimaging Ltd and the Ministry of Education through the MedTech CoRE. The authors would like to acknowledge the Medipix2, Medipix3 and Medipix4 collaborations. Also, we would like to take this opportunity to acknowledge the generous support of the MARS Collaboration as well as Marco Schneider and Thor Besier from the Auckland Bioengineering Institute. European MARS Collaboration: S. A. Adebileje, S. D. Alexander, M. R. Amma, M. Anjomrouz, F. Asghariomabad, A. Atharifard, S. T. Bell, F. O. Bochud, P. Carbonez, C. Chambers, A. I. Chernoglazov, J. A. Clark, J. S. Crighton, S. Dahal, A. Denys, N. J. A. deRuiter, D. Dixit, R. M. N. Doesburg, K. Dombroski, N. Duncan, S. P. Gieseg, A. Gopinathan, B. P. Goulter, J. L. Healy, L. Holmes, K. Jonker, T. Kirkbride, C. Lowe, V. B. H. Mandalika, A. Matanaghi, M. Nowak, B. Paulmier, D. Racine, P. Renaud, D. Rundle, N. Schleich, E. Searle, R. Senzig, J. S. Sheeja, A. Smith, L. Vanden Broeke, F. R. Verdun, V. Vitzthum, Vivek V. S., E. P. Walker, M. Wijesooriya, W. R. Younger.

References

1. Zainon, R., et al.: Spectral CT of carotid atherosclerotic plaque: comparison with histology. Eur. Radiol. **22**(12), 2581–2588 (2012). https://doi.org/10.1007/s00330-012-2538-7
2. Vanden Broeke, L., Grillon, M., Yeung, A.W., Wu, W., Tanaka, R., Vardhanabhuti, V.: Feasibility of photon-counting spectral CT in dental applications-a comparative qualitative analysis. BDJ Open **7**(1) (2021). https://doi.org/10.1038/S41405-021-00060-X
3. Lowe, C., et al.: Molecular imaging of pulmonary tuberculosis in an ex-vivo mouse model using spectral photon-counting computed tomography and micro-CT. IEEE Access **9**, 67201–67208 (2021). https://doi.org/10.1109/ACCESS.2021.3076432
4. Lau, L.C.M., et al.: Multi-energy spectral photon-counting computed tomography (MARS) for detection of arthroplasty implant failure. Sci. Rep. **11**(1) (2021). https://doi.org/10.1038/S41598-020-80463-2
5. Ostadhossein, F., et al.: Hitchhiking probiotic vectors to deliver ultra-small hafnia nanoparticles for 'Color' gastrointestinal tract photon counting X-ray imaging. Nanoscale Horiz. **7**(5), 533–542 (2022). https://doi.org/10.1039/D1NH00626F
6. Moghiseh, M., et al.: Discrimination of multiple high-Z materials by multi-energy spectral CT-a phantom study. JSM Biomed. Imaging Data Pap. **61**, 1007 (2016)
7. Tao, S., Rajendran, K., McCollough, C.H., Leng, S.: Feasibility of multi-contrast imaging on dual-source photon counting detector (PCD) CT: an initial phantom study. Med. Phys. **46**(9), 4105–4115 (2019). https://doi.org/10.1002/MP.13668
8. Si-Mohamed, S., et al.: Spectral photon-counting computed tomography (SPCCT): in-vivo single-acquisition multi-phase liver imaging with a dual contrast agent protocol. Sci. Rep. **9**(1), 1–8 (2019). https://doi.org/10.1038/s41598-019-44821-z
9. Mory, C., Sixou, B., Si-Mohamed, S., Boussel, L., Rit, S.: Comparison of five one-step reconstruction algorithms for spectral CT. Phys. Med. Biol. **63**(23), 235001 (2018). https://doi.org/10.1088/1361-6560/AAEAF2

10. Abascal, J.F., et al.: Material decomposition in spectral CT using deep learning: a Sim2Real transfer approach. IEEE Access **9**, 25632–25647 (2021). https://doi.org/10.1109/ACCESS.2021.3056150
11. Wu, X., et al.: Multi-material decomposition of spectral CT images via fully convolutional DenseNets. J. X-Ray Sci. Technol. **27**(3), 461–471 (2019). https://doi.org/10.3233/XST-190500
12. Zhu, J., et al.: Feasibility study of three-material decomposition in dual-energy cone-beam CT imaging with deep learning. Phys. Med. Biol. **67**(14), 145012 (2022). https://doi.org/10.1088/1361-6560/AC7B09
13. Raja, A., et al.: Measuring identification and quantification errors in spectral CT material decomposition. Appl. Sci. **8**(3), 467 (2018)
14. Trevethan, R.: Sensitivity, specificity, and predictive values: foundations, pliabilities, and pitfalls in research and practice. Front. Public Health **5**, 307 (2017). https://doi.org/10.3389/FPUBH.2017.00307/BIBTEX
15. Hunt, B.R., Kaloshin, V.Y.: Prevalence. In: Handbook of Dynamical Systems, vol. 3, no. C, pp. 43–87, May 2022. https://doi.org/10.1016/S1874-575X(10)00310-3
16. Deeks, J.J., Altman, D.G.: Diagnostic tests 4: likelihood ratios. BMJ **329**(7458), 168–169 (2004). https://doi.org/10.1136/BMJ.329.7458.168
17. Bateman, C.J.: Methods for material discrimination in MARS multi-energy CT. Ph.D. thesis, University of Otago, Christchurch (2015). http://hdl.handle.net/10523/5888
18. Van Den Ende, J., Moreira, J., Basinga, P., Bisoffi, Z.: The trouble with likelihood ratios [6]. Lancet **366**(9485), 548 (2005). https://doi.org/10.1016/S0140-6736(05)67096-1

Explainable Network Pruning for Model Acceleration Based on Filter Similarity and Importance

Jinrong Wu[(✉)], Su Nguyen, and Damminda Alahakoon

Research Centre for Data Analytics and Cognition, La Trobe University, Melbourne, Australia
{melody.wu,p.nguyen4,d.alahakoon}@latrobe.edu.au

Abstract. Filter-level network pruning has effectively reduced computational cost, as well as energy and memory usage, for parameterized deep networks without damaging performance, particularly in computer vision applications. Most filter-level network pruning algorithms focus on minimizing the impact of pruning on network performance using either importance-based or similarity-based pruning approaches. However, no study has attempted to compare the effectiveness of the two approaches across different network configurations and datasets. To address these issues, this paper compares two explainable network pruning methods based on importance-based and similarity-based approaches to understand their key benefits and limitations. Based on the analysis findings, we propose an innovative hybrid pruning method and demonstrate its effectiveness using several models and datasets. The comparisons with other state-of-the-art filter pruning methods show the superiority of the new hybrid method.

Keywords: Model compression · Network pruning · Explainable AI (XAI) · Visualization · Convolutional neural networks

1 Introduction

Recent years have seen significant improvement in the accuracy of deep neural networks, achieving huge success in computer vision natural language processing, stock prediction, and forecasting [1]. However, over-parameterization is a widely recognized problem with deep neural networks (DNNs). It leads to high computational costs and high memory footprints, impeding the model deployment that concerns the availability of energy, computation, and memory resources [2]. To pursue a satisfactory tradeoff between processing efficiency and prediction accuracy, model compression and acceleration methods have received much attention from researchers and have shown great progress in the past few years [3].

Methods for DNNs model compression can be generally categorized into network pruning, quantization, low-rank factorization, and knowledge distillation. Recent developments in these domains contribute dramatically to model efficiency [3, 4]. However, recent works [4] have highlighted the importance but a lack of explainability in the existing model compression approaches, leading to model robustness issues. Explainable and

W. Q. Yan et al. (Eds.): IVCNZ 2022, LNCS 13836, pp. 214–229, 2023.
https://doi.org/10.1007/978-3-031-25825-1_16

robust compression can minimize the effort to re-evaluate the compressed model, and thus reliable and predictable in production [5]. However, it is difficult to provide explanations for knowledge distillation and low-rank factorization methods because the former essentially treats the entire network model as a black box [6], and the latter focuses on the matrices level which can be hard to explain due to the millions of matrices in a large network. As such the limited explainable model compression studies have focussed mainly on network pruning and quantization. While the effect of network pruning and quantization can be combined without any compatibility issues, this study focuses on the explainable methods for network pruning, specifically filter/channel pruning, because of its easy integration with general hardware and Basic Linear Algebra Subprograms (BLAS) libraries [7].

The existing filter pruning methods can be generally categorized into importance-based and similarity-based approaches. The importance-based approach prunes filters by ranking their importance to the network classification. It identifies less important filters by locating filters with fewer zeros or smaller norms in the filter weights or filter outputs [10, 29], or by measuring the change in the final model response before and after removing a filter from the network [16, 28]. The second approach, similarity-based pruning, locates the redundant filters based on their similarity. This approach aims at maintaining a diverse but small set of filters by removing redundant filters that are similar to others. Most of the methods in this category use distance measures to calculate the similarity between filters with either accumulated pairwise similarity scores [9, 30] or filter clustering [14].

However, both importance-based and similarity-based methods have limitations. To obtain satisfactory results from a single importance measure, such as a norm-based measure, the distribution of values needs to be wide enough to retain some values but contain enough values close to zero such that a smaller network organization is still accurate [13]. In addition, retaining filters with higher importance scores but capturing the same information can lead to redundancy in the pruned network [15]. Similarly, the selection process using similarity can only ensure the uniqueness of filters, as such without an additional measure on their contribution to the network, the selected irreplaceable filters can still be non-essential.

Therefore, it is beneficial to consider both filter importance and similarity in network pruning. Very recently, a study by Shao et al. [16] proved the advantage of taking both importance and similarity into account in network pruning with their proposed combination method. However, their work with pure quantitative comparisons on the post-pruned model accuracy, could not help understand the advantage and limitations of each approach, nor about when and where one approach can be better than another. As such, in this study, we aim to critically analyze the two pruning approaches using our proposed explainable similarity-based and importance-based pruning methods and evaluate their performance in pruning different layers on different networks at various pruning ratios and further explain their pruning behaviors with visualization. Based on the analysis, we propose a hybrid pruning method with combined use of importance-based and similarity-based methods, to enable flexibility in catering to different requirements and thus deliver a better pruning outcome.

The contribution of our study presented in this paper is summarized as follows:

- To our knowledge, this is the first study to critically analyze the strengths and weaknesses of both importance-based and similarity-based filter pruning, via an explainable technique supported by empirical performance evaluation.
- This study uses clustering and visualization in combination for explainability of the entire pruning process.
- Based on the critical analysis of the two approaches, a novel hybrid pruning method is proposed, with experimental results demonstrating its superiorness to other state-of-the-art pruning methods.

2 Related Work

2.1 Filter Level Network Pruning

Network pruning, specifically for Convolutional Neural Networks, is generally categorized into structured pruning and unstructured pruning [9, 14]. Unstructured pruning refers to individual weight pruning without considering the structure of the model [18]. It can result in creating sparsities in node weight connections, which require dedicated hardware/libraries for compression and speedup [18]. Structured pruning, on the other hand, prunes networks at the level of channels, filters, or layers. These techniques can preserve the original model structure without dedicated hardware or libraries [17]. Filter or channel pruning (which in certain instances are interlaced), is the most popular structured pruning method, as it is operated at the most fine-grained level while still fitting in conventional deep learning frameworks.

There are two different types of pruning processes, namely static pruning and dynamic pruning [19]. The pruning processes include preparing a model, either a pre-trained model or a scratch model, locating pruning candidates, pruning the identified candidates, and finetuning the model. Static pruning has all pruning steps performed offline before inferencing, whereas dynamic pruning is performed during runtime.

2.2 Approaches for Filter Pruning

To identify the filters that are not crucial and thus can be pruned from the network, filters are located based on their importance and irreplaceability to the network. The former is referred to as the importance-based approach and the latter is referred to as the similarity-based approach.

Importance-Based Approach. The importance-based approach aims at identifying the filter pruning candidates with the lowest importance by ranking their contribution to the network [13]. There are several ways to measure filter contributions. One measure focuses on filter activation and if a filter generates a small output, it is less likely to be activated and thus less important. Methods that use this type of measure identify less important filters by ranking the filter norms or the percentage of zeros in the filter weight or output. For example, the norm-based criteria, including ℓ_1-norm [20] and ℓ_2-norm [21], prune filters with lower norm values. Average Percentage of Zeros (APOZ) [8] and

Zero-Keep [9] filter pruning methods prune the filters with less proportion of zeros in the filter weights or filter outputs. Another measure to identify less important filters is by comparing the network middle layer or final layer output before and after removing some filters, and the removed filters leading to smaller changes are considered less important. For example, measuring the feature reconstruction error for each filter to identify the less important filter group is one popular technique. Filters with higher reconstruction errors on the next layer [11] or the final output layer [10] are considered less important to the network and are thus pruned. Similarly, the Taylor criterion [22] has also been proposed to approximate the change in loss caused by removing a feature map, and the removed filters leading to small changes are pruned. Gradient-based criterion proposed by Liu and Wu [23], prunes filter-related channels by calculating the mean gradient of output feature maps in each layer. Alqahtani et al. evaluate the degree of alignment between the semantic concept and individual hidden unit representations to detect important filters [24].

Similarity-Based Approach. The similarity-based approach aims at identifying the filters that are the best representatives for the groups of similar filters with similar weight structures or generating similar outputs. The goal of this approach is to reduce redundancy by removing the replaceable filters and retaining the most representative filters [25]. Similarity score ranking and clustering-based ranking are the two popular ways to identify replaceable filters in the network. Similarity score ranking identifies replaceable filters using accumulated pairwise filter similarity scores, and the filters with higher scores are more replaceable. Zhu and Pei [25] and Yao et al. [15] scored filter similarities by comparing higher-level extracted features from filter outputs, for example, color, texture, brightness, contrast, and structure. He et al. [13] identified the most replaceable filters by measuring the filter geometric median, so that filters with similar values within the layer are pruned. Clustering methods, such as k-means [14], measure the filter similarities by identifying the filter clusters. Only the most representative filters from clusters are reserved and the rest of the filters are pruned.

2.3 Existing State of Explainability in Network Pruning

To identify the right channels or filters to be pruned for a given network, it is important to understand the network structure and network components in decision-making. However, due to the complex structures in deep neural networks (DNNs), it is difficult to develop such an understanding. Furthermore, most existing pruning methods are evaluated purely via post-prune model accuracy. This may lead to a less valid and general network that can fail to provide correct classification when deployed in different datasets in the real world, because the pruning decisions can be made based on spurious in the experiment data [6, 12], and Such a network is likely to, where spurious or artifactual correlations may not be present [6]. Thus, explainable AI (XAI), has been applied recently in network pruning to deal with the black-box nature of DNNs while providing explanations of pruning procedures [26].

Among the limited number of existing explainable network pruning studies, Layer-wise Relevance Propagation (LRP) [12], and visualization [27, 30] are the most popular methods. Apart from these, SHAP [27], DeepLift [28] and Degree of information presented in feature maps [24] have also shown their effectiveness in identifying the redundant hidden units in the network. However, these existing applications of XAI in network pruning focus on the individual interpretation of filters, thus overlooking the understanding of the grouping of filters. Since filters in the network work collaboratively rather than individually, explanations from filter groups or filter clusters can help identify the diversity and variance of network filters. As such, cluster-based filter interpretation can be used for explainable network pruning. Although there are some existing clustering-based network pruning methods [14], none of them attempts to explain the filter grouping to assist understanding.

3 Method

To take advantage of both filter similarity and importance in network pruning, we first develop two benchmarking methods for both similarity-based and importance-based approaches with built-in visualization. This enables critical analysis of the strengths and weaknesses of both approaches. Based on the analysis findings, we propose a hybrid method that can make use of the advantages of both approaches.

3.1 Similarity-Based Filter Pruning

For a convolutional neural network model M, each convolutional layer C_i has N_{C_i} filters. And for each filter $f_{C_i,j}$, $j \in [1, N_{C_i}]$ in the layer, filter output $o_{f_{C_i,j}} \in \mathbb{R}^{D \times S}$ (D is the number of observations in the dataset and S is the flattened size of the output from each filter) are used to determine filter similarity.

The proposed similarity-based method prunes the filters in a layer-wise manner, including two main steps, namely GSOM-based filter clustering and distance-based similarity measurement.

GSOM-Based Filter Clustering. Clustering is commonly used to identify patterns in the data, via which, data with similar patterns are grouped. As such, it has been used to identify similar filters in the network [14]. To assist human understanding of clustering results, visualization is frequently applied in a two-dimensional space. Therefore, in this study, we apply an automatic clustering algorithm growing self-organizing maps (GSOM) [29] to identify similar filters. Since its natural topology enables it to make better and unbiased natural grouping than other clustering methods which requires a pre-defined number of clusters, such as HDBSCAN and K means [29], GSOM is able to select the most representative filters. In addition, because the topographic clustering map generated by GSOM is two-dimension, filter grouping results can be easily visualized for better understanding.

To cluster the filters in layer C_i, GSOM takes all filter outputs $O_{C_i} = o_{f1}, o_{f2}, \cdots o_{fN_{C_i}-1}, o_{fN_{C_i}}$ to generate a topological node map, where each node is considered as one cluster, and filters assigned in the same GSOM node have relatively similar filter outputs. As shown in Eq. (1), the output of GSOM is the pairs of filters and their assigned nodes. Suppose GSOM creates t nodes, by aggregating the filters for each node, a list of nodes $\{G_{C_i,1}, \cdots, G_{C_i,t}\}$ is generated, and each node includes a list of filters $\left\{F_{G_{C_i,1}}, \cdots, F_{G_{C_i,t}}\right\}$ that is assigned to it. One important parameter that needs to be adjusted in the experiment, is the spread factor controlling the spreading of the GSOM map. As a default, it is 0.98 based on the original experiment results on different datasets [29].

$$GSOM\left(O_{C_i}\right) = \left(f_{C_i,1}, G_{f_{C_i,1}}\right), \left(f_{C_i,2}, G_{f_{C_i,2}}\right), \ldots, \left(f_{C_i,N_{C_i}}, G_{f_{C_i,N_{C_i}}}\right) \quad (1)$$

Distance-Based Filter Similarity Measurement. Since the GSOM node center is a calculated centroid based on all assigned filters and might not represent any real filter, to identify the filter best represent the filters in the node which is closest to the node center, we use Euclidean distance to measure the distance between each filter $f_{G_{C_i,p},k}$ in $F_{G_{C_i,p}}, p \in [1, t]$ in the node $G_{C_i,p}$ and the node center $Center\left(G_{C_i,p}\right)$ with $w_{Centre\left(G_{C_i,p}\right)} - f_{G_{C_i,p},k}$. By ranking the distance of all filters in the node, we identify the most representative filter, which has the shortest distance to the center, and the least representative filter which has the longest distance to the node center (which could also be abnormal or outliers in that node).

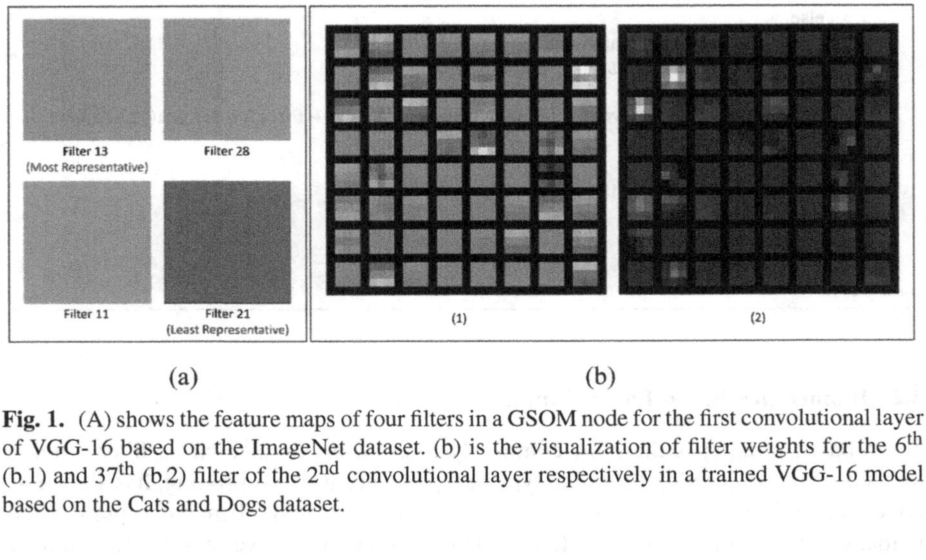

(a) (b)

Fig. 1. (A) shows the feature maps of four filters in a GSOM node for the first convolutional layer of VGG-16 based on the ImageNet dataset. (b) is the visualization of filter weights for the 6th (b.1) and 37th (b.2) filter of the 2nd convolutional layer respectively in a trained VGG-16 model based on the Cats and Dogs dataset.

We give an example of why the least representative filter is also considered to be reserved in some circumstances in our method. As shown in Fig. 1(a), we visualized four filter output feature maps assigned to the same GSOM node in the first convolutional layer of the VGG-16 model. Filter 13 is selected as the most representative, since it is closest to the node center. Filter 21 is considered the least representative, because it is the farthest to the node center. In this case, filter 11 and filter 28 are redundant, because they are similar to filter 13. But filter 21 presents a significantly different pattern from the rest of the filters, thus should be reserved during pruning. Therefore, the least representative filters could also be reserved when there is a need for additional filters to capture more variety of features from inputs.

The detailed algorithm for similarity identification procedures is as follows:

R represents the number of filters to be reserved for a convolutional layer in the network. *TL* is the filter list of the most representative filter of each GSOM node ranked by how many filters are assigned to the node. OL_0 is the filter list of the least representative filters of each node ranked by their number of assigned filters. A node is skipped if there are no more filters in it. The same rule applies to the n_{th} least representative list OL_n. The complete pruning steps are shown in Algorithm 1.

ALGORITHM 1: Similarity-based Filter Pruning

```
for each layer in the network do
    N ← number of filters in the layer; R ← number of filters to be reserved
    if R < N then
        generate GSOM map based on filter outputs of the layer
        filters to be reserved ← {}
        most representative filters (TL) ← filters that are closest to each GSOM node center
        least representative filters round (i) ← 0
            if R ≤ length(TL) then
                filters to be reserved ← first R filters from TL
            else
                get the least representative filter list OL_i
                while R > length(TL) + Σ_1^i length(OL_i) do
                    i ← i + 1
                    filters to be reserved ← filters to be reserved + TL + first (R − length(TL)) filters
                                                  from Σ_1^i OL_i
            end if
        prune the filters that are not the filters to be reserved
    else
        skip the layer
    end if
end for
```

3.2 Importance-Based Filter Pruning

Norm-based pruning is the most commonly referred importance-based filter pruning method. ℓ_1-norm pruning [20], one typical type of norm-based pruning, has shown its effectiveness in various filter pruning tasks and is considered a benchmarking method in many pruning studies [14, 25]. In this method, filters with lower norm values in filter weights are pruned, since they can have limited contribution to network decisions during convolution calculations due to their low value in outputs [20]. Because of its easy-to-understand working mechanism, we adopt this method as the basis for the importance-based filter pruning method. But for the purpose of further analyzing and comparing the similarity-based and importance-based filter pruning methods, we further visualize each filter to help understand the decision-making process.

In this importance-based pruning method, the importance of each filter $f_{C_i,j}$ in the layer C_i is calculated via Eq. (2), where $W\left(f_{C_i,j}\right)$ is the weight of the filter $f_{C_i,j}$. By ranking the importance scores of all filters in the layer, the filters with the lowest importance ranking are pruned.

$$Importance\left(f_{C_i,j}\right) = W\left(f_{C_i,j}\right)_1 \tag{2}$$

Figure 1(b1) is the visualization of the filter that is identified as the most important among all the filters in the 2nd convolutional layer of the VGG-16 model, while Fig. 1(b2) is the filter selected to be pruned. Each 3 × 3 square pixel in the figure is a visualization of the filter kernel for each filter channel. If the weight is close to −1 or 1, the color is brighter, and if the weight is close to 0, it is darker. As such, although two kernels in the filter in Fig. 1(b2) are bright, the majority of the kernels in the filter are darker than those in Fig. 1(b1). As a result, it is visually understandable that the filter presented in Fig. 1(b1) is more important than the filter presented in Fig. 1(b2).

3.3 Hybrid Filter Pruning

As discussed in Sect. 1, since both importance-based and similarity-based methods have limitations, we are motivated to propose a hybrid method that can selectively use both pruning methods. After critically analyzing the advantages and limitations of both methods in the experiment with visualizations (in Sect. 5.1), we confirmed and explained the effectiveness of the hybrid method, where importance-based pruning method performed better at pruning early convolutional layers, while the similarity-based pruning method performed better at pruning later convolutional layers.

To identify the proper layers to apply importance-based method and similarity-based method respectively, we further analyzed the impact of layer sequence number, the filter number, and the dimension of filter weights and filter outputs on the pruning performance. The results shows layer sequence number has the most significant influence. Before the first six to seven layers, the importance-based method shows better performance, while the similarity-based method shows better performance in pruning the last seven layers. Two methods have comparable performance in pruning the middle layers. In our experiment, the change in model types and data inputs does not have much influence on this finding. Therefore, there are three simple ways to selectively use the two methods in pruning: 1) apply the importance-based method when pruning the first seven layers, and apply the similarity-based method when pruning the rest of the layers; 2) apply the importance-based method when pruning all the layers, before the last seven layers where similarity-based pruning applies; 3) apply importance-based method when pruning the first half of the layers, and similarity-based pruning for the rest of the layers. By investigating the individual layer pruning results in Sect. 5.1, the first way shows better performance, particularly when the layer pruning ratio is below 60%. Therefore, we apply the first way in the pruning experiment setting.

4 Experiment Setup

To analyze the advantages and limitations of the similarity-based method and the importance-based method, we first set up a single-layer filter pruning task (in Sect. 5.1)

[15]. In this task, we pruned layers individually at different pruning ratios without fine-tuning the model, and evaluated how the method performs in pruning different layers in the network. The datasets used in this task were two benchmark datasets, Cats and Dogs [15] and Oxford Flower 102 [30]. We applied two typical types of convolutional networks, VGG [31] and ResNet [32] with a setup of fewer layers, but still high accuracy, for easy visualization. To be specific, we used the VGG-16 and Resnet-34 models pre-trained on the two datasets with more than 97% accuracy.

In the second experiment task (in Sect. 5.2), we validated our proposed hybrid method with a model-level filter pruning setup and compared the pruning results with the other state-of-the-art methods. Two other benchmark datasets were applied in this task, namely CIFAR-10 and ILSVRC2012 (ImageNet) datasets [20]. VGG-16, ResNet-56, and ResNet-110 models were used for the CIFAR-10 dataset, whereas VGG-16 and ResNet-34 were used for the ImageNet dataset.

To evaluate the effectiveness of the proposed pruning method, similar to other static structured pruning methods, we applied a pre-defined pruning ratio (percentage of parameters to be pruned) for each filter before pruning, and compared the model predicting accuracy after pruning and fine-tuning. The pruning ratio applied in the comparison analysis is adopted from a well-cited pruning study by Li et al. [20] which sets a smaller pruning ratio for the more important convolutional layers and a higher pruning ratio for the less important layers. During the model re-training, a.k.a finetuning, the initial learning rate was set as 0.001, and momentum and weight decay were 0.9 and 0.0001, respectively. The training batch size was 64 for training and 128 for testing, and a total number of 100 epochs were used to re-train the model.

5 Experiment Results

In the first experiment, we analyzed the similarity-based and importance-based filter pruning by comparing the after-pruning model accuracy after applying these two methods and visualizing the decision-making process to understand the underlying reasons for pruning the selected filters. In the second experiment, we compared the effectiveness of our hybrid pruning method with other state-of-the-art methods.

5.1 Comparisons of Similarity-Based Filter Pruning and Importance-Based Filter Pruning

In this experiment, we first compared the similarity-based and importance-based filter pruning methods based on their effectiveness in a single-layer pruning setup. We pruned the filters of each convolutional layer individually at different pruning ratios from 10% to 90% without finetuning. Then, we analyzed both methods with the generated to explain the pruning behavior.

Similarity-Based Filter Pruning VS Importance-Based Filter Pruning. Based on the single-layer pruning results for the VGG-16 model on the Cats and Dogs dataset in Fig. 2, we found that the importance-based method is performing well in pruning the early layers from layer 1 to layer 6, before the filter number increases. But starting from layer 7 to layer 13, the similarity-based method began to show its advantage in pruning. Similar findings were found in the Flower dataset as shown in Fig. 3, where pruning before layer 7, the importance-based method performed better, and pruning after layer 7, the similarity-based method performed better.

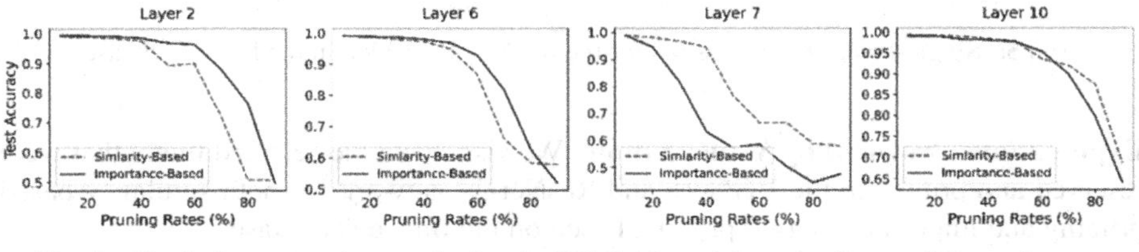

Fig. 2. Single-layer pruning results for the VGG-16 model on the Cats and Dogs dataset.

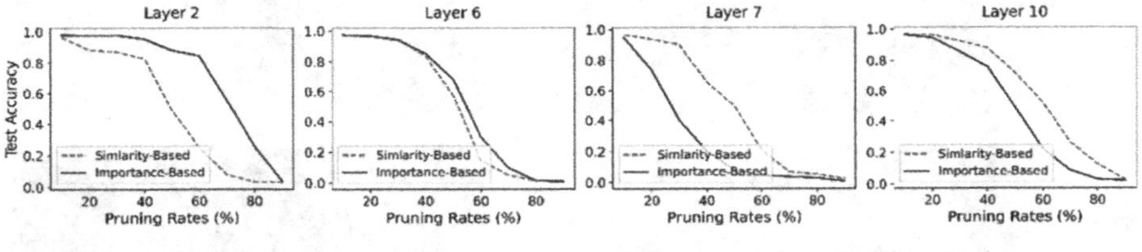

Fig. 3. Single-layer pruning results for the VGG-16 model on the Flowers dataset.

Figure 4 and Fig. 5 are the results of pruning the ResNet-18 model on the Cats and Dogs dataset and the Flowers dataset respectively. In pruning 17 convolutional layers (not including the three convolutional layers for residual) in the ResNet-18 model, the importance-based pruning methods outperformed the similarity-based method in the early layers, from layer 1 to layer 6, and starting from layer 7, the similarity-based method outperformed importance-based method.

Therefore, from the above evaluations of the two pruning methods based on different pruning tasks, we found that the importance-based method performed better at pruning the first six to seven layers, while the similarity-based method performed better at pruning the last seven layers in the network. Two methods have comparable performance in pruning the middle layers.

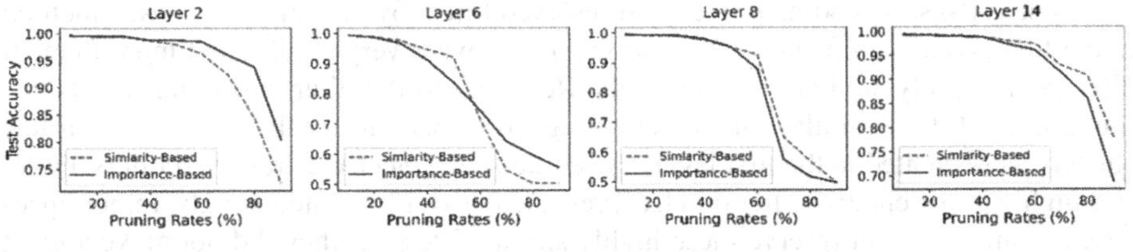

Fig. 4. Single-layer pruning results for the ResNet-18 model on the Cats and Dogs dataset.

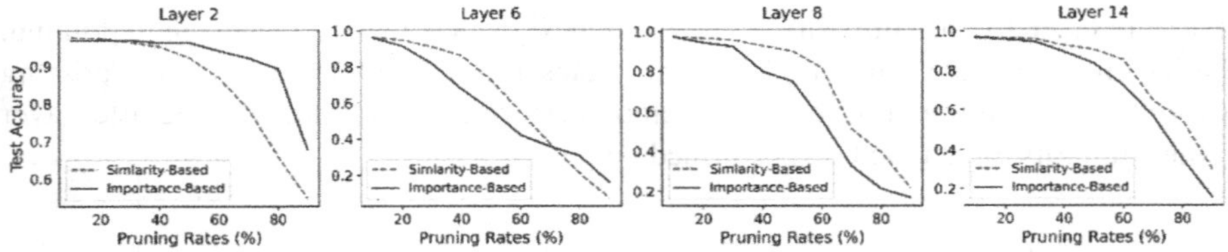

Fig. 5. Single-layer pruning results for the ResNet-18 model on the Flowers dataset.

Explanations Provided by Visualization. We visualized the selected filters that were reserved and pruned in the VGG-16 and ResNet-34 network for both similarity-based pruning and importance-based pruning based on the ImageNet dataset.

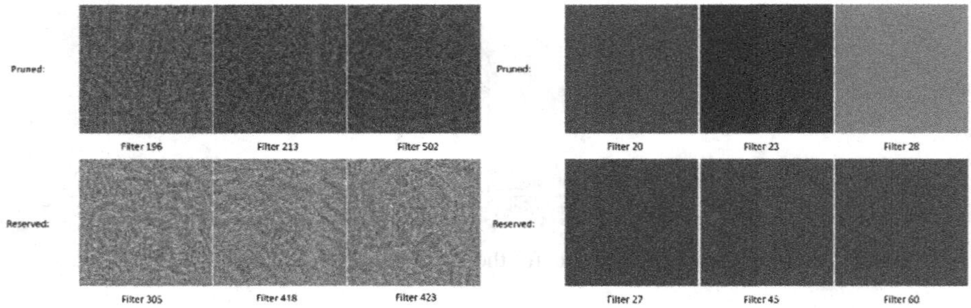

(a) (b)

Fig. 6. Filter selection for (a) important-based pruning and (b) similarity-based pruning.

Based on the visualization analysis, we found the similarity-based method underperformed when the filter structure is simpler, whereas the importance-based method underperformed when the filter structure is complex. For example, Fig. 6(a) shows some filters from the last convolutional layer of ResNet34. Filters in the first row are pruned and the filters in the second row are reserved. It is significant that, although the first row has a lower norm with a darker visualized filter image, they captured unique patterns from the input data when compared with other filters in the layer. As such, they are irreplaceable and should not be pruned in the network. Interestingly, similarity-based pruning performed well in this layer and did not prune these filters.

Figure 6(b) shows some pruned and reserved filters by the similarity-based method in the first layer. Significantly, the reserved filters were very similar and more than 20 filters were highly similar to these filters. Referring to the filter clustering results, we found that all these similar filters were assigned to one cluster. But, since only a few clusters were created in the first layer, these filters are selected as reserved filters, either as a highly representative filter or a least representative filter. Interestingly, importance-based pruning did not reserve these highly similar filters, as they did not make a great contribution to the network classification.

From this visualization analysis, we explain the reasons why the similarity-based method performs better at pruning later layers, and the importance-based method performs better at pruning early layers. These observations suggest that each pruning method has its own limitations and combing the two methods can potentially improve the pruning performance. In the next experiment, we further explored this potential.

5.2 Comparisons of Our Hybrid Filter Pruning with the State-of-the-Art Filter Pruning Methods

Based on the findings from Sect. 5.1, we proposed a novel hybrid filter pruning method described in Sect. 3.3. To validate the effectiveness of the hybrid filter pruning method, we compared it with the state-of-art methods in model-level pruning, based on the retrained accuracy of the pruned models using the CIFAR-10 dataset and the ImageNet dataset. Since it can be biased when comparing different methods directly from their reported results in the papers due to the differences in experimental environments and settings, in our experiments, we only included the state-of-the-art methods that provided implementation codes in public. We run and compared their methods in the same environment with the same experiment settings.

Experiments on the CIFAR-10 Dataset. In the experiment, we aimed to compare our hybrid filter pruning method with different types of state-of-the-art filter pruning methods. Apart from the most commonly applied pruning procedures with single-time static pruning, where all the filter pruning candidates were identified and pruned at one single time, we also included iterative pruning and an innovative pruning from the scratch method which in nature is to train a small network from scratch. We used norm-based importance pruning method [20] to present importance-based pruning in a single-time pruning manner. Filter pruning based on geometric median (FPGM) [13] was used to present similarity-based pruning. Thinet [11] was used to present the iterative pruning, whereas Prune from Scratch [17], presented the pruning from the scratch method. Except for the Prune from Scratch method which does not need pre-trained weights, the rest of the methods applied the same pre-trained model. As such, no further training was required thus saving time and resources during training.

We followed the pre-defined pruning ratio setting provided by Li et al. [20], since they provided very detailed justifications in their study. Their study described why these settings are reasonable in presenting performance analysis at pruning each layer and identifying the layers that are sensitive to pruning.

Table 1 shows the pruning performance of different methods on the CIFAR-10 dataset. The values presented are based on average testing results of 10 runs. The results show that our hybrid method achieves the highest pruning performance for VGG-16, ResNet-56, and ResNet-110 models under different pruning ratios. It is surprising to find that the pruning performance is quite poor for the Prune from Scratch method, but a similar underperformance result has also been identified in another study by Gale et al. [18], and they suggested that this may be due to the poor learning rate setting applied in the Prune from Scratch method.

Table 1. Pruning results on the CIFAR-10 dataset

Methods	VGG-16 (Original accuracy 93.27%)	ResNet-56 (Original accuracy 93.50%)		ResNet-110 (Original accuracy 93.88%)	
	Prune ratio 45%	Prune ratio 10%	Prune ratio 14%	Prune ratio 2%	Prune ratio 34%
Norm-based [20]	93.26%	93.54%	93.45%	93.65%	93.43%
Prune from scratch [17]	92.53%	92.39%	91.74%	91.45%	91.38%
Thinet [11]	93.15%	93.30%	93.30%	93.60%	93.55%
FPGM [13]	93.03%	93.35%	93.49%	93.40%	93.39%
Hybrid pruning	93.39%	93.59%	93.52%	93.75%	93.59%

Experiments on the ImageNet Dataset. We applied a standard preprocessing and augmentation process [11]: re-sizing images to have a small dimension of 256, randomly cropping a 224 $\times$ 224 patch, randomly applying horizontal flips, and normalizing images by subtracting a per-dataset mean and dividing by a per-dataset standard deviation. During testing, we used the central crop of size 224 $\times$ 224.

Table 2. Pruning results on the ImageNet (ILSVRC2012) dataset

Methods	VGG-16 (Original accuracy 73.36%)	ResNet-34 (Original accuracy 73.31%)	
	Prune ratio 45%	Prune ratio 8%	Prune ratio 10%
Norm-based [20]	68.98%	72.55%	72.15%
Prune from scratch [17]	68.00%	71.50%	71.02%
Thinet [11]	68.05%	71.58%	71.85%
FPGM [13]	68.85%	72.30%	72.10%
Hybrid pruning	70.11%	72.60%	72.25%

Same as the predefined settings applied in the CIFAR-10 experiment, the pruning ratios setting for each convolutional layer are provided by Li et al. [20], who identified that pruning the first layer of the residual block is more effective at reducing the overall FLOP than pruning the second layer. Thinet was not implemented for ResNet-34 in the original paper, so we implemented it by ourselves in the experiment.

Table 2 shows the pruning performance of our hybrid pruning method and other state-of-the-art pruning methods on ImageNet data. From the results, it is evident that our hybrid method achieves the best pruning performance for both VGG-16 and ResNet-34 in different pruning settings.

6 Conclusions

In this paper, we critically analyzed the strengths and weaknesses of two types of widely used filter pruning methods, namely, similarity-based pruning and importance-based pruning, through the quantitative evaluation based on the post-prune model re-trained accuracy and the qualitative explanations to analyze and understand the pruning processes. The research was carried out to address the important gap in the literature with a comprehensive study and analysis of the applicability of the two different neural network pruning approaches. Based on the analysis and findings, we proposed a novel hybrid method to take the advantage of both types of methods to improve the network pruning performance with explainability. The comparison results demonstrate that the proposed hybrid pruning method outperforms state-of-the-art pruning methods in different testing scenarios.

References

1. Choudhary, T., Mishra, V., Goswami, A., Sarangapani, J.: A comprehensive survey on model compression and acceleration. Artif. Intell. Rev. **53**(7), 5113–5155 (2020). https://doi.org/10.1007/s10462-020-09816-7
2. Ba, L.J., Caruana, R.: Do deep nets really need to be deep? In: Advances in Neural Information Processing Systems. Neural Information Processing Systems Foundation, pp. 2654–2662 (2014)
3. Deng, B.L., Li, G., Han, S., et al.: Model compression and hardware acceleration for neural networks: a comprehensive survey. Proc. IEEE **108**, 485–532 (2020)
4. Xu, C., Zhou, W., Ge, T., et al.: Beyond preserved accuracy: evaluating loyalty and robustness of BERT compression. In: EMNLP 2021-2021 Conference on Empirical Methods on Natural Language Processing Procceeding, pp. 10653–10659 (2021)
5. Xu, C., McAuley, J.: A survey on model compression for natural language processing. arXiv (2022)
6. Du, M., Mukherjee, S., Cheng, Y., et al.: What do compressed large language models forget? Robustness challenges in model compression. arXiv (2021)
7. Zhang, Y., Lin, M., Lin, C.W., et al.: Carrying out CNN channel pruning in a white box. IEEE Trans. Neural Networks Learn. Syst. 1–10 (2022)
8. Hu, H., Peng, R., Tai, Y.-W., Tang, C.-K.: Network trimming: a data-driven neuron pruning approach towards efficient deep architectures. arXiv (2016)
9. Woo, Y., Kim, D., Jeong, J., et al.: Zero-keep filter pruning for energy/power efficient deep neural networks†. Electron **10**, 1238 (2021)
10. Yu, R., Li, A., Chen, C.F., et al.: NISP: pruning networks using neuron importance score propagation. In: Proceedings of the IEEE Computer Society Conference on Computer Vision and Pattern Recognition, pp. 9194–9203. IEEE Computer Society (2018)
11. Luo, J.H., Wu, J., Lin, W.: ThiNet: a filter level pruning method for deep neural network compression. In: Proceedings of the IEEE International Conference on Computer Vision, pp. 5068–5076. IEEE Computer Society (2017)

12. Yeom, S.K., Seegerer, P., Lapuschkin, S., et al.: Pruning by explaining: a novel criterion for deep neural network pruning. Pattern Recognit **115**, 107899 (2021)
13. He, Y., Liu, P., Wang, Z., et al.: Filter pruning via geometric median for deep convolutional neural networks acceleration. In: Proceedings of the IEEE Computer Society Conference on Computer Vision and Pattern Recognition, pp. 4335–4344. IEEE Computer Society (2019)
14. Qi, J., Yu, Y., Wang, L., et al.: An effective and efficient hierarchical K-means clustering algorithm. Int. J. Distrib. Sens. Networks **13**, 1–17 (2017)
15. Yao, K., Cao, F., Leung, Y., Liang, J.: Deep neural network compression through interpretability-based filter pruning. Pattern Recognit. **119**, 108056 (2021)
16. Shao, M., Dai, J., Wang, R., Kuang, J., Zuo, W.: CSHE: network pruning by using cluster similarity and matrix eigenvalues. Int. J. Mach. Learn. Cybern. **13**(2), 371–382 (2021). https://doi.org/10.1007/s13042-021-01411-8
17. Liu, Z., Sun, M., Zhou, T., et al.: Rethinking the value of network pruning. In: 7th International Conference on Learning Representations, ICLR 2019. International Conference on Learning Representations, ICLR, pp. 4335–4344 (2019)
18. Gale, T., Elsen, E., Hooker, S.: The state of sparsity in deep neural networks. CoRR (2019)
19. Liang, T., Glossner, J., Wang, L., et al.: Pruning and quantization for deep neural network acceleration: a survey. Neurocomputing **461**, 370–403 (2021)
20. Li, H., Samet, H., Kadav, A., et al.: Pruning filters for efficient convnets. In: 5th International Conference on Learning Representations, ICLR 2017-Conference Track Proceedings. International Conference on Learning Representations, ICLR (2017)
21. He, Y., Dong, X., Kang, G., et al.: Asymptotic soft filter pruning for deep convolutional neural networks. IEEE Trans. Cybern. **50**, 3594–3604 (2020)
22. Molchanov, P., Tyree, S., Karras, T., et al.: Pruning convolutional neural networks for resource efficient inference. In: 5th International Conference on Learning Representations, ICLR 2017-Conference Track Proceedings. International Conference on Learning Representations, ICLR (2017)
23. Liu, C., Wu, H.: Channel pruning based on mean gradient for accelerating Convolutional Neural Networks. Signal Process. **156**, 84–91 (2019)
24. Alqahtani, A., Xie, X., Jones, M.W., Essa, E.: Pruning CNN filters via quantifying the importance of deep visual representations. Comput. Vis. Image Underst. **208**, 103220 (2021)
25. Zhu, J., Pei, J.: Filter pruning via structural similarity index for deep convolutional neural networks acceleration. In: Proceedings of IEEE 14th International Conference on Intelligent Systems and Knowledge Engineering, ISKE 2019. Institute of Electrical and Electronics Engineers Inc., pp. 730–734 (2019)
26. Sabih, M., Hannig, F., Teich, J.: DyFiP: explainable AI-based dynamic filter pruning of convolutional neural networks. In: EuroMLSys 2022 - Proceedings of the 2nd European Workshop on Machine Learning and Systems. Association for Computing Machinery, Inc., pp 109–115 (2022)
27. Lundberg, S.M., Lee, S.I.: A unified approach to interpreting model predictions. In: Advances in Neural Information Processing Systems, pp. 4766–4775 (2017)
28. Shrikumar, A., Greenside, P., Kundaje, A.: Learning important features through propagating activation differences. In: 34th International Conference on Machine Learning, ICML 2017, pp. 4844–4866 (2017)
29. Alahakoon, D., Halgamuge, S.K., Srinivasan, B.: Dynamic self-organizing maps with controlled growth for knowledge discovery. IEEE Trans. Neural Networks **11**, 601–614 (2000)
30. Nilsback, M.E., Zisserman, A.: Automated flower classification over a large number of classes. In: Proceedings-6th Indian Conference on Computer Vision, Graphics and Image Processing, ICVGIP 2008, pp. 722–729 (2008)

31. Simonyan, K., Zisserman, A.: Very deep convolutional networks for large-scale image recognition. In: 3rd International Conference on Learning Representations, ICLR 2015-Conference Track Proceedings. International Conference on Learning Representations, ICLR (2015)
32. He, K., Zhang, X., Ren, S., Sun, J.: Deep residual learning for image recognition. In: Proceedings of the IEEE Computer Society Conference on Computer Vision and Pattern Recognition, pp. 770–778 (2016)

Outlier Detection for Visual Odometry in Vegetated Scenes Using Local Flow Consistency

Sam Schofield(✉), Andrew Bainbridge-Smith, and Richard Green

Computer Science and Software Engineering, University of Canterbury, Christchurch, New Zealand
sam.schofield@pg.canterbury.ac.nz

Abstract. We present a novel outlier detector for visual odometry algorithms operating in dynamic vegetated environments. The outlier detector utilises the difference in optical flow patterns caused by camera motion and plants moving in the wind. The proposed method is compared to existing methods on real and synthetic data containing wide ranges of scene motion. Our results show that the proposed method works well as a pre-processing step to RANSAC, improving pose estimation accuracy in dynamic vegetated environments with minimal computation overhead.

Keywords: Visual odometry · Outlier detection · Dynamic environments

1 Introduction

Visual odometry (VO) is the process of estimating the motion of a camera using images that the camera is capturing. VO is often used in robotics to provide the robot with a pose estimate when GPS is unavailable or does not provide enough precision. Traditionally, VO relied on the assumption that it operated in a static environment. However, most modern visual odometry/visual-inertial odometry (VIO) algorithms typically use either RANSAC [22,23,26] or a Chi-squared test [1,9,23] to provide robustness to outliers. As a result, most algorithms can handle small amounts of scene motion, but struggle in highly dynamic scenes, as shown in [24].

More recently, there has been an abundance of work aiming to improve the robustness of VO in scenes where these standard methods fail. With the popularity of autonomous driving and the abundance of driving datasets, much of this research targets urban environments [5,11,15,17,27]. Unlike urban environments, dynamic vegetated environments (such as forests) remain relatively unstudied, yet are essential for expanding robotics (specifically UAVs in this work) into areas such as agriculture and forestry.

The nature of scene motion in vegetated environments differs from that in urban areas. Urban settings typically contain a static background with objects

W. Q. Yan et al. (Eds.): IVCNZ 2022, LNCS 13836, pp. 230–245, 2023.
https://doi.org/10.1007/978-3-031-25825-1_17

(a) Example image from KITTI dataset. (b) Forest example image.

Fig. 1. Example images highlighting the differences between urban and forest environments. In the urban environment (Fig. 1a), the moving objects (bus, pedestrians and trees) can be easily identified and ignored leaving a static background (roads, buildings, etc.) to use for pose estimation. The forest environment (Fig. 1b), however, has no distinguishable static background as the entire scene is potentially moving.

(such as cars or people) moving through it (Fig. 1a). This configuration has led to many scene-motion-robust VO algorithms that work by separating the moving objects from the static background, then using the background for pose estimation [5,11,15,17,27]. This segmentation of foreground and background often relies on assumptions about the scene's geometry, e.g. that the background consists of lines or planes [17,27,29], or about the moving objects, e.g. identifying (and ignoring) cars and pedestrians using machine learning [5,11,15]. However, these methods are unlikely to work in a vegetated environment such as a forest where the scene's geometry is relatively unpredictable. Furthermore, even if the foreground and background can be segmented, the background is also likely to move (Fig. 1b), which may cause significant problems for existing methods.

A more promising area of research focuses on directly analysing 3D landmarks to determine whether they are static or dynamic. Scene flow analysis, for example, examines the 3D flow vectors between landmarks viewed across consecutive images (within the camera's reference frame). One problem with scene flow analysis is that rotation of the camera induces a larger scene flow for features with a greater depth (see Fig. 2). This inconsistency can be dealt with in several ways. For example, pre-rotating the point cloud using angular velocity measurements from a gyroscope [16], or clustering the features based on their position to ensure similar scene flow [19]. Another alternative is to look at the consistency of the distances between landmarks over time [6,31] as these distances are invariant to where they are viewed from (see Fig. 3). Unfortunately, this approach can become computationally inefficient depending on the sampling strategy. A downside to all methods that analyse 3D data is that if the depth is generated by a stereo camera (as in this work), then they have to deal with the quadratic and anisotropic nature of stereo error (Fig. 4). Typically, this is done using the Mahalanobis distance [21] instead of the Euclidean distance. While the Mahalanobis distance deals with the problems associated with stereo errors, it requires the covariance of each landmark to be estimated and maintained. If the covariance is not already estimated for other purposes, this can worsen the system's complexity and efficiency.

An alternative approach is to analyse the optical flow of features in the scene. Since optical flow is measured in image space, it does not suffer from the effects of stereo error. As such, there is no need to use the Mahalanobis distance or estimate landmark covariance, improving computational performance. Additionally, operating in image space aligns well with reprojection-error minimisation, often used for pose estimation in VO. The primary problem with using optical flow analysis to identify moving features is that optical flow is not consistent across the whole image. This inconsistency is mainly due to the relationship between optical flow and scene depth when the camera is translating (although other factors are also involved). The inconsistency can be addressed by forming neighbourhoods of features with similar 3D positions as these features will have similar camera-motion-induced optical flows (Fig. 7b). Although the optical flow vectors for static features are locally consistent, this is typically not the case for features on moving trees or bushes (Fig. 7c). We utilise this discrepancy to identify flow vectors contaminated with scene motion. We show that the proposed approach, named Local Flow Consistency (LFC), is a good pre-processing step to RANSAC, improving pose estimation accuracy while adding little processing overhead and, in some scenarios, actually reducing the processing time of state-of-the-art RANSAC.

2 Related Work

The use of optical flow to detect scene motion is not new; however, it is far more common in the literature for moving object detection than visual odometry. The optical-flow-based moving object detection literature can be divided into three broad categories.

The first category uses machine learning. These methods often work by first identifying potentially moving objects such as cars or people, then analysing the optical flow of the object and its surroundings to determine if it is moving [3,20, 34]. This approach works well for large objects that move coherently. However, they are likely to struggle with the more chaotic motion of plants. Other methods such as [7,28] feed both the images and the optical flow field into a neural network to segment the image into moving and static regions. While these methods could work with moving vegetation, they are unlikely to operate in real-time onboard a computationally constrained UAV.

The second category uses a motion prior to compensate for the camera's movement between frames. The motion-compensated optical flow residuals can then be used to identify which image regions are influenced by scene motion. Many of these methods estimate the camera motion between consecutive images using the estimated optical flow [4,10,13]. These approaches rely on the assumption that most of the scene is static, so the camera's motion can be estimated accurately. However, this assumption does not hold in a forest environment where

(a) Camera translation (and movement of single landmark) in world reference frame.

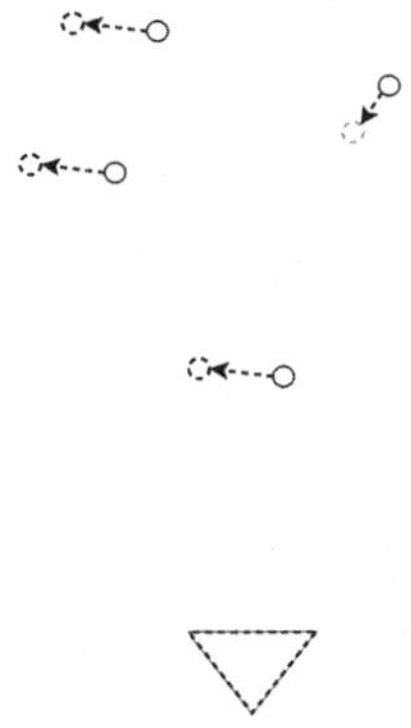

(b) Scene flow as a result of the motion in Figure 2a.

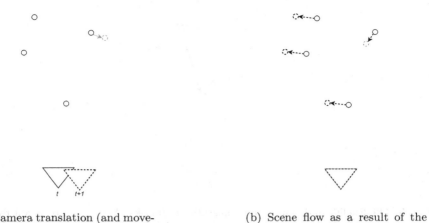

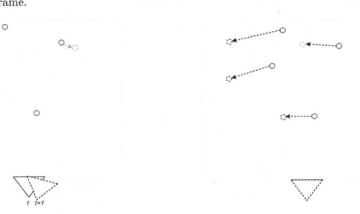

(c) Camera translation and rotation (and movement of single landmark) in world reference frame.

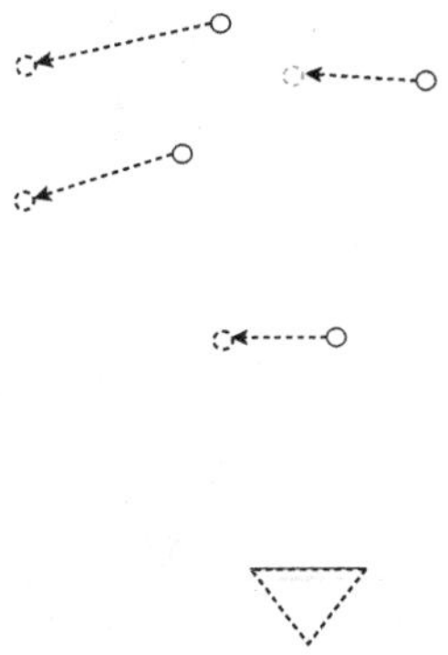

(d) Scene flow as a result of the motion in Figure 2c.

Fig. 2. Camera motions and the resulting scene flows. Figure 2b shows that scene flow analysis can easily detect the moving landmark when the camera undergoes pure translation. However, when rotation is introduced (Fig. 2d) it becomes challenging to distinguish between static and moving landmarks.

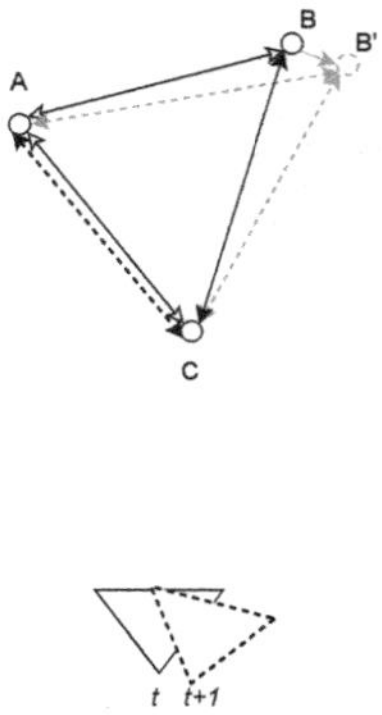

Fig. 3. The Euclidean distance between static landmarks (A, C) stays constant regardless of viewpoint. However, the distance between landmarks when one or both of them are moving typically changes over time (AB, BC).

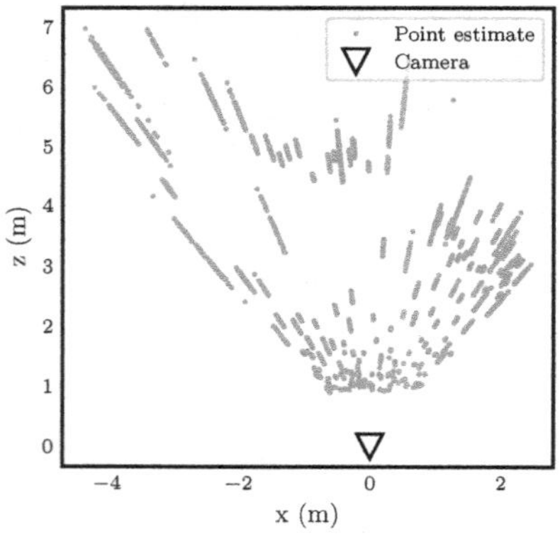

(a) Landmark position estimates from a stationary stereo camera in a static scene for one hundred samples.

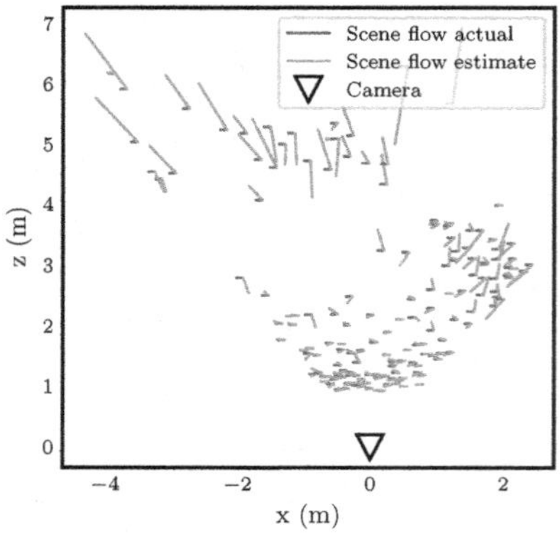

(b) Scene flow estimates in the presence of stereo error for a camera motion of approximately 0.1 m along the x axis.

Fig. 4. The influence of stereo matching error on landmark estimation and scene flow analysis.

the entire scene can be moving. A plausible alternative is to use a motion prior from other sensors such as wheel odometry [18], or an IMU [12] that are not affected by scene motion. Similarly, a constant-velocity motion model can be used to predict the camera's motion for the motion compensation step (provided the prediction is accurate enough).

The final category segments the image into static and dynamic regions by analysing the flow field directly. These methods [14,30,32] typically work by modelling the flow of the background, then comparing that approximated flow to the measured flow field to identify regions of scene motion. While these methods can accurately segment the image into static and dynamic regions without a motion prior, they typically require dense optical flow estimates, which results in slower processing time. The fastest algorithm in this category [14] requires a desktop-grade GPU to process frames in 50 ms—making it ill-suited for use on a UAV. These methods may also struggle when the background is not static, as the flow becomes more difficult to model.

The proposed approach combines the second and third categories. We analyse local neighbourhoods of sparse optical flow vectors in conjunction with a motion prior to identify features undergoing scene motion. This approach improves performance over the naive use of a motion prior with little processing overhead (as shown in Sect. 4).

3 Method

At a high level, the proposed approach works as follows. Given a set of 3D points at time t and a set of corresponding vectors representing the optical flow from time t to $t+1$, we first assign each point a neighbourhood of other points likely to have a similar camera-motion-induced optical flow. The expected variation in the optical flow for each neighbourhood (in the absence of scene motion) is then calculated by predicting the flow using a motion prior. Features whose measured flow varies from their neighbours by more than the expected amount are deemed contaminated by scene motion and discarded as outliers. Figure 5 gives a visual overview of this process. This strategy relies heavily on the fact that camera-motion-induced optical flow is locally consistent, while scene-motion-induced optical flow is not (particularly that caused by moving vegetation).

3.1 Neighbourhood Formation

The first step in the pipeline is to associate each feature with a neighbourhood of other features likely to have a similar (camera-motion-induced) optical flow. Camera-motion-induced optical flow varies between image features for multiple reasons. For example, the features' depth when translating parallel to the image plane (Fig. 6a), depth and feature position relative to the centre of expansion

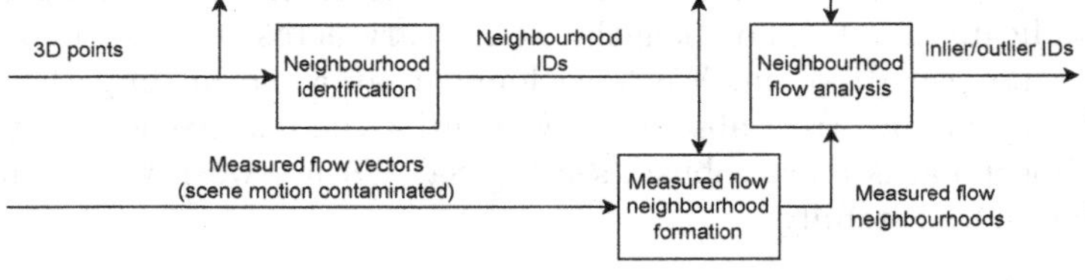

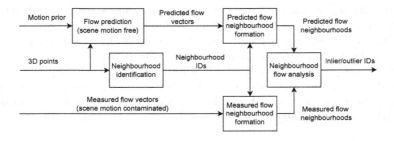

Fig. 5. System diagram of the LFC process.

when moving along the optical axis (Fig. 6c), and the features' distance from the centre of rotation when rotating about the optical axis (Fig. 6d). This work deals with these factors by assigning each feature a neighbourhood consisting of the n features closest to it in 3D space, as this typically ensures similar flow vectors across the group. The exception is when the centre of rotation or expansion falls within a group of features, in which case the flow directions can vary significantly. Figure 6a shows some example feature clusters when $n = 5$. The optimal neighbourhood size depends on how close (on average) the 3D points are to their neighbours. A larger neighbourhood will be less affected by outliers. However, increasing neighbourhood size also increases the distance between points—reducing the expected similarity in camera-motion-induced optical flow.

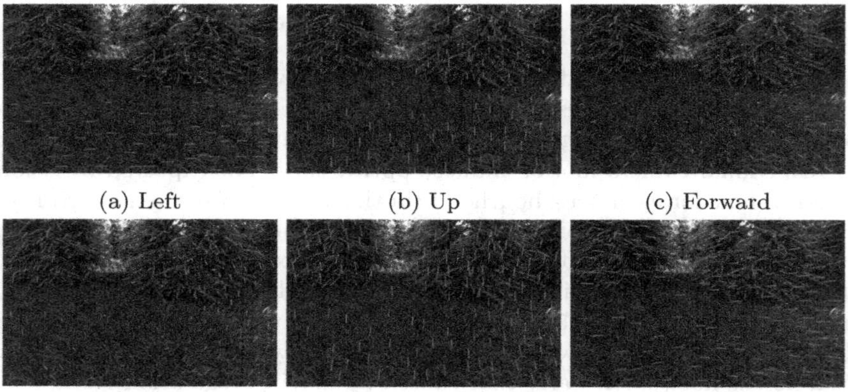

(a) Left (b) Up (c) Forward

(d) Roll (e) Pitch (f) Yaw

Fig. 6. Examples of optical flow vectors for different camera motions.

3.2 Outlier Identification

Once the neighbourhoods have been formed, their flow lengths are analysed to detect outliers. We choose to analyse only the length of the vectors and not the direction, as the flow direction of short flow vectors is very sensitive to noise and can vary significantly if the centre of expansion or rotation fall within a neighbourhood—this is not the case for flow length.

First, the optical flow for each neighbourhood is predicted using a motion prior to indicate how the flow lengths would vary across the neighbourhood if there were no scene motion. We have found that while an imperfect motion prior will generate incorrect absolute flow lengths, the relative flow length across the neighbourhood is more robust (see Fig. 8a). Future work will examine this observation more carefully.

(a) (b) (c)

Fig. 7. Example feature neighbourhoods (Fig. 7a) along side the ideal optical flow (as if there were no scene motion) (Fig. 7b) and the actual optical flow (Fig. 7c). The introduction of scene motion in Fig. 7c makes a significant difference to the deviation in flow direction and magnitude. Note that the same colour is used per cluster across the different sub-figures.

We utilise the relative flow length's robustness to motion prior error by constructing a flow "template" that captures the relative flow lengths of the neighbourhood. Features contaminated by scene motion are identified using the template (which is free from scene motion error) in a hypothesise-then-verify type fashion. A hypothesis template is formed for each feature in the neighbourhood by aligning the *nth* flow length in the template with the *nth* measured flow length (see Fig. 8a). Each template hypothesis is then scored using a sum of squared truncated errors,

$$s = \sum_i \rho(e_i^2),$$

where the squared truncated error, $\rho(e^2)$, is

$$\rho(e^2) = \begin{cases} e^2 & e^2 < t^2 \\ t^2 & e^2 \geq t^2 \end{cases},$$

e_i is the difference between the template hypothesis and measured flow length for the ith feature in the neighbourhood, and t is the threshold for e_i to determine if a feature is an outlier. The template with the lowest score is selected, and any feature whose flow length differs from the template by more than t is identified as an outlier and discarded (Fig. 8b). Note that the ideal value for t depends on the quality of the motion prior (more erroneous motion priors require a higher threshold), but typically a value of 0.3–1 px worked well during our testing.

Figure 9 shows an example of feature classification using the proposed method. Qualitatively, the method exhibits the desired behaviour, i.e. the outliers are predominantly detected on the tree closest to the camera (those with the most scene motion), while the inliers are primarily located on the grass (least scene motion). A quantitative analysis of the proposed method is given in Sect. 4.

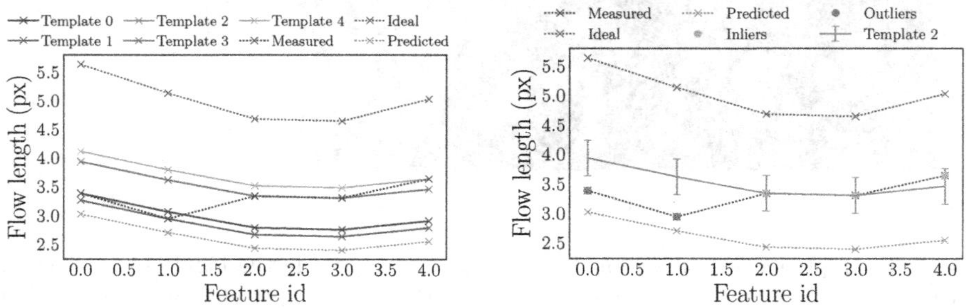

(a) Candidate flow length templates. Template n corresponds to aligning the predicted flow length template with the nth length in the measured flow neighbourhood.

(b) Outlier identification process using the chosen template. Note that while the predicted absolute flow length is incorrect (by 2.5 px), the shape of the predicted flow curve (dashed red) is similar to that of the ideal flow curve (dashed green).

Fig. 8. Outlier identification process used by the LFC approach. First the template with the lowest sum of truncated distances is found, then any features with a flow length outside a pre-defined tolerance are identified as outliers. (Color figure online)

4 Experiments

The efficiency and accuracy of the proposed method (LFC) are compared to two commonly used outlier detection methods. Specifically, RANSAC and a motion prior residual threshold (MP). The three outlier detection methods are evaluated as part of a rudimentary visual odometry pipeline using both synthetic and real images of moving vegetation. Additional experiments were performed to examine how the methods responded to increased levels of measurement error, outlier rate and the number of features. Those experiments largely agreed with the results given in Sects. 4.4 and 4.5 so were excluded for brevity.

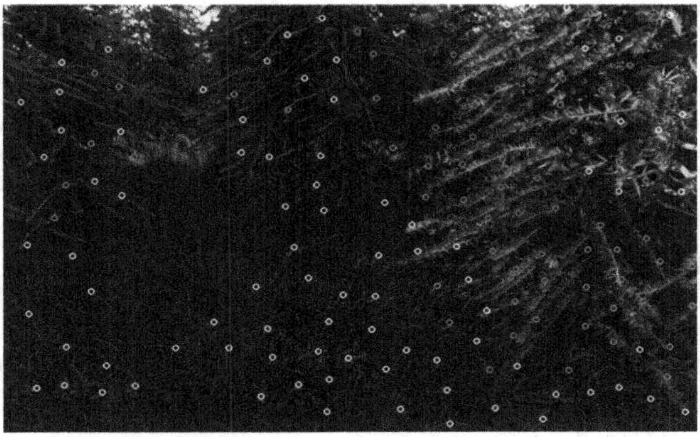

Fig. 9. Example of feature classification using Local Flow Consistency. Red features indicate outliers while green features indicate inliers. (Color figure online)

4.1 Methods

The three outlier detection methods (LFC, RANSAC and MP) are tested both on their own and as a pre-processing step for RANSAC resulting in six variants total (see Fig. 10). Each variant is used in a trajectory estimation pipeline that works as follows. First, a set of 3D points captured at time t and a set of corresponding 2D points captured at time $t+1$ are passed to the outlier detection variant (more detail on how this data is generated is given in Sect. 4.2). The outlier detector then segments those feature correspondences into inliers and outliers. The inliers are passed to an iterative perspective-n-point algorithm to estimate the camera's motion. Finally, the motion estimates are concatenated together to form an estimate of the camera's trajectory.

The MP method works by projecting the 3D points from time t onto the image plane at time $t+1$ using a motion prior. The distance between the projected points and the measured features is calculated, and any feature whose distance is above the specified threshold is discarded. Note, the MP method without an additional RANSAC step performed poorly in most of the experiments, so was omitted from the following figures for clarity.

RANSAC works by forming motion hypotheses from random, minimal samples from the data, then checking how well the remaining data fits the hypothesis [8]. In this work, the OpenCV USAC_PARALLEL RANSAC implementation was used because, in testing, it reliably provided good accuracy (relative to the other OpenCV RANSAC variants) and had a low processing time.

Method Parameters. A grid search was used per experiment to find the optimal settings for each method. The LFC methods varied both the group size and the outlier threshold, while the RANSAC methods changed the reprojection error threshold. Additionally, the maximum iterations parameter of the

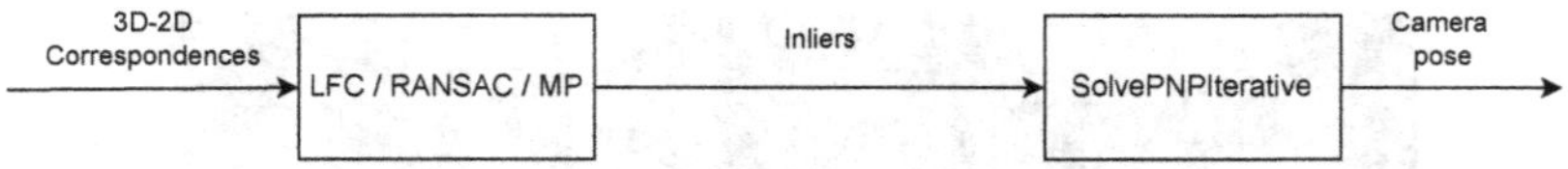

(a) Pipeline for testing outlier detection methods on their own.

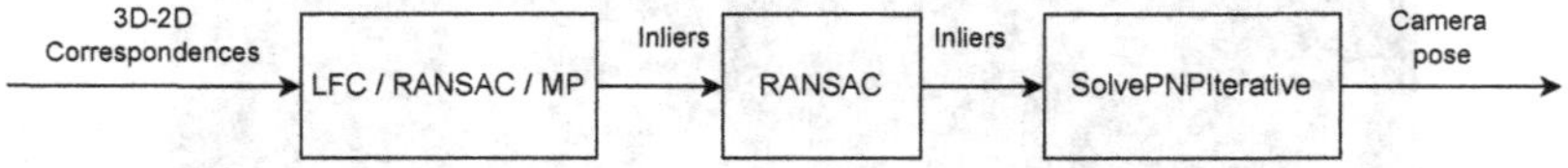

(b) Pipeline for testing outlier detection methods as a pre-processing step to RANSAC.

Fig. 10. System diagrams depicting the outlier detection variants.

RANSAC methods was limited to ensure they terminated in less than 0.05 s. The MP method also varied the reprojection error threshold. The parameters and their ranges are displayed in Table 1. The ranges were selected by finding the boundary where changing the parameters further either degrade performance substantially, or no longer resulted in significant changes in performance.

Note that the LFC and MP methods require a motion prior for optical flow prediction. For simplicity, a constant-velocity motion model was used to provide this motion prior. In practice, a better alternative would be to incorporate information from an IMU to give a better prediction.

Table 1. Parameters used during the parameter search for each method.

Method	Parameter	Range
LFC	Template threshold (px)	0.1–1 px
LFC	Group size	5–13
RANSAC	Reprojection error threshold (px)	0.6–1.6 px
MP	Reprojection error threshold (px)	0.1–10 px

4.2 Test Procedure

The data for the visual odometry experiments is generated as follows. For every sequence evaluated, two hundred features are extracted in every frame, triangulated, and then tracked to the next frame. Features are detected using OpenCV's implementation of the Shi-Tomasi feature detector [25]. These features are tracked, both between consecutive frames and between the left and right image of the stereo pair using sparse optical flow [2]. Any features that could not be tracked were discarded. This data extraction is performed as a pre-processing step to ensure each method is tested on the same data.

The 3D-2D correspondences extracted from each sequence are then used to evaluate the six outlier detection methods. The resulting pose estimates are concatenated together to form a trajectory that is compared to the ground truth trajectory to calculate the absolute trajectory error (ATE) [33].

4.3 Dataset

The Windy Forest dataset [24] was used to evaluate the outlier detection methods. The dataset consists of fifteen synthetic sequences comprised of three trajectories moving through a forest scene at five different wind levels. The dataset also includes five sequences containing real images captured by a static camera facing trees and bushes undergoing different amount of scene motion. Stereo data is provided at 20 Hz for both the synthetic and real data at a resolution of 640 × 480 and 1280 × 1024 respectively.

4.4 Synthetic Data

The first experiment examines how well the Local Flow Consistency method performs on the synthetic sequences from the Windy Forest dataset compared to RANSAC and motion prior residual thresholding. The results shown in Fig. 11a, Fig. 11b and Fig. 11c show that LFC on its own outperforms the other variants on V1_01 (Fig. 11a). However, it performs poorly on the other sequences (Fig. 11b, Fig. 11c). Upon further inspection, this decline in performance appears to be caused by the inaccuracy of the constant-velocity motion model resulting in the misclassification of the majority of the features. As mentioned previously, this could potentially be remedied by incorporating IMU measurements into the motion prior. The LFC+RANSAC variant, however, performed much better than LFC alone. Additionally, it outperformed the other variations consistently for the full-wind sequences and typically performed similarly or better on the other sequences.

The processing time for the LFC methods (see Fig. 11d, Fig. 11e and Fig. 11f) is similar to that of RANSAC and typically much better than RANSAC+RANSAC—only being outperformed by the MP method. Additionally, the results show that the processing time for the LFC-based methods is consistent across the different wind levels.

4.5 Real Data

We verify the synthetic data results using real data sequences from the Windy Forest dataset. The results in Fig. 12 show that both LFC methods outperform the other methods on the "dynamic" sequence (which has the highest levels of scene motion) for both accuracy and processing time. In fact, the LFC+RANSAC requires less processing time than pure RANSAC on the two most challenging sequences. This result is likely because LFC is able to filter many of the potential outliers—reducing the iterations required by RANSAC.

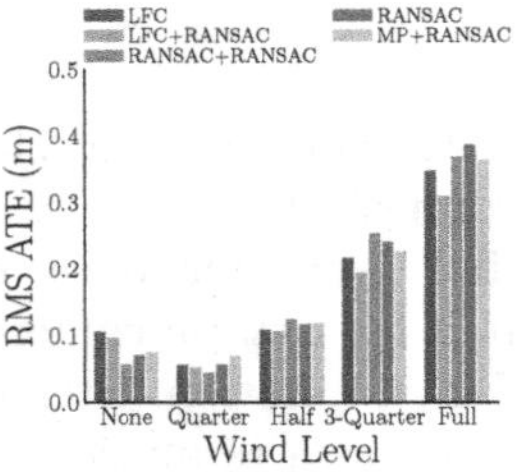

(a) V1_01 - Trajectory accuracy.

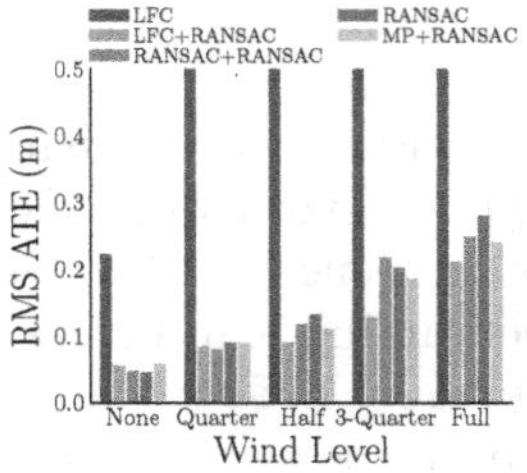

(b) V1_02 - Trajectory accuracy.

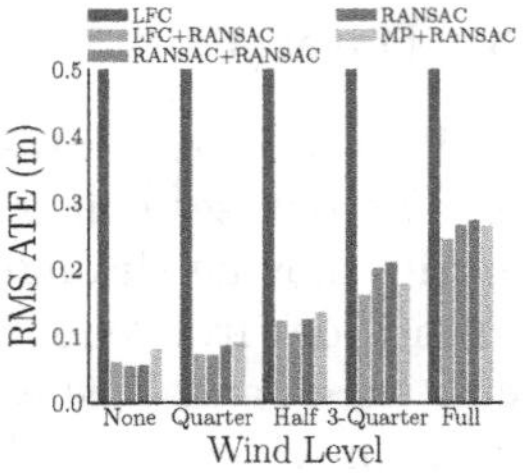

(c) V1_03 - Trajectory accuracy.

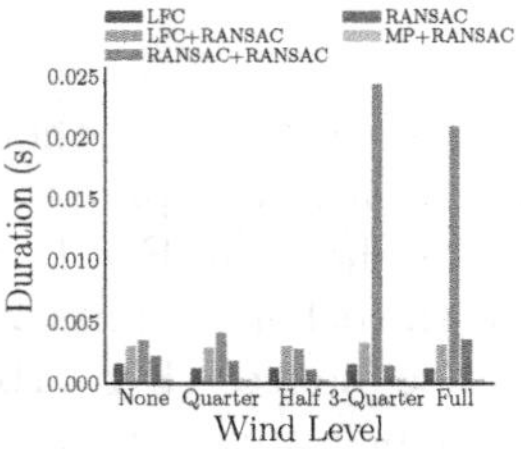

(d) V1_01 - Processing time.

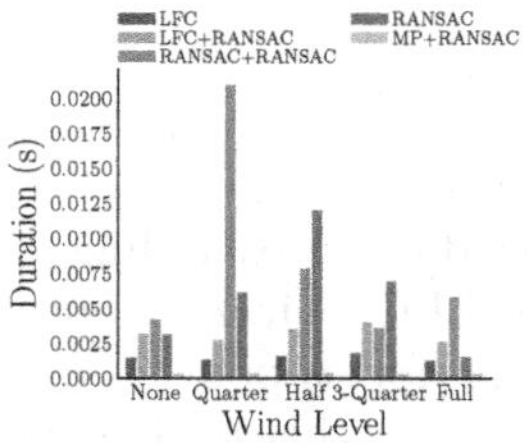

(e) V1_02 - Processing time.

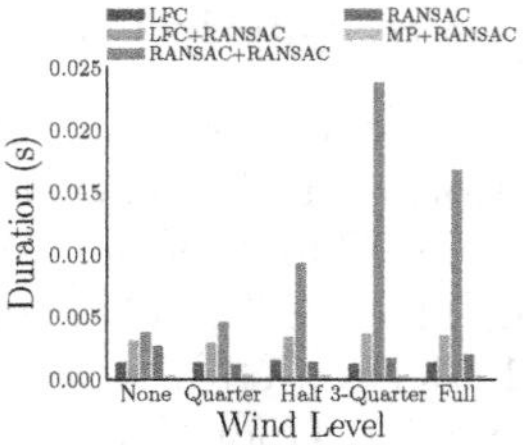

(f) V1_03 - Processing time.

Fig. 11. Graphs showing how the positional accuracy and processing time of the different methods compare on the Windy Forest Dataset.

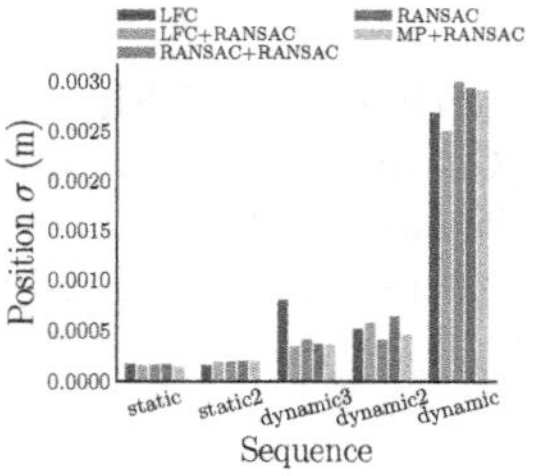

(a) Position accuracy.

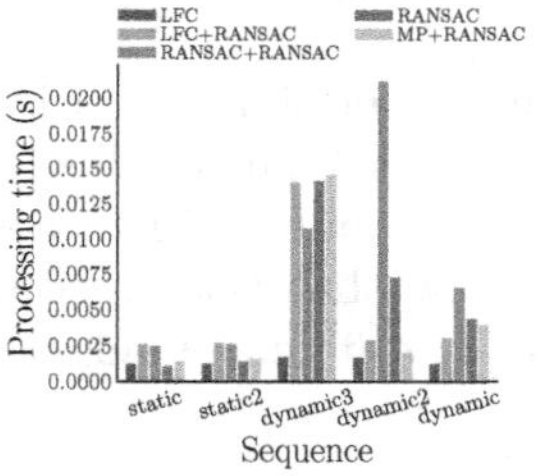

(b) Processing time.

Fig. 12. Graphs showing how the positional accuracy and processing time of the different methods compare on the static camera dataset.

5 Conclusion

This work presented a novel outlier detector for visual odometry named Local Flow Consistency (LFC). The performance of LFC was compared to a state-of-the-art RANSAC implementation and an alternative method that also utilised a motion prior to identify outliers (both independently and as a pre-filtering step to RANSAC). The results showed that using LFC in conjunction with RANSAC gave notable accuracy improvements with minimal additional processing time compared to alternative methods. This improvement is likely because LFC is able to filter the majority of outliers before the RANSAC step and is more robust to motion-prior error than a naive motion-prior-filtering approach.

Future work will investigate different methods of feature grouping, for example, through the use of flow correlation. Additionally, we will investigate whether the error threshold can be set adaptively based on the geometry/correlation of each group. Finally, we will investigate how LFC compares to other dynamic-scene robustification methods such as scene-flow analysis or 3D consistency.

Acknowledgements. The research reported in this article was conducted as part of "Enabling unmanned aerial vehicles (drones) to use tools in complex dynamic environments UOCX2104", which is funded by the New Zealand Ministry of Business, Innovation and Employment.

References

1. Bloesch, M., Omari, S., Hutter, M., Siegwart, R.: Robust visual inertial odometry using a direct EKF-based approach. In: 2015 IEEE/RSJ International Conference on Intelligent Robots and Systems (IROS), vol. 2015-December, pp. 298–304. IEEE, September 2015. https://doi.org/10.1109/IROS.2015.7353389, http://ieeexplore.ieee.org/document/7353389/
2. Bouguet, J.Y., et al.: Pyramidal implementation of the Affine Lucas Kanade feature tracker description of the algorithm. Intel Corporation **5**(1–10), 4 (2001)
3. Chen, T., Lu, S.: Object-level motion detection from moving cameras. IEEE Trans. Circ. Syst. Video Technol. **27**(11), 2333–2343 (2017). https://doi.org/10.1109/TCSVT.2016.2587387, http://ieeexplore.ieee.org/document/7505596/
4. Cheng, J., Sun, Y., Meng, M.Q.H.: Improving monocular visual SLAM in dynamic environments: an optical-flow-based approach. Adv. Robot. **33**(12), 576–589 (2019). https://doi.org/10.1080/01691864.2019.1610060
5. Cheng, J., Wang, Z., Zhou, H., Li, L., Yao, J.: DM-SLAM: a feature-based SLAM system for rigid dynamic scenes. ISPRS Int. J. Geo-Inf. **9**(4), 202 (2020). https://doi.org/10.3390/ijgi9040202, https://www.mdpi.com/2220-9964/9/4/202
6. Chiu, A.: Probabilistic Outlier Removal for Stereo Visual Odometry. Ph.D. thesis, Stellenbosch University, Stellenbosch (2017)
7. Dave, A., Tokmakov, P., Ramanan, D.: Towards segmenting anything that moves. In: 2019 IEEE/CVF International Conference on Computer Vision Workshop (ICCVW), pp. 1493–1502. IEEE, November 2019. https://doi.org/10.1109/ICCVW.2019.00187, https://ieeexplore.ieee.org/document/9022103/
8. Fischler, M.A., Bolles, R.C.: Random sample consensus: a paradigm for model fitting with applications to image analysis and automated cartography. Commun. ACM **24**(6), 381–395 (1981). https://doi.org/10.1145/358669.358692, http://portal.acm.org/citation.cfm?doid=358669.358692
9. Geneva, P., Eckenhoff, K., Lee, W., Yang, Y., Huang, G.: OpenVINS: a research platform for visual-inertial estimation. In: 2020 IEEE International Conference on Robotics and Automation (ICRA), pp. 4666–4672. IEEE, Paris, France, May 2020. https://doi.org/10.1109/ICRA40945.2020.9196524 , https://ieeexplore.ieee.org/document/9196524/

10. Ghahremannezhad, H., Shi, H., Liu, C.: Real-time hysteresis foreground detection in video captured by moving cameras. In: IST 2022 - IEEE International Conference on Imaging Systems and Techniques, Proceedings. Institute of Electrical and Electronics Engineers Inc. (2022). https://doi.org/10.1109/IST55454.2022.9827719
11. Henein, M., Zhang, J., Mahony, R., Ila, V.: Dynamic SLAM: the need for speed. In: 2020 IEEE International Conference on Robotics and Automation (ICRA), pp. 2123–2129. IEEE, May 2020. https://doi.org/10.1109/ICRA40945.2020.9196895, https://ieeexplore.ieee.org/document/9196895/
12. Huang, C., Chen, P., Yang, X., Cheng, K.T.: REDBEE: a visual-inertial drone system for real-time moving object detection. In: 2017 IEEE/RSJ International Conference on Intelligent Robots and Systems (IROS), pp. 1725–1731 (2017). 10.0/Linux-x86_64
13. Huang, J., Zou, W., Zhu, J., Zhu, Z.: Optical flow based real-time moving object detection in unconstrained scenes, July 2018. http://arxiv.org/abs/1807.04890
14. Huang, J., Zou, W., Zhu, Z., Zhu, J.: An efficient optical flow based motion detection method for non-stationary scenes. In: Proceedings of the 31st Chinese Control and Decision Conference (2019 CCDC) (2019)
15. Irmisch, P., Baumbach, D., Ernst, I.: Robust visual-inertial odometry in dynamic environments using semantic segmentation for feature selection. ISPRS Ann. Photogrammetry Remote Sens. Spatial Inf. Sci. **2-2020**, 435–442 (2020). https://doi.org/10.5194/isprs-annals-V-2-2020-435-2020, https://www.isprs-ann-photogramm-remote-sens-spatial-inf-sci.net/V-2-2020/435/2020/
16. Kim, D.-H., Han, S.-B., Kim, J.-H.: Visual odometry algorithm using an RGB-D sensor and IMU in a highly dynamic environment. In: Kim, J.-H., Yang, W., Jo, J., Sincak, P., Myung, H. (eds.) Robot Intelligence Technology and Applications 3. AISC, vol. 345, pp. 11–26. Springer, Cham (2015). https://doi.org/10.1007/978-3-319-16841-8_2
17. Kim, D.H., Kim, J.H.: Effective background model-based RGB-D dense visual odometry in a dynamic environment. IEEE Trans. Rob. **32**(6), 1565–1573 (2016). https://doi.org/10.1109/TRO.2016.2609395
18. Kundu, A., Krishna, K.M., Sivaswamy, J.: Moving object detection by multi-view geometric techniques from a single camera mounted robot. In: 2009 IEEE/RSJ International Conference on Intelligent Robots and Systems (IROS), pp. 4306–4312. IEEE, October 2009. https://doi.org/10.1109/IROS.2009.5354227, http://ieeexplore.ieee.org/document/5354227/
19. Lenz, P., Ziegler, J., Geiger, A., Roser, M.: Sparse scene flow segmentation for moving object detection in urban environments. In: 2011 IEEE Intelligent Vehicles Symposium (IV), pp. 926–932. IEEE, June 2011. https://doi.org/10.1109/IVS.2011.5940558, http://ieeexplore.ieee.org/document/5940558/
20. Liu, H., Liu, G., Tian, G., Xin, S., Ji, Z.: Visual SLAM based on dynamic object removal. In: 2019 IEEE International Conference on Robotics and Biomimetics (ROBIO), pp. 596–601. IEEE, December 2019. https://doi.org/10.1109/ROBIO49542.2019.8961397, https://ieeexplore.ieee.org/document/8961397/
21. McLachlan, G.J.: Mahalanobis distance. Resonance **4**(6), 20–26 (1999)
22. Qin, T., Li, P., Shen, S.: VINS-mono: a robust and versatile monocular visual-inertial state estimator. IEEE Trans. Robot. **34**(4), 1004–1020 (2018). https://doi.org/10.1109/TRO.2018.2853729, https://ieeexplore.ieee.org/document/8421746/
23. Rosinol, A., Abate, M., Chang, Y., Carlone, L.: Kimera: an open-source library for real-time metric-semantic localization and mapping. In: IEEE International Conference on Robotics and Automation (ICRA), October 2019. http://arxiv.org/abs/1910.02490

24. Schofield, S., Bainbridge-Smith, A., Green, R.: Evaluating visual inertial odometry using the windy forest dataset. In: 2021 36th International Conference on Image and Vision Computing New Zealand (IVCNZ), pp. 1–6. IEEE, December 2021. https://doi.org/10.1109/IVCNZ54163.2021.9653391, https://ieeexplore.ieee.org/document/9653391/
25. Shi, J., et al.: Good features to track. In: 1994 Proceedings of IEEE Conference on Computer Vision and Pattern Recognition, pp. 593–600. IEEE (1994)
26. Sun, K., et al.: Robust stereo visual inertial odometry for fast autonomous flight. IEEE Robot. Autom. Lett. **3**(2), 965–972, April 2018. https://doi.org/10.1109/LRA.2018.2793349, http://arxiv.org/abs/1712.00036, https://ieeexplore.ieee.org/document/8258858/
27. Sun, Y., Liu, M., Meng, M.Q.H.: Motion removal for reliable RGB-D SLAM in dynamic environments. Robot. Auton. Syst. **108**, 115–128 (2018). https://doi.org/10.1016/j.robot.2018.07.002
28. Tokmakov, P., Alahari Inria, K., Schmid, C.: Learning motion patterns in videos. In: Proceedings of the IEEE Conference on Computer Vision and Pattern Recognition (CVPR), pp. 3386–3394 (2017). http://thoth.inrialpes.fr/research/mpnet
29. Yao, E., Zhang, H., Xu, H., Song, H., Zhang, G.: Robust RGB-D visual odometry based on edges and points. Robot. Auton. Syst. **107**, 209–220 (2018). https://doi.org/10.1016/j.robot.2018.06.009, https://linkinghub.elsevier.com/retrieve/pii/S0921889018300770
30. Zhang, T., Zhang, H., Li, Y., Nakamura, Y., Zhang, L.: FlowFusion: dynamic dense RGB-D SLAM based on optical flow. In: 2020 IEEE International Conference on Robotics and Automation (ICRA), pp. 7322–7328. IEEE, May 2020. https://doi.org/10.1109/ICRA40945.2020.9197349, https://ieeexplore.ieee.org/document/9197349/
31. Zhang, Y., Dai, W., Peng, Z., Li, P., Fang, Z.: Feature regions segmentation based RGB-D visual odometry in dynamic environment. In: IECON 2018–44th Annual Conference of the IEEE Industrial Electronics Society, vol. 1, pp. 5648–5655. IEEE, October 2018. https://doi.org/10.1109/IECON.2018.8591053, https://ieeexplore.ieee.org/document/8591053/
32. Zhang, Z., Forster, C., Scaramuzza, D.: Active exposure control for robust visual odometry in HDR environments. In: 2017 IEEE International Conference on Robotics and Automation (ICRA), pp. 3894–3901. IEEE, May 2017. https://doi.org/10.1109/ICRA.2017.7989449, http://ieeexplore.ieee.org/document/7989449/
33. Zhang, Z., Scaramuzza, D.: A tutorial on quantitative trajectory evaluation for visual(-Inertial) odometry. In: 2018 IEEE/RSJ International Conference on Intelligent Robots and Systems (IROS), pp. 7244–7251. IEEE, October 2018. https://doi.org/10.1109/IROS.2018.8593941, https://ieeexplore.ieee.org/document/8593941/
34. Zhao, L., Liu, Z., Chen, J., Cai, W., Wang, W., Zeng, L.: A compatible framework for RGB-D SLAM in dynamic scenes. IEEE Access **7**, 75604–75614 (2019). https://doi.org/10.1109/ACCESS.2019.2922733

M3T: Multi-class Multi-instance Multi-view Object Tracking for Embodied AI Tasks

Mariia Khan[1(✉)], Jumana Abu-Khalaf[1], David Suter[1], and Bodo Rosenhahn[2]

[1] Edith Cowan University, Perth, WA 6027, Australia
{mariia.khan,j.abukhalaf,d.suter}@ecu.edu.au
[2] Leibniz University Hannover, 30167 Hannover, Germany
rosenhahn@tnt.uni-hannover.de

Abstract. In this paper, we propose an extended multiple object tracking (MOT) task definition for embodied AI visual exploration research task - multi-class, multi-instance and multi-view object tracking (M3T). The aim of the proposed M3T task is to identify the unique number of objects in the environment, observed on the agent's way, and visible from far or close view, from different angles or visible only partially. Classic MOT algorithms are not applicable for the M3T task, as they typically target moving single-class multiple object instances in one video and track objects, visible from only one angle or camera viewpoint. Thus, we present the M3T-Round algorithm designed for a simple scenario, where an agent takes 12 image frames, while rotating 360° from the initial position in a scene. We, first, detect each object in all image frames and then track objects (without any training), using cosine similarity metric for association of object tracks. The detector part of our M3T-Round algorithm is compatible with the baseline YOLOv4 algorithm [1] in terms of detection accuracy: a 5.26 point improvement in AP_{75}. The tracker part of our M3T-Round algorithm shows a 4.6 point improvement in HOTA over GMOTv2 algorithm [2], a recent, high-performance tracking method. Moreover, we have collected a new challenging tracking dataset from AI2-Thor [3] simulator for training and evaluation of the proposed M3T-Round algorithm.

Keywords: Scene Understanding · Multiple Object Tracking · Embodied AI

1 Introduction

Recent research trends in deep learning, reinforcement learning, computer vision, and robotics have led to a shift from "internet AI" towards "embodied AI". Internet AI focuses on learning from data collected from internet datasets of photos, videos, and text. On the other hand, embodied AI allows artificial agents to learn from interactions with their surroundings [4].

Embodied AI tasks can include, but are not limited to: visual exploration and visual navigation, instruction following, embodied question answering and room rearrangement. The environment exploration phase is common for all embodied AI research

W. Q. Yan et al. (Eds.): IVCNZ 2022, LNCS 13836, pp. 246–261, 2023.
https://doi.org/10.1007/978-3-031-25825-1_18

tasks. It is critical for an agent exploring an environment to be able to understand its surroundings. This is referred to as scene understanding, which might be a simple task for humans, but remains very challenging for embodied AI agents [5].

Typically, an agent's goal is to obtain (at the end of the exploration phase) a 2D or 3D map of the environment, based on the inputs from the available sensors. For the map reconstruction an intelligent agent must first perform object detection, or in other words, understand, what objects are in the environment, and where they are located. Object detection is an important topic in computer vision and has been applied for various embodied AI tasks. However, object association and tracking algorithms have not been widely applied for better scene understanding in embodied AI tasks. Object association between image frames over the whole path of the agent in the environment can improve scene understanding. Knowing the amount of unique tracks in the environment leads to identifying the number of unique objects in the 3D scene (over all image inputs). This information can be further leveraged for environment map reconstruction. Hence, applying re-identification and tracking approaches can accelerate research in embodied AI tasks.

The Multiple-Object Tracking (MOT) computer vision task is not new to the research community. It is mainly used to analyse videos to detect and track object instances belonging to one or more categories, without prior knowledge about the appearance and number of targets [6].

The main difference between classic MOT scenarios and embodied AI tasks is in the input type. The input for a classic tracking algorithm is a video sequence, from which a large number of image frames is extracted with high frame rates. Therefore, the appearance (size and shape) and the motion (position) of moving objects (like pedestrians or cars) are changing gradually within the following image frames (Fig. 1). As for embodied AI tasks, there are constraints in terms of memory usage, agent rotation degree and step of movement. Accordingly, the input for such tasks is a relatively small set of image frames, where the appearances of static objects (like furniture items) can differ drastically. The object can appear to have a different size, may be visible from a close or far viewpoint, or can change shape, depending on which angle it was observed from.

To overcome limitations of the traditional MOT algorithms, discussed above, we propose a spatial-temporal independent model, where objects are compared based only on their appearance features, using a cosine similarity metric. The aim of the proposed M3T method is to identify the unique number of objects in the environment by associating and tracking objects, visible from far or close view, from different angles or visible only partially within the consecutive image frames. Therefore, we propose to build on classic MOT algorithms, but concentrate on appearance feature extraction, based on the known object location in the image (detected bounding boxes) instead of the object location prediction in the next frame, using an appearance or motion model.

Figure 2 illustrates the main four stages of the proposed M3T tracking method. Initially, the robotic agent moves around a household environment and captures RGB images alone the way. Then, each input image frame is passed through the object detector in order to identify target objects. The output of this stage is a set of bounding boxes and object labels (classes). Next, a ROI align layer is utilized to extract object appearance features during the feature extraction stage. In the affinity stage, collected object features are used to generate a cosine similarity matrix between pairs of detections and trackers. Finally, the similarity metrics are used to associate detection and tracker sets by assigning the same ID to detections that identify the same target [5].

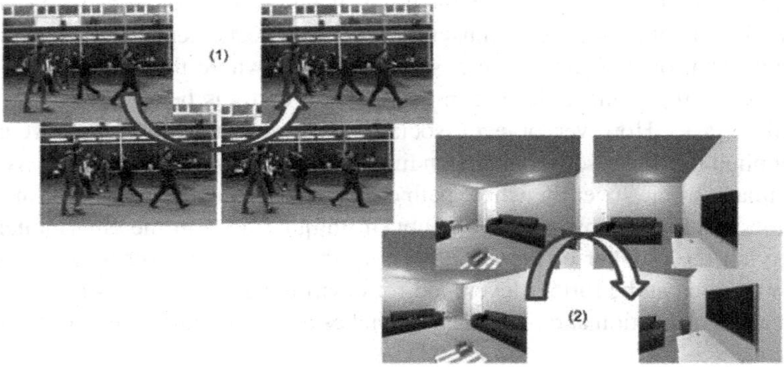

Fig. 1. An illustration of the input image frames for classic MOT task and for embodied AI scene understanding task: (1) MOT-15 challenge [7] image frames extracted from TUD-Campus video sequence; (2) image frames extracted from AI2-Thor simulator on the agent's path around the scene. Notice the significant differences in object appearance and angle of view between image frames from (2).

We evaluate our M3T-Round tracking model using the AI2-Thor embodied AI simulator. For algorithm evaluation we use 42 different object categories (Table 2). The task is to track objects in all frames that the agent observes, while navigating in a scene, and identify classes of all objects in the output tracks.

In general, **our contributions** can be summarized as follows:

- An extended MOT task definition for embodied AI visual exploration research tasks - multi-class, multi-instance, multi-view, visual object tracking (M3T).
- An object association and tracking pipeline for embodied AI tasks to identify the number of the unique objects in the embodied environment – M3T-Round.
- A new challenging dataset for object association, collected within AI2-Thor simulator, which is better suited for embodied AI research tasks.

2 Related Work

2.1 Object Detection

Object detection is a common computer vision task for identification of objects within images or videos. Object detection models can be divided into the following groups: CNN-based one-stage dense prediction algorithms (YOLO series [1, 8]), CNN-based two-stage sparse prediction object detectors (R-CNN series [9–11]) and transformer-based approaches (DETR [12], Deformable DETR [13]).

Object detection is an essential part of the scene understanding stage for embodied AI tasks. Although most of existing object detection algorithms can be directly trained and applied on images collected within embodied AI simulators, they typically do not leverage agents' interaction with the environment. In contrast, several algorithms were designed specifically for embodied AI agents, navigating through various embodied environments: Interactron [14] and Embodied Mask R-CNN [15] algorithms.

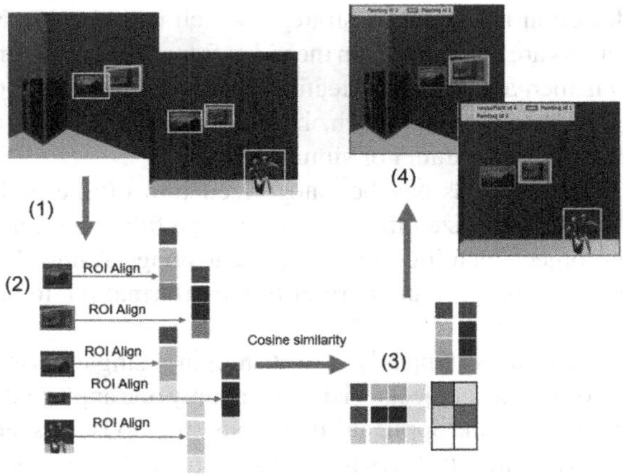

Fig. 2. Our proposed M3T approach follows the classic MOT algorithm stages: (1) an object detector is applied on the image frames to detect sets of objects bounding boxes and labels. (2) The appearance features are computed for each recognised object. (3) The probability of two objects belonging to the same target is computed (the affinity computation step). (4) A numerical track ID is set to each object during a final association step.

2.2 Object Re-identification

The object re-identification task has been extensively researched in literature. However, it has generally been described as a problem of person [16, 17] or vehicle re-identification [18, 19]. The goal of such tasks is to re-identify objects in the images by utilizing visual search, to find alike images for a given query image of the target object.

Existing re-identification methods can be divided into two main types: appearance-based and motion-based techniques [20]. Motion-based approaches are not really applicable in indoor embodied AI tasks, as such methods aim to identify each object based on a motion model. This usually fails, when the same object instance is observed from a different viewpoint on the following image frame. This is due to the large degree of rotation that the robotic agent undergoes from one view to the next.

2.3 Multiple Object Tracking

Multiple Object Tracking is a critical computer vision problem that has received a lot of interest within the research community, due to its academic and commercial possibilities. MOT algorithms can be divided into fully CNN-based (SORT [21], DeepSORT [22]), transformer backbone-based (ViTT [23]) and CNN-transformer mix algorithms, where CNN is necessary to extract features from image frames. These features then become inputs for the tracking transformer (TransTrack [24], TrackFormer [25] and MOTR [26]).

Tracking-by-detection is a standard strategy, which is utilized in MOT algorithms, where a set of detections are collected from the video frames and used for further tracking. The same track ID is then assigned to detections associated together [6]. In contrast, in the joint-tracking-and-detection approach, both detection and tracking processes are performed together: either in parallel or simultaneously.

Additionally, MOT methods can be categorized into offline and online trackers. Offline tracking algorithms utilize information from the following image frames, while trying to detect the object identities for the current image frame. In contrast, online tracking algorithms only use past and current frame information, to predict the current frame tracks [6].

Classic MOT algorithms typically target moving single-class multiple object instances in one video and track objects, visible from only one angle or camera viewpoint. These two assumptions limit the applicability of these methods for scene understanding in the Embodied AI Research tasks, where objects of multiple classes are static and can be visible from continuously changing viewpoints. Thus, for the best of our knowledge, our M3T method is a first tracking algorithm designed specifically for the scene understanding stage of the embodied AI research tasks.

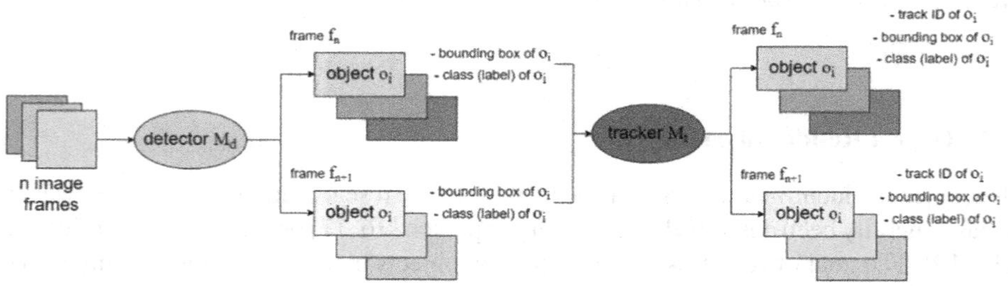

Fig. 3. The M3T task definition: the input is an image set F (n image frames, collected by an agent in scene s), the output is a set of a bounding box, a class label and a numerical track ID for every object in each image frame from F

3 Multi-class Multi-instance Multi-view Object Tracking

We first introduce our definition of a multi-class, multi-instance, multi-view visual object tracking (M3T) task for indoor embodied environments (Fig. 3 and 4). M3T task definition is broader than the typical MOT task, which can be referred to as a single-class, multi-instance, single-view visual tracking. Typical MOT algorithms are based on the appearance or motion estimation model and assume minor changes in objects' positions and sizes in subsequent video frames. In contrast, the proposed M3T task is not limited to such spatial-temporal restrictions, as objects in subsequent frames are compared only based on their general appearance features. Therefore, objects from close or far views, which may be visible from different angles, can be correctly associated.

3.1 Task Definition

The proposed M3T visual object tracking task is suitable for various indoor interactive environments such as AI2-Thor [3] or Habitat [27]. The aim of the task is to correctly predict a bounding box, a class (label) and a numerical track ID for all objects within each image frame on the path of the robotic agent. In other words, the aim is to identify all unique objects in the interactive embodied environment.

Formally, we are given a scene $s \in S$, an agent initial position p and an agent initial rotation r, then asked to identify every object $o \in O_S$ in this scene (Fig. 4). There are m actions available for an agent from the action set A, defined by policy P. An agent is allowed to record observed n image frames (image set F). Next, an object detector model M_d is applied for F. The outputs from M_d are a bounding box and a class for all objects in O_S. The detection should be performed for all n RGB frames from F.

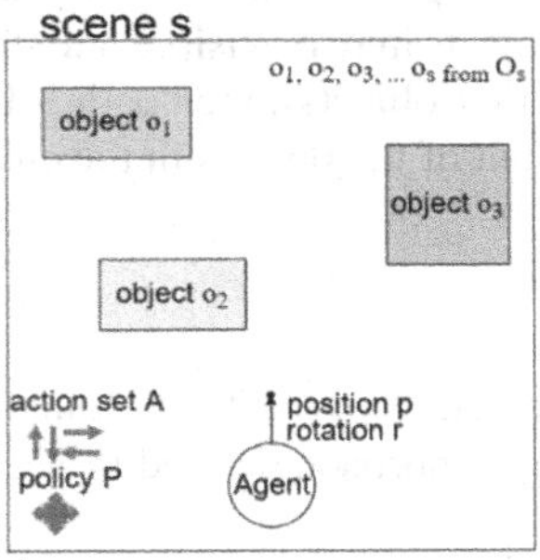

Fig. 4. Notation of M3T visual object tracking task for the given scene s

In the next stage, we apply a tracking model M_t, which takes the bounding boxes and the class labels for each object $o \in O_S$, identified by M_d, as input; to predict a numerical track ID for each object $o \in O_S$ (Fig. 3).

Therefore, the final output of the task is the set of: a bounding box, a class label and a numerical track ID for every object in each RGB image frame from F. To evaluate the M3T task, standard MOT metrics can be used: HOTA [28], MOTA [29] and IDF1 [30]. Higher Order Tracking Accuracy (HOTA) is a geometric mean of a detection accuracy and an association accuracy, averaged across localization thresholds. Multi-Object Tracking Accuracy (MOTA) combines three error sources: false positives, missed targets and identity switches. IDF1 Score is the ratio of correctly identified detections over the average number of ground-truth and computed detections.

We further define two approaches for the proposed M3T visual object tracking task with a different level of complexity. The first one is a simple "turning around" approach, where the agent doing a full rotating circle from the initial starting point in the scene. The second approach is more complex: an agent follows some path in the environment, simultaneously changing its location and rotation in space.

3.2 Policy Definition

The simplest approach to solve the M3T tracking problem is by using a turn-around policy P to move an agent around the initial position p with a pre-defined rotation degree d in one direction (left or right), until reaching the initial rotation r in space again (360° closed loop). We propose to train an off-the-shelf object detector M_d directly on the visual data, extracted from an interactive environment (embodied AI simulator). During the movement of an agent, n image frames will be collected around the starting position p: making it possible to detect and track the objects in the environment, with an appearance-based tracking model M_t.

The advantage of the described approach lies in the small amount of collected image frames. It allows to keep the consistent track IDs of objects in the scene, without a big number of track ID switches. We implement "the turning around" approach in the M3T-Round Model. Implementation details are further discussed in Sect. 4.

A second approach can be referred to as a "path following" approach, where an agent simultaneously changing its position and rotation alone the predefined by policy P path in the scene. As some objects may be visible only from a certain view point in the environment (hidden behind other objects), the "path following" approach may detect and track possibly bigger amount of the objects in the environment.

4 M3T Model

In this section, we describe the details of the M3T-Round model: its main stages, and a strategy for the "turning around" approach defined in Sect. 3.

4.1 Stages of M3T Model

There are four main stages in the M3T model, namely: detection, feature extraction, affinity stage and association stage. The overview of the algorithm is present in Fig. 5. The user sets two parameters for the M3T model: the cosine similarity threshold (or a minimum value to match objects) and the maximum number of frames, to keep a track without associated detections.

The Detection Stage. Any off-the-shelf object detection model can be applied to this stage. We specifically use a state-of-the-art YOLOv4 [1] algorithm for the M3T-Round model. YOLOv4 utilizes a CSPDarknet53 backbone, trained on ImageNet for extraction of image feature maps, and a YOLOv3 head for bounding box coordinates and object labels prediction for each object in all image frames, collected by an embodied agent. Although we use a CNN-based detector elsewhere in our model, it is not a requirement of this architecture.

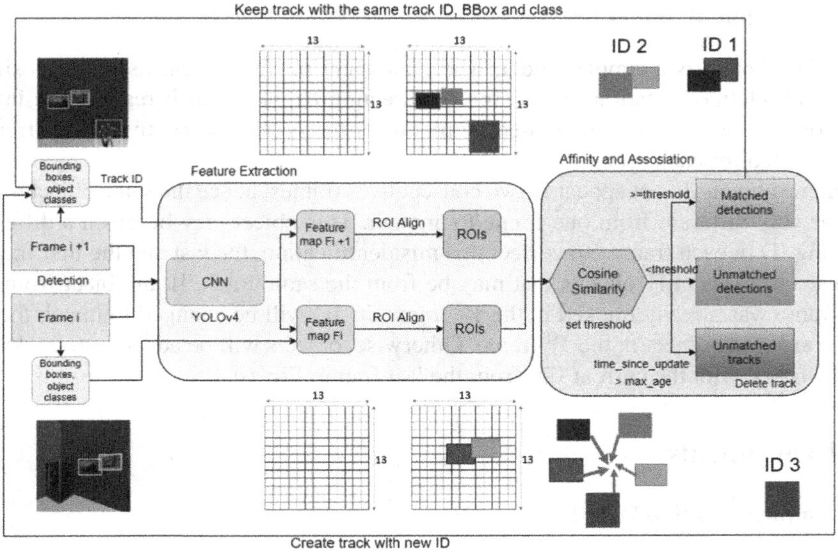

Fig. 5. M3T algorithm pipeline overview

The Feature Extraction Stage. To extract object appearance features from feature maps, we use a ROI Align layer from [10]. The M3T-Round algorithm takes in each step bounding boxes and classes of objects from two consecutive image frames: a current frame F_i and the next (future) frame F_{i+1}, which is used to initialize tracks of objects. The ROI Align layer extracts a small feature map using each bounding box, which was found during the detection stage, organizing the so-called ROI sets for both current and future frames.

The Affinity Stage. The object appearance feature vectors (ROIs) are further leveraged to compute the cosine similarity matrix between the current and future image frames. A cosine similarity metric aims to associate the appearance feature vectors of objects between two image frames, where two extracted features corresponding to the same object are likely to be closer, than the features from different objects.

The Association Stage. During this final stage, the cosine similarity matrix values are used to link detections of the same target object with the specific track ID. The cosine similarity value is checked to pass the pre-defined by user threshold. If the value is greater or equal to the threshold, the objects belong to one track and are considered as matched detections. The track will be kept with the same track ID. If the value is smaller than the threshold, the pair is treated as an unmatched detection. Thus, one more track is associated with a new track ID. In case, if the time passed since track update is less than the the maximum number of frames, set to keep a track without associated detections, the track is marked as unmatched and should be deleted.

4.2 M3T-Round Model

The M3T-Round is a simple model, where an agent moves around its initial position with a pre-defined rotation of 30°, in one direction (right), until it reaches the initial position in space again (360° closed loop). Accordingly, 12 consecutive image frames are collected for each scene.

Some objects do not appear in two consecutives frames, hence the same object might appear very different from one frame to another. This object may be given a different tracking ID in each frame. To correct this misidentification, the last and the first frames are used to re-identify objects that may be from the same track. If an object from the last frame was already tracked in the 1st frame, its ID will be changed to match the ID of the matching object in the 1st frame. Otherwise, objects will be added into tracks for the 1st frame with the current IDs from the last frame (Fig. 6).

5 Experiments

5.1 Implementation Details

To run our experiments, we have used the AI2-Thor [3] simulator, as it offers four different types of indoor scenes, fast rendering and a rich ground truth metadata about objects in the scenes. For our task, we consider 12 RGB image frames saved by the agent rotating in the scene. We use the action RotateRight and a 30° rotation angle to move the agent in space. The evaluation of the model is done with 416 × 416 resolution images to reduce the computational complexity and increase the performance speed.

5.2 Dataset

As there are currently no existing datasets, collected for the M3T task from embodied AI environments, we collect our dataset from AI2-Thor simulator. We concentrate on objects from living room type scenes only, as there is the largest amount of object types (classes). The total amount of 42 object classes were chosen for the YOLOv4 detection model re-training. The classes' list can be observed in Table 2.

All AI2-Thor scenes were divided into 80 training (rooms 1–20) and 20 testing (rooms 21–25) sets. For initial scene set up, we use the following information from the Room Rearrangement dataset [31]: scene id S, agent strating position p and rotation r, as well as starting positions p_{obj} and rotations r_{obj} of all objects in the scene. For training of the object detector part of the algorithm, annotations in YOLOv4 format were collected automatically through the simulator for each image frame.

The tracking part of our algorithm doesn't require any training and can be used on-the-go, however annotations are still needed to perform evaluation of the tracking part of the algorithm. Therefore, an additional set of annotations was collected using the MOT-15 Challenge [7] format with an additional value for object class ID. The ground truth file format contains one object instance per line. Each line must contain 11 values: frame ID, track ID, class ID, upper left x and y coordinates of a bounding box, its width and height, detection confidence score and x, y, z 3D coordinates, which are ignored, as we work with 2D data, and are filled with -1.

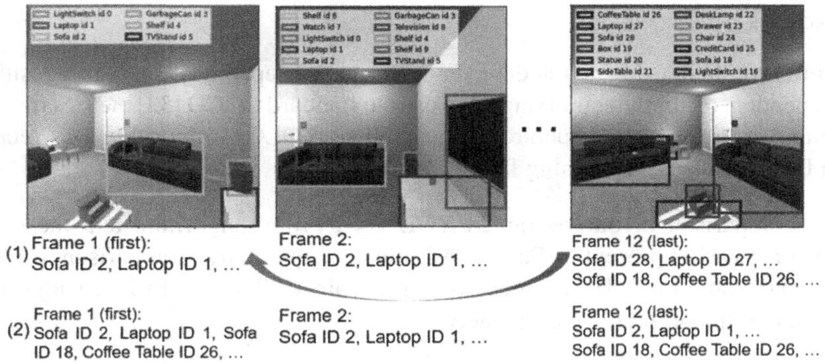

Fig. 6. Illustration of the closed loop re-identification. (1) On the 1st image frame only objects, appeared in the 2nd frame, are tracked therefore, some objects are missing: second Sofa or Coffee Table. As soon as the algorithm reaches the last image frame, 1st and last frames are compared again to re-identify objects. (2) Objects from the last frame correspondent with the 1st frame objects will change their IDs: Sofa changes ID from 28 to 2, Laptop changes ID from 27 to 1. Objects which were not tracked from the 1st frame will be added into tracks: Coffee Table ID 26, second Sofa ID 18.

For evaluation of the tracking part of the algorithm, an additional testing set was collected. For scene set up, we use the following information from the ProcTHOR-10K [32] dataset: scene id S, agent strating position p and rotation r, as well as starting positions p_{obj} and rotations r_{obj} of all objects in the scene. From all 10000 houses, only 1048 houses with 1 room set-up were used, for simplicity of the testing set creation.

Fig. 7. Qualitative comparison results between YOLOv4 baseline model (left) and our re-trained YOLOv4-PyTorch algorithm (right) on data collected from AI2-Thor simulator

5.3 Results

Detector. We, first, evaluate the detector performance separately, as tracker results are fully dependent on the detection stage. We use the standard COCO [33] metrics for results evaluation. Note, that the structure of the original YOLOv4 algorithm was recreated within PyTorch machine learning framework.

The comparison of our re-trained YOLOv4-PyTorch algorithm and the original YOLOv4 network is present in Table 1. Additionally, qualitative comparison results can be observed from Fig. 7. The accuracy of our re-trained YOLOv4-PyTorch algorithm is compatible to the original YOLOv4 network.

Table 1. Comparison of the our YOLOv4-PyTorch and the original YOLOv4 algorithms

Model	Size	Dataset	Batch size	mAP@0.50 ↑	mAP@0.75↑
YOLOv4-PyTorch (our)	416 × 416	Our from AI2-Thor	16	62,42%	52,56%
Original YOLOv4	608 × 608	COCO	16	65,7%	47,3%

Tracker. Next, we evaluate the tracker's performance separately, using the ground truth tracking annotations that are extracted from the simulator. We use the standard MOT metrics for results evaluation: HOTA [28], MOTA [29] and IDF1 [30]. The evaluation metric scores for each class are shown in Table 2: the left side of the table shows the results for large size objects, which are easier to track, and the right side of the table shows the results for medium and small size objects, which are more challenging for tracking. The average scores over all 42 classes are: 73.41 HOTA, 60.86 MOTA and 75.72 IDF1.

As our M3T algorithm cannot be directly compared to MOT state-of-the-art algorithms, we present the scores from the four top ranked MOT algorithms from MOT-15 [2] test dataset in Table 3. We chose MOT-15 dataset, as it is one of the most widely used challenging datasets for MOT algorithms evaluation and contains video sequences in unconstrained environments filmed with both static and moving cameras. Additionally, Table 4 contains the average scores over all classes for our M3T-Round algorithm on test set collected from Ai2-Thor simulator. As in classic MOT challenges, there is usually only 1 class of objects under evaluation, we also include in Table 4 scores that are calculated for the most frequent large and small size object classes separately, to make a fair comparison.

Detector and Tracker. Finally, we evaluate the joint performance of the whole M3T-Round pipeline: detector and tracker together. We use the standard MOT metrics for the results evaluation as well: HOTA [28], MOTA [29] and IDF1 [30]. The results of the evaluation for the most frequent big size object class (Painting) are in Table 3.

Table 2. MOT metric scores for all classes under evaluation for the M3T-Round tracker: all objects are sorted in order of occurrence frequency in the dataset (from largest to smallest)

Class	Scores for large size object classes			Class	Scores for small and medium size object classes		
	HOTA↑	MOTA↑	IDF1↑		HOTA↑	MOTA↑	IDF1↑
Painting	88.914	87.808	89.3	Drawer	73.606	63.986	75.5
Shelf	73.715	63.803	75.8	Book	72.293	60.828	76.29
Chair	80.16	71.609	81.7	Pillow	76.515	68.938	81.3
DiningTable	79.513	68.912	78.99	Statue	81.704	75.481	85.2
HousePlant	85.244	80.674	87.72	CellPhone	60.731	37.594	64.4
Dresser	77.724	67.359	77.86	Bowl	72.265	62.123	76.97
Television	90.312	87.987	89.8	RemoteControl	61.437	41.618	65.2
GarbageCan	86.515	83.22	89.29	Vase	77.018	68.829	80.7
ShelvingUnit	79.164	69.961	80.19	Pencil	43.045	1.914	43.9
Stool	85.977	81.797	88.5	Pen	50.403	17.782	52.3
ArmChair	81.36	76.152	84.01	Plate	70.885	59.339	75.9
SideTable	80.49	70.806	81.7	Laptop	77.016	68.248	80.6
Sofa	76.664	70	78.4	Box	77.793	68.309	80.78
TVStand	72.391	60.778	74.4	CreditCard	49.742	12.327	48.83
FloorLamp	89.659	87.285	90.71	KeyChain	55.989	27.063	58.06
Desk	78.27	67.805	79.02	DeskLamp	78.77	71.171	81.59
Cabinet	75.922	69.31	77.5	Watch	58.544	33.886	62.6
DogBed	76.829	71.739	82.6	Candle	69.57	54.932	73.6
Safe	79.088	73.81	84.1	TissueBox	73.124	62.704	77.4
CoffeeTable	73.34	55.102	73.68	Newspaper	60.399	42.581	65.25
Ottoman	38.188	0	33.3	Boots	92.932	90.909	95.24
Average	**78.544**	**69.805**	**79.93**	**Average**	**68.27**	**51.931**	**71.5**

6 Discussion and Limitations

One of the benefits of using our M3T-Round embodied model versus a standard off-the-shelf MOT algorithm for the M3T task is that we only need to train the detector part of the model and use the tracker part on-the-go (without training).

Table 1 shows that the detector part of our model (YOLOv4-PyTorch), which is re-trained directly on the images extracted from simulator, is compatible with the baseline YOLOv4 algorithm in terms of detection accuracy. The bounding boxes in Fig. 7 represent detections with a confidence score greater than 0.5 for both our model and the baseline model. Our model is able to detect significantly larger number of objects: both small (KeyChain or Box) and large size (FloorLamp or CoffeeTable) objects. Some of the object classes are still undetected even after re-training: RemoteControl or CreditCard on the table. This is due to the quite small object sizes and not enough occurrence in the training dataset collected from AI2-Thor.

Despite the differences in the complexity of the tracking task and the number of classes under evaluation in our M3T tracking model and typical MOT problem, our experiments show that the averaged performance of our tracking model is compatible. This is despite using the simple Turning Around M3T approach. The results from Table 2 shows, that the average tracking scores (using ground truth detections) for large size object classes are higher than for medium and small size objects: by 10 points in HOTA, 18 points in MOTA and 8 points in IDF1. Thus, small objects are more challenging to track by our algorithm.

The tracking performance comparison of the state-of-the-art MOT algorithms and our M3T-Round model (in Table 3) shows competitive results: 51.1 HOTA of the GMOTv2 algorithm versus 55.705 HOTA for our M3T-Round model (for the most frequent large size object class evaluated on the detections from YOLOv4-PyTorch).

Additionally, the tracking evaluation results for M3T-Round model from Table 3 on the ground truth detections are much higher, than with the detections results from YOLOv4-Pytorch retrained algorithm: 88.914 HOTA versus 55.705 HOTA for Painting object class. We can conclude, that the detector part of our model requires further improvement to be able to detect correctly more objects (especially small size objects).

Table 3. Results of four top MOT algorithms in MOT-15 challenge

Algorithm	Metric scores			
	HOTA↑	MOTA↑	IDF1↑	Dataset
GMOTv2 [2]	**51.1**	**64.5**	**67.0**	MOT-15
GGDA [2]	46.8	57.5	60.6	MOT-15
HMM [2]	47.3	56.9	62.3	MOT-15
NEUT [2]	46.2	55.1	61.1	MOT-15

6.1 Limitations

One of the limitations of our M3T-Round model is that it works with 360° round turn of the agent only, therefore not all objects can be detected and tracked in a scene. Alternative approaches for tasks that require an agent to explore an environment in a more complex way can be used to address this limitation. Thus, this model can be extended for the "following path" M3T approach using frames collected by the agent, as it moves in a longer path in the environment, allowing to identify larger number of objects in the scene.

Another limitation is in the evaluation of the model: the standard MOT metrics can be further adjusted to include the evaluation of the predicted track class and concentrate

Table 4. Results of our M3T algorithm on dataset collected from AI2-Thor simulator

Algorithm	Metric scores			
	HOTA↑	MOTA↑	IDF1↑	Dataset
M3T (ours) on ground truth detections (average over all 42 classes)	73.41	60.86	75.72	Our from AI2-Thor
M3T (ours) on ground truth detections (the most frequent large object-Painting)	**88.914**	**87.808**	**89.3**	Our from AI2-Thor
M3T (ours) on ground truth detections (the most frequent small object - Drawer)	73.606	63.986	75.5	Our from AI2-Thor
M3T (ours) with YOLOv4-Pytorch detector (the most frequent large object class - Painting)	55.705	31.686	51.79	Our from AI2-Thor

on the results evaluation not only on separate object classes, but also on the fair results aggregation for all object classes defined for the task.

Additionally, we have used the off-the-shelf YOLOv4 object detector algorithm in our model, which does not leverage agents' interaction with the environment, and the detection results can be improved. We can further design an algorithm specific for embodied AI agents navigating through various embodied environments or use one of the existing ones [14, 15]. This will potentially improve the performance of the detection stage of the model thus, strengthen the tracking part further as well.

7 Conclusions

This paper makes two contributions to solving a multi-class, multi-instance and multi-view object tracking problem for embodied interactive indoor environments. First, we introduce M3T-Round, a method for embodied visual object tracking in an interactive indoor environment. The method's main aim is to correctly identify tracks of unique objects in the scene from image frames that are collected by a moving robotic agent. The object detection model first detects all objects in image frames, and then the tracking model associates object tracks (without any training) using the cosine similarity metric. Moreover, we propose a novel embodied AI dataset collected from Ai2-Thor simulator for objects tracking in indoor interactive environments. We evaluate our approach using classic MOT metrics and show that it is compatible with the state-of-the-art MOT trackers.

References

1. Bochkovskiy, A., Wang, C.-Y., Liao, H.-Y.M.: YOLOv4: optimal speed and accuracy of object detection. arXiv preprint arXiv:2004.10934 (2020)

2. MOT15 results. https://motchallenge.net/results/MOT15/. Accessed 20 Sept 2022
3. Kolve, E., et al.: AI2-THOR: an interactive 3D environment for visual AI. arXiv preprint arXiv:1712.05474 (2017)
4. Batra, D., et al.: Rearrangement: a challenge for embodied AI. arXiv preprint arXiv:2011.01975 (2020)
5. Hall, D., et al.: The robotic vision scene understanding challenge. arXiv preprint arXiv:2009.05246 (2020)
6. Ciaparrone, G., et al.: Deep learning in video multi-object tracking: a survey. Neurocomputing **381**, 61–88 (2020)
7. Leal-Taixé, L., et al.: MOT challenge 2015: towards a benchmark for multi-target tracking. arXiv preprint arXiv:1504.01942 (2015)
8. Redmon, J., Ali, F.: YOLOv3: an incremental improvement. arXiv preprint arXiv:1804.02767 (2018)
9. Girshick, R.: Fast R-CNN. In: Proceedings of the IEEE International Conference on Computer Vision, pp. 1440–1448 (2015)
10. He, K., et al.: Mask R-CNN. In: Proceedings of the IEEE International Conference on Computer Vision, pp. 2961–2969 (2017)
11. Ren, S., et al.: Faster R-CNN: towards real-time object detection with region proposal networks. In: Advances in Neural Information Processing Systems, vol. 28 (2015)
12. Carion, N., Massa, F., Synnaeve, G., Usunier, N., Kirillov, A., Zagoruyko, S.: End-to-end object detection with transformers. In: Vedaldi, A., Bischof, H., Brox, T., Frahm, J.-M. (eds.) ECCV 2020. LNCS, vol. 12346, pp. 213–229. Springer, Cham (2020). https://doi.org/10.1007/978-3-030-58452-8_13
13. Zhu, X., et al.: Deformable DETR: deformable transformers for end-to-end object detection. arXiv preprint arXiv:2010.04159 (2020)
14. Kotar, K., Mottaghi, R.: Interactron: embodied adaptive object detection. In: Proceedings of the IEEE/CVF Conference on Computer Vision and Pattern Recognition, pp. 14860–14869 (2022)
15. Yang, J., et al.: Embodied visual recognition. arXiv preprint arXiv:1904.04404 (2019)
16. Li, A., Liu, L., Wang, K., Liu, S., Yan, S.: Clothing attributes assisted person reidentification. IEEE Trans. Circ. Syst. Video Technol. **25**(5), 869–878 (2015)
17. Farenzena, M., Bazzani, L., Perina, A., Murino, V., Cristani, M.: Person re-identification by symmetry-driven accumulation of local features. In: Proceedings IEEE Conference on Computer Vision and Pattern Recognition, pp. 2360–2367. IEEE (2010)
18. Zhao, J., et al.: Heterogeneous relational complement for vehicle re-identification. In: Proceedings of the IEEE/CVF International Conference on Computer Vision, pp. 205–214 (2021)
19. Yu, J., et al.: Camera-tracklet-aware contrastive learning for unsupervised vehicle re-identification. In: 2022 International Conference on Robotics and Automation (ICRA), pp. 905–911. IEEE (2022)
20. Bansal, V., Foresti, G.L., Martinel, N.: Where did i see it? Object instance re-identification with attention. In: Proceedings of the IEEE/CVF International Conference on Computer Vision, pp. 298–306 (2021)
21. Bewley, A., et al.: Simple online and realtime tracking. In: 2016 IEEE International Conference on Image Processing (ICIP), pp. 3464–3468. IEEE (2016)
22. Wojke, N., Bewley, A., Paulus, D.: Simple online and realtime tracking with a deep association metric. In: 2017 IEEE International Conference on Image Processing (ICIP), pp. 3645–3649. IEEE (2017)
23. Zhu, X., et al.: ViTT: vision transformer tracker. Sensors **21**(16), 5608 (2021)
24. Sun, P., et al.: TransTrack: multiple object tracking with transformer. arXiv preprint arXiv:2012.15460 (2020)

25. Meinhardt, T., et al.: TrackFormer: multi-object tracking with transformers. In: Proceedings of the IEEE/CVF Conference on Computer Vision and Pattern Recognition, pp. 8844–8854 (2022)
26. Zeng, F., et al.: MOTR: end-to-end multiple-object tracking with transformer. arXiv preprint arXiv:2105.03247 (2021)
27. Savva, M., et al.: Habitat: a platform for embodied AI research. In: Proceedings of the IEEE/CVF International Conference on Computer Vision, pp. 9339–9347 (2019)
28. Luiten, J., et al.: HOTA: A higher order metric for evaluating multi-object tracking. Int. J. Comput. Vis. **129**(2), 548–578 (2021). https://doi.org/10.1007/s11263-020-01375-2
29. Bernardin, K., Stiefelhagen, R.: Evaluating multiple object tracking performance: the clear MOT metrics. EURASIP J. Image Video Process. **1**, 1–10 (2008). https://doi.org/10.1155/2008/246309
30. Ristani, E., Solera, F., Zou, R., Cucchiara, R., Tomasi, C.: Performance measures and a data set for multi-target, multi-camera tracking. In: Hua, G., Jégou, H. (eds.) ECCV 2016. LNCS, vol. 9914, pp. 17–35. Springer, Cham (2016). https://doi.org/10.1007/978-3-319-48881-3_2
31. Weihs, L., et al.: Visual room rearrangement. In: Proceedings of the IEEE/CVF Conference on Computer Vision and Pattern Recognition, pp. 5922–5931 (2021)
32. Deitke, M., et al.: ProcTHOR: large-scale embodied AI using procedural generation. arXiv preprint arXiv:2206.06994 (2022)
33. Lin, T.-Y., Maire, M., Belongie, S., Hays, J., Perona, P., Ramanan, D., Piotr Dollár, C., Zitnick, L.: Microsoft COCO: common objects in context. In: Fleet, D., Pajdla, T., Schiele, B., Tuytelaars, T. (eds.) ECCV 2014. LNCS, vol. 8693, pp. 740–755. Springer, Cham (2014). https://doi.org/10.1007/978-3-319-10602-1_48

A VR Tool for Labelling 3D Data Sets

Leo Venn and Steven Mills(✉)

University of Otago, Dunedin, New Zealand
venle611@student.otago.ac.nz, steven.mills@otago.ac.nz

Abstract. Point clouds are a common representation of 3D scenes, and labelled point clouds are necessary as input to many machine learning systems. Current labelling tools, however, are predominantly 2D. A 3D interface would be more natural fit to the task, and we investigate Virtual Reality as a mechanism for point cloud labelling. In contrast to previous studies we find that the choice of 2D or 3D interface is not the determining factor for labelling speed or accuracy. The nature of the task is important, with some tasks being better suited to the 2D or 3D tools.

Keywords: Point cloud labelling · Virtual reality · 3D data

1 Introduction

Point clouds are a ubiquitous representation of 3D data. They are used in environment visualisation [8], mesh generation [6], and for neural network based object recognition [2]. With the rise of supervised machine learning (notably deep neural networks), an important task is labelling these point clouds.

Most of the point-cloud labelling software uses a traditional 2D desktop interface [1,5,7]. These interfaces tend to be complex and difficult to use, as the 3D data is only seen in projection, and six-degree of freedom navigation in a 2D interface is often unintuitive for naïve users. Virtual reality (VR) offers an alternative, where users can interact with 3D data in a 3D environment.

In this work we investigate the use of VR for point cloud labelling. We find that both 2D (screen-based) and 3D (VR) interfaces have their advantages, and that task completion time depends not only on the interface, but on the type of labelling task being undertaken.

2 Point Cloud Labelling Tools

Many tools already exist for labelling point clouds. Most of these use traditional 2D interfaces, but there are some 3D VR-based options.

2.1 Point Cloud Labelling with 2D Interfaces

Many 2D software packages exist already to label point clouds. Pointly [5], BasicAI [1], and Supervise.ly [7] are just a few of the popular examples. These tools

W. Q. Yan et al. (Eds.): IVCNZ 2022, LNCS 13836, pp. 262–271, 2023.
https://doi.org/10.1007/978-3-031-25825-1_19

share many labelling features, such as bounding-box selection, but also add tools for either general or task-specific labelling requirements.

In order to navigate the 3D space, multiple views may be displayed of a selected region (e.g. front, side, top facing) to assist the user in visualising the data points within a bounding box, as shown in Fig. 1. Synthesising these views can be difficult, and requires some experience to become used to.

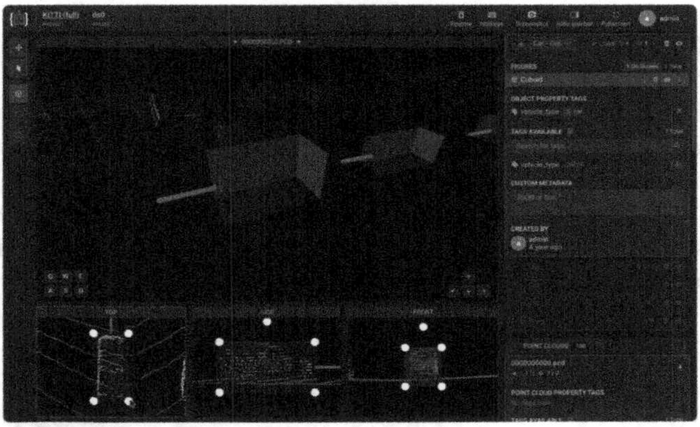

Fig. 1. Example of a bounding box being used to label a car in Supervise.ly [7]. The main interface window shows a 3D perspective view, but smaller orthographic views from the top, side, and front are also provided to aid in 3D labelling.

Bounding boxes are a common tool found in most 2D labelling software but can be difficult to use for intricate labelling tasks, such as labelling fingers on a hand. Other tools such as brush painting (using the mouse to label points near it on the screen) are difficult to use in 2D as each screen location represents a ray into the scene rather than a specific depth. This problem means points behind where the user thinks that they are clicking may be labelled incorrectly.

The main issues with current 2D labelling software packages lies in two areas. Firstly, the tools and interfaces are often overly complex to compensate for the projection from 3D to 2D. And second, the tools used in this software tend to be difficult to use and less accurate when it comes to more complex shaped objects or scenes.

2.2 Point Cloud Labelling in VR

In order to address the deficiencies of 2D point cloud labelling interfaces, some VR tools have been developed. Fanzleubbers *et al.* [3] propose a VR tool for the specific task of labelling plants. Comparing their VR labelling tool using the motion controller as a brush and a basic 2D interface, Franzleubbers *et al.* found that participants were faster to reach up to 90% label accuracy, and also at counting tasks in VR. Labelling to 98% accuracy, however, was not faster in VR, and these results were on a very specific task.

A similar experiment was conducted in PointAtMe [9], labelling cars in street scenes. Again, VR labelling was found to be faster and more accurate, but was limited to bounding box tools, rather than natural gesture-painting in VR.

Overall, there is some evidence that VR is more effective than 2D interfaces for point cloud labelling, but there is scope for further investigation.

3 A VR Point Cloud Labelling Tool

Although there have been several VR point labelling tools reported in the literature, they are not readily available. As a result, we have implemented a basic labelling tool using Unreal Engine and a mix of Blueprints and C++. Our VR Labelling Tool allows the user to load point clouds, navigate through them, assign labels to points, and save the resulting output. An example of the tool in use is shown in Fig. 2.

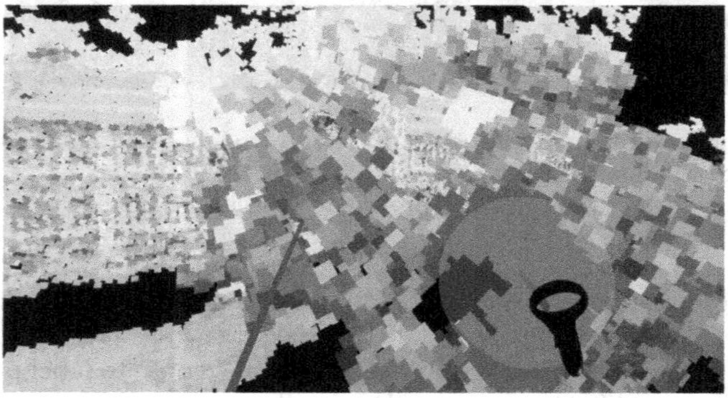

Fig. 2. Our VR tool in use. Here the user is labelling a tree red. (Color figure online)

The only tool currently implemented is a brush. This allows the user to hold down the trigger on the right controller to label points near their hand with the currently selected label. This was chosen as a simple tool that takes advantage of the natural interaction provided in the VR environment.

The user has the following controls available to them in the VR environment:

- Right Joystick: Move forwards, backwards, side to side
- Right Joystick: (left/right movement) + Right Grip Button: Scale brush size
- Right Trigger: Apply current label
- Left Joystick (up/down): Move up/down
- Left Joystick (up/down movement): + Left Grip Button: scale world
- A Button: Change Labels
- B Button: Enter/exit undo mode

For the purposes of the experiments reported here, rotation of the scene was not enabled. This is because the street scene used (see below) has a natural vertical orientation and we did not want to confuse the user if they turned their world upside down.

4 Experimental Evaluation

We compare our VR labelling tool to the cloud-based annotation tool Pointly [5]. Pointly provides three main labelling tools:

1. Bounding box selection, where a cuboid can be placed, scaled, and rotated in all three axes. All points within the box are selected.
2. Lasso selection, where a polygon is drawn on screen and all points inside it are selected.
3. Semi-automatic segmentation, where an attempt is made to divide the point cloud into sub-parts that can be selected individually.

These tools are shown in Fig. 3, and the user can assign different labels to the selected points.

Fig. 3. Pointly annotation tools: selecting a bounding box (left), a polygonal lasso (centre), and semi-automatic segmentation (right).

4.1 Labelling Tasks

The test data for our experiment is a sample model provided with Pointly. This is a street scene with a mix of buildings, cars, vegetation, etc., and is shown in Fig. 4.

Participants were asked to label three different objects, which were chosen to give a variety of tasks. The labelling tasks are:

1. A car, which is moderately complex, but has a compact 3D extent.
2. A flat rooftop, which is quite simple but extended in space.
3. A detailed road marking, which requires intricate selection.

The areas of the model that need to be labelled are illustrated in Fig. 5.

Fig. 4. The example scene from Pointly.

Fig. 5. The three tasks in our experiment: labelling a car, a flat rooftop area, and a painted pattern on the ground.

4.2 Method

Each participant is taken through a fixed regime. Participants are randomly allocated to use VR or Pointly first, to avoid ordering effects, and the following procedure is used:

1. The purpose of the experiment is explained, and participants complete consent and are asked if they have experience in VR or Pointly.
2. The first tool is demonstrated, participants are taught the controls of the software and given 5 min to explore, during which they can ask questions.
3. Participants undergo a 10-minute test where they are to label the three specified objects as quickly and accurately as they can. The time spent on each task is recorded.
4. Participants answer the NASA TLX user experience questionnaire [4] about their experience with the first tool.
5. Steps 2 to 4 are repeated for the second tool.
6. A final interview is performed to compare the two labelling tools.

The NASA Task Load Index (TLX) questionnaire [4] is widely used to measure task workload. Participants rank their experience on 7-point scales measuring

- Mental Demand – How mentally demanding was the task?
- Physical Demand – How physically demanding was the task?
- Temporal Demand – How hurried or rushed was the pace of the task?
- Performance – How successful were you in accomplishing what you were asked to do?
- Effort – How hard did you have to work to accomplish your level of performance?
- Frustration – How insecure, discouraged, irritated, stressed, and annoyed were you?

Each point on the scale is subdivided into low, medium, and high estimates, for a total of 21 gradations.

The final interview consists of four questions:

1. Which software did you prefer using and why?
2. Are there any situations where you would choose the less preferred software and why?
3. What improvements do you think should be made to the Virtual Reality Labelling tool?
4. Do you have any other feedback?

Ethical approval was the University of Otago Ethics Committee (ref D22/228) and participants were informed that they could leave the study at any stage at no disadvantage. In particular they were told of the risk of cybersickness and to remove the VR headset if they felt unwell.

4.3 Participants

Ten participants were recruited for this experiment. Their ages ranged from 19 to 64, but the majority (eight) were aged 19 to 23. There was an equal split of male and female. No participants had any experience with Pointly, and all had either no or minimal experience with VR.

5 Results and Discussion

We present results on the task completion time, accuracy, and survey results.

5.1 Task Completion Time

The time taken for each participant to complete the tasks is shown in Fig. 6. Times are given for labelling each individual object (Car, Roof, and Pattern), and the overall time spent labelling is also reported. Summary statistics of the timings are given in Table 1.

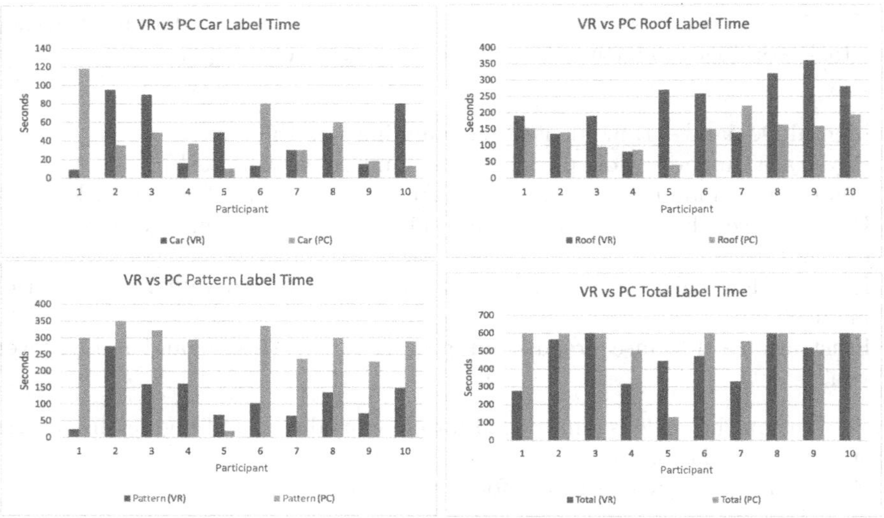

Fig. 6. Task completion times in seconds for each participant using the VR tool (VR) and Pointly (PC). Times are shown for each individual object to be labelled, as well as overall time spent using the tool.

Table 1. Times (in seconds) to complete various parts of the labelling task. All times are averages across the cohort ± one standard deviation.

Task	VR tool	Pointly
Car	44.5 ± 31.72	45 ± 31.88
Roof	222.1 ± 85.12	139.8 ± 50.42
Pattern	120.8 ± 67.61	267.5 ± 90.2
Non-labelling	84.4 ± 30.81	77 ± 29.88
Total	471.8 ± 120.43	529.3 ± 138.34

Conducting two-tailed unpaired t-tests with a threshold of $p = 0.05$ shows that the VR tool is significantly faster for labelling the Pattern ($p < 0.01$), but Pointly is significantly faster for the Roof task ($p = 0.027$). If a Bonferroni correction is applied, the significance level drops to 0.01 (as there are five tests) and we find that only the VR tool on Pattern labelling has any significant difference. Overall the choice of tool does not have a significant effect on task completion time.

5.2 Task Accuracy

To measure the accuracy of the labelling we produce a ground truth labelling by careful hand annotation. This allows us to categorise the final labelling from each participant into true positives (TP), false positives (FP), true negatives

(TN), and false negatives (FN). Since the majority of the scene is not labelled, then proportion of true negatives is always very high, so we exclude them and compute accuracy as

$$A = \frac{TP}{TP + FP + FN}.$$

The accuracy on each task and overall is summarised in Table 2. We note that one participant accidentally removed the labels from the Car, so was excluded from the Car and Overall accuracy computations.

Table 2. Accuracy on each task and overall. Values are reported ± one standard deviation.

Task	VR tool	Pointly
Pattern	0.607 ± 0.117	0.497 ± 0.095
Car	0.848 ± 0.063	0.951 ± 0.039
Roof	0.854 ± 0.091	0.916 ± 0.08
Overall	0.768 ± 0.069	0.794 ± 0.042

Statistical analysis as before shows that Pointly is significantly more accurate for the Car ($p = 0.004$), and the VR Tool is significantly more accurate for the Pattern ($p = 0.005$). Both of these results hold after a Bonferroni correction. Neither has a significant advantage for the Roof or overall.

5.3 Survey Results

The results of the NASA TLX scores are shown in Fig. 7. Participants generally found the VR Tool to be more physically demanding, but faster, more performant, and less frustrating.

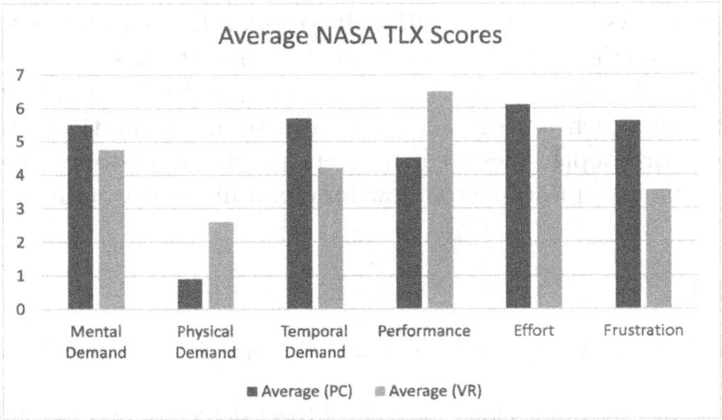

Fig. 7. Average NASA TLX scores for Pointly and the VR tool

In the final survey, eight of the ten participants preferred the VR Tool. They reported that it was more fun, intuitive, and more accurate. Using VR for a long period was a concern, with some participants feeling slightly dizzy (although not enough to discontinue the tasks). Pointly's advantages were the greater range of options.

5.4 Other Observations

While observing the participants, a number of observations about the use of VR for point cloud labelling were made. The first, which is common with novice users of VR, is that participants did not physically move through the space, preferring to move the scene with the controllers. Many were also reluctant to turn around, making it difficult to orient themselves in the scene.

Another common problem was a lack of depth perception. Despite the stereoscopic projection of the VR headset, many users had difficulty judging the distance to the object. As a result, they became frustrated when they thought they were labelling a surface, even though their hands were several meters away in the virtual space. This may be due to conflicting depth cues from the structure of the scene itself. The depth perception problem was the most common improvement suggested by the participants.

6 Conclusions and Future Work

In contrast to earlier research we did not find a consistent benefit to using VR over traditional 2D interfaces. Rather, we found that the nature of the task was important for both accuracy and speed. Our VR Tool did prove top be faster and more accurate for fine-detailed labelling (such as the Pattern task), but less efficient for other tasks. This may be due to the wider range of tools available in the more mature 2D technology, but hybrid tools could also be worth exploring.

Participants generally found the VR tool more engaging and intuitive, which is a benefit for novice users. None of our participants had prior experience with the task, however. As a result, it is difficult to generalise our results to large-scale point cloud labelling efforts without more extensive studies.

Other technical advancements to the current VR Tool would be to add more labelling methods (such as the box select and semi-automatic segmentation of Pointly); add more depth cues to help users judge the distance to the surface; and to extend the user interface to allow for customisation of labels and controls.

References

1. BasicAI Inc.: BasicAI: Multisensory Data Platform for Machine Learning (2022). https://www.basic.ai/
2. Fernandes, D., et al.: Point-cloud based 3D object detection and classification methods for self-driving applications: a survey and taxonomy. Inf. Fusion **68**, 161–191 (2021)

3. Franzluebbers, A., Li, C., Paterson, A., Johnsen, K.: Virtual reality point cloud annotation. In: 2022 IEEE Conference on Virtual Reality and 3D User Interfaces Abstracts and Workshops (VRW), pp. 886–887. IEEE (2022)
4. Hart, S.G., Staveland, L.E.: Development of NASA-TLX (task load index): results of empirical and theoretical research. In: Advances in Psychology, vol. 52, pp. 139–183. Elsevier (1988)
5. Pointly GmbH: 3D point cloud classification: automatic & manual. Pointly (2022). https://pointly.ai/
6. Remondino, F.: From point cloud to surface: the modeling and visualization problem. Int. Arch. Photogrammetry Remote Sens. Spat. Inf. Sci. **34** (2003)
7. Supervisely OÜ: Supervisely 3D sensor fusion labelling (2022). https://supervise.ly/labeling-toolbox/3d-lidar-sensor-fusion/
8. Varga, R., Costea, A., Florea, H., Giosan, I., Nedevschi, S.: Super-sensor for 360-degree environment perception: point cloud segmentation using image features. In: 2017 IEEE 20th International Conference on Intelligent Transportation Systems (ITSC), pp. 1–8. IEEE (2017)
9. Wirth, F., Quehl, J., Ota, J., Stiller, C.: PointAtMe: efficient 3D point cloud labeling in virtual reality. In: 2019 IEEE Intelligent Vehicles Symposium (IV), pp. 1693–1698. IEEE (2019)

Determining Realism of Procedurally Generated City Road Networks

Alex Shaw, Burkhard C. Wünsche(✉), Joachim Yde, Peter Vergerakis, Lance Chaney, Yuqiang Jin, and Neha Sarwate

School of Computer Science, University of Auckland, Auckland, New Zealand
l.shaw@auckland.ac.nz, burkhard@cs.auckland.ac.nz, {jyde833,pver135,lcha770,yjin139,nsar492}@aucklanduni.ac.nz

Abstract. Cities are both expansive and relatively homogeneous, making them well suited to procedural generation. Road networks are one of the most important components of generated cities, and rely on both exogenous and endogenous factors such as the environment and population. These factors significantly affect the characteristics of the generated city, and thus the evaluation of the realism or plausibility of the generated city is a non-trivial task. Previous work largely relied of subjective assessments of realism.

In this paper, we present an extensible approach for procedurally generating city road networks, utilizing real world data and intended to mimic the characteristics of a real city. We also present a set of metrics such as entropy and orientation order suitable for objectively assessing the generated road network in comparison to the real city it is mimicking. We find that employing these metrics can provide a reasonable baseline for future projects exploring realism in procedural city generation.

Keywords: Procedural generation · City generation · Realism · Metrics · Road network

1 Introduction

Procedural generation is a popular approach for generating content in computer graphics [9] such as cities: the many buildings and complicated road networks are arduous to create by hand, but sufficiently consistent that they can be produced algorithmically. With the ever-increasing level of computational power available in modern computers, the boundaries for what is possible in procedural modelling expands, and so does the potential use cases for procedurally modelled cities which range from entertainment products to urban planning. In such applications, realism of content becomes a bigger priority.

A key component of a city is its road network. The road network both significantly influences the appearance of the city layout and content distribution (whether the network is generated around the city contents or vice-versa), and is often the primary means of navigating the city or orienting within it. Achieving

W. Q. Yan et al. (Eds.): IVCNZ 2022, LNCS 13836, pp. 272–287, 2023.
https://doi.org/10.1007/978-3-031-25825-1_20

realism in the road network of a procedurally generated city is thus an important step in achieving realism for the city overall.

In general, existing solutions focus on first constructing the road network of a city, and subsequently adding one or more types of buildings. However, an algorithmically generated city is not necessarily a realistic, life-like city. Accordingly, a number of different approaches have been proposed in pursuit of increasing realism: In their pioneering paper, Parish and Mueller account for terrain specifications and population density when constructing road networks using L-systems inspired by four architecturally distinct cities around the world [14]. Cullen and O'Sullivan implement object-oriented shape grammars in conjunction with various geographic information system data detailing, amongst others, urban planning information [4]. Lechner et al. introduce agent-based systems working with predefined residential, commercial, and industrial land [11–13], and Groenewegen et al. account for different historical backgrounds and districts [8].

Despite these efforts, the subsequent evaluation of the generated cities usually relies on subjective, manual determinations: does the generated city look and feel real? While all the aforementioned papers express intentions of enhancing realism in their future work sections - e.g. by expanding the types of roads supported in their networks, implementing L-systems mirroring different types of cities from around the world, simulating growth, including railways, and similar—none directly discusses metrics to help determine realism. Aerial images can be an objective metric for measuring correctness if the goal is to completely replicate a real city. However, in many applications, such as games, we want generate cities inspired by one or multiple real ones, without necessarily replicating them. For this purpose additional metrics are required capturing the "look and feel" of the city. Similarly, different types of input might also be necessary depending on which city one wishes to generate. All this, ultimately, poses a number of challenges in first understanding which type of input parameters produces which type of output, and, secondly, in determining which metrics should be applied under which circumstances to best evaluate the generated city. Therefore, our research question is two-fold:

How can we, given a collection of real-world data sets, implement a system capable of procedurally generating realistic cities and how can we use objective metrics to determine the validity of generated cities?

In this paper we present two primary contributions: a system capable of generating cities based on input from real-world cities and an approach to evaluate the realism of the cities produced by our generator. By using objective metrics we will determine the realism of a generated city by comparing it to the real-world city from which data was used as input in the generation process.

2 Related Work

A number of different approaches have been proposed in pursuit of increasing the realism of procedurally generated cities. In their pioneering paper, Parish and Mueller propose a system which, based on a hierarchical set of comprehensible rules implemented using extensible L-systems, procedurally generates urban environments from scratch [14]. The authors use socio-statistical and geographical input such as terrain specifications and population density to invoke two distinct L-systems: one to generate a road network, and another to generate buildings. A major contribution is their self-sensitive road creation L-system, which considers both global goals and local constraints as to enhance realism. Parish and Mueller evaluate the correctness of their generated city through an inherently subjective, visual inspection.

Given the importance of road networks in city generation, several papers emphasise their generation. Accordingly, numerous efforts towards enhancing road networks have been published. Several techniques rely either entirely or partially on L-systems akin to Parish and Mueller [10,16], though exceptions do exist. Aliaga et al. employ an example based approach to generating a road network using imagery synthesised from aerial photographs to attain a wider variety of road configurations [1]. The space-colonization method employed by Fernandez and Fernandez grows a road network based on a set of attraction points [5]. The placement and density of these attraction points helps ensure interconnectedness of relevant locations.

Yu et al. combine the example-based approach by Aliaga et al. with Parish and Mueller's notion of local constraints, and propose a more flexible method to organically grow road networks perceptually similar to supplied examples [20]. Despite apparent increased realism, Yu et al. do not account for global patterns; something which is otherwise likely to influence how streets in a particular city intersect between various districts or neighbourhoods.

One approach to handling districts, which does not require external input, is proposed by Groenewegen et al. [8]: Appreciating how different districts are distinguishable by residential and social characteristics, their model is based on two distinct urban land use models - Western Europe and North America - and supports 18 different types of districts. They generate cities by feeding four city-specific parameters (diameter, continent, historical background, and number of highways passing through) to their model which then determines an appropriate mix of district types. Groenewegen et al. do not propose any metrics for evaluating their output, though they do consult with experts on urban design and planning who approved their model and its ability to generate plausible city layouts.

Merely generating realistic city-bounded road networks in accordance with district specific rules is not necessarily sufficient either. As argued by Jan Benes et al., a city inherently cannot be modelled realistically without also accounting for nearby cities [2]. The underlying logic being that surrounding areas influence how major roads are placed, and thus also, by natural extension, how smaller roads within a city are placed. Two earlier works lay the foundation for Jan Benes et al.'s work: Galin et al. first propose a method for generating individual roads across different terrains [7], and later extend this to a road network generation model connecting possible cities and settlements [6]. Jan Benes et al. then extend this further to also account for both water transportation and neighbouring cities as well as the fact that cities evolve over time.

An additional consideration in terms of generating realistic road networks ties back to the population distribution: In any real-life city, people move back and forth throughout the day. Any generated road network should, therefore, be able to support the expected traffic. Accordingly, Weber et al. augment the original Parish and Mueller approach by (a) including traffic simulation; and (b) supporting a sequence of urban configurations allowing the city to evolve over time [19]. Taking as input a terrain elevation map, a list of neighbouring cities, and a set of control functions (e.g., level of traffic, rate of growth), they grow a road network and associated city. They also pioneer the discontinuation of relying on rigid grids in favour of realistic geometric configurations. Weber et al. affirm realism by evaluating the similarity of their generated land use patterns to real-world data on a pixel-by-pixel basis.

Similarly, Vanegas et al. propose a method based on geometrical and behavioural modelling which also allows them to model the development of cities over time, including parcels and streets using current aerial imagery and GIS data [17]. Their main focus, however, is visualisation, and since they generate a 2030-version of Seattle, evaluating the realism of their generation is largely impossible. In subsequent work Vanegas et al. focus on procedurally generating realistic parcel distributions capable of accommodating change [18].

A different approach to procedurally generating cities is proposed by Lechner et al., and entails a highly controllable agent-based system which, given a geographical map, models the development of five fundamental types of land use (residential, commercial, industrial, roads, and parks) [12]. The authors associate each land use with distinct, so-called developer agents, which have distinct responsibilities and behaviour. A primary developer for roads, for instance, ensures that residents may move quickly through the generated city; somewhat akin to what Parish and Mueller achieve using global goals and local constraints in their L-systems. Lechner et al. propose a unique method of verifying correctness based on (a) the composition of land cover by use; and (b) the spatial configuration of land use inspired by urban geography literature.

In summary, previous work focuses on different aspects of procedurally generating realistic looking city and do so using a wide range of different methods and techniques. However, how realism or correctness is actually defined, as well how it is measured, varies greatly and is more often than not entirely subjective.

3 Design

The primary focus of our research is being able to evaluate the city generated by our system using predefined metrics. This means procedurally generating a city and outputting it in an intermediate representation which can be employed for both statistical analysis and visualisation purposes. A design like this allows us to employ objective evaluation metrics as opposed to subjective ones currently dominating related work.

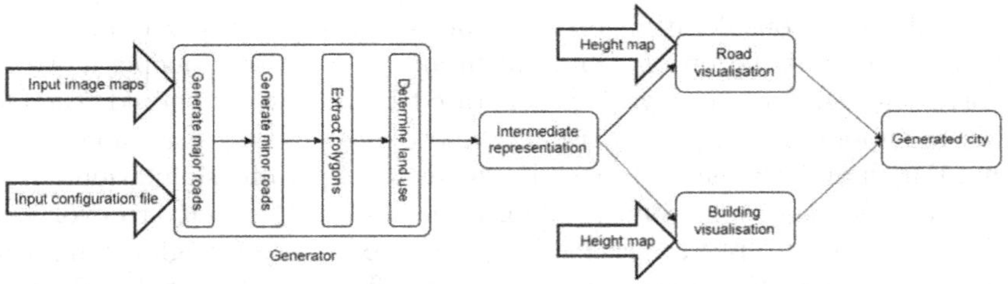

Fig. 1. A high-level overview of our system. The large arrows represent input manually supplied to the system, whereas the small arrows represent the individual steps throughout the pipeline-like structure. Note how each box represents an independent module or subsystem in accordance with our extensibility requirement. Our system starts by accepting image maps and a configuration file as input. These effectively determine how the road network is generated. The road network is outputted in an intermediate representation by the generator. This representation is subsequently employed in the road and building visualisation modules, both of which require the same height map as input. The output from both modules are combined at the end to visualise the generated city.

The only way in which we can objectively evaluate a city is if we have a representation which actually allows such evaluation. We argue that a visual representation similar to that implemented in related work is generally not sufficient for this purpose, except in the case where the generated city is directly mirroring one or more real-life cities, in which case comparing aerial imagery is possible [17,19].

Defining metrics in order to measure the correctness of a generated city is the crux of our research. However, what makes a city correct is a subjective matter dependent on the specific context in which the city is generated. This means that different metrics are required for different use cases. In one case, producing a road network capable of supporting large flows of traffic might be important, whereas the exact placement of buildings might be more important in another. Since we cannot reasonably anticipate every possible metric, we consider extensibility to be an important requirement of our system, such that future iterations can accommodate additional metrics. Similarly, by prioritising modularity we separate the process of generating, evaluating, and visualising a city.

3.1 System Overview

Based on the requirements established so far, we create a system using the architecture seen in Fig. 1 which consists of three major parts: The city generator subsystem, the road visualisation subsystem, and the building visualisation. In this paper we focus primarily on the generation components of the system as our contributions are not related to the visualization.

Our system takes as input a selection of real-world maps, and a configuration file. The current design specifically requires a waterways map, land usage map, population density map, and an elevation map. We source these from Koordinates.com. Additionally, we require an image map specifying road layout rules, which we provide. The water map, population density map, and road patterns are used to generate the road network, and the population density map and land use map are used to determine the population and land use in the generated city. We will describe each input maps in more detail and how they are used in the subsequent sections detailing each part of the generator subsystem. In addition to the above-mentioned maps, the visualisation subsystems also rely on an input height map such that the road network and buildings conform to the height of the ground. Such data should arguably also be included as input to the generator subsystem as it may impact how the road network is generated. However, for the current state of the system, height maps are used only for visualisation.

The system also relies on a configuration file containing parameters for various aspects of the generation process which includes length of road segments, angles when generating a turn, and which land uses to consider for city blocks, among other things. This ensures some flexibility in our system because the parameters can be tuned to change the output of the generator in order to achieve better results.

3.2 Road Network

We use a graph-based approach to generate the road network starting with the generation of major roads, which define the overall structure of the road network. We consider major roads equivalent to arterial roads or high capacity collector roads. Reading from the configuration file, we start with a seed. This seed will usually correspond to a single road segment, though multiple segments may be supplied. Using the seed, we generate a list of candidate segments that will be placed such that they obey our global and local constraints as well as any constraints specified in the configuration file. We suggest up to three types of candidate segments; a segment going straight from the current segment or seed; a segment turning left from the current segment or seed; and a segment turning right from the current segment or seed.

In generating candidate segments, we use different growth rules corresponding to distinct road network patterns similarly to previous work [1,10,14,19]:

- **Organic**, in which segments are allowed a larger degree of freedom in terms of the angle of their turns (e.g., London).
- **Grid**, in which segments are either perpendicular or parallel i.e. straight segments are placed with an angle of 180° from the current segment, whereas turning segments are placed with an angle of 90° (e.g., Manhattan).
- **Radial**, in which segments either follow a radial track around an identified centre, or grow towards the centre, (e.g. city centre of Paris).

Using the work by Boeing [3], we have devised each rule by analysing real-world cities and the pattern(s) exhibited by their road networks. A city may exhibit different patterns, e.g., old parts vs. new parts, and in our program this information can be added manually by preparing image maps depicting areas with different road network patterns.

The generator determines which pattern to employ when generating candidate segments by checking the supplied image map specifying which growth rule applies where. Specifically, different colours correspond to the different growth rules. The difference between the three growth rules is the minimum and maximum allowed angle a road can turn as well as the minimum and maximum allowed segment distance. In all three growth rules, the population density further impacts the probability of a road turning. That is, the closer to a dense area, the higher the probability of the road turning. We argue that since there are more roads in city centres than rural areas, there are more intersections and, by natural extension, more roads which turn. All parameters are specified in the configuration file.

Once a list of candidate segments has been generated, we proceed to verify each segment. This is done by enforcing a set of global and local constraints similar to those presented in [14]. We first check whether the segment breaks the boundaries of our canvas. If it does not, we then check whether the end of the segment is in water. As our system currently does not support bridges, any segment found to be in water is dismissed. At this stage, the candidate segment has satisfied our global constraints.

We thus proceed to analyse the exact placement of each candidate segment and enforce our local constraints where necessary. Specifically, we check whether the segment:

1. intersects an existing segment and is not close to another vertex or segment
2. intersects an existing segment and is close to another vertex
3. intersects an existing segment and is close to another segment
4. does not intersect an existing segment but is close to another segment
5. does not intersect an existing segment but is close to another vertex

If any of these cases occurs the candidate segment is connected with the existing segment or vertex to form a branch or intersection. Otherwise the segment is placed and added to a queue. We add the verified segment to a queue because we now consider the end of that segment a position from which new segments can grow. In the following iteration, a segment is popped from the queue and the cycle repeats. This process continues until either the queue is empty, or the maximum allowed iterations is reached.

We generate minor roads using the same approach as the major roads. The only difference is that we start by generating seeds based on the existing road network. This is because we deliberately do not supply any seeds for minor roads; we argue that doing so would not make sense as they inherently depend on the location of the major roads. In generating seeds for minor roads, any point where only two major road segments connect has a probability of becoming a minor road seed. We impose this limitation with the argument that we do not consider it realistic for an intersection with major roads to also spawn a minor road. Similar to the probability of a road turning, the probability of eligible major road segments becoming minor road seeds also increases based on the population density.

3.3 City Blocks

Using the road network as input, city blocks can be determined by first extracting the polygons which are enclosed by roads in the road network. Once the polygons of a city blocks have been extracted, the system checks the following attributes for each city block: Population density, total population, and land use. In order to do this, the system uses the population density image map, and the land use image map. The population density image map contains values with people per square meter and the land use map contains values associated with a specific land use such as residential and industrial. In order to determine the attributes of a city block, the system chooses random points in the city blocks polygon and samples into the two maps in order to determine which value is present at that point. The population density attribute of a city block is then the average value of sampled values. The average value is multiplied by the size of the polygon in order to get the total population of the city block. The land use of the block is the land use which was sampled the most times from the land use map. In the current version of the system, we only consider three types of land uses: Residential, commercial, and industrial. However, as many types as the land use image map contains can be modelled.

Having determined the attributes of each city block the system creates the intermediate representation of the generated city. This contains all placements of road segments and the attributes and placement of each city block.

4 Implementation

The generator subsystem shown in Fig. 1 has been implemented in Python. The Python Imaging Library fork, `Pillow` and `NumPy` are used to convert the input image maps to arrays and the `GDAL` library is used to open `tif` files which we have prepared in QGIS. In addition to converting input images, `NumPy` is used extensively in the road network generation to compute angles and unit vectors which are used to determine the position of new segments. In order to determine distances between end points of segments used for the third and fourth case of the road network constraints, we use a kd-tree implementation in the `SciPy` library. This structure allows the system to look up distances between points in the road network in order to find the point that is closest to the end point of a suggested segment. The `Shapely` library is used to compute the area of city block polygons, and the `Path` class in the `Matplotlib` library is used as part of our implementation to find all the points inside a polygon which is required by the generator subsystem in order to sample points that lie within polygons in the input image maps. Our implementation of the radial road network pattern and polygon extraction is based on the implementation made by Sauder et al. [15]. For the radial pattern, we have used the `label` function in the `morphology` module of the `scikit-image` library in order to find clusters of the colour used to represent the radial pattern in the input road pattern image.

5 Evaluation

In order to evaluate the capabilities of our system we will be measuring correctness in terms of realism by using statistical metrics. We present a set of metrics focused on analysing the intermediate representation independent of the visualization. In keeping with the concept of road layout types being a distinguishing characteristic of cities, our metrics are based on road density, layout, and interconnectivity.

5.1 Methodology

As no prior related work with the exception of those attempting to replicate existing cities, e.g. [17,19], has established statistical evaluation metrics, we start by proposing the following metrics inspired by Boeing's analysis of real-world cities [3]. Note that since we are representing our generated city as an undirected graph in the intermediary representation, we will henceforth be using the term node when describing the start or the end of a road segment.

The metrics that we use to evaluate our results are:

- **Entropy** - Using a binned histogram of road orientations in the generated city's street network, we compute the entropy of the street network as:

$$H = -\sum_{i=1}^{n} P(o_i) * log_e P(o_i)$$

  where n is the total number of bins (currently defaulted to 36 such that one bin represents 10°), i is the indices of the bins, and $P(o_i)$ is the proportion of orientations in the i-th bin [3]. We employ the natural logarithm such that the value of H is in nats, the dimensionless natural unit of information. The maximum entropy H_{max} that a city can have thus equals the logarithm of the total number of bins. In our case, this corresponds to 3.584 with a perfectly uniform distribution of street bearings across all bins. In theory, if all bearings fell into the same bin, H_{min} would equal 0. However, because we are working with an undirected graph, the theoretical minimal entropy our road network can have is 0.693 in the case where all roads run either north-south, or east-west, and thus fall into two bins. Such a road network is hardly plausible in the real-world, and we thus consider a road network exhibiting a perfect grid pattern (north-south-east-west in equal proportions across four bins) to be our H_{min}. In other words; we assume $H_{min} = 1.386$.
- **Orientation Order** - Using the entropy, we can compute the normalised orientation order of the generated road network [3]. We do this using:

$$\Phi = 1 - (\frac{H - H_{min}}{H_{max} - H_{min}})^2$$

where H_{min} is the lowest possible entropy of a plausible road network perfectly adhering to the grid-pattern, and H_{max} is the highest possible entropy in which the road network has a perfectly uniform distribution. Φ thus allows us to describe where a generated road network is positioned on the continuum from perfectly disordered to perfectly ordered. Or, in other words; how organic or how grid-based the generated road network is.
- **Average node degree** - The average node degree (that is, average number of edges connected to a given node) of the road network including dead ends and intersections.
- **Proportion of dead ends** - The proportion of nodes in the road network which connect only to one other node (dead end roads).
- **Proportion of three-way intersections** - The proportion of nodes with three edges in the road network.
- **Proportion of four-way intersections** - The proportion of nodes with four edges in the road network.
- **Intersection count** - The total number of intersections in the generated road network. Note that the current implementation does not support roundabouts or intersections with more than four edges.
- **Total road length** - The total length of the generated road network in kilometers.

Whilst the above is not an exhaustive list of metrics, we do consider the proposed metrics rather fundamental and thus likely applicable to most use-cases in which determining some level of realism is desirable. In accommodating potential future work by others, it would be helpful to have a set of standardised metrics to compute as to enable the comparison of different solutions. We hope that the above metrics can help establish that.

In evaluating our generator module, we will be relying on input data from Auckland, New Zealand. Specifically, we will focus on the central part of Auckland within the bounding box −36:83N, −36:94S, 174:82E, and 174:68W per OpenStreetMap2. Using Koordinates.com, we download waterways, land usage, and population density input maps covering this area. This, in turn, allows us to (a) test the impact of changing road rules and modifying individual probabilities; and (b) extract the corresponding real-world version of Auckland and apply the same set of metrics for subsequent comparison. We additionally input a manually created road rule map which determines for any given area whether roads will be laid out using an organic, grid, or radial pattern.

Thus, if a specific growth rule should be applied to the entire canvas, the rule map would just feature the corresponding colour. Accordingly, we propose the following six distinct configurations. Note that by default, we allow 100 major road iterations, and 10000 minor road iterations. These numbers were chosen largely arbitrarily as it seemed to suffice in covering the area within the bounding box. Additionally, and unless otherwise stated, minor roads have a 90% probability of using the grid-based pattern and a 10% probability of using the organic pattern.

1. **Organic** configuration with an organic road rule map covering the entire canvas and a default 15% probability of the road turning. Based on analysing map data we hypothesise that this configuration best matches the real-world version of Auckland.
2. **Organic (0.5 Turn)** configuration with an organic road rule map covering the entire canvas and an increased 50% probability of the road turning.
3. **Radial** configuration with a radial road rule map covering the entire canvas and a default 10% probability of the road turning.
4. **Grid (Perfect)** configuration with a grid road rule map covering the entire canvas and 10% probability of the road turning. Minor roads are upwards 33% shorter and 100% grid-based. We hypothesise that this configuration will match the real-world version of Auckland the least.
5. **Grid (Major)** configuration with a grid road rule map covering the entire canvas and 10% probability of the road turning. Minor roads maintain 10% probability of using the organic pattern.
6. **Grid (Both)** configuration with a grid road rule map covering the entire canvas and 1% probability of the road turning. Minor roads are 100% grid-based.

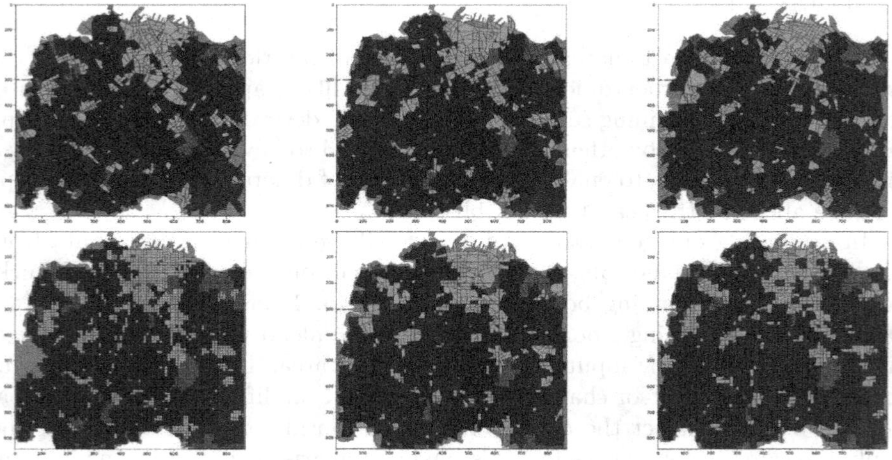

Fig. 2. Auckland configurations output. Blue lines represent major roads, black lines represent minor roads, purple polygons are residential, yellow polygons are commercial, blue polygons are industrial, and red polygons are areas such as parks with no types of buildings. Top row: organic, organic (0.5 turn), and radial configurations. Bottom row: grid configurations: perfect, major, and both. (Color figure online)

In addition to computing the nine metrics outlined above, we time the generation process. Although performance is not a design priority in our system, we include the metric to highlight the correlation between the number of intersections and total road length and the time it takes to generate. The computations

Table 1. An overview of the statistical metrics applied to the six configurations. The Auckland column represents data from the real city of Auckland over the same region. For each metric, the configuration most closely matching the real world value is in bold.

	Organic	Organic (0.5)	Radial	Grid (perfect)	Grid (major)	Grid (both)	Auckland
Entropy	3.5571	**3.5362**	3.5641	1.3863	3.1511	2.755	3.4957
Orientation order	0.0238	**0.0426**	0.0157	0.9999	0.3549	0.612	0.0783
Avg. node degree	**3.5728**	3.6196	3.5678	3.7824	3.6095	3.6579	2.7823
Prop. dead-ends	**0.0545**	0.0493	0.0524	0.0144	0.0399	0.031	0.1761
Prop. 3-ways	0.2638	0.2355	**0.2752**	0.1746	0.2715	0.25	0.6955
Prop. 4-ways	0.6817	0.7136	**0.6721**	0.8108	0.6883	0.7188	0.123
# intersections	2864	3124	2872	**5255**	2960	3167	4876
Total length (km)	1000	**1073**	1004	1394	1022	1081	1056
Generation time (s)	110.6	121.9	217.5	212.1	114	132.2	N/A

were performed using an AMD Ryzen 7 2700X 3.7 GHz CPU, 32 GB memory, and a NVidia GeForce GTX 1080 Ti GPU.

5.2 Results

The computed metrics for all six configurations alongside the real-world Auckland are summarised in Table 1. A visual representation of the intermediary representation generated by each respective configuration is shown in Fig. 2.

It can be seen that the organic- and radial-based configurations generate cities are closer to the real-world version of Auckland than the cities generated by grid-based configurations. However, our hypothesis that the organic configuration would most closely match the real-world version of Auckland is only partially correct: It does score highest for average node degree (3.5728 compared to 2.7823) and the proportion of dead-ends (0.0545 compared to 0.1761), though the organic configuration with an increased road turn probability would have been a better choice. This scores the best of both entropy (3.5362 compared to 3.4957), orientation order (0.0426 compared to 0.0783), and total road length (1,072,566 m compared to 1,055,638 m). In either case, the entropy of the real-world Auckland is very uniform and thus confirms that the majority of the road network can be considered organic.

Our hypothesis that the Grid (Perfect) configuration would match the real-world version of Auckland the least is correct on all metrics except the number of intersections. Interestingly, with the exception of this configuration, all other configurations have approximately 38% fewer intersections than the real-world

Auckland. The main reason for this is that, unlike all six of our configurations, the real-world version of Auckland has a significantly larger proportion (0.6955) of three-way intersections than the proportion of four-way intersections (0.123). The numbers are more or less flipped in our case. This could suggest that our local constraints are too keen on generating four-way intersections, though it is more likely that our average segment distance is too short.

When examining the rose diagrams in Fig. 3 depicting the various compass bearings, it becomes even clearer that our generated cities differ from the real world version of Auckland. Observe how in the real city, the real orientation distributions are heavily skewed towards east and west with the remaining bins appearing somewhat uniform. None of our rose diagrams depict a similar distribution. This indicates that the generator system is not yet able to completely mimic the characteristics of a real world city. This is likely due to the fact that our generator only considers factors within its borders, while the real city's road network is influenced by the surrounding area and terrain.

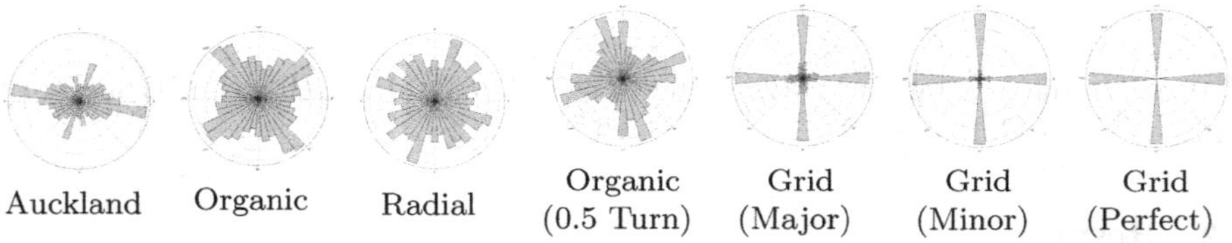

Fig. 3. Binned histograms for the real Auckland area used as input and the six different configurations.

Our results suggests that the generation time is correlated with intersection count and total road length, as the configurations with the lowest generation time are also among the configurations with the least amount of intersections and shortest road network.

Figure 4 shows a simple visualization of the city generated from the intermediate format output for the organic configuration.

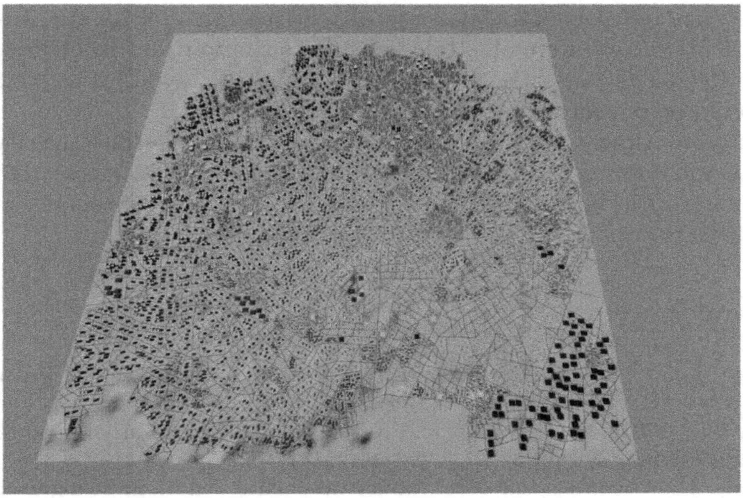

Fig. 4. A simple visualization of the generated city. Building colours indicate land use: apartments in green, houses in blue, industrial buildings in black, shops in pink, and large-scale shopping centres in yellow. (Color figure online)

6 Conclusion and Future Work

We have presented a modular and extensible system capable of accepting a selection of real-world input maps and outputting a procedurally generated city in an intermediate representation. Our system constructs road networks using three distinct growth rules derived from real-world cities and accounts for population density and land use

Our results suggest that the presented city generation techniques is able to create a large variety of plausible road layouts. However, it is still to naive to generate realistic real world cities such as Auckland. We believe that one key factor for this is insufficient input data such as the lack of terrain data. Major roads tend to follow ridges and valleys. Also, our system does not simulate the growth of a city and that land usage can change over time (e.g., areas initially designed as industrial become residential).

We proposed a selection of objective evaluation metrics which can help determine the correctness or realism of a generated city by evaluating its intermediate representation. These metrics may be applied on both real-world cities as well as procedurally generated cities, thereby allowing the comparison between real-world cities and procedurally generated cities from different solutions.

Our analysis indicates that the metrics are a useful tool to evaluate and quantify characteristics of cities and are hence a valuable addition to visual analysis. For example, while the outputs of configurations one and five are easy to visually compare given their widely different use of road patterns, comparing outputs of configuration one and two visually becomes a lot more difficult given that they are based on the same use of road patterns. With the use of our statistical metrics we were able to reach the conclusion that configuration one was not able to output satisfactory results, as we had hypothesised, and more work should be put into the implementation.

In future work we would like to incorporate terrain maps into the city generation and add information about neighbouring high density centres, which effect traffic and growth direction. Also we would like to automate the generation of road growth rule maps, e.g., by analysing existing cities.

References

1. Aliaga, D.G., Vanegas, C.A., Benes, B.: Interactive example-based urban layout synthesis. ACM Trans. Graph. (TOG) **27**(5), 1–10 (2008)
2. Beneš, J., Wilkie, A., Křivánek, J.: Procedural modelling of urban road networks. Comput. Graph. Forum **33**(6), 132–142 (2014)
3. Boeing, G.: Urban spatial order: street network orientation, configuration, and entropy. Appl. Netw. Sci. **4**(1) (2019). Article number: 67. https://doi.org/10.1007/s41109-019-0189-1
4. Cullen, B., O'Sullivan, C.: A caching approach to real-time procedural generation of cities from GIS data. J. WSCG **19**(3), 119–126 (2011)
5. Fernandes, G.D., Fernandes, A.R.: Space colonisation for procedural road generation. In: 2018 International Conference on Graphics and Interaction (ICGI), pp. 1–8. IEEE (2018)
6. Galin, E., Peytavie, A., Guérin, E., Beneš, B.: Authoring hierarchical road networks. Comput. Graph. Forum **30**(7), 2021–2030 (2011)
7. Galin, E., Peytavie, A., Maréchal, N., Guérin, E.: Procedural generation of roads. Comput. Graph. Forum **29**(2), 429–438 (2010)
8. Groenewegen, S., Smelik, R.M., de Kraker, K.J., Bidarra, R.: Procedural city layout generation based on urban land use models. In: Eurographics (Short Papers), pp. 45–48 (2009)
9. Hendrikx, M., Meijer, S., Van Der Velden, J., Iosup, A.: Procedural content generation for games: a survey. ACM Trans. Multimed. Comput. Commun. Appl. **9**(1), 1:1–1:22 (2013). https://doi.org/10.1145/2422956.2422957
10. Kelly, G., McCabe, H.: CityGen: an interactive system for procedural city generation. In: Fifth International Conference on Game Design and Technology, pp. 8–16 (2007)
11. Lechner, T., Ren, P., Watson, B., Brozefski, C., Wilenski, U.: Procedural modeling of urban land use. In: ACM SIGGRAPH 2006 Research Posters, SIGGRAPH 2006, p. 135-es. Association for Computing Machinery, New York (2006). https://doi.org/10.1145/1179622.1179778
12. Lechner, T., Watson, B., Ren, P., Wilensky, U., Tisue, S., Felsen, M.: Procedural modeling of land use in cities. Technical report, NWU-CS-04-38, Northwestern University (2004)
13. Lechner, T., Watson, B., Wilensky, U.: Procedural city modeling. In: 1st Midwestern Graphics Conference. Citeseer (2003)
14. Parish, Y.I., Müller, P.: Procedural modeling of cities. In: Proceedings of the 28th Annual Conference on Computer Graphics and Interactive Techniques, pp. 301–308 (2001)

15. Sauder, J.: procedural_city_generation: procedural city generation program implemented in python and visualized with blender (2015). https://github.com/josauder/procedural_city_generation
16. Smelik, R.M., De Kraker, K.J., Tutenel, T., Bidarra, R., Groenewegen, S.A.: A survey of procedural methods for terrain modelling. In: Proceedings of the CASA Workshop on 3D Advanced Media in Gaming and Simulation (3AMIGAS), vol. 2009, pp. 25–34 (2009)
17. Vanegas, C.A., Aliaga, D.G., Benes, B., Waddell, P.: Visualization of simulated urban spaces: inferring parameterized generation of streets, parcels, and aerial imagery. IEEE Trans. Vis. Comput. Graph. **15**(3), 424–435 (2009)
18. Vanegas, C.A., Kelly, T., Weber, B., Halatsch, J., Aliaga, D.G., Müller, P.: Procedural generation of parcels in urban modeling. Comput. Graph. Forum **31**(2pt3), 681–690 (2012)
19. Weber, B., Müller, P., Wonka, P., Gross, M.: Interactive geometric simulation of 4D cities. Comput. Graph. Forum **28**(2), 481–492 (2009)
20. Yu, Q., Steed, A.: Example-based road network synthesis. In: Eurographics (Short Papers), pp. 53–56 (2012)

Video Quality Assessment Considering the Features of the Human Visual System

Anastasia Mozhaeva[1,2], Vladimir Mazin[2], Michael J. Cree[1], and Lee Streeter[1(✉)]

[1] The University of Waikato, Hamilton, New Zealand
lee.streeter@waikato.ac.nz
[2] Moscow Technical University of Communications and Informatics, Moscow, Russia

Abstract. Nowadays, numerous video compression quality assessment metrics are available. Some of these metrics are "objective" and only tangentially represent how a human observer rates video quality. On the other hand, models of the human visual system have been shown to be effective at describing spatial coding. In this work we propose a new quality metric which extends the peak signal to noise ratio metric with features of the human visual system measured using modern LCD screens. We also analyse the current visibility models of the early visual system and compare the commonly used quality metrics with metrics containing data modelling human perception. We examine the Pearson's linear correlation coefficient of the various video compression quality metrics with human subjective scores on videos from the publicly available Netflix data set. Of the metrics tested, our new proposed metric is found to have the most stable high performance in predicting subjective video compression quality.

Keywords: Video quality metric · The model of visibility · Acceptable visual quality · Artefact perception

1 Introduction

Video quality metrics are a critical component in modern streaming video processing algorithms. Alongside compression and data transmission, accurate quality estimation is key to maximising use of bandwidth while also maximising the user experience. Video quality methods can be divided into two categories: subjective and objective quality assessment criteria [1]. The human user is typically the final recipient in typical video processing applications, so subjective quality criteria that reflects human visual perception is arguably the more important method of assessing video quality [2]. Objective quality criteria are the most common and popular method of evaluating video quality since the assessment is performed algorithmically. Objective metrics avoid expensive research with user participation. At the present stage of the development of media content transmission technologies, objective algorithms for evaluating video quality

W. Q. Yan et al. (Eds.): IVCNZ 2022, LNCS 13836, pp. 288–300, 2023.
https://doi.org/10.1007/978-3-031-25825-1_21

have achieved high optimization and simplicity. Unfortunately, popular objective prediction models correlate poorly with subjective perceptions of quality by the human visual system (HVS) and depend on the systems or processes involved [2]. On the other hand, there are algorithmically complex video quality metrics (VQM) based on models of the human visual system [1,3], and an open question is whether complex video quality metrics based on models of the human visual system provide significantly better predictions than objective metrics. Another problem when used of visual models, is in developing video quality metrics we must represent the HVS in software, a task impeded by the limited new fundamental knowledge of the HVS perception of video content using modern equipment.

In visual perception studies, the primary focus is most often on studying the physical aspects of HSV, of which video quality is a secondary question and often not discussed [4]. In addition, in most psychophysical experiments, participants enter a controlled laboratory environment in order to stabilise tests and control any confounding factors. In other words, a reduction in the number of experiments, consequently, errors, is made [5]. The laboratory approach limits many of the problems that arise in the "real world". Therefore, developers of video quality assessments and video compression algorithms need to determine how psychophysical results relate to quality in today's video content presentation environment, without a full understanding of the processes of HVS. Modern knowledge and tests about the relationship between HVS and quality metrics are needed.

Human visual sensitivity can be characterised by a distinct region of spatio-temporal frequencies. Region boundaries determine the visibility of artefacts in the displayed information [6]. More than five years ago, a linear model of spatio-temporal contrast sensitivity [7], which determines the visibility of visual information artefacts by the human eye, was presented. However, the viewing conditions in the research differed from modern computer monitors and television screens. In particular, cathode ray tube (CRT) screens were used, which are not in common use today, and the viewing angle was controlled to be small or normal to the screen, which is not necessarily true in everyday use. Consequently, the results do not guarantee a proper description of the conditions in which media content is presently consumed. At present, the problem of imitation in the evaluations of the quality of the video of the HVS has led to the creation in 2022, of a model HVS [8] that considers stimulus parameters: spatial and temporal frequency, eccentricity, luminance, and viewing area. However, to create the model, there were data from 11 publications, where the viewing conditions were also different from modern computer monitors and television screens. At the current stage of technological development, there are no theoretical obstacles to creating a sufficiently comprehensive video quality assessment correlated with HVS. However, there exists a practical obstacle, that large-scale experiments may be extremely time-consuming and expensive. In the previous work, we presented a methodology [9] for conducting large-scale experiments and presented initial measurements of the characteristics of the HVS [1].

While many video scoring metrics are available, this paper represents the first practical consideration of a comprehensive solution to adapting a simple objective video quality metric to a typical video user experience. The objective of this paper is to demonstrate the need to research and collect new HVS data under modern circumstances of providing information. We present a comparison of commonly used video quality metrics with HVS based metrics. Our hypothesis is that video quality metrics containing HVS data perform better than other currently used metrics. Consequently, we aim to show that there is a significant opportunity to develop and extend existing video quality metrics using new HVS based data.

2 Related Work

Considering all video quality assessment methods, the most popular ones are peak signal-to-noise ratio (PSNR) and structural similarity image metric (SSIM) [10]. PSNR is used more often than other methods to assess similarities between original and reconstructed images and videos. PSNR is calculated on a logarithmic scale by amplitude (in decibels). The benefit being that the HVS also perceives brightness on a logarithmic scale. In PSNR, the signal-to-noise ratio based on standard deviation never gives overestimated results [10], which is why the method is the most widely used. However, PSNR is poorly correlated with visual quality estimation and does not consider spatial and temporal psychovisual models. PSNR gives significantly underestimated results, even a slight shift of the reference and estimation frame in space or desynchronisation of video sequence in time can degrade the performance of PSNR estimation [11]. Another common metric is structural similarity of the image (SSIM) [10]. However, it has been shown analytically and experimentally that, for images with fragments of large or small mean luminance values, the local estimates of the metric are unstable [4]. SSIM does not consider different levels of absolute luminance, temporal aspects or viewing distances, and poorly correlates with human perception.

A more fundamental approach to creating video quality metrics includes low-level visual modelling based on psychophysical models, such as the contrast sensitivity function (CSF) [3]: the threshold at which a human observer can detect change in a given brightness pattern as a function of spatial and temporal frequency. The artefacts visible to the human visual system in early vision are regulated by the function of visual sensitivity. Early vision involves three processes: filtering, encoding, and interpretation [12]. Human visual sensitivity can be characterised as a reference filter in terms of spatial and temporal frequencies and there is no need for rendering beyond the limitations of the region [6].

The existing HVS models do not consider all the necessary spatio-temporal variations of the stimuli. Popular CSF models, such as Barten [13] or Daly [14], do not include the temporal frequency. Kelly's spatio-temporal CSF [15] consider spatial and temporal frequencies without luminance. In 2016, Watson and

Ahumada presented a linear model of spatio-temporal contrast sensitivity, called the "visibility pyramid" [7]. The authors constructed an exhaustive description of spatial and temporal contrast sensitivity, and its dependence on retinal illumination, and also derived a number of strong relations, all from the modest results of much older studies [16–18]. StelaCSF, unlike previous studies, tries not to get a perfect fit for individual datasets but creates a single model that can explain all datasets without overfitting [8]. To model the 5-dimensional contrast sensitivity space, the authors combined data from 11 publications on CSF. The main interest of modelling is datasets. However, the equipment and, therefore, viewing conditions in older studies differed from the conventional computer monitors and television screens used today.

In previous work, we presented a new video quality metric (PSNR-M) [1] that considered data for the first part of early vision, namely the filtering stage that determines which spatial and temporal fluctuations in stimuli the HVS responds to. PSNR-M was created using HVS data from modern screens, but peripheral vision effects were not considered. The participants controlled the spatial aspects, flicker amplitude, and the brightness of the round sinusoidal lattice pattern [9]. Contrast thresholds were measured at 8 different spatial and 15 temporal frequencies at 3 different brightness levels (L). The resultant HVS model is reproduced in Fig. 1 [1].

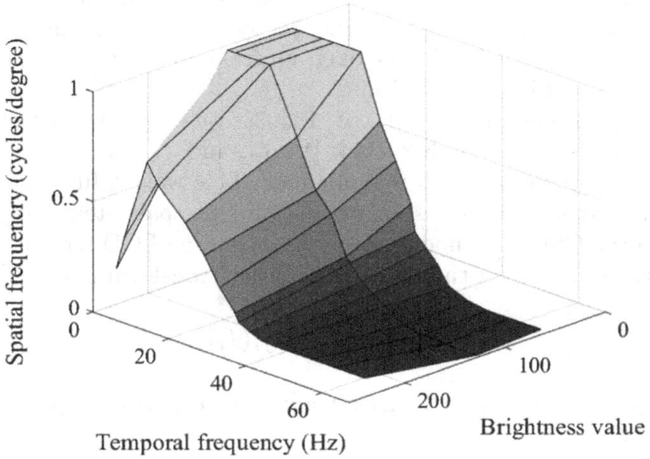

Fig. 1. The time-frequency response of the HVS with respect to brightness [1].

3 Video Quality Assessment Considering the Features of the Human Visual System: Adaptation Aspects, Spatial and Temporal Frequency, Eccentricity, Luminance

We propose a PSNR based video quality assessment method that also incorporates measurements of the HVS, which we call PSNR-M+. The HVS factors included are spatial and temporal frequency, eccentricity (distance from the centre of fovea in visual degrees), luminance, adaptation aspects of the HVS, and stimulus size. This new metric is an extention of our earlier PSNR-M [1] which considered a small dataset on the dependence of separate spatial and temporal HVS characteristics on luminance, which was proven effective in comparison to PSNR [1].

3.1 Video Quality Assessment Metric

The proposed method centres around a weighted PSNR calculation which we proceed to elucidate,

$$\mathrm{PSNR}'(I(t), I_R(t), t) = \mathrm{PSNR}(I(t)K(t), I_R(t)K(t), t), \tag{1}$$

where I is a compressed frame, I_R is the reference uncompressed frame, and $K(t)$ is a weight coefficients matrix for t frames. The quality of the distorted video is measured incorporating both the spatio-temporal-luminance component and a peripheral component. A flow diagram framework of the methodology for calculating $K(x, y, t)$ is shown in Fig. 2.

In the spatio-temporal block of Fig. 2, we use the weight function $H_{L,f_t}(f_x, f_y)$ measured in earlier work [1], f_x, and f_y are spatial frequencies, f_t is temporal frequency, and L is luminance. This weight function models the ability of the human visual system to respond to spatio-temporal change, as measured via the CSF on a modern in-plane-switching LCD screen. The impulse response may be found via the following Fourier transform pair

$$h_{L,f_t}(x, y) \overset{F}{\Longleftrightarrow} (H_{L,f_t}(f_x, f_y)), \tag{2}$$

where $\overset{F}{\Longleftrightarrow}$ is the (invertible) Fourier transform. Filtering is performed in the spatial domain via.

$$I_R^{'}(x, y) = I_R(x, y) * h(x, y), \tag{3}$$

where $*$ is convolution. The spatio-temporal-luminance weighting factor, K_{stL}, is then computed as

$$K_{\mathrm{st}L}(x, y) = \frac{I_R^{'}(x, y)}{I_R(x, y)}. \tag{4}$$

The task of the region of interest (ROI) sampling is to identify objects that are more significant for the HVS. In this work, the ROI distinguishes individual objects using a variant of the watershed algorithm [20]. No more than five objects

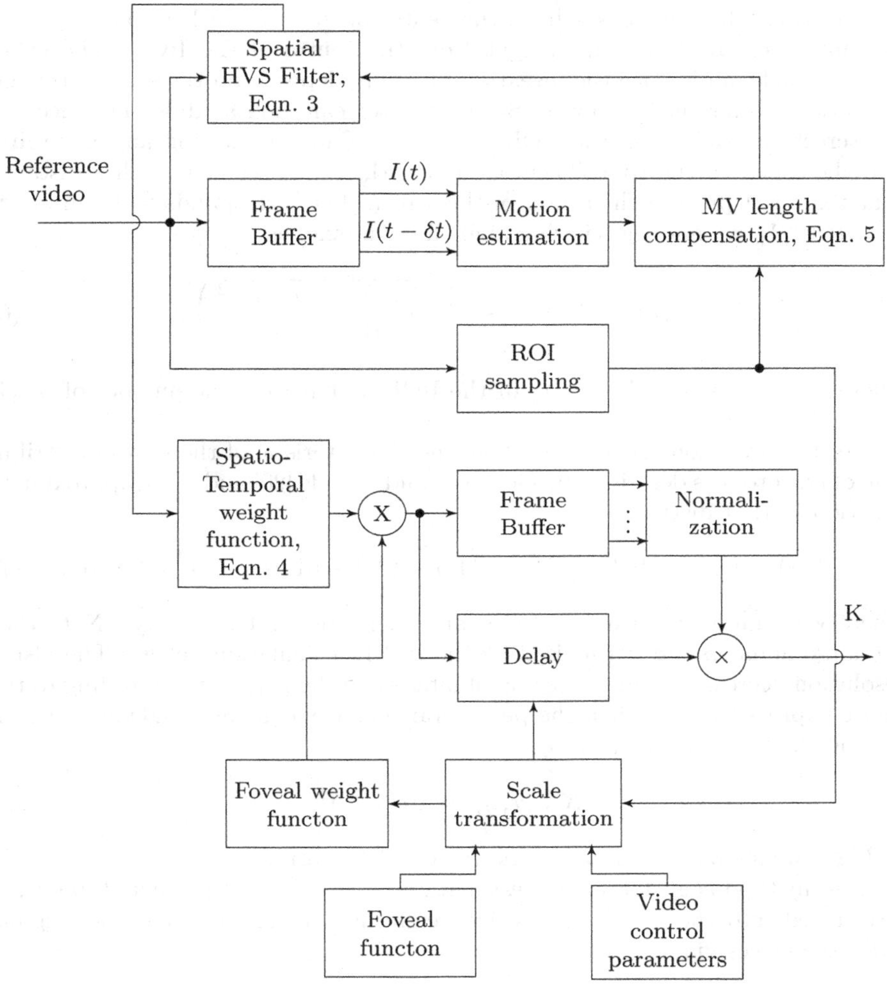

Fig. 2. The framework of the methodology for weight estimate.

are selected close to the centre, with fewer objects selected if the total area of the selected clustered object is greater than a user determined threshold [1].

The motion estimation utilises an adaptation of the MPEG block matching technique [19], which uses a 16 × 16 pixel block size, method with 32 × 32 pixel search area. The motion vectors, v(x, y, t), are "compensated" by subtracting the average within the ROI, v_{ROI}, viz.

$$v_{\text{comp.}}(x, y, t) = v(x, y, t) - v_{\text{ROI}}, \tag{5}$$

The peripheral coefficient is an approximate foveation function, which models the decrease of focus from the centre of the ROI outwards. (Foveation being the blurring which increases from the centre of vision, which in our metric is accounted for by decreasing weight from the centre of the ROI). The exact viewing angle, and hence the foveation function, depends on the screen size and resolution of the display device. Since the user can look at different places on the screen depending on the ROI, in order to find the viewing angle, we first find the centre of the ROI. Then, assuming that the user is at such a distance that there is 1 pixel in the center of the screen that corresponds to 1 min of arc of vision [21], we find the viewing angle for the pixel as

$$\alpha(x, y) = \arctan\left(\frac{\pi\sqrt{(x - x_c)^2 + (y - y_c)^2}}{1080}\right), \tag{6}$$

where, x_c, and y_c are the centre of the ROI, and 1080 is the number of pixels width of the screen.

For the foveation function, we take the characteristic of the spatial distribution of receptors as described by Gonzalez and Woods [20] and fit an approximate function to that distribution, viz.

$$A(\alpha) = \left(1.26 \times 10^5\right) e^{-0.71\alpha} + \left(1.5 \times 10^4\right) - 512\alpha + 11\alpha^2 - 0.08\alpha^3, \tag{7}$$

where we temporarily exclude the spatial dependency for brevity. (Note that the large numbers are normalised by the PSNR′ calculation below.) The visual resolution decreases from the region of interest to the periphery according to the above expression [22]. Then the peripheral coefficient in the weight function in the metrics is determined by

$$K_{pr}(x, y) = A(\alpha(x, y)). \tag{8}$$

In Fig. 3 we show an example of the $K_{\text{pr}}(x, y)$ pattern.

The methodology for weight estimation developed and presented above, is introduced into the PSNR metric by weighting the function and the original video sequences via

$$\text{PSNR}' = 20\log_{10}\frac{\sqrt{\sum_{t_n=-\frac{n}{2}}^{\frac{n}{2}}\sum_{x,y}K_{stL}^2(x, y, t_0 + t_n)\,K_{pr}^2(x, y, t_0 + t_n)}}{\sqrt{(n+1)\sum_{x,y}(I - I_R)^2 K_{stL}^2(x, y, t_0)\,K_{pr}^2(x, y, t_0)}}, \tag{9}$$

where t_0 is the current frame, and t_n is the number of frames from the current frame, $n = \Delta t f_v$, Δt is the reaction time of the eye to frames (without considering the cognitive factors) [1], and f_v is the number of frames per second. In this work we took $\Delta t = 0.8$ s [23,24].

4 Methods

We compare the proposed metric to a range of metrics found in the literature and in current use. These metrics include: the most popular metrics at the current

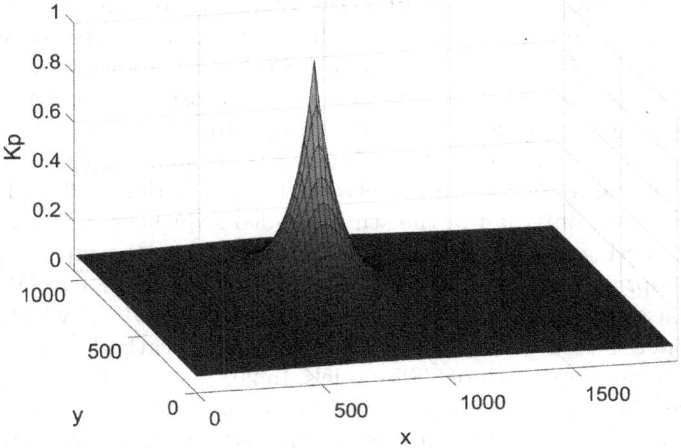

Fig. 3. An example of K_{pr}, where the frame size is 1920 × 1080 and the centre of the ROI is set to (760, 640).

moment (PSNR, SSIM); video multimethod assessment fusion (VMAF) [25], a method based on machine learning on databases of real users' evaluations of the quality of training videos [26]; spatio-temporal reduced reference entropy difference (STRRED) [27] and HDR-VQM [28], which are metrics that address temporal aspects; and metrics which considered the HVS visibility models such as PSNR-M and FovVideoVDP.

The LIVE-NFLX [29,30] was used to test and compare the video quality metrics. The LIVE-NFLX data set consists of 112 compressed (hence distorted) videos. In this work we use 12 of these videos. The LIVE-NFLX database is selected because it represented highly realistic content with quality of experience responses to various design dimensions, including varying compression rates in the form of simulated varying transmission video bit rates over the course of each video.

5 Results and Discussion

The mean opinion scores (MOS), or subjective score, provided by the LIVE-NFLX database were used to compare the performance of PSNR-M+ with the presented above VQM. The scatter plots of the VQM algorithms under comparison are shown in Fig. 4. In Fig. 5, the metrics and their Pearson's linear correlation coefficient (PLCC) with the subjective quality scores are compared. As can be seen Fig. 5, the proposed metric gives the most consistent positive correlation. The gain in performance is mostly due to PSNR-M+'s ability to generalize predictions across video sequences. The strong compression distortion represented in this dataset and the difficulty of replicating subjective scores are apparent in the correlations herein. For example, in the scatter plots in Fig. 4, HDR-VQM predicts a comparable correlation between the prediction and real subjective score estimates for only half of the studied video sequences. HDR-VQM does a good job predicting spatial processing but does not model temporal processing.

It is appropriate to compare PSNR-M+, in which the model of HVS is derived from data based on modern LCD screens, to FovVideoVDP [3] which was based on much older data derived from experiments using cathode ray tube (CRT) screens. Both metrics consist of CSF models. In 2022, a visibility model StelaCSF (an improved FovVideoVDP model, the exact model we test herein) was presented based on 11 data sets that also use the older CRT screen data [8]. Their results show that stelaCSF can explain current data sets better than other previously existing models. FovVideoVDP, as well as PSNR-M, show variable results in low contrast objects and low bitrate. When comparing FovVideoVDP with PSNR-M+, the metrics show approximately similar results for video sequences with a normal bitrate. However, for videos containing sufficient motion, PSNR-M+ gives an average of 15% higher correlation.

The primary purpose of the analysis is to observe the behaviour of current VQM measures when deployed for predicting the perceptual quality of video content. From Fig. 5, the new HVS data based model, taken on modern equipment, gives an increase in the stability of the correlation of the obtained metric value with the real subjective score when evaluating video quality. Figure 5 compares the correlation intervals of video quality metrics with the real subjective score on video sequences. PSNR-M+ has a better correlation interval than the VQM algorithms under comparison. Or in other words, the metric more effectively predicts the perception of videos, with different content and distortions, by the human visual system.

Despite the objectively better stability of the developed method, there is still potential room for improvement. For example, the motion compensation is coarse and could be refined to better reflect the motion of individual moving objects. The size of objects selected by the ROI stage could also be refined. The use of more videos from more disparate data sets will also enhance our confidence in our correlation scores. Moreover, our HVS model is limited and does not simulate certain aspects of vision such as inter-channel masking, eye movement, and a model of peripheral vision.

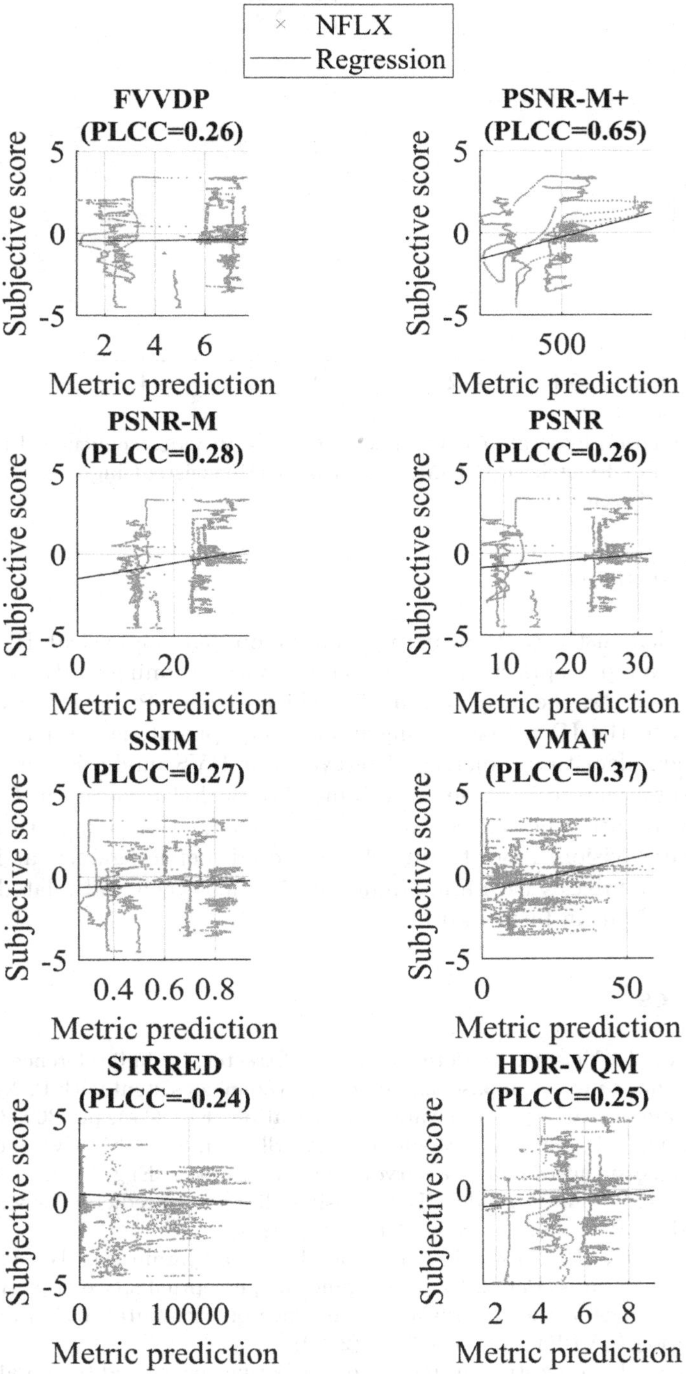

Fig. 4. Visualisation of LIVE-NFLX database results.

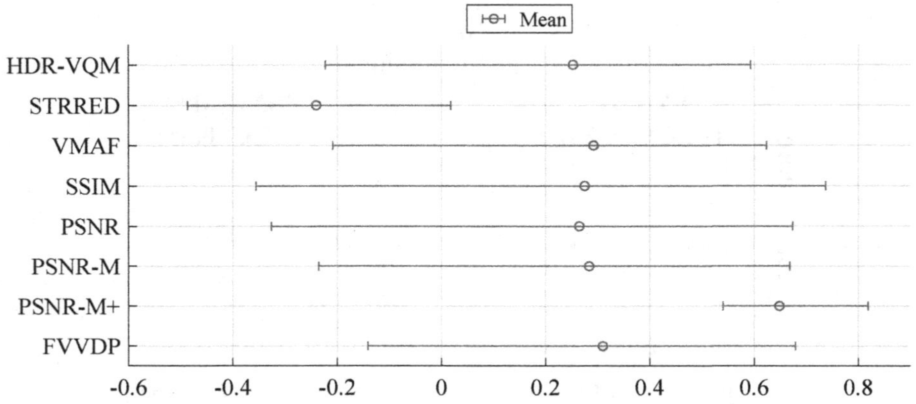

Fig. 5. Correlation interval of video quality metrics on video sequences LIVE-NFLX. The new proposed metric, PSNR-M+, has the most consistent high correlation of the metrics tested herein.

6 Conclusion

Our work demonstrates that metrics based on psychophysical HVS models explain human perception of video quality, outperforming statistically based metrics. In this test the proposed metric, which extends PSNR by incorporating recent data of the HVS taken using modern equipment, is comparable to the best complex algorithmic metrics. However, our HVS model is still somewhat limited and we plan to increase the volume of data that informs our CSF model, and to incorporate other aspects such as inter-channel masking, eye movement and peripheral vision. Also, testing the proposed metric against, at least, four independent databases with a minimum of 15 videos each and related statistics will be carried out in future work.

References

1. Mozhaeva, A., Vlasuyk, I., Potashnikov, A., Streeter, L.: Full reference video quality assessment metric on base human visual system consistent with PSNR. In: 2021 28th Conference of Open Innovations Association (FRUCT), pp. 309–315 (2021)
2. Mohammadu, P., Ebrahimi-Moghadam, A., Shirani, S.: Subjective and objective quality assessment of image: a survey. Majlesi J. Electr. Eng. **9**(1), 55–83 (2015)
3. Mantiuk, R., et al.: FovVideoVDP: a visible difference predictor for wide field-of-view video. ACM Trans. Graph. **40**, 1–19 (2021)
4. Ying, Z., Niu, H., Gupta, P., Mahajan, D., Ghadiyaram, D., Bovik, A.: From patches to pictures (PaQ-2-PiQ): mapping the perceptual space of picture quality. In: Proceedings of the IEEE/CVF Conference on Computer Vision and Pattern Recognition (CVPR), pp. 3575–3585 (2020)
5. Kulikowski, J.: Some stimulus parameters affecting spatial and temporal resolution in human vision. Vis. Res. **11**, 83–93 (1971)

6. Watson, A.: High frame rates and human vision: a view through the window of visibility. SMPTE Motion Imaging J. **122**, 18–32 (2013)
7. Watson, A., Ahumada, A.: The pyramid of visibility. J. Vis. **16**(12), 567 (2016)
8. Mantiuk, R., Ashraf, M., Chapiro, A.: stelaCSF - a unified model of contrast sensitivity as the function of spatio-temporal frequency, eccentricity, luminance and area. ACM Trans. Graph. (Proceedings of SIGGRAPH 2022) **41**(4), 1–19 (2022). Article no. 145
9. Mozhaeva, A., Vlasuyk, I., Potashnikov, A., Cree, M.J., Streeter, L.: The method and devices for research the parameters of the human visual system to video quality assessment. In: 2021 Systems of Signals Generating and Processing in the Field of Onboard Communications, pp. 1–5 (2021)
10. Wang, Z., Bovik, A.: Mean squared error: love it or leave it? A new look at signal fidelity measures. IEEE Sig. Process. Mag. **26**(1), 98–117 (2009)
11. Seshadrinathan, K., Bovik, A.: Motion tuned spatio-temporal quality assessment of natural videos. IEEE Trans. Image Process. **19**(2), 335–350 (2009)
12. National Research Council: Human Performance Models for Computer Aided Engineerings. The National Academies Press (1989)
13. Barten, P.: Formula for the contrast sensitivity of the human eye. In: Proceedings Volume 5294, Image Quality and System Performance, pp. 231–238 (2003)
14. Daly, S.: Visible differences predictor: an algorithm for the assessment of image fidelity. In: Proceedings of SPIE 1666, Human Vision, Visual Processing, and Digital Display III (1992)
15. Kelly, D.: Motion and vision. II. Stabilized spatio-temporal threshold surface. J. Opt. Soc. Am. **10**(10), 1340–1349 (1979)
16. de Lange, H.: Research into the dynamic nature of the human fovea-cortex systems with intermittent and modulated light. Attenuation characteristics with white and colored light. J. Opt. Soc. Am. **48**(11), 777–784 (1958)
17. Campbell, F., Cooper, G., Robson, J.: Application of Fourier analysis to the visibility of gratings. J. Physiol. **179**(3), 551–566 (1968)
18. van Nes, F., Bouman, M.: Spatial modulation transfer in the human eye. J. Opt. Soc. Am. **57**(3), 401–406 (1967)
19. Potashnikov, A., Vlasuyk, I., Augstkaln, I.: Analysis of methods for detecting moving objects of different types on video image. In: Fundamental Problems of Radio Electronic Instrumentation, pp. 1201–1204 (2017)
20. Gonzalez, R., Woods, R.: Digital Image Processing. Technosphere, Moscow (2012)
21. Poynton, K.: Digital Video and HD. Algorithms and Interfaces, 2nd edn. Morgan Kaufmann Publishers, San Francisco (2012)
22. Vlasuyk, I.: Development of a model of the human visual system for the method of objective image quality control in digital television systems. Telecommun. Transp. **51**, 189–192 (2009)
23. Hubel, D.H.: Eye, Brain, Vision. Mir, Moscow (1990)
24. Mozhaeva, A., Potashnikov, A., Vlasuyk, I., Streeter, L.: Constant subjective quality database: the research and device of generating video sequences of constant quality. In: International Conference on Engineering Management of Communication and Technology (EMCTECH), Vienna, Austria, pp. 1–5 (2021)
25. Liu, T.-J., Lin, Y.-C., Lin, W., Kuo, C.-C.J.: Visual quality assessment: recent developments, coding applications and future trends. APSIPA Trans. Sig. Inf. Process. **2**(1), 20 (2013)
26. Li, Z., Aaron, A., Katsavounidis, I., Moorthy, A., Manohara, M.: Image quality assessment: from error visibility to structural similarity. IEEE Trans. Image Process. **13**(4), 600–612 (2004)

27. Soundararajan, R., Bovik, A.C.: Video quality assessment by reduced reference spatio-temporal entropic differencing. IEEE Trans. Circ. Syst. Video Technol. **23**(4), 684–694 (2012)
28. Narwaria, M., Da Silva, M.P., Le Callet, P.: HDR-VQM: an objective quality measure for high dynamic range video. Sig. Process. Image Commun. **35**, 46–60 (2015)
29. Bampis, C.G., Li, Z., Moorthy, A.K., Katsavounidis, I., Aaron, A., Bovik, A.C.: Study of temporal effects on subjective video quality of experience. IEEE Trans. Image Process. **26**(11), 5217–5231 (2017)
30. Bampis, C.G., Li, Z., Moorthy, A.K., Katsavounidis, I., Aaron, A., Bovik, A.C.: LIVE Netflix Video Quality of Experience Database. http://live.ece.utexas.edu/research/LIVE_NFLXStudy/index.html

Small Visual Object Detection in Smart Waste Classification Using Transformers with Deep Learning

Jianchun Qi(✉), Minh Nguyen, and Wei Qi Yan

Auckland University of Technology, Auckland, New Zealand
yhy5508@autuni.ac.nz

Abstract. Smart object waste classification is relatively essential for protecting the environment and saving resources. This is considered a vital pathway towards sustainability. In waste classification, we see that it is challenging to detect waste of small visual objects with low resolutions that directly affect the overall performance of waste classification. While current visual object detection algorithms focus on the exploration of larger objects, the development of small object detection is being expanded relatively slowly due to the inability to acquire more visual information. In this paper, we propose a novel method combining contextual information and multiscale learning to improve small object detection performance in waste classification by enabling small object detection to obtain more feature information at high resolution. Furthermore, based on the advantages of parallel computing in Transformers, we utilize the DETR model to explore our method. The experimental results show that our method achieves high accuracy in the detection of a small object in waste.

Keywords: Small object detection · Transformers · Waste detection · Waste classification

1 Introduction

Waste classification generally refers to the conversion of waste into a public resource by classifying wastes into storage and transportation according to classification standards. The purpose is to increase the economic and resource value of waste, promote the recycling of resources, reduce the cost of waste disposal and the consumption of land resources so as to protect our environment. Besides, conventional waste disposal, such as landfilling and stacking, may produce harmful chemicals, which contaminate soil and groundwater resources and lead to reduced crop yields [4]. Therefore, it is necessary to develop efficient and accurate waste classification methods.

The development of computer vision has made pattern classification and visual object detection easy. Visual object detection occupies a vital position in the research field of computer vision [10, 20, 27, 33]. It can solve problems such as pedestrian tracking, visual object segmentation, and smart driving, etc. By applying deep learning [22, 23, 29, 32] to waste classification, exploring automated and efficient waste classification methods

W. Q. Yan et al. (Eds.): IVCNZ 2022, LNCS 13836, pp. 301–314, 2023.
https://doi.org/10.1007/978-3-031-25825-1_22

has ecological, social, and economic significance. Meanwhile, we find that the accuracy of detecting small waste objects could be improved, such as broken nut shells and button batteries. These objects are small in size compared to plastic bottles and carton boxes. If they are detected in an image, a fewer of pixels are occupied in the image than other objects. This makes the waste classification task as a challenging problem.

Similarly, small visual object detection is also abundant and broadly applied to ordinary life, such as traffic sign detection in automated driving, etc. Small object detection has always been a challenging task in visual object detection, because the visual features of small objects need to be accurately detected. For example, a small object may have less than 32×32 pixels while a standard image resolution is 1024×1024.

In recent years, the performance of small object detection has also gradually improved [11, 13], but the performance is still inferior to that of large objects. The feature maps of small objects do not have high resolution, resulting in less visual information to be detected by deep neural networks. Currently, too many downsampling operations and too big receptive field are all the factors that could affect small object detection. Furthermore, solving the problem of small object detection also requires both shallow representational information and deep semantic information.

To sum up, we make use of both context learning [19] and multiscale learning [34] to improve the small object detection performance in waste classification. Also, owing to the advantage that Transformer models can be computed in parallel, there are fewer studies on small object detection based on Transformer, we choose to study the detection of small objects in waste using Transformer. In this paper, we choose the DETR model. The main contributions of this paper are as follows:

(1) Transformers for detecting small objects of waste are trained, the results are high in accuracy.
(2) Multiscale learning and context learning are combined together for the improvement of small object detection for waste classification.
(3) A dataset including the small waste object is created.

In this paper, our related work is presented in Sect. 2, while Sect. 3 shows the methods, after the results is stated in Sect. 4. Finally, Sect. 5 contains our conclusions.

2 Related Work

2.1 Small Object Detection

Currently, the detection of small visual objects is significantly different from that of large objects, in many cases, it has only half size of large objects. However, small object detection has important research significance. For example, in autonomous vehicles, it is important to accurately detect small visual objects that can trigger traffic accidents to preserve the road safety. There are a slew of solutions for the shortcomings of small object detection as follows.

2.1.1 Data Augmentation

Data augmentation is a simple and effective method to improve the performance of small object detection. It enhances the generalization ability and robustness of the model by expanding the size of small object samples. In recent years, a plenty of data enhancement methods for regular object detection have been broadly employed, such as random cropping [12], translation [31], adjusting image saturation [17, 21], and mosaic enhancement.

Similarly, data augmentation methods for small object detection have also emerged. For example, the number of small objects is increased by repeatedly copying and pasting the small objects in the image to improve the model performance [11]. There is also an adaptive learning method proposed to enhance the performance of the small object detection. The data augmentation has solved the problems of small number of samples and lack of features in small object detection which improved the generalization ability of the model.

2.1.2 Contextual Information

Contextual information can improve the performance of small object detection because there is a group of informational correlations between the object and the background. For example, while a small object is flying in sky, we may not be able to see exactly what the object is, but with the background of the sky and the size of the object, we will associate a bird flying over our heads. Using this informational correlation will assist us to improve the detection of small visual objects.

Currently, a spate of studies explored and exploited this research issue. A method based on contextual feature enhancement is proposed [14], which firstly generates image proposal regions, and then produces multiscale windows around the targets for object feature enhancement. There are also recurrent neural networks proposed to encode and concatenate contextual information. However, though these methods improved the performance of small object detection, they are still affected by the size of the receptive field, resulting in partial loss of contextual information.

2.1.3 Multiscale Learning

Multiscale learning allows small visual object detection to take into account in both the need for representing shallow information and deep semantic information, avoiding the loss of location and feature information of small objects as the depth of the network increments. There are various ideas for using multiscale detection. For example, using dilated convolution to obtain various receptive field sizes, image pyramids [6], multiscale object detection [15], deconvolution layers [2], and feature pyramids [18]. These methods improve the resolution of small object feature maps, but some of them also have the problems with too much computational cost.

Overall, multiscale learning can effectively betterment the performance of small object detection, but the huge computational costs and unstable feature fusion process are also the reasons that hinder the further development of multiscale learning.

2.2 Visual Object Detection

2.2.1 Convolutional Neural Network

Convolutional neural networks (CNN) can be trained using the corresponding feature maps from a large number of visual object samples and reduce the complexity of the model by using downsampling, weight sharing, and local receptive fields [17]. At present, the existing CNN models applied to object detection can be classified into two categories: One-stage network and two-stage network.

One-Stage Network. It directly returns the class and position information of visual object through backbone network without using Region Proposal Network (RPN), which is fast in visual object detection but low in accuracy [26]. At present, the classical one-stage object detection networks are YOLO [24, 25], YOLOv4 [1], and Single Shot MultiBox Detector (SSD) [15].

Two-Stage Network. It mainly extracts features through convolutional neural networks, trains the RPN network, then conducts fine-tuning the network with the proposal regions, which has high accuracy but lower detection speed than one-stage models. Currently, classical algorithms include Region-CNN (R-CNN) [8], Faster R-CNN [18], and Mask R-CNN [9].

2.2.2 Transformer Models

In recent years, Transformer models [28] have become popular. Compared with CNNs, Transformers have better computational complexity and solve the problem of time consumption. With the advancement of Transformers in the field of Natural Language Processing (NLP), Transformers applied in the computer vision field are also emerging.

Vision Transformer [7], which was applied to pattern classification, cuts the 3D data of an image into patches, arranges them in sequence, converts them into serialized data, and uses the Transformer model for processing. Similarly, there is also a Transformer model applied to visual object detection, namely DETR [3]. It firstly extracts feature maps by using CNN to form a patch sequence, then Robject queries and a new loss function are formed. The model is very succinct and concise.

3 Our Method

In this paper, the proposed model structure is shown in Fig. 1. The detection of small visual objects is often difficult, visual features extracted from the proposed regions have weak discriminative ability. After considering multiple methods for small object detection [12, 14, 16], inspired by R-CNN [8] and context-based small object detection methods [5, 34], based on the DETR model [3], we introduce a new combination of neural networks, which can provide important feature maps for small object detection. We input the target image and its context into different neural networks through the target channel and context channel, respectively; we aggregate the contextual information for fusion, and input the obtained visual features into the Transformer [28] encoder and decoder. In this case, the context channel is modified according to the CAB model [5]. The model is mainly split into three stages, its details are represented in the following sections. It is worth noting that our proposed method can also be applied in other detectors such as Faster R-CNN.

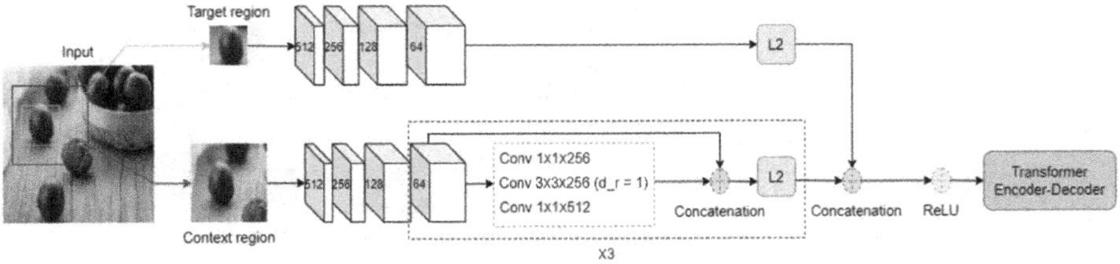

Fig. 1. The proposed model structure.

Backbone Network

In our experiments, we are use of ResNet-50 as the backbone network to extract feature information from the images. Firstly, we adjust the input image to 512 × 512. After that, we extract the visual features by using downsampling. Following this step, we assign the model settings of ResNet-50 for the experiments. The target channel and contextual channel parameters are consistent that have the same structure. In this stage, we only kept the four layers: Conv1, Conv2, Conv3, and Conv4, in the ResNet-50 net as shown in Fig. 1. Retaining the shallow feature map not only reduces the loss of small object features but also preserves the receptive field of small object detection and strengthens the accuracy of border regression [30].

Target Channel. We firstly crop the proposed region in the image as the input of the target channel. As known from the backbone network part, after input the target image region, it is convolved by Conv1, Conv2, Conv3, and Conv4. Hence, we add an L_2 normalization layer.

Contextual Channel. The structure of the context channel is different from the target channel as shown in Fig. 1. At the first step, we crop the contextual region with the proposed region in the image as the input of the context channel. Again, as shown in the backbone network section, we keep the first four convolutional layers. After that, in this channel, we take use of multiple stacked dilated convolution layers, which is consistent with CAB model [5], we expand the convolution kernel by adding 0 to the convolution kernel to achieve the goal of expanding the receptive field and obtain multiscale contextual information without losing resolution [34].

Specifically, a 1 × 1 convolution layer is added, after multiple stacked dilated convolution layers are employed to obtain more contextual information. This reduces the computational effort by not introducing additional parameters. Besides, regarding the dilated convolution, we choose three parallel stacked dilated convolutions, each stacked dilated convolution has an increasing dilation rate as shown in Fig. 2. Thus, more contextual information from different angles can be obtained. Then, we concatenate each

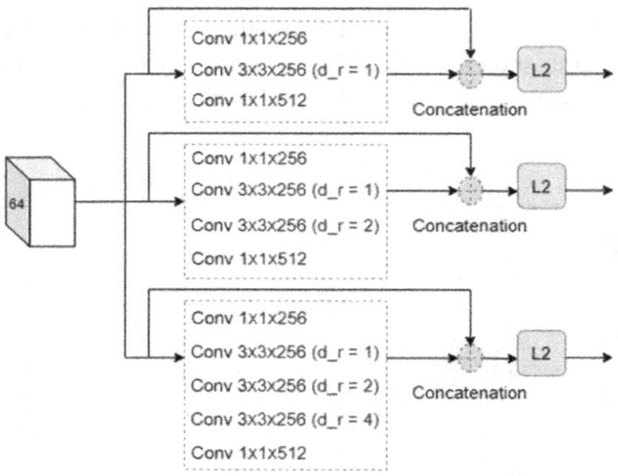

Fig. 2. The architecture of context channel.

input layer with its corresponding dilated convolution features for optimization. Finally, after each output is added to an L2 normalization, the three outputs are concatenated. In this model, the dilated convolution is calculated as Eq. (1).

$$f_k = \mathrm{dr} \times (\mathrm{k} - 1) + 1 \tag{1}$$

where f_k is the convolution kernel size after expansion, *dr* is the expansion coefficient, and *k* is the convolution kernel size. In this model, *n* is the number of convolutions, we set the selection rule of *dr* as 1, 2, and 4, that means, the computational complexity is $1 + n(n - 1)/2$. The size of receptive field is determined by the kernel size and stride size. Therefore, the computational method of the receptive field is shown in Eq. (2).

$$RF_{n+1} = RF_n + (f_k - 1) \times \prod_{i=1}^{n} s_i \tag{2}$$

where RF_n is the size of the receptive field corresponding to the *n*-th convolutional layer, s_i is the stride size of layer *i*. Because the stride length of the convolution kernel represents the extraction accuracy, we set the stride size as 1.0 to avoid losing the information of the original image [5].

If $dr = 1$, the convolution kernel size is 3×3. If $dr = 2$, the size of the convolution kernel after adding the hole is 5×5, the receptive field is 7×7. If $dr = 4$, the convolution kernel size is 9×9, the receptive field increases to 15×15. Although the output dimensions of all dilated convolutions are the same, we see that the receptive fields are distinct. It is also worth noting that the parameter quantities do not change after dilated convolution. Increasing the receptive field does not group the size of the convolution kernel, even in multiple stacked dilated convolutions [34].

Finally, the output of the target channel after L_2 normalization is concatenated with the three outputs of the contextual channel, then a layer of ReLU is added to feed the result into the encoder-decoder structure of the Transformer model, so that the visual feature can show a better balance between semantic and spatial aspects, and achieve the ideal combination of multiscale learning and contextual learning.

Transformer Detection

In DETR [3], Transformer and feedforward network (FFN) are combined to form the net architecture for visual object detection. Regarding the Transformer, its structure is almost identical to the original one of encoder-decoder architecture. The encoder consists of a multi-head self-attention and an FFN with the addition of positional encoding to obtain the attention results of each target. While the decoder retains the original multi-head self-attention, multi-head attention, and FFN, we decode the targets in parallel and queries these targets together. The results are fed into a fixed number of FFNs in the form of embedding [28]. Finally, the predicted classification and bounding box corresponding to each target are obtained by parallel calculation. In our model, finding the optimal bipartite matching [3] is also employed to determine the bounding box of each object, as shown in Eq. (3).

$$\hat{\sigma} = arg \min_{\sigma \in \mathfrak{S}N} \sum i^N \mathcal{L}\text{match}(yi, \hat{y}\sigma(i)) \tag{3}$$

where N is the number of predictions, y is the ground truth set, and $\hat{y}$ is the set of predictions, $\hat{y}$ contains the predicted category and bounding box. The loss between each y and $\hat{y}$ is L_i. Therefore, Eq. (3) finds a permutation that can map the predicted indices to the indices of the ground truth, avoiding getting the same loss in different ranking predictions. Besides, regarding the calculation of L_i, we adopt the same method as DETR, which is a linear combination of L_1 loss and GIOU loss [29] for ground truth and predicted values.

4 Result Analysis

The model takes use of AdamW optimizer with an initial learning rate of 10^{-4}. The backbone network has a learning rate of 10^{-5} and a weight decay of 10^{-4}. Additionally, a dropout of 0.1 was adopted, 300 epochs were selected for model training.

4.1 Our Dataset

In this paper, we collected the waste dataset with 1,053 images of small objects, including batteries, fruit cores, nut shells, egg shells, and bottle caps. Furthermore, we selected small waste objects with an area of less than 32×32 pixels in the image.

We merge these five types of wastes into four classes according to the waste classification criteria, i.e., batteries belong to the class "Hazardous", fruit cores and egg shells are "Wet" class, the nut shells are classified into "Dry" class, and the bottle caps are the "Recyclable" class. Figure 3 shows the small waste images in the dataset and Table 1 illustrates the number of samples of each class.

Table 1. The number of samples of each class

Classes	Numbers of samples
Battery	202
Egg shell	221
Bottle cap	220
Nut shell	207
Fruit core	203
Total	1,053

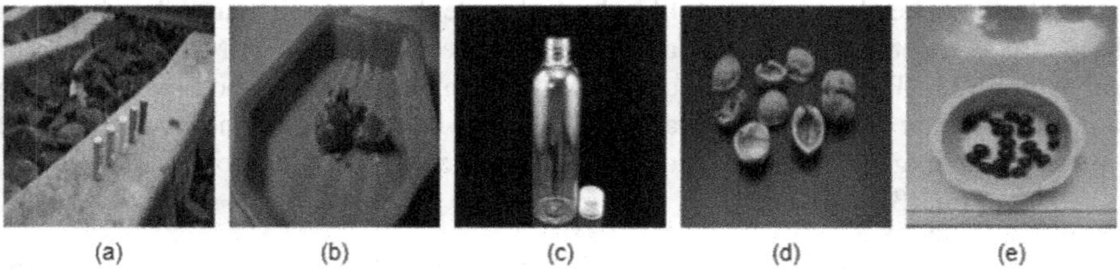

Fig. 3. The samples in the waste dataset. The images, (a), (b), (c), (d), and (e) show battery, egg shell, bottle cap, nutshell, and fruit core, respectively.

4.2 Evaluation Methods

In order to verify the performance of the model, we take use of a series of evaluation metrics: Average Precision (AP) and Mean Average Precision (mAP). The range of thresholds is [0.5:0.05:0.95]. In addition, we also adopted the Precision-Recall curve (PR curve) for the performance evaluations.

4.3 Result Analysis

Figure 4 shows us an example of small waste object detection. We see that there are various classes of waste samples in the image, each class has a color bounding box and label.

In Fig. 5, we see the PR curves of three small object waste classifications by using multiple models. AP values are calculated by calculating the area under the curve. In Fig. 5(a), AP values of the battery of our proposed model, DETR, and Faster R-CNN are 12%, 9%, and 8%, respectively. For bottle cap, its AP value is the highest, reaching about 28% in our proposed model, 23% and 21% in DETR and Faster R-CNN, respectively. Finally, in Fig. 5(c), the AP values of fruit core of our model, DETR and Faster R-CNN are 11%, 10%, and 7%, respectively.

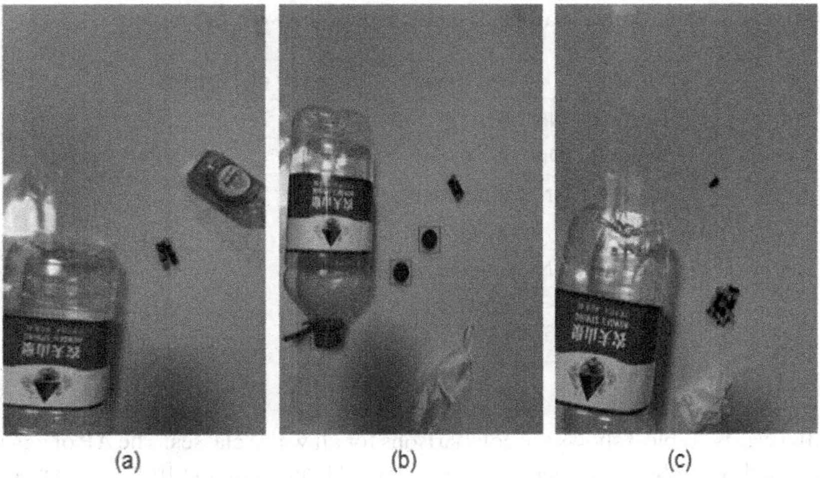

Fig. 4. Visual object detection results (a) the results of classifying batteries, which belong to "Hazardous" class, (b) the classification results of bottle cap and battery, which are from "Recyclable" class and "Hazardous" class (c) the classification results of fruit core, which belong to "Wet" class.

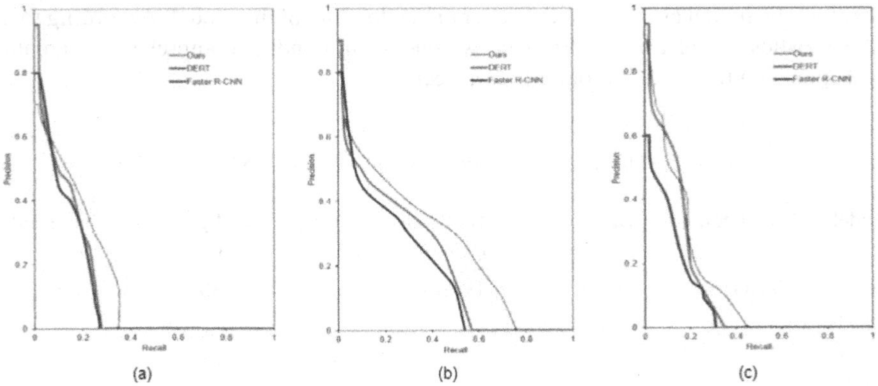

Fig. 5. PR curves of three small object classifications, comparing our model to other two advanced models (a) the PR curve of battery (b) the PR curve of bottle cap (c) the PR curve of fruit core.

We quantitatively compare our model with other models by using our own dataset. Table 2 shows the comparison results. The mAP of DETR is 28.8%, which is slightly lower than that of our model by 0.9%. Then, the mAP of Faster R-CNN (ResNet-50) and Mask R-CNN is 20.5% and 26.2%, respectively. Finally, SSD has 23.7% mAP values, which is 5.1% higher than Faster R-CNN (VGG16). It is evident that our model is more vibrant for small waste objects.

Table 2. Mean average precision results between five models

Models	Backbone	mAP (%) (small)
Faster R-CNN	VGG16	18.6
Faster R-CNN	ResNet-50	20.5
Mask R-CNN	ResNet-101	26.2
SSD	VGG16	23.7
Mask R-CNN	Swin transformer	27.8
DETR	ResNet-50	28.3
Ours	ResNet-50	29.2

Afterwards, Table 3 shows the comparisons for all waste classes. The AP of egg shell and nut shell is higher than that of other classes. Meanwhile, the fruit core has almost the lowest AP value among the five classes.

4.4 Ablation Experiments

Pertaining to the overall performance of these models, the components were employed to explore the model and facilitate a better understanding of the model. According to the characteristics of our proposed model, we choose to conduct comprehensive ablation experiments on the model through four aspects.

Table 3. Average precision results between five models for each class

Models	Backbones	Battery (%)	Bottle cap (%)	Fruit core (%)	Egg shell (%)	Nut shell (%)
Faster R-CNN	VGG16	6.9	19.6	6.3	30.6	29.7
Faster R-CNN	ResNet-50	8.1	21.4	7.2	31.6	34.7
Mask R-CNN	ResNet-101	11.2	24.9	9.6	43.3	42.2
SSD	VGG16	9.8	22.5	9.7	37.6	39.1
Mask R-CNN	Swin transformer	11.5	25.4	9.8	47.6	44.9
DETR	ResNet-50	9.6	23.7	10.3	49.4	48.8
Ours	ResNet-50	12.1	28.0	11.6	45.1	49.6

4.4.1 Number of Channels

Our model has a target channel and a contextual channel. Firstly, we cut off the target channel and make use of only one input image for visual object detection, as shown in Table 4. If only the target channel is kept, it is impossible to perform detection better. After keeping the context channel, though the method of context information cannot be used, the mAP of the multiscale learning of the receptive field reaches 23.4%, which is 5.8% lower than the way of retaining both. The speed also decreased from 3.1 to 1.6 FPS, indicating that the effect of the target channel on the model is also present. Furthermore, this brings us to a future direction on how to make the model improve FPS with guaranteed accuracy.

Table 4. Influence of target channel on mAP and FPS

Target channel	Context channel	mAP % (small)	Speed (FPS)
√		–	–
	√	23.4	1.6
√	√	29.2	3.1

4.4.2 Number of Convolutional Layers

In the paper, Conv1, Conv2, Conv3, and Conv4 layers are preserved. Therefore, in our ablation experiments, we keep Conv2, Conv3, and Conv4 layers, respectively, the results are shown in Table 5. By retaining the feature map of Conv3 for detection, the FPS is only 1.0 FPS, while using Conv4 normally, the model will get a speed of 3.1 FPS. Regarding mAP, the results are also 7.3% lower (from 29.2% to 21.9%), with only the convolutional layers retained to Conv3 than with Conv4.

Table 5. Influence of convolutional layers on mAP and FPS

Layer	mAP % (small)	Speed (FPS)
Conv2	–	–
Conv3	21.9	1.0
Conv4	29.2	3.1
Conv5	28.6	2.7
Conv6	26.3	2.4

4.4.3 Application of Dilated Convolutions

In our model, we employ the dilated convolution in the context channel. Therefore, we also applied the dilated convolution in the target channel as well by conducting ablation experiments. The experimental results are shown in Table 6. We see that the mAP value of using the dilated convolution in both channels is only 0.2% lower than that of the original model. This shows the importance of the dilated convolution to the model. But on the contrary, the speed is only 0.8 FPS.

Table 6. Influence of the application of dilated convolutions on mAP and FPS

Dilated convolution	mAP % (small)	Speed (FPS)
In context channel	29.2	3.1
In both channels	29.0	0.8

4.4.4 Number of Dilated Convolutions

In the model, we applied three dilated convolutions. Therefore, we increase the number of dilated convolutions to evaluate the model performance. The number of dilated convolutions is denoted as d_c, the experimental results are shown in Table 7.

Table 7. Influence of the number of dilated convolutions on mAP values

Num (d_c)	mAP % (small)	Speed (FPS)
2	22.7	3.8
3	29.2	3.1
4	30.0	1.9
5	30.2	1.8

We see that the mAP value increases with the increase of Num(d), from 22.7% to 30.0%. However, if four dilated convolutions are employed, the FPS value is only 1.9. Since the mAP at Num(d) of 4 is only 0.8% more than that at Num(d) of 3, on balance, we choose to set Num(d) as 4.

5 Conclusion

In this paper, we improve the performance of Transformer models for small visual object detection in waste classification with the DETR model. One is to expand the receptive field to obtain more feature information for small waste object detection whilst ensuring high resolution. Secondly, the contextual information of small objects is enhanced by extracting target regions and contextual regions. The experimental results show that the proposed model achieves small object classification for the wastes. In future, we will improve the model in three directions: Improving FPS, simplifying the model and reducing the computation, and expanding the small waste object dataset.

References

1. Bochkovskiy, A., Wang, C.Y., Liao, H.Y.M.: YOLOv4: optimal speed and accuracy of object detection. arXiv (2020)
2. Cai, Z., Fan, Q., Feris, R.S., Vasconcelos, N.: A unified multi-scale deep convolutional neural network for fast object detection. In: Leibe, B., Matas, J., Sebe, N., Welling, M. (eds.) ECCV 2016. LNCS, vol. 9908, pp. 354–370. Springer, Cham (2016). https://doi.org/10.1007/978-3-319-46493-0_22
3. Carion, N., Massa, F., Synnaeve, G., Usunier, N., Kirillov, A., Zagoruyko, S.: End-to-end object detection with transformers. In: Vedaldi, A., Bischof, H., Brox, T., Frahm, J.-M. (eds.) ECCV 2020. LNCS, vol. 12346, pp. 213–229. Springer, Cham (2020). https://doi.org/10.1007/978-3-030-58452-8_13
4. Chen, S.S., et al.: Carbon emissions under different domestic waste treatment modes induced by garbage classification: case study in pilot communities in Shanghai, China. Sci. Total Environ. **717**, 137193 (2020)
5. Cui, L., et al.: Context-aware block net for small object detection. IEEE Trans. Cybern. **52**(4), 2300–2313 (2022)
6. Dalal, N., Triggs, B.: Histograms of oriented gradients for human detection. In: IEEE CVPR, pp. 886–893 (2005)
7. Dosovitskiy, A.,et al.: An image is worth 16 × 16 words: transformers for image recognition at scale. arXiv (2020)
8. Girshick, R., Donahue, J., Darrell, T., Malik, J.: Rich feature hierarchies for accurate object detection and semantic segmentation. In: IEEE CVPR, pp. 580–587 (2014)
9. He, K.M., Gkioxari, G., Dollár, P., Girshick, R.: Mask R-CNN. In: IEEE ICCV, pp. 2961–2969 (2017)
10. He, K.M., Zhang, X.Y., Ren, S.Q., Sun, J.: Deep residual learning for image recognition. In: IEEE CVPR, pp. 770–778 (2016)
11. Kisantal, M., Wojna, Z., Murawski, J., Naruniec, J., Cho, K.: Augmentation for small object detection. arXiv (2019)
12. Krizhevsky, A., Sutskever, I., Hinton, G.E.: ImageNet classification with deep convolutional neural networks. In: NIPS, pp. 1–9 (2012)
13. Li, J., et al.: Attentive contexts for object detection. IEEE Trans. Multimed. **19**(5), 944–954 (2016). https://doi.org/10.1109/TMM.2016.2642789
14. Liu, W., et al.: SSD: single shot multibox detector. In: Leibe, B., Matas, J., Sebe, N., Welling, M. (eds.) ECCV 2016. LNCS, vol. 9905, pp. 21–37. Springer, Cham (2016). https://doi.org/10.1007/978-3-319-46448-0_2
15. Liu, Z., Mao, H.Z., Wu, C.Y., Feichtenhofer, C., Darrell, T., Xie. S.N.: A ConvNet for the 2020s. arXiv (2022)
16. Li, Z., Zhou, F.: FSSD: feature fusion single shot multibox detector. arXiv:1712.00960 (2017)
17. Luo, Z., Nguyen, M., Yan, W.: Sailboat detection based on automated search attention mechanism and deep learning models. In: IEEE IVCNZ (2021)
18. Nie, Z.F., Duan, W.J., Li, X.D.: Domestic garbage recognition and detection based on Faster R-CNN. In: Journal of Physics: Conference Series (2021)
19. Oliva, A., Torralba, A.: The role of context in object recognition. Trends Cogn. Sci. **11**(12), 520–527 (2017)
20. Pan, C., Yan, W.: A learning-based positive feedback in salient object detection. In: IEEE IVCNZ (2018)
21. Pan, C., Yan, W.Q.: Object detection based on saturation of visual perception. Multimed. Tools Appl. **79**(27–28), 19925–19944 (2020). https://doi.org/10.1007/s11042-020-08866-x

22. Pan, C., Liu, J., Yan, W., Zhou, Y.: Salient object detection based on visual perceptual saturation and two-stream hybrid networks. IEEE Trans. Image Process. **30**, 4773–4787 (2021)
23. Qi, J., Nguyen, M., Yan, W.: Waste classification from digital images using ConvNeXt. In: PSIVT (2022)
24. Redmon, J., Divvala, S., Girshick, R., Farhadi, A.: You only look once: unified, real-time object detection. In: IEEE CVPR, pp. 779–788 (2016)
25. Redmon, J., Farhadi, A.: YOLO9000: better, faster, stronger. In: IEEE CVPR, pp. 7263–7271 (2017)
26. Rezatofighi, H., Tsoi, N., Gwak, J., Sadeghian, A., Reid, I., Savarese, S.: Generalized intersection over union: a metric and a loss for bounding box regression. In: IEEE CVPR, pp. 658–666 (2019)
27. Shen, D., Xin, C., Nguyen, M., Yan, W.: Flame detection using deep learning. In: ICCAR (2018)
28. Vaswani, A.,et al.: Attention is all you need. In: NIPS (2019)
29. Wan, L., Zeiler, M., Zhang, S., Le Cun, Y., Fergus, R.: Regularization of neural networks using DropConnect. In: ICML, pp. 1058–1066 (2013)
30. Yin, X., Goudriaan, J.A.N., Lantinga, E.A., Vos, J.A.N., Spiertz, H.J.: A flexible sigmoid function of determinate growth. Ann. Bot. **91**, 361–371 (2002)
31. Xiao, B., Nguyen, M., Yan, W.Q.: Apple ripeness identification using deep learning. In: Nguyen, M., Yan, W.Q., Ho, H. (eds.) ISGV 2021. CCIS, vol. 1386, pp. 53–67. Springer, Cham (2021). https://doi.org/10.1007/978-3-030-72073-5_5
32. Yan, W.Q.: Computational Methods for Deep Learning – Theoretic, Practice and Applications. Springer, Heidelberg (2021). https://doi.org/10.1007/978-3-030-61081-4
33. Yan, W.Q.: Introduction to Intelligent Surveillance - Surveillance Data Capture, Transmission, and Analytics, 3rd edn. Springer, Heidelberg (2019). https://doi.org/10.1007/978-3-030-107 13-0
34. Yu, F., Koltun, V.: Multiscale context aggregation by dilated convolutions. In: ICLR (2016)

A Hybrid Human-Machine System for Image-Based Multi-weather Detection

Harsh Bhandari[1], Soumajit Chowdhury[2], and Sarbani Palit[1(✉)]

[1] Indian Statistical Institute, Kolkata, India
sarbanip@isical.ac.in
[2] University of Sheffield, Sheffield, UK

Abstract. Accurate determination of weather from images or video is of prime significance in applications such as autonomous vehicles drive or atmospheric pollution estimation. In contrast to image classification approaches generally adopted in the literature based on neural networks acting directly on the input, we propose a novel joint learning approach combining features extracted from the input image, sensitive to the human visual system (HVS) and those generated by the CNN model trained on benchmark datasets. The features work in a joint collaboration that is able to detect the presence of weather features during learning. The novel approach outperforms many state-of-the-art methods which use only CNN-based extracted features in order to classify the images. Experimental results with publicly available benchmark datasets establish the robustness and effectiveness of the proposed method in multi-weather classification.

Keywords: SKYNET · Visibility · Weather · Multi-class latent SVM

1 Introduction

Accurately predicting the weather conditions from an image is of immense importance not only for meteorological forecasting but also for many other applications including automated driver-less cars, air quality monitoring, etc. Correctly determining the environment is crucial for the safe and proper functioning of an autonomous vehicle, a task which is drastically affected by bad weather conditions such as haze and rain. Even snow and strong sunlight may cause unwanted reflections. Since an autonomous vehicle might have to travel through different kinds of weather, accurate and fast determination of the prevalent weather based on the output of the camera fixed on the vehicle is essential.

Another important application of weather detection is in the estimation of the concentration of particulate matters of varying sizes, from an image of the locality. We have observed that the concentration of particulate matter in air, especially of size 2.5 µm, is greatly affected by the type of weather. Figure 1 shows the daily average of $PM_{2.5}$ for the year 2019 at two different monitoring stations of Kolkata, India. The high values of $PM_{2.5}$ can be seen during the months from

W. Q. Yan et al. (Eds.): IVCNZ 2022, LNCS 13836, pp. 315–329, 2023.
https://doi.org/10.1007/978-3-031-25825-1_23

December to February when it is winter at Kolkata, while low values can be seen during summer from March to June and lower still during the monsoon season from July to August. It is reasonable to conclude that information regarding weather conditions is vital for good prediction of the concentration of $PM_{2.5}$.

In this paper, we propose a novel approach that also provides technical insight into the problem of multi-weather classification. Given an outdoor image, the system will be able to judge the type of weather - "Sunny", "Cloudy", "Rainy", "Hazy" and "Snow" as shown in Fig. 3.

Not much research has been conducted in the field of Weather Detection using single images. Most of the research has mainly focused on images with a few selected static scenes for weather classification leading to limited performance since such classifiers would fail for images with varying background scenes. While [16,19] proposed weather recognition from vehicles, they rely on information from vehicle-mounted sensors and operate only on a chosen region of interest. Understanding a scene for categorizing it into a particular class of weather requires structural information which may be based on illumination-invariant features like SIFT or HOG [6,12].

The approach proposed in this paper makes the following major contributions:

1. We propose image extracted features, emulating the human visual system that are used to detect weather. The detection of illumination has been employed in order to detect sunlight in an image. Further, a feature based on the visibility of a scene has been proposed.
2. We propose a diverse foreground-background sky mask dataset for developing the SKYNET model to segment sky and non-sky areas in an image.
3. A combination of handcrafted and Resnet-101 extracted 512-D feature vectors are used to make the final classification.
4. A Multi-class Latent Joint Support Vector Machine (MLJSVM) classifier to effectively classify the images into five weather conditions - "Sunny", "Cloudy", "Haze", "Snow" and "Rainy".

The flow diagram of the proposed approach is provided in Fig. 2. The rest of the paper has been structured into five sections. Section 2 deals with the literature survey of the existing works done in the allied fields. Section 3 elaborates upon the various human visual features emulated in the proposed work. Section 4 illustrates the various datasets used in the work along with the proposed sky mask dataset. It also provides a detailed comparison of results obtained by the proposed approach along with other state-of-the-art techniques. Section 5 finally concludes the article provides the concluding remarks and the guide to future scope relevant to the work.

2 Related Work

This section provides an overview of the related work by classifying them into two broad categories with respect to their area of application.

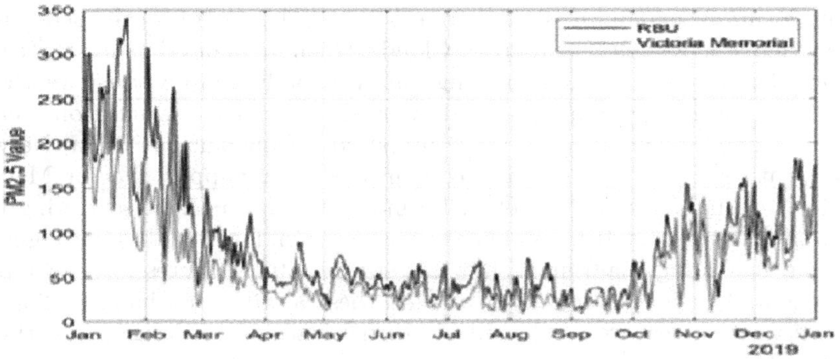

Fig. 1. Concentration of $PM_{2.5}$ correlated with the change in weather conditions.

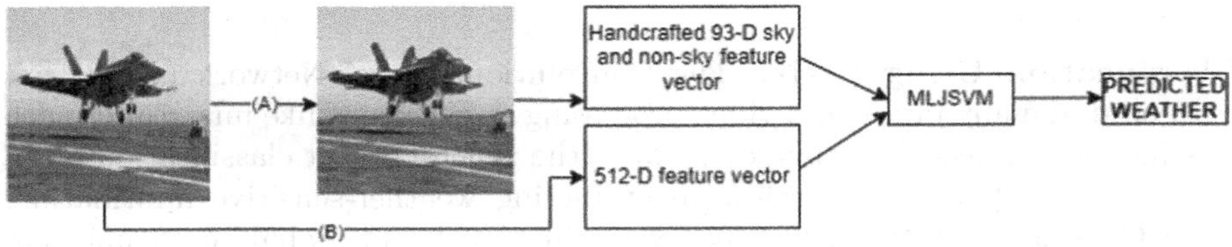

Fig. 2. The proposed weather detection method. (A) represents the proposed SKYNET to perform sky detection and (B) represents the state-of-the-art RESNET-101 feature extractor. Handcrafted features are extracted from the image are combined with the machine-based features. The combined features are fed to the MLJSVM to finally yield the output.

2.1 Weather Classification

Classification Using Image-Based Features. Understanding weather plays a crucial role in multiple real-world applications such as self-driving cars by controlling the speed of the vehicle during adverse weather situations [16,19]. Kurihata in [9] used an in-vehicle camera to detect rainy weather by detecting raindrops on the windshield. A discriminative raindrop framework was learned to detect the weather condition. Martin in [16] performed classification by distinguishing weather situations using a monocular color image. But the method suffered from a major drawback as it is based on images for driver assistance. Yan in [19] developed a weather recognition system to be used in a vehicle on the basis of features extracted from images and employing Real Adaboost classifier. The aforementioned techniques are only able to detect rainy weather with further limitation of being restricted to similar target scenes. Li in [17] emphasized two-class weather classification - Sunny and Cloudy, of popular tourist spots from images of the same scene. Multiple images are required to estimate the illumination of a given location, making it limited to a few sites. Lu in [14] developed a technique to label images either Sunny or Cloudy. Chen in [4] developed a three class weather classification method covering Sunny, Cloudy and Overcast weather of the same scene image. Song in [18] developed a weather classification method used for a fixed scene only. An in [1] proposed a Convolution Neural Network based weather image classification combined with a Support Vector Machine (SVM). Narasimhan in [15] developed a model to capture light scattering from light source to the lens. It works in cases such as rain, fog, and, haze at night due to strong scattering. Laffront in [10] has studied 40 attributes for a given single image and designed a model to predict these attributes. Zhang in [20] used multiple weather features and multiple kernel learning to design a multi-class weather classification model. Though the methods outlined above have shown good performance in their designed applications, conditions or assumptions were often required which may be perceived as significant drawbacks.

Classification Using CNN. The Convolution Neural Network (CNN) has been used in many image and video processing applications like image and video classification, object detection, etc. Unlike the general object classification problem, weather classification relies on extracting weather-sensitive information. Deep CNN are robust models for extracting local and global scene information to be incorporated in order to make the recognition successful rather than fine-grained information required for weather classification. Elhoseiny in [7] has directly used VGGNet in classifying weather. Lu in [14] has used a combination of both image and CNN based features to classify weather images. An inherent limitation in these approaches is that the network based models may avoid some weather information which is inherent in the image. Further, the models have been used to predict only two types of weather viz "Sunny" and "Cloudy", making their functionality limited. Image based weather observation has tremendous potential for developing into a powerful computer vision application owing to the availability of low-cost smartphones and surveillance cameras. This motivates us to look for smart and efficient solutions to the problem.

Fig. 3. The five weather conditions - rainy, cloudy, sunny, haze and snow being read from left to right.

3 Overall Weather Features

We segment each input image into two mutually exclusive sky and non-sky segments. The non-sky segment is further over-segmented using Mean Shift clustering [2]. The complete sky segment generated after this segmentation is used to extract features from the sky area. The overall weather feature vector consists of two parts, a sky feature f_{sky} and five non-sky weather features. Of these five non-sky features, visibility, denoted by f_{vis} is computed patch-wise while the rest are computed directly from the segments of the image. Concatenation of six components yields the feature vector which is described as:

$$[f_{il}, f_{vis}, f_{Col}, f_{ce}, f_{sat}, f_{sky}] \tag{1}$$

where the features $f_{il}, f_{vis}, f_{Col}, f_{ce}, f_{sat}$, and f_{sky} namely represent illumination, visibility, colorfulness, contrast energy, saturation and sky respectively.

3.1 Illumination

Images taken under sunny weather exhibit a clear presence of sunlight in the image. All the image-based weather detection algorithms have tried to detect the presence of shadow in the image. Many a time, shadow detection fails for images with differing weather conditions where dark regions are often mis-classified as the shadow. It is also common to come across sunny images that either do not contain any shadow of the objects or have very thin and blurry shadows. The proposed method looks at the impact of the presence of sunlight on the scene color. The presence of sunlight increases the vibrancy of the object's color, making the object brighter and colorful.

With the objective of accurately detecting the presence of sunlight, a pool P_{NS} of the twenty largest non-sky segments (having sunlight) each from M images belonging to the sunny image dataset is selected. The rationale behind the selection of the number of segments is explained in Sect. 4.3. Thus, a pool size of 20*M segments is created. A strongly white color segment may get misclassified as a sunny segment due to very high "perceptual lightness" which is given by the L component of the CIE L*a*b image format of the selected segments. To overcome the problem of mis-classification, we look for non-white and non-black segments to establish the presence of perceptually light segments.

Given an unknown segment, we measure it's feasibility of being a segment of a sunny image by computing the average euclidean distance to its K-nearest neighbour ($K = 5$) in P_{NS}. This generates the 20-D illumination feature vector f_{il} which mainly helps in differentiating Sunny weather from the other weather conditions as in Fig. 4, we display the results of detecting segments with presence of light in a given color image.

Fig. 4. Sunlight detection results from non-sky region for images showing the impact of illumination

3.2 Colorfulness

The colorfulness of a given image denotes the visual perception of an image, as perceived by the human eye. The colorfulness of a segment depends not only on its spectral reflectance but also on the power of the illumination and increases with illumination unless the brightness is too high. Given a color image, the proposed Sky-Detection algorithm, segments the image into the sky and non-sky regions marked by S and NS respectively. Colorfulness of an image, is computed as follows:

$$ColorFulness = \lambda_1 \cdot \sqrt{\sigma_a^2 + \sigma_b^2} + \lambda_2 \cdot \sqrt{\mu_a^2 + \mu_b^2} \tag{2}$$

σ_a and σ_b refers to sample standard deviations and μ_a and μ_b to sample means of the channels a and b of CIE L*a*b colorspace respectively. λ_1 and λ_2 set to 1 are weights assigned to the standard deviation and mean components respectively. For each of the largest 20 segments of the non-sky area of image I, we define

$$\begin{aligned} f_{col}(i) &= w_i * ColorFulness_i(I) \\ w_i &= N_i/N_I \end{aligned} \tag{3}$$

where w_i is the weight assigned to $ColorFulness_i$ of the segment i of the image I, N_i is the number of pixels of segment i and N_I is the total number of pixels of the segment. This generates a 20-D f_{col} Colorfulness feature vector.

3.3 Presence of Haze

Much study has been carried out on haze in Computer Vision, with dark channel prior (DCP) being considered as a useful haze feature. However, it has been found in real-world images that the DCP feature often fails to effectively capture the impact of haziness. The reason for this was on account of the assumption made in DCP, that at least one of the color channels in [R, G, B] are zero or close to zero in the original, non-hazy image, but haze being white leads to higher value in all the color channels. This assumption led to the choice of DCP as one of the distinguishing feature for identifying a hazy image from a set of hazy and non-hazy images. However, we have observed that in actual practice, the values of the color channels are not always very high, since real-world images often show bluish tinge of haze, as in Fig. 5, where it may be observed that the haze being non-white gives rise to DCP values much lower than the images with white haze. Reddish and yellowish tinges of haze have also been observed elsewhere.

Visibility of the scene is the most important factor in the hazy image. Haze often occurs when dust and smoke particles accumulate in relatively dry air found mainly during winter and with lower temperatures. It settles down at lower heights, impairing the visibility of the region. We try to estimate the visibility of the image by estimating the blur in the image as in (4) which is further used to postulate a measure of visibility as in (5).

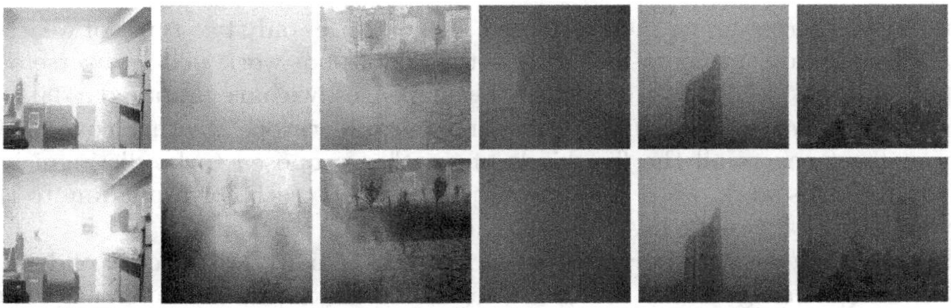

Fig. 5. DCP maps shown in the second row pertaining to first three columns for images with *white haze* and the last three columns for images with *non-white haze*.

$$\begin{aligned} I_{blur} &= I(x,y) \otimes G_p(x,y,\sigma_1) \\ V_1 &= \sqrt{(I(x,y) - I_{blur}(x,y))^2} \\ V_2 &= rgb2gray(V_1) \end{aligned} \tag{4}$$

$$Visibility(x,y) = \sum_{i=x-\frac{N}{2}}^{x+\frac{N}{2}} \sum_{j=y-\frac{N}{2}}^{y+\frac{N}{2}} \frac{V_2(i,j) - \mu}{\mu^{\alpha+1}} \tag{5}$$

$G_p(x,y,\sigma)$ is a 2D-Gaussian circular symmetric function of size $p \times p$ and σ referring to its standard deviation. Operator $\otimes$ is a 2D convolution operation. $V_2(i,j)$ is the gray intensity value at pixel position (i, j). α is a visual coefficient empirically set to 0.65 and μ is the mean of the gray values in the window. Most of the haze regions are smooth in nature with the smoothness increasing with the haze level. Each image is uniformly partitioned into overlapping regions with patch sizes N set to 5, 7, 9, 11, 13, 15 and 17. Median of median values of the visibility map in these patch sizes form a 7-D f_{vis} feature vector. The Visibility map of the hazy and the haze-free images have been shown in Fig. 6. The measure is observed to work well even in the case of white haze.

Fig. 6. Scenes with clear and natural haze. The top row contains haze and haze-free images with the bottom row being their corresponding Visibility Map.

3.4 Sky Segmentation

The sky segmentation method proposed in [3,14], uses only two types of weather while performing segmentation. The methods does not work well in diverse and complex scenes. In order to overcome the challenges, we have proposed a dataset consisting of images containing outdoor scenes in various weather conditions. The dataset covers all the five kinds of weather proposed to be distinguished between in this work. Most importantly, the ground truth corresponding to the sky region i.e. sky-mask has also been included.

We have used a U-Net architecture network to achieve the objective of detecting and segmenting the sky from the image. The major point of difference of our proposed network from the conventional U-Net architecture is the addition of a 3×3 tail convolution to the last convolution layer to refine the output. The tail convolution effectively helps in creating a segmented foreground-background output with the sky being foreground (white pixels) and the non-sky region (black pixels) being the background area. The model has been extensively trained on various benchmark datasets and has yielded results with very high accuracy for all weather conditions except for haze as shown in Fig. 7.

Loss Function. Given the color image and the corresponding ground truth mask, we intend to determine the parameters of the network by minimizing the spatial loss function as stated in Eq. 6. It uses the standard L_2 loss function which guides the network for the proper classification of pixels in the spatial domain and is defined as:

$$L_{sd} = \frac{1}{N} \sum_{i=1}^{N} \| x_i^{gt} - SN(x_i^{img}) \|^2 \tag{6}$$

where $SN(I)$ is the output obtained from the SKYNET model for an image I, I^{gt} is the corresponding reference image and N is the total number of images.

3.5 Sky Features

The sky color has an important contribution in determining the weather conditions of an area. Color of the sky during a sunny day is dominated by the blue component while during a cloudy or rainy day it leans towards a gray-black color. Further, scattered or absence of clouds in sunny images causes a large smooth region which leads to a lower gradient. The sky in cloudy and rainy images, shows the excessive presence of the clouds making the sky look grayish white as clouds are mainly composed of water droplets. Similarly, the sky during the winter season predominantly remains clear, transparent with color tilt towards white whereas haze adds granularity with brownish effect to the sky. This causes higher average gray intensity but low gradient in amplitude.

We calculate the mean value of each of the R, G and B color channels for the sky segment(s) denoted by f_{skyR}, f_{skyG} and f_{skyB} respectively in order to capture the color information. In order to take into account the texture of the sky, we consider the smoothness of the sky through computation of the average gradient amplitude in the sky segment. The image gradient, for a channel c, is defined as,

$$f_{grad}(I_S) = \frac{\partial I_S}{\partial x}\hat{x} + \frac{\partial I_S}{\partial y}\hat{y} \tag{7}$$

where I_S stands for the sky region of the image, $\partial I_S/\partial$x is the gradient in the X direction and $\partial I_S/\partial$y is the gradient in the Y direction. The average of the gradient amplitude for a channel c is defined as,

$$f_{\nabla I_S,c} =. \frac{1}{N_{sky}} \sum_{x,y \in Sky} \sqrt{(\frac{\partial I_S}{\partial x})^2 + (\frac{\partial I_S}{\partial y})^2} \tag{8}$$

Thus the sky feature f_{sky} can be expressed as:

$$f_{sky} = [f_{sky,c}, f_{\nabla I_S,c}] \quad c \in \{R, G, B\} \tag{9}$$

generating a 6-D sky feature vector.

3.6 Contrast Energy

The Contrast Energy (CE) computes the contrast for a given image at local level. The segments generated from non-sky region of an image I are decomposed using Gaussian filters of second order. The generated responses are rectified and normalized to control the non-linearity in contrast-gain, followed by a threshold to reduce the noise. For each component L, $a*$ and $b*$ of CIEL*a*b colorspace of the image I, CE is computed as follows:

$$Z(I_c) = \sqrt{(I_c \otimes h_h)^2 + (I_c \otimes h_v)^2} \tag{10}$$

$$CE(I_c) = \frac{\alpha.Z(I_c)}{Z(I_c) + \alpha.\kappa} - \tau_c \tag{11}$$

where $\kappa = 0.1$ is the contrast-gain and noise threshold denoted by τ_c for each color channel c $\in$ [L, a*, b*], α being the highest value of Z(I_c). $\otimes$ denotes convolution, h_h and h_v being horizontal and vertical Gaussian functions of second-order respectively using a 20×20 filter. The thresholds set for the L, *a, and *b channels are 0.2353, 0.2287 and 0.0528 respectively [8].

For each image, we select the normalized value as stated below, of the largest twenty segments from the non-sky area.

$$f_{ce}(i) = w_i * CE_i(I) \qquad (12)$$
$$w_i = N_i/N_I$$

where w_i is the weight assigned to contrast energy CE_i of the segment i of the image I. N_i is the number of pixels of segment i and N_I is the total number of pixels of the largest 20 segments of the non-sky area of image I. This generates a 20-D Contrast Energy feature vector f_{ce}.

3.7 Snow Feature

Detecting the presence of snow in a single image is a challenging task. Snow is light and soft and in the majority of the cases, it is uniformly white in appearance. For detection of the presence of snow, we take into consideration the lower 30% of the image frame. Snow, mainly uniform, continuous, and white in color leads to low saturation and very high gray intensity. Since the lower 30% may contain areas without any presence of snow, this region is divided into multiple segments using the Mean Shift [5]. For each of the generated segments S, we calculate the mean normalized saturation as,

$$f_{sat,S} = Mean\{\frac{S_i - \min(S)}{\max(S) - \min(S)}\} \quad pixel\ i \in S \qquad (13)$$

where max(S) is the maximum saturation and min(S) is the minimum saturation of segment S. This generates a 20-D Saturation feature vector f_{sat} which primarily helps in discriminating snow from other weather conditions.

(a) (b) (c) (d) (e)

Fig. 7. Sky detection for (a) cloudy image (b) rainy image (c) sunny image (d) snow image and (e) hazy image

3.8 CNN Model

A CNN model with the backbone of Resnet-101 on a five-class image set is used to analyze the performance of our proposed composite model consisting of hand-crafted and network-based extracted features. The loss function is defined by cross-entropy loss. We optimized the CNN parameters by maximizing the average of the log-probability of the correct label over the training examples.

4 Weather Dataset

Benchmark datasets available publicly on the internet have been used viz the Two Weather Dataset [14] for Sunny and Cloudy weathers, Desnow Dataset [13] for snow images and MWI Dataset [20] for Sunny, Hazy, Rainy, and Snow images. In total, we have used 30855 images collated from all the mentioned benchmark datasets.

4.1 Sky Dataset and SKYNET

A sky dataset has been proposed incorporating all the weather conditions being studied in this work. The dataset contains 30855 images, pertaining to different weather conditions along with their corresponding ground truth with white pixels indicating sky area and black being non-sky area. Figure 8 shows the weather images with their corresponding sky mask. For training the SKYNET, we have used 24684 images, each of size 256 × 256, while the remaining were used for testing the model on an Nvidia GeForce RTX 2060 Super. The batch size has been fixed at 32 with ADAM optimizer to optimize the loss function with the learning rate set at 0.0001 with total epochs set to 200 having 100 iterations each.

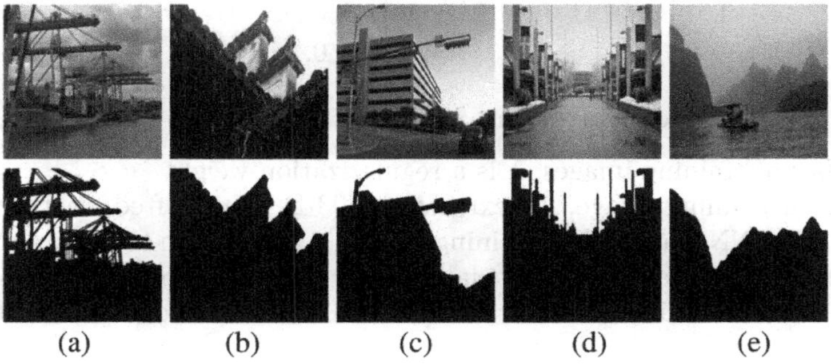

Fig. 8. Sky Mask for (a) cloudy image (b) rainy image (c) sunny image (d) snow image (e) hazy image

4.2 Multi-class Latent Joint Support Vector Machine (MLJSVM)

We define the Multi-class Latent Joint Support Vector Machine (MLJSVM) as follows:

Let $Y \equiv \{1, 2, \cdots, 5\}$ be a set of 5 classes. We define ϕ, ψ and matrix W as

$$\begin{aligned}\psi(x_i, \beta) &= \beta x_i^T \in \mathbb{R}^{m \times d} \\ \phi(x_i, y, \beta) &= \left[0 \cdots 0 \; \psi(x_i, \beta) \; 0 \cdots 0\right]^T \in \mathbb{R}^{5m \times d} \\ W &= \left[W_1^T \; W_2^T \cdots W_5^T\right] \in \mathbb{R}^{5m \times d}\end{aligned}$$

where matrix W_y contains m d-dimensional models and $\beta \in \mathbb{R}^m$ is a latent variable. Here, $d = 605$, is the dimension of the input feature vector and m is the number of training images. The prediction made by the technique is given by

$$\begin{aligned}y &= \psi(x_i, \beta) \\ y_i &\equiv s_w(x_i, y), \\ s_w(x_i, y) &\equiv \max_{\beta \in \Omega_p} f_w(x_i, y, \beta), \\ f_w(x_i, y, \beta) &\equiv W \odot \phi(x_i, y, \beta)\end{aligned}$$

where $y \in Y$, Ω_p is a p-norm unit ball, $\Omega_p \equiv \{\beta \in \mathbb{R}^m : \|\beta\|_p \leq 1, \beta_i \geq 0, \forall i = 1, \cdots, m\}$, $\odot$ indicates element-wise multiplication and β can be interpreted as a model weight factor. Finally, the objective function is given as:

$$\min_{W,\xi} \frac{\lambda}{2}\|W\|_F^2 + \sum_{i=1}^{n} \xi_i \tag{14}$$

such that

$$\begin{aligned}1 - (s_w(x_i, y_i) - n_w(x_i, y_i)) \leq \xi_i, i = 1, 2, \cdots, n \\ \xi_i \geq 0, i = 1, \cdots, n\end{aligned}$$

where $\|.\|_F$ is the Frobenius norm, $n_w(x_i, y_i) \equiv \max_{y \in Y/\{y_i\}, \beta \in \Omega_p} f_W(x_i, y, \beta)$, n is the number of training images, λ is a regularization weight.

For each training image, we extract the 93-D handcrafted weather feature and a 512-D CNN feature. For training the SVM model, the training and testing composition from the benchmark dataset has been 80% and 20% respectively. We have used 24684 images for training and 6171 images for testing the model.

4.3 Experimental Results

The classification results have been reported under different parameter settings. The Resnet based classifier referred to as CNN model has been trained end-to-end on the benchmark datasets.

Ablation Study. We have used MLJSVM to evaluate each of the individual features. Table 1[a] tabulates the classification results. The learning rate, decay and momentum used in CNN Model have been set at 0.01, 0.9, and 0.9 respectively. We conducted an ablation study where we analyse the performance of the system when one of the features is excluded. Table 1[b] shows that all the extracted handcrafted weather features are required in accounting for accurate weather detection. We note that the feature extracted from CNN model plays an important role in comparison to other individual features, but the combination of all the hand-crafted features overpower the CNN-based model.

Comparison of Performance. We present our result for multi-weather detection problem compared with other state-of-the-art weather systems. We implemented MLJSVM with linear kernel on the 605-D feature vector extracted from an image. The hyper-parameters of the CNN model have been fine-tuned with the best result being obtained for learning rate, momentum and weight decay being tuned to 0.0001, 0.9 and 0.005 respectively.

Table 1. (a) Classification results obtained using only the stated feature. (b) Classification results obtained by leaving out the stated feature and the features extracted from Resnet-101. (c) Classification results of different methods.

(a)

Feature	*Accuracy*
Illumination	34.3
Colorfulness	41.8
Sky	73.0
Visibility	83.0
Contrast Energy	35.2
Saturation	39.7
CNN Model	73.15

(b)

Feature	*Accuracy*
Illumination	89%
Colorfulness	90%
Sky	73.0
Visibility	74.0
Contrast Energy	90.0
Saturation	88.7

(c)

Model	*Accuracy*
Ours	94.19%
Li [11]	83.79%
Resnet-101	73.15%

Comparison with Related Methods. Most of the research work has been limited to two or three kinds of weather. In [11], images with five different weathers are classified based upon fusion of some weather features with different neural networks with highest accuracy being obtained in Resnet-50 with comparison presented in Table 1[c]. It may be noted, that Liu in [14] has worked *only* on Sunny and Cloudy weather and we have tested our approach with same classes using the code available in the public domain. The accuracy obtained in [14] was found to be 91.4%, while the proposed method yielded 97.63%.

5 Concluding Remarks and Scope for Future Work

A major contribution of this work is in devising a strategy which rises above a simple classification approach with the objective of achieving higher accuracy. Our proposed algorithm tries to emulate the characteristics used by the our brain in order to predict the weather. However, extracting all such weather features in handcrafted space are impossible and thus we combine the effect of Neural Network based features. This complements the handcrafted features used in the method and makes the model robust. The experimental results obtained on highly diverse datasets have demonstrated its effectiveness. The proposed approach rules out the existence of multiple weather conditions which would naturally be reflected in an image acquired at such a juncture. Thus, further research is required to make the system more robust to such weather patterns.

References

1. An, J., Chen, Y., Shin, H.: Weather classification using convolutional neural networks. In: 2018 International SoC Design Conference (ISOCC), pp. 245–246 (2018). https://doi.org/10.1109/ISOCC.2018.8649921
2. Ancuti, C., Ancuti, C.O., De Vleeschouwer, C.: D-hazy: a dataset to evaluate quantitatively dehazing algorithms. In: 2016 IEEE International Conference on Image Processing (ICIP), pp. 2226–2230 (2016). https://doi.org/10.1109/ICIP.2016.7532754
3. Bhandari, H., Palit, S., Chowdhury, S., Dey, P.: Can a camera tell the weather? In: 2021 36th International Conference on Image and Vision Computing New Zealand (IVCNZ), pp. 1–6 (2021). https://doi.org/10.1109/IVCNZ54163.2021.9653246
4. Chen, Z., Yang, F., Lindner, A., Barrenetxea, G., Vetterli, M.: Howis the weather: automatic inference from images. In: 2012 19th IEEE International Conference on Image Processing, pp. 1853–1856 (2012). https://doi.org/10.1109/ICIP.2012.6467244
5. Comaniciu, D., Meer, P.: Mean shift: a robust approach toward feature space analysis. IEEE Trans. Pattern Anal. Mach. Intell. **24**(5), 603–619 (2002). https://doi.org/10.1109/34.1000236
6. Derpanis, K.G., Lecce, M., Daniilidis, K., Wildes, R.P.: Dynamic scene understanding: the role of orientation features in space and time in scene classification. In: 2012 IEEE Conference on Computer Vision and Pattern Recognition, pp. 1306–1313 (2012). https://doi.org/10.1109/CVPR.2012.6247815
7. Elhoseiny, M., Huang, S., Elgammal, A.: Weather classification with deep convolutional neural networks. In: 2015 IEEE International Conference on Image Processing (ICIP), pp. 3349–3353 (2015). https://doi.org/10.1109/ICIP.2015.7351424
8. Groen, I.I.A., Ghebreab, S., Prins, H., Lamme, V.A.F., Scholte, H.S.: From image statistics to scene gist: evoked neural activity reveals transition from low-level natural image structure to scene category. J. Neurosci. **33**, 18814–18824 (2013)

9. Kurihata, H., et al.: Rainy weather recognition from in-vehicle camera images for driver assistance. In: IEEE Proceedings Intelligent Vehicles Symposium 2005, pp. 205–210 (2005). https://doi.org/10.1109/IVS.2005.1505103
10. Laffont, P.Y., Ren, Z., Tao, X., Qian, C., Hays, J.: Transient attributes for high-level understanding and editing of outdoor scenes. ACM Trans. Graph. **33**, 149:1–149:11 (2014). https://doi.org/10.1145/2601097.2601101
11. Li, Z., Li, Y., Zhong, J., Chen, Y.: Multi-class weather classification based on multi-feature weighted fusion method. IOP Conf. Ser. Earth Environ. Sci. **558**(4), 042038 (2020). https://doi.org/10.1088/1755-1315/558/4/042038
12. Lin, D., Lu, C., Liao, R., Jia, J.: Learning important spatial pooling regions for scene classification. In: 2014 IEEE Conference on Computer Vision and Pattern Recognition, pp. 3726–3733 (2014). https://doi.org/10.1109/CVPR.2014.476
13. Liu, Y.F., Jaw, D., Yeh, W.C., Hwang, J.N.: DesnowNet: context-aware deep network for snow removal. IEEE Trans. Image Process. **27**(6), 3064–3073 (2018). https://doi.org/10.1109/TIP.2018.2806202
14. Lu, C., Lin, D., Jia, J., Tang, C.K.: Two-class weather classification. IEEE Trans. Pattern Anal. Mach. Intell. **39**(12), 2510–2524 (2017). https://doi.org/10.1109/TPAMI.2016.2640295
15. Narasimhan, S., Nayar, S.: Shedding light on the weather. In: 2003 IEEE Computer Society Conference on Computer Vision and Pattern Recognition 2003 Proceedings, vol. 1, p. I (2003). https://doi.org/10.1109/CVPR.2003.1211417
16. Roser, M., Moosmann, F.: Classification of weather situations on single color images. In: 2008 IEEE Intelligent Vehicles Symposium, pp. 798–803 (2008). https://doi.org/10.1109/IVS.2008.4621205
17. Shen, L., Tan, P.: Photometric stereo and weather estimation using internet images. In: 2009 IEEE Conference on Computer Vision and Pattern Recognition, pp. 1850–1857 (2009). https://doi.org/10.1109/CVPR.2009.5206732
18. Song, H., Chen, Y., Gao, Y.: Weather condition recognition based on feature extraction and K-NN. In: Sun, F., Hu, D., Liu, H. (eds.) Foundations and Practical Applications of Cognitive Systems and Information Processing. AISC, vol. 215, pp. 199–210. Springer, Heidelberg (2014). https://doi.org/10.1007/978-3-642-37835-5_18
19. Yan, X., Luo, Y., Zheng, X.: Weather recognition based on images captured by vision system in vehicle. In: Yu, W., He, H., Zhang, N. (eds.) ISNN 2009. LNCS, vol. 5553, pp. 390–398. Springer, Heidelberg (2009). https://doi.org/10.1007/978-3-642-01513-7_42
20. Zhang, Z., Ma, H.: Multi-class weather classification on single images. In: 2015 IEEE International Conference on Image Processing (ICIP), pp. 4396–4400 (2015). https://doi.org/10.1109/ICIP.2015.7351637

A Novel CNN-Based Approach for Distinguishing Between COVID and Common Pneumonia

Somrita Bakshi[1], Sarbani Palit[1](✉), Ujjwal Bhattacharya[1], Kimia Gholami[2], Nushrat Hussain[1], and Debasis Mitra[2]

[1] Indian Statistical Institute, Kolkata, India
bakshi.somrita@gmail.com, hnushrat@gmail.com, {sarbanip,ujjwal}@isical.ac.in
[2] Florida Institute of Technology, Melbourne, USA
fgholami2021@my.fit.edu, dmitra@fit.edu

Abstract. It is well known that the symptoms of Coronavirus disease (COVID) and common pneumonia (CP) disease are very similar though the first one often leads to severe complications and may even be fatal. Hence, it is of vital importance to be able to correctly distinguish between the two. This paper attempts to achieve this task using whole 3-D CT scans of lungs. A number of models have been experimented with, using convolutional and radiomic features as well as their concatenations, and different classifiers (MLP and Random Forest) with two different sizes of input CT images ($50 \times 128 \times 128$ and $25 \times 256 \times 256$) and their performances have been compared. The most significant contribution of this work is the postulation of a 3-D dual-scale framework using CT scans, employing both intra-scale and inter-scale information, thereby achieving performance scores which are much higher than the state of the art methods to distinguish between COVID-19 and CP using lung CT scans. Specifically, Accuracy of 98.67% and Receiver Operating Characteristics-Area Under The Curve (AUC) of 99% are worth mentioning.

Keywords: 3-D dual scale CNN · 3-D CT scan · Random forest · Radiomic features

1 Introduction

COVID is an infectious disease, which requires fast identification and isolation of affected persons/patients. World Health Organisation has declared it to be a world pandemic. Although reverse transcription-polymerase chain reaction (RT-PCR) is considered as the gold standard for the diagnosis of COVID-19, medical imaging techniques have also been tried for the same recently, towards increased robustness of the detection procedure. Moreover, this is a viable option since artificial intelligence based image analysis is faster than RT-PCR testing and is also cost effective. Additionally, CT scan or any imaging technique is non-invasive which is an added advantage.

W. Q. Yan et al. (Eds.): IVCNZ 2022, LNCS 13836, pp. 330–344, 2023.
https://doi.org/10.1007/978-3-031-25825-1_24

This paper attempts to classify between COVID and CP diseases using lung CT scans of a publicly available dataset, the Mendeley dataset [24] (which contains scans of COVID and common pneumonia patients only). Healthy patients are not included in this dataset. Figure 1a shows a Lung CT image of a COVID patient and Fig. 1b shows that of a CP patient.

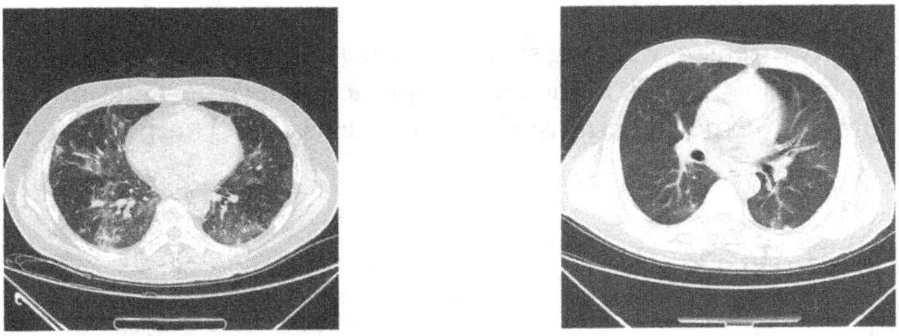

(a) A CT scan slice of lungs of a COVID affected patient

(b) A CT scan slice of lungs of a common pneumonia (CP) affected person

Fig. 1. Lung CT images.

Instead of adopting the conventional approach of using 2-D slices [1,2,5], we have stacked all the CT slices of each patient into respective 3-D scan level images. We, then used these whole scans of patients lungs to detect the disease. This scan level classification reduces the no. of data samples, (each scan has variable number of slices) making classification more challenging for the AI system. However its significant advantages are -

- It uses the 3-D voxels containing both intra-slice/(spatial) and inter-slice/(depth) information, compared to its 2-D counterpart.
- Whole scan level classification is clinically more useful than slice based classification.
- In the present dataset, the ground truth is given by RT-PCR test which is a whole scan level information and slice level ground truth is not available.

The highlights and hence important contributions made in this paper are:

1. Comparison has been performed between the results obtained using volumes containing smaller sized slices but larger in number versus volumes containing larger sized slices which are less in number.
2. Both high intra-slice and high inter-slice information have been combined using a 3D dual-scale CNN, where one CNN emphasizes intra-slice information while the second CNN emphasizes inter-slice information.
3. Three dimensional interpolation has been employed in order to get different sizes of the input scans, namely different slice sizes as well as number of slices. Each scan being fed to our CNN's, corresponds to the whole original scan.

4. While using CNNs we did not use segmentation and yet we got satisfactory results. Thus, on the whole our system is faster and less prone to errors compared to most other works.
5. Since in the existing literature, radiomics features had been used, we explored the use of convolutional features concatenated with the radiomic features and observed that such a concatenation fails to improve the detection performance of convolutional features.

Section 2 briefly discusses existing studies of the problem. The proposed approach have been detailed in Sect. 3. Some more approaches have been detailed in Sect. 4. Section 5 presents experimental results and discussion. The paper is concluded in Sect. 6.

2 Related Work

A new dataset (Mendeley dataset) of CT scan image samples of the lungs of COVID and CP patients was introduced in [1]. A multi-scale convolutional neural network (MSCNN), ensembled by three different CNN's which have the same structure but different input scales are used to extract scale specific information. Efficient Net B0 pre-trained on the ImageNet dataset has been used as the backbone architecture of the 3 CNNs. Performance of classification has been computed both at the slice level and scan level with the top three highest scores of all slices of a scan being averaged as the scan level score. The scan level classification is derived from the slice level classification. However, the neural network has been trained using 2-D slice level images only. Their performance scores were satisfactory for the per-slice classification. However, the scan level accuracy dropped significantly.

One of the first papers on detecting corona virus using chest CT scans was proposed by Wang et al. in [2]. Data of a total of 259 patients were used from three hospitals namely Xi'an Jiaotong University First Affiliated Hospital (center 1), Nanchang University First Hospital (center 2), and Xi'an No.8 Hospital of Xi'an Medical College (center 3). The input images were then pre-processed using image processing techniques. GoogleNet InceptionIt v3 CNN pre-trained on ImageNet was used for transfer learning. It may be noted that this operated only at the slice level.

Vaidyanathan et al. in [3] used 169 COVID cases, 60 pneumonia cases and 76 infection cases for their study. The data was collected from two hospitals, namely CHU Sart-Tilman and CHU Notre Dame des Bruyères. A lung segmentation model was used. Then a lung abnormality segmentation model was used. Different sets of 48 consecutive axial slices with an overlap of 10 slices between two consecutive sets were obtained. Each data point containing the 48 consecutive axial slices were processed in three different ways. An inflated 3-D Inception model was trained on 48 consecutive axial slices. The overall patient-wise prediction was calculated as follows: - if $\geq 20\%$ of the predictions correspond to the class COVID, then the patient is assigned to that class. If the probability for influenza/pneumonia is $\geq 20\%$ then the patient is assigned to that class. Otherwise it is classified as no infection. This suffers from the drawback of deriving results from 48 slices at a time instead of using the entire volume.

Hasan et al. in [4] used 1110 scans in total for both binary (COVID versus normal) and multiclass classification. They used the MosMedData dataset. They segmented the lungs using image processing techniques. Geometric augmentation and intensity based augmentation are applied to increase the database size. As, their dataset was imbalanced, they added extra CT volumes from another dataset and weighted the loss function for penalizing the over-represented class. A base 3D-CNN model was built on the extracted 3-D patches from whole chest CT scans. Then, the patch size was doubled and the base model's input size was also doubled and a modified base network was trained. The base model's trained weights for smaller patches were also kept. This is repeated till the patch size is as large as the original image size. Unfortunately the imbalanced dataset led to a lower specificity value.

Abdar et al. in [5] used CT images - (131 COVID patients and 150 healthy patients). The images were collected from Alinasab Hospital of Tabriz. The lung part of the images were extracted using image processing techniques. They used VGG-16 as the base model for transfer learning. This approach too suffered from the limitation of being only slice-based and also did not include CP patients.

Chen et al. in [6] included 326 CT exams. A 3D-UNet was used to segment detected consolidation and ground glass opacity lesions. The segmentation results were used to extract quantifying CT characteristics and radiomics features. Four groups of features were included for model building. Grid searching was used for feature selection. Support Vector Machine (SVM) models were built on the four groups of features individually and their combination. The approach however did not yield very good results.

de Moura LV et al. in [7], used 2611 COVID and 2611 non-COVID chest X-Rays. They used the BIMCV dataset. First and second order radiomics features were used by them. Lung masks were generated using a U-Net inspired segmentation model. SHAP-RFECV (Recursive Feature Elimination with Cross Validation) was used for feature selection. XGBoost (XGB) and Random Forest (RF) models were used for classification. RF and XGB models were applied with 10-fold cross-validation. However, the use of X-Ray images as input possibly led to lower output scores.

A 3-D deep learning framework was developed in [9] and called COVNet. They used 4563 chest CT scans and the dataset was acquired from six hospitals. It extracts both 2-D local and 3-D global features employing the RESNET50 which takes as input CT image slices which have been segmented using U-Net in order to extract the lung region. A Max-Pooling operation combines the entire set of extracted features which is then fed to a fully connected layer with softmax activation. The output is a probability score for COVID, CP, and non-pneumonia. Though its overall performance was good, it still left room for improvement.

A dual sampling approach was adopted in [10] to deal with the problem of imbalanced distribution infected areas in both Covid and CP patient scans. A 2-stage strategy was followed where in the first stage two different sampling strategies are used to train two 3-D RESNET34 models. Attention maps have been used on the feature maps produced by the 3-D RESNET34 networks. In the second stage an ensemble learning layer is trained to integrate the predictions obtained from the two models. However, some of the scores achieved still needed to be improved upon.

The approach followed in [11] was to build a diagnostic system comprising of two models: a lung-lesion segmentation model and a model to produce the final diagnosis. They used the China Consortium of Chest CT Image Investigation (CC-CCII) Dataset. The Deep Lab v3 formed the backbone of the segmentation network to produce the lung-lesion maps which was used by the diagnosis classifier for diagnosis prediction. The method suffers from the twin drawbacks of requiring a preliminary manual segmentation and also that lesion segmentation is involved which makes the task difficult.

Multiscale networks for detecting and classifying lung nodules have been used in [12,13] while a multi-scale rotation-invariant model has been used for classifying lung tissue patterns in [14]. In this paper the concept of multi-scale has been introduced while operating upon 3-D image volumes.

In recent years Random forest classifiers have been used in [22] for string instrument recognition classification algorithms and very high performance was achieved in their work.

Various complex image features after being subjected to a feature selection procedure and other screening strategies were used to build radiomics features as outlined in [17]. van Griethuysen et al. [8] developed an open-source platform, called PyRadiomics, for extraction of hand engineered radiomic features from medical images. These authors also demonstrated the application of PyRadiomics in characterizing lung lesions. In [18] semiautomatic volumetric segmentation approach was implemented on the 3D Slicer platform. They observed that a major challenge of the use of Radiomics in analysis of medical images is segmentation of the object. They used a publicly available 3D-Slicer platform implementing a semiautomatic region growing volumetric segmentation method and extracted fifty six 3D-radiomic features from CT lung images for their study. The superiority and robustness of radiomic features was observed with respect to hand crafted features. This technique was applied in [15] for analysing tumour phenotype through the extraction of features used as imaging biomarkers. Radiomic features were also used in [21] to predict outcomes of neuroblastoma. The utility of radiomic features in predicting outcome of treatments was reported in [16]. van Griethuysen et al. [8] developed an open-source platform, called PyRadiomics, for extraction of hand engineered radiomic features from medical images. Recent studies of Radiomics in medical image analysis tasks have been reviewed in [19]. The main thrust of this study is the field of oncology towards development of personalized medicine. A more recent survey of the use of radiomics can be found in [20]. The authors of this later study described some of the promising and important results achieved in various applications.

3 Proposed Approach

Two chief stages constitute the approach proposed in this paper. The first one comprises of a sequence of operations which perform a preprocessing of the input. This is followed by the network which carries out the task of classification.

- **Preprocessing**
  1. The non-lung images were manually excluded from each scan. Figure 2a shows an example of a non-lung CT of a patient suffering from COVID while Fig. 2b shows a corresponding image for a CP patient.
  2. All slices of a scan were stacked into a 3-D array forming a volumetric scan.
  3. The volumetric scans were interpolated to fixed sizes of $(50 \times 128 \times 128)$ and $(25 \times 256 \times 256)$ using a 3-D nearest neighbour interpolation technique.
  4. The intensity of each voxel was normalized in the range (0, 1) for classification of COVID and CP patients.

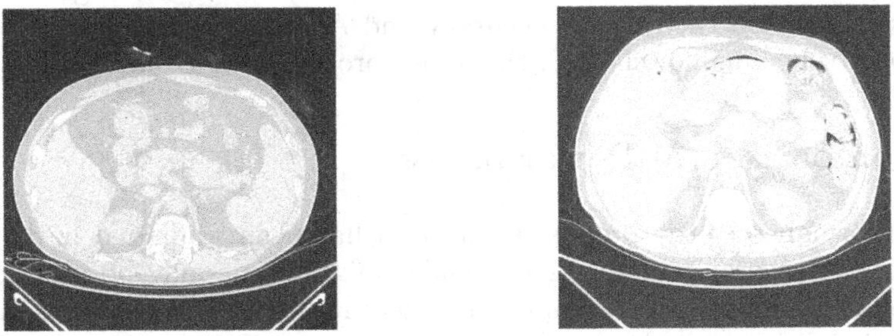

(a) CT scan slice with COVID (b) CT scan slice with CP

Fig. 2. Non-lung CT images

- **Network implementation**
  We use a 3-D dual-scale CNN to classify between COVID and CP patients. The first CNN extracts features from our volume scans of size $(25 \times 256 \times 256)$ which emphasizes more on intra-slice information. The second CNN extracts features from our volume scans of size $(50 \times 128 \times 128)$ which emphasizes more on inter-slice information. Each CNN uses two blocks of convolution and Max-Pooling layers. This was followed by a Global Max-Pooling and a flatten layer in each CNN. Then the features of both CNN were concatenated. These concatenated features were used as input to the hidden layer with 32 units which was added to it. Finally a output layer with 2 units were added. Each layer except the final output layer uses ReLU activation. The final output

layer uses a softmax activation. Adam optimizer with a constant learning rate of 0.001 was used. All codes were written in Python using Visual Studio Code. All models were built using tensorflow.keras package. All models were trained using NVIDIA quadro RTX 24 GB GDDR 6 GPU.

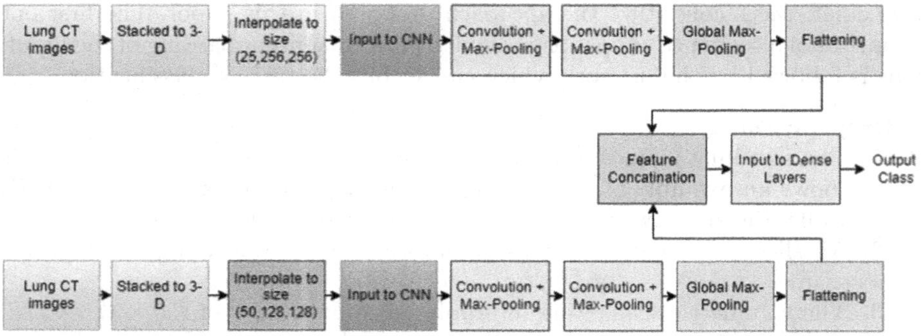

Fig. 3. Flow of 3-D dual-scale convolutional features based classification

A batch-size of 4 was used for training the model. After testing it with 5-fold cross validation the average accuracy and AUC score were 98.67% and 0.99 respectively. The flowchart of the above procedure is given in Fig. 3.

4 Comparison of Performance

The performance of the proposed approach has been examined in comparison with that of several configurations. The configurations may be grouped under three main heads viz. those employing solely radiomic features, those based on convolutional features alone and finally those utilizing a combination of both features. These are now elaborated upon:

4.1 Classification Using Radiomic Features

The use of radiomic features for classification is examined for two different scenarios: one in which the entire set of features is employed and the other in which a subset of the features is considered.

4.1.1 The General Procedure

The main steps for classification using radiomic features is described below.

■ **Preprocessing** -

1. From each scan the non-lung images were manually filtered out.
2. A U-Net was used to segment the lungs of each slice.

3. For all scans, all slices of the scan were stacked into a 3-D array.
4. Similarly, all segmented images of a scan were stacked into a 3-D array. This was done for all scans.

- **Radiomic feature extraction -**
Both the 3-D CT scans and the 3-D segmentation maps were fed into the Pyradiomics package [8] in order to extract radiomic features. 112 3-D radiomic features were extracted from each lung CT scan.
- **Classification based on Radiomic features -**
  1. A Multilayer Perceptron (MLP) with 2 hidden layers using rectified linear units (RELU) activation (64 and 32 units each) was used to classify COVID versus CP patients. The entire set of radiomic features was fed to its input layer. The learning strategy involved the Adam Optimizer and a batch size of 100 samples. A stopping criterion of 30 epochs was used. After testing it with 5-fold cross validation, the average accuracy and AUC score (described later in Sect. 5 were 57.32% and 0.57 respectively. The flowchart of this procedure is given in Fig. 4. It may be noted that in the last box, the MLP option applies for this case.

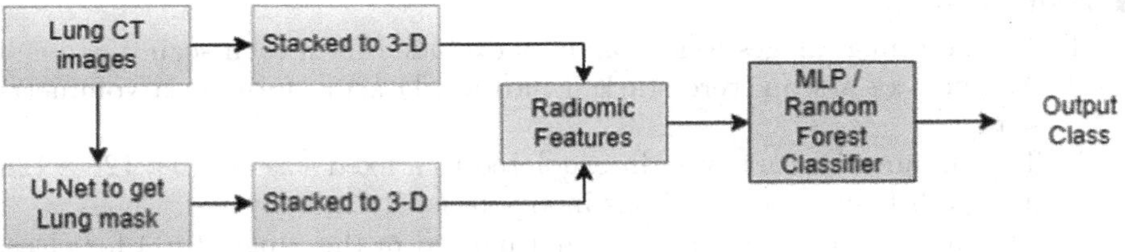

Fig. 4. Flow of radiomic features based classification

  2. Next, a Random Forest classifier (RF) with 256 decision trees was used for the classification task instead of the MLP, as portrayed in Fig. 4 with the Random Forest option in the last box. After testing it with 5-fold cross validation the average accuracy and AUC score were 87.42% and 0.87 respectively.

4.1.2 Classification Based on Selected Radiomic Features

The mutual information [23] between each extracted radiomic feature and the target label (given ground truth) were calculated. The radiomic features were then sorted according to their calculated mutual information with the feature having maximum information being ranked 1. We have considered two different subsets of features S_1 and S_2 consisting of top ranked 56 and 10 features respectively.

1. An approach exactly similar to that for all features was adopted. Using the top 56 radiomic features, and testing with 5-fold cross validation the average accuracy and AUC score were 62.03% and 0.62 respectively. The average accuracy and AUC scores were 82.108% and 0.82 respectively, using the top 10 features. Reducing number of features further to 5 the accuracy reduces to 70.62%.
2. Following the procedure shown in Fig. 4 a Random Forest (RF) classifier was used for the classification based on the selected 56 features. After testing it with 5-fold cross validation the average accuracy and AUC score were 87.67% and 0.88 respectively. On the other hand, using the top 10 features yielded an average accuracy and AUC score were 84.77% and 0.84 respectively. Reducing number of features further to 5 the accuracy reduces to 83.07%.

4.2 Classification Using Convolutional Features

Preprocessing of the input scans is described below. Several different variants of the basic network used for classifying the preprocessed images are considered and their performances computed.

- **Preprocessing -**
  1. The non-lung images were manually excluded from each scan.
  2. All slices of a scan were stacked into a 3-D array forming a volumetric scan.
  3. The volumetric scans were interpolated to a fixed size of $(50 \times 128 \times 128)$ using a 3-D nearest neighbour interpolation technique.
  4. The intensity of each voxel was normalized in the range (0, 1) for classification of COVID and CP patients.
- **Network implementation -**
  1. We built a 3-D CNN model (Model-1) to distinguish between COVID and CP. The entire volume scans of size $(50 \times 128 \times 128)$ were passed as input to the CNN. Four blocks of convolution, each having 32 filter kernels of size $(3 \times 3 \times 3)$, followed by Max-Pooling layers with filter size $(2 \times 2 \times 2)$ was used. This was followed by a Global Max-Pooling and a flatten layer. Two hidden layers were then added each with 32 units. Finally a output layer with 2 units were added. While the final layer uses a softmax activation, all other layers used for training the model use ReLU. There was no padding and a stride of 1 was used. The learning strategy employed the Adam optimizer with a constant learning rate of 0.001 and a batch-size of 25. After testing it with 5-fold cross validation the average accuracy and AUC score were 96.02% and 0.96 respectively. The flowchart of the above procedure is given in Fig. 5.
  2. Next we extracted the convolutional features from the flatten layer of the above model - (model 1) and used a Random Forest classifier for the final classification task. After testing it with 5-fold cross validation the average accuracy and AUC score were 96.13% and 0.96 respectively.

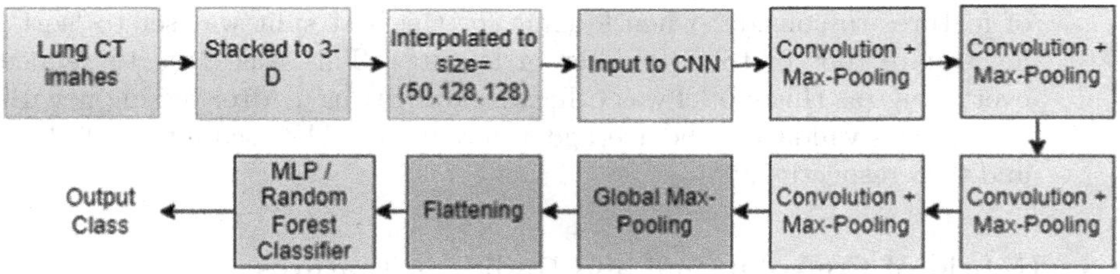

Fig. 5. Flow of Convolutional features based classification model - 1

3. We next examined if larger images with less number of slices yields better results than smaller images with more number of slices. That is we tried to check if emphasizing on intra-slice information with less inter-slice information is more important, or vice versa. For this, we interpolated our volume scans to size (25 × 256 × 256). We now built another 3-D CNN model (model-2) to classify COVID versus CP patients using our new interpolated volume scans of size (25 × 256 × 256). Two blocks of convolution followed by Max-Pooling layers were used. Each convolution layer had 32 filter kernels each of size (3 × 3 × 3). The filter size in each of the Max-Pooling layer was (2 × 2 × 2). This was followed by a Global Max-Pooling and a flatten layer. One hidden layer was then added with 32 units. Finally a output layer with 2 units were added. Each layer except the final output layer uses ReLU activation. The final layer uses a softmax activation. There was no padding and a stride of 1 was used. Adam optimizer with a constant learning rate of 0.001 was used. A batch-size of 2 was used for training the model. After testing it with 5-fold cross validation the average accuracy and AUC score were 97.59% and 0.97 respectively. The flowchart of the above procedure is given in Fig. 6.

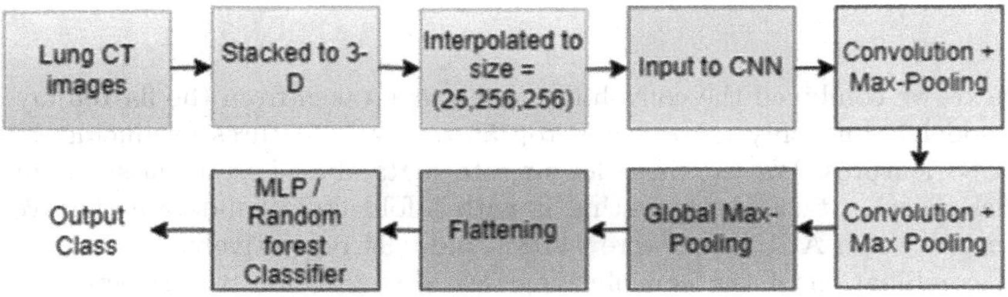

Fig. 6. Flow of Convolutional features based classification - model - 2

4. Next we extracted the convolutional features from the flatten layer of the above model - (model 2) and used a Random Forest classifier for the final classification task. The maximum depth of our tree was 10 and number

of features to consider when looking for the best split was set to 'sqrt', i.e. square root of total number of features. This was done to reduce overfitting (as this model was otherwise overfitting). After testing it with 5-fold cross validation the average accuracy and AUC score were 93.10% and 0.93 respectively.

4.3 Hybrid of Convolutional and Radiomic Features

Preprocessing of the input scans is accomplished following the procedures described in Sects. 4.1 and 4.2. The strategies for combining the two sets of features is now elaborated upon.

1. A combination of both convolutional and radiomic features was considered as the input. The convolutional features from the flatten layer of (model-1) were extracted and concatenated with the radiomic features leading to a total of 144 features. A Random Forest classifier was used for the final classification task. After testing it with 5-fold cross validation the average accuracy and AUC score were 95.77% and 0.96 respectively. The flowchart of the above procedure is given in Fig. 7.

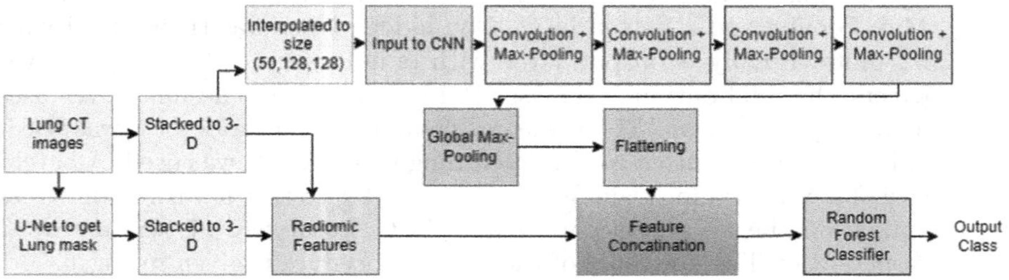

Fig. 7. Flow of convolutional and all radiomic features based classification

2. Next, we combined the convolutional features taken from the flatten layer of (model-1) and only the selected top 56 radiomic features producing a total of 88 features. We fed these features to a Random Forest classifier for the classification task. After testing it with 5-fold cross validation the average accuracy and AUC score were 94.08% and 0.94 respectively.
3. A combination of the convolutional features and only the selected top 10 radiomic features was also considered. The 42 features obtained after concatenation were fed to a Random Forest classifier. After testing it with 5-fold cross validation the average accuracy and AUC score were 92.75% and 0.93 respectively.

5 Results

The performances of all the configurations described in the last section using standard metrics is now presented systematically, including those of some standard classification approaches.

5.1 Evaluation Metrics

In the present study, the following metrics have been used for evaluation.

1. Accuracy (Acc) $= \frac{TP+TN}{TP+TN+FP+FN}$
2. Sensitivity (Sen) $= \frac{TP}{TP+FN}$
3. Specificity (Sp) $= \frac{TN}{TN+FP}$
4. Precision (Pre) $= \frac{TP}{TP+FP}$
5. AUC score (AUC) = AUC is the area under the entire Receiver Operating Characteristic curve (ROC) curve, where the ROC curve plots sensitivity versus (1 - specificity).
6. F1 score (F1) $= \frac{2*(precision*recall)}{precision+recall} = \frac{TP}{TP+\frac{1}{2}(FP+FN)}$,
   where TP = True Positive, TN = True Negative, FP = False Positive and FN = False Negative.

5.2 Average Performance

The performance metrics for each of the approaches studied here are reported in Table 1 The highest value in each column is shown in bold.

It may be observed that the proposed algorithm employing dual-scale 3D CNN produces the best result with respect to every metric. For sensitivity the best performance is also achieved by 3-D CNN with input size 25 × 256 × 256.

The proposed approach of 3-D dual scale CNN is compared with past works in Table 2. It can be seen that the proposed approach performs much better than all other approaches with respect to every metric except Sensitivity. Our Sensitivity is only marginally lower than the work of Hassan et al. but, their specificity is drastically low implying a biased training in their work. Our AUC score is also much higher than theirs.

Table 1. Average performance after 5-fold cross validation

	Accuracy (%)	Sensitivity	Specificity	Precision	AUC score	F1-score
MLP based on All Radiomic Features	57.32	0.76	0.38	0.67	0.57	0.62
Random Forest based on All Radiomic Features (All)	87.42	0.86	0.89	0.89	0.87	0.87
MLP based on 56 Radiomic Features	62.03	0.47	0.95	0.80	0.62	0.52
Random Forest based on 56 Radiomic Features	87.67	0.87	0.88	0.88	0.88	0.88
MLP based on 10 Radiomic Features	82.11	0.81	0.83	0.83	0.82	0.82
Random Forest based on 10 Radiomic Features	84.77	0.83	0.86	0.86	0.85	0.84
CNN-input size (50 × 128 × 128)	96.02	0.97	0.95	0.95	0.96	0.96
CNN-input size (50 × 128 × 128)+ Random Forest	96.13	0.96	0.96	0.96	0.96	0.96
CNN-input size (50 × 128 × 128) + All Radiomic Features + Random Forest	95.77	0.97	0.94	0.95	0.96	0.96
CNN-input size (50 × 128 × 128) + 56 Radiomic Features + Random Forest	94.08	0.95	0.93	0.94	0.94	0.94
CNN-input size (50 × 128 × 128) + 10 Radiomic Features + Random Forest	92.75	0.92	0.94	0.94	0.93	0.93
CNN-input size (25 × 256 × 256)	97.59	**0.98**	0.96	0.97	0.97	0.98
CNN-input size (25 × 256 × 256) + Random Forest	93.10	0.93	0.93	0.93	0.93	0.93
3-D dual scale CNN	**98.67**	**0.98**	**0.99**	**0.99**	**0.99**	**0.99**

Table 2. Comparison with other works

	Accuracy (%)	Sensitivity	Specificity	Precision	AUC score	F1-score
Yan et al. [1]	87.50	0.89	0.86	–	0.93	–
Wang et al. [2]	89.5	0.88	0.87	–	0.93	0.77
Vaidyanathan et al. [3]	–	0.88 (Covid)	0.87 (Covid)	–	0.91 (Covid)	–
	–	0.83 (CP)	0.89 (CP)	–	0.89 (CP)	–
	–	0.78 (normal)	0.98 (normal)	–	0.98 (normal)	–
Hasan et al. [4]	–	**0.99**	0.66	–	0.91	–
Abdar et al. [5]	90.14	0.88	0.90	0.91	–	0.91
Chen et al. [6]	84.30	0.82	0.92	–	0.92	–
de Moura LV et al. [7]	82.00	0.82	–	0.82	–	0.82
Lin Li et al. [9]	–	0.90 (Covid)	0.96 (Covid)	–	0.96 (Covid)	–
	–	0.87 (CP)	0.92 (CP)	–	0.95 (CP)	–
	–	0.94 (normal)	0.96 (normal)	–	0.98 (normal)	–
Ouyang X et al. [10]	87.90	0.87	0.91	–	0.95	0.82
Zhang K et al. [11]	90.71	0.92	0.90	–	0.98	–
Proposed 3-D dual scale CNN	**98.67**	0.98	**0.99**	**0.99**	**0.99**	**0.99**

6 Conclusions and Scope for Future Work

The performance of the 3-D CNN models are observed to be good while the overall best results have been achieved by the dual-scale 3-D CNN model. Here both intra-slice as well as inter-slice information is employed with each CNN giving more importance to one kind of information, intra-slice or inter-slice. Another reason for the dual-scale to work better is that medical images are scale invariant [1]. It may be noted that incorporation of radiomic features did not help in enhancing performance.

The ground truth of our dataset was determined by RTPCR test. Although RTPCR test is still considered to be the gold standard [6], yet it sometimes gives false negative results. Hence further testing of performance of the proposed dual-scale 3-D CNN is necessary with data having clinical confirmation of COVID.

New variants of Corona virus keep appearing. Our methods may not be able to detect cases of these new variants. Although, intuitively CT scans of the newer variants would be visually more similar to the older variants of Corona virus than pneumonia but that is yet to be checked experimentally, a direction of future work could be to detect even new strains of the virus.

As the proposed algorithm is seen to work well in distinguishing between COVID and CP, we intend to apply it for obtaining distinctions between less characterisable type of pneumonia.

References

1. Yan, T., Wong, P.K., et al.: Automatic distinction between COVID-19 and common pneumonia using multi-scale convolutional neural network on chest CT scans. Chaos, Solitons Fractals **140**, e110153 (2020)
2. Wang, S., et al.: A deep learning algorithm using CT images to screen for Corona virus disease (COVID-19). Eur. Radiol. **31**(8), 6096–6104 (2021). https://doi.org/10.1007/s00330-021-07715-1
3. Vaidyanathan, A., Guiot, J., et al.: An externally validated fully automated deep learning algorithm to classify COVID-19 and other pneumonias on chest computed tomography. ERJ Open Res. **8**(2) (2022)
4. Hasan, K.M., Jawad, T.M., et al.: COVID-19 identification from volumetric chest CT scans using a progressively resized 3D-CNN incorporating segmentation, augmentation, and class-rebalancing. Inform. Med. Unlocked **26**, e100709 (2021)
5. Abdar, A.K., Sadjadi, S.M., et al.: Automatic detection of coronavirus (COVID-19) from chest CT images using VGG16-based deep-learning. In: 2020 27th National and 5th International Iranian Conference on Biomedical Engineering (ICBME), pp. e212–e216 (2020)
6. Chen, H.J., Mao, L., et al.: Machine learning-based CT radiomics model distinguishes COVID-19 from non-COVID-19 pneumonia. BMC Infect. Dis. **21**(1), e1–e13 (2021). https://doi.org/10.1186/s12879-021-06614-6
7. de Moura, L.V., Mattjie, C., et al.: Explainable machine learning for COVID-19 pneumonia classification with texture-based features extraction in chest radiography. Front. Digit. Health **3**, 662343 (2021)
8. van Griethuysen, J.J.M., Fedorov, A., et al.: Computational radiomics system to decode the radiographic phenotype. Can. Res. **77**(21), e104–e107 (2017)
9. Li, L., Qin, L., et al.: Using artificial intelligence to detect COVID-19 and community-acquired pneumonia based on pulmonary CT: evaluation of the diagnostic accuracy. Radiology **296**(2), e65–e71 (2020)
10. Ouyang, X., Huo, J., et al.: Dual-sampling attention network for diagnosis of COVID-19 from community acquired pneumonia. IEEE Trans. Med. Imaging **39**(8), e2595–e2605 (2020)

11. Zhang, K., Xiaohong, L., et al.: Clinically applicable AI system for accurate diagnosis, quantitative measurements, and prognosis of COVID-19 pneumonia using computed tomography. Cell **181**(6), e1423–e1433 (2020)
12. Liu, K., Kang, G.: Multiview convolutional neural networks for lung nodule classification. Int. J. Imaging Syst. Technol. **27**(1), e12–e22 (2017)
13. Kim, B.-C., Yoon, J.S., et al.: Multi-scale gradual integration CNN for false positive reduction in pulmonary nodule detection. Neural Netw. **115**, e1–e10 (2019)
14. Wang, Q., Zheng, Y., et al.: Multiscale rotation-invariant convolutional neural networks for lung texture classification. IEEE J. Biomed. Health Inform. **22**(1), e184–e195 (2017)
15. Dong, D., Zhang, F., et al.: Development and validation of a novel MR imaging predictor of response to induction chemotherapy in locoregionally advanced nasopharyngeal cancer: a randomized controlled trial substudy (NCT01245959). BMC Med. **17**(1), e1–e11 (2019)
16. Song, J., Shi, J., et al.: A new approach to predict progression-free survival in stage IV EGFR-mutant NSCLC patients with EGFR-TKI therapy prediction of EGFR-TKI treatment outcome in stage IV NSCLC. Clin. Can. Res. **24**(15), e3583–e3592 (2018)
17. Lambin, P., Rios-Velazquez, E., et al.: Radiomics: extracting more information from medical images using advanced feature analysis. Eur. J. Can. **48**(4), e441–e446 (2012)
18. Parmar, C., Rios Velazquez, E., et al.: Robust radiomics feature quantification using semiautomatic volumetric segmentation. PLoS ONE **9**(7), e102107 (2014)
19. Liu, Z., Wang, S., et al.: The applications of radiomics in precision diagnosis and treatment of oncology: opportunities and challenges. Theranostics **9**(5), e1303–e1322 (2019)
20. Scapicchio, C., Gabelloni, M., Barucci, A., Cioni, D., Saba, L., Neri, E.: A deep look into radiomics. Radiol. Med. (Torino) **126**(10), 1296–1311 (2021). https://doi.org/10.1007/s11547-021-01389-x
21. Liu, G., Poon, M., et al.: Incorporating radiomics into machine learning models to predict outcomes of neuroblastoma. J. Digit. Imaging **35**(3), e605–e612 (2022). https://doi.org/10.1007/s10278-022-00607-w
22. Banerjee, A., Ghosh, A., Palit, S., et al.: A novel approach to string instrument recognition. In: International Conference on Image and Signal Processing, pp. e165–e175 (2018)
23. Gierlichs, B., Batina, L., Tuyls, P., Preneel, B.: Mutual information analysis. In: Oswald, E., Rohatgi, P. (eds.) CHES 2008. LNCS, vol. 5154, pp. 426–442. Springer, Heidelberg (2008). https://doi.org/10.1007/978-3-540-85053-3_27
24. Yan, J.: COVID-19 and common pneumonia chest CT dataset. Mendeley Data (2020)

Improving Masked Face Recognition Using Dense Residual Unit Aided with Quadruplet Loss

Muhammad Aasharib Nawshad(✉) and Muhammad Moazam Fraz

School of Electrical Engineering and Computer Science (SEECS), National University of Sciences and Technology (NUST), Islamabad, Pakistan
{mnawshad.mscs20seecs,moazam.fraz}@seecs.edu.pk

Abstract. All over the world, people are wearing face masks and practising social distancing to protect themselves against the Coronavirus disease (COVID-19). The need for contactless biometric systems has increased to avoid the common point of contact. Among contactless biometric systems, facial recognition systems are the most economical and effective ones. Conventional face recognition systems rely heavily upon the facial features of the eyes, nose, and mouth. But due to people wearing face masks, the important facial features of the nose and mouth get hidden under the mask, resulting in degraded performance by the facial recognition systems on masked faces. In this paper, we propose a Dense Residual Unit (DRU) aided with Quadruplet loss on top of existing facial recognition systems. This solution tries to unveil the masked faces by producing embeddings for masked faces, which are similar to embeddings of unmasked faces of the same identity but different from embeddings of different identities. We have evaluated our method using two pre-trained facial recognition models' backbones, i.e. Resnet-101 and MobileFaceNet, and upon two datasets, among them, one is a real-world dataset, i.e. MFR2, and one is a simulated masked face dataset i.e. masked version of LFW. We have achieved improvement in the performance of masked face recognition in terms of False Match Rate, False Non-Match Rate, Fisher Discriminant ratio, and Equal Error Rate.

Keywords: Biometric · Masked face recognition · Masked face verification · Quadruplet loss

1 Introduction

The Coronavirus disease (COVID-19) has affected the lives of people around the world in several ways as people were advised to cover their faces, avoid points of frequent contact and practice social distancing to avoid the negative impact of the pandemic. These restrictions severely impacted the biometric systems. For example, fingerprint scanner systems that require contact were left useless to avoid the point of common contact. In these special circumstances, contactless

W. Q. Yan et al. (Eds.): IVCNZ 2022, LNCS 13836, pp. 345–360, 2023.
https://doi.org/10.1007/978-3-031-25825-1_25

biometric systems came to the rescue. Among all available contactless biometric systems, face recognition systems are the most efficient and economical. Due to this, facial recognition systems are being used in border control, surveillance, person verification, and authentication scenarios. The majority of face recognition systems are based upon computer vision techniques and require all facial features like the forehead, eyes, nose, and mouth to be visible for accurate working. According to the COVID-19 Standard Operating Procedures (SOPs), people are wearing face masks, and it is advised not to remove masks even for a short period due to the chances of infection during that period. However, due to facial masks, important facial features of the nose and mouth get hidden, thus making it difficult for facial recognition systems to work accurately. The National Institute of Standards and Technology (NIST) 2020 reports on the performance of facial recognition systems in pre [17] and post-COVID era [18] have reported the degradation in the performance of facial recognition systems due to the presence of mask on faces. Moreover, according to Damer et al. [6], there is more error in genuine pair decisions than imposter pair decisions. Several other studies have shown adverse effects on facial recognition performance due to masked faces [5,8]. So, given the findings of the studies mentioned above, urgent improvement is required in performance of existing facial recognition algorithms.

We have proposed a novel approach for improving the masked face recognition performance of existing facial recognition algorithms in this paper. We have designed our solution to utilise the results (embeddings) from existing facial recognition models and do some post-processing on masked faces to improve the overall system's performance. The basic idea for this approach is that when a person wears a face mask, the facial feature gets hidden under the mask, and the conventional face recognition systems produce corrupted facial embeddings. Our system tries to make the facial embeddings for masked faces similar to the facial embeddings of unmasked faces of the same person, whereas different from facial embeddings of unmasked faces of another person. Further, we have trained this system using Quadruplet loss which is a novel idea, as previously, Triplet loss or its variations have been used for facial recognition tasks. Also, it can be seen in the experimental evaluations section that there is an improvement in masked recognition performance in terms of False Non-Match Rate (FNMR) over Triplet loss by using Quadruplet loss.

In the rest of the paper, we shall first discuss related work in Sect. 2. Then we shall present our proposed methodology and Quadruplet loss in Sect. 3. After it, the implementation will be discussed in Sect. 4, followed by results in Sect. 5. Then we shall discuss the results in Sect. 6. In the end, we shall present the conclusion of our proposed approach and findings in Sect. 7.

2 Literature Review

Face Recognition is quite a mature field but still there are challenging scenarios like illumination, occlusion, and pose where its performance degrades. Several studies have been conducted on improving face recognition performance in occlusion scenarios in recent years. However, these studies do not specifically handle face masks; they handle general occlusion like sunglasses, sun shades, and other tiny objects which come in front of face [10,19,20].

The COVID-19 outbreak compelled people to wear face masks, resulting in degraded performance by conventional facial recognition algorithms. NIST 2020 report on the performance of face recognition algorithms on masked faces concluded that the False Non-Match Rate has increased in all the evaluated algorithms [17]. In addition, Damer et al. reported the presence of greater error in genuine pair decisions compared to imposter pair decisions [5,6]. The biggest challenge with masked face recognition is the absence of a large masked face dataset. Several mask augmentation techniques [1,17] have been proposed to solve this problem. These techniques generate a masked version of the unmasked dataset. However, Anwar et al. [1] have developed a small real-world masked face recognition dataset, i.e. MFR2, which has 269 unmasked and masked images of 53 identities. Wang et al. have developed another dataset, i.e. RMFRD, which contains masked and unmasked images of Asian faces. But the problem with this dataset is that the ratio of masked images to unmasked images of the same person is very small. Further, some identities do not have masked faces [21].

Maharani et al. [13], and Muhi et al. [15] have proposed a transfer-learning based approach for masked face recognition. But their solution is limited to just the evaluation of simulated masked face datasets or very small self-collected real-world masked face datasets, which cannot be generalized to real-world scenarios. Li et al. [12] have proposed a combination of attention mechanism and a cropping-based approach to focus only on the non-occluded region of the face. In this way, the occluded region of the face is discarded, and the network is trained to process only non-occluded regions. Boutros et al. [2], have proposed a variation of the Triplet loss function, i.e. Self Restraint Triplet Loss (SRTL), to train a three-layer neural network. Nawshad et al. [16], have proposed a skip connection based dense unit trained with SRTL. [2] and [16] have conducted several ablation studies using different backbones and datasets. They have reported improvement in the separability of genuine and imposter pair scores. Malakar et al. [14] have proposed a reconstruction-based approach using Principal Component Analysis (PCA) for regenerating the masked occluded face region. Then, a deep learning algorithm is applied to recognize the identity. They have achieved a 15% improvement in performance.

3 Methodology

In this section, we present our proposed methodology for improvement in the masked face verification performance of the existing face recognition algorithms. Our proposed approach operates upon the embeddings resulting from the existing state-of-the-art (SoTA) face recognition models. The advantage of using this approach is, no need to modify the architecture of existing face recognition models. In addition to this, we do not require retraining of already trained existing facial recognition models. We have developed a Dense Residual Unit (DRU) architecture that takes masked face feature embeddings as input from existing face recognition models. Then it tries to unveil the face covered with the mask by making the masked face feature embeddings similar to the unmasked face feature embeddings of the same person and dissimilar to the unmasked face feature embeddings of a different person. Then it outputs refined feature embeddings.

We have also proposed the Quadruplet loss for training the DRU. Quadruplet loss is similar to Triplet loss as it shares similar learning objectives. It tries to minimize the distance between unmasked and masked face embeddings of the same person (minimizing the intra-class variation) and maximize the distance between unmasked and masked face embeddings of a different person (maximizing the inter-class variation). But, it is different from Triplet loss because, instead of a single negative example, it takes two negative examples and tries to maximize the inter-class distance. We shall discuss the motivation of Quadruplet loss and its functional formula later in this section.

3.1 Dense Residual Unit

Dense Residual Unit architecture takes M-dimensional feature embeddings from the backbone, i.e. pre-trained face recognition network. The value of M is 512 for the Resnet and 128 for the MobileFacenet backbone. The M-dimensional input is passed through Dense Unit Blocks repeated N times. N is the hyper-parameter; in our case, N is kept equal to 3. The value of N is determined through rigorous experimentation. Each Dense Unit block has L number of fully connected layers consisting of M number of neurons each (the value of M depends upon the Resnet or MobileFacenet used). The L is also a hyper-parameter; in our case, L is kept equal to 2. The value of L is determined through rigorous experimentation. A skip connection is introduced to the output of each Dense Unit Block from the input of that Dense Unit Block. Therefore, in summary, the input embedding is passed through N number of Dense Unit Blocks followed by an M-dimensional fully connected layer in the end. The output of DRU is M-Dimensional embedding for the original masked face embedding, which is similar to the unmasked embedding of the same person and different from the unmasked facial embedding of an other person. The complete architecture is shown in Fig. 1.

3.2 Triplet Loss

A Triplet loss helps in learning the discriminative facial embeddings. Let B is set of all training batches, and $b \in B$ is a batch of training images. $e(b)$ is the embedding generated from the pre-trained face recognition model backbone. A Triplet loss requires a triplet of sample in the form of $\{b_k^a, b_k^p, b_k^n\} \in B$, where b_k^a is anchor, b_k^p is a positive example of the same identity as anchor and b_k^n is the negative example of a different identity. For a mini-batch of size S, mathematically Triplet loss is defined as:

$$L_t = \frac{1}{S} \sum_{k}^{S} max([d(e(b_k^a), e(b_k^p)) - d(e(b_k^a), e(b_k^n)) + m], 0) \tag{1}$$

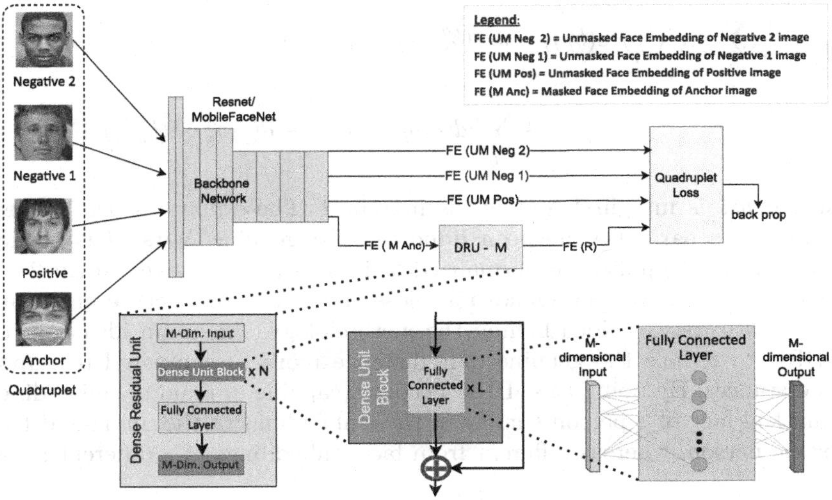

Fig. 1. The workflow of the proposed DRU with pre-trained face recognition model backbone and Quadruplet loss. The DRU processes the M-dimensional masked face embeddings to produce embeddings similar to the unveiled face embeddings of same person and dissimilar from unveiled face embeddings of different person. The value of M is 512 for Resnet and 128 for MobileFaceNet backbone. The Dense Unit Block inside DRU has N times repetitions and this is hyper-parameter. In our case, the value of N is kept equal to 3. Further, each Dense Unit Block has L number of fully connected layer and it is also hyper-parameter whose value is kept equal to 2.

where m is the margin which introduces separability between the genuine and the imposter pairs. d is the euclidean distance between two points i.e. a and b. It is mathematically defined as,

$$d(a_k, b_k) = ||a_k - b_k||^2 \tag{2}$$

3.3 Quadruplet Loss

Quadruplet loss helps in learning discriminative facial embeddings. Let B is set of all training batches, and $b \in B$ is a batch of training images. *e(b)* is the embedding from the pre-trained face recognition model. Quadruplet loss requires a quadruplet of sample in the form of $\{b_k^a, b_k^p, b_k^{n1}, b_k^{n2}\} \in B$, where b_k^a is anchor, b_k^p is a positive example of the same identity as the anchor, b_k^{n1} is the first negative example, and b_k^{n2} is second negative example of a different identity. The learning objective of the Quadruplet loss function is to reduce the distance of anchor e(b_k^a) and positive e(b_k^p), i.e. genuine pair and maximize the distance of anchor e(b_k^a) and first negative example e(b_k^{n1}), i.e. first imposter pair as well as anchor e(b_k^a) and second negative example e(b_k^{n2}), i.e. second imposter pair. Mathematically, Quadruplet loss L_q for a mini-batch of size S is defined as:

$$L_q = \frac{1}{S}(\sum_{k}^{S}[d(e(b_k^a), e(b_k^p)) - d(e(b_k^a), e(b_k^{n1})) + m1] + \sum_{k}^{S}[d(e(b_k^a), e(b_k^p)) - d(e(b_k^a), e(b_k^{n2})) + m2]) \quad (3)$$

where *m1* is margin 1 and *m2* is margin 2. These margins are applied to increase the separability between imposter and genuine pairs. d is euclidean distance which is applied on normalized facial feature. d is given in Eq. 2

Now that we have seen Quadruplet loss and DRU architecture in detail, we can look into the core idea behind the methodology. The main idea is to train the DRU to reduce the genuine pair distance more as compared to imposter pairs distances. By doing this, DRU becomes capable of making embeddings of the masked face of a person similar to the embeddings of the unmasked face of the same person, whereas different from face embeddings of a different person.

4 Materials and Implementation

In this section, we present the implementation details for the conducted experiments.

4.1 Pre Trained Face Recognition Model Backbone

First, we compute and save the embeddings of face images using a pre-trained face recognition model backbone. We have evaluated the performance of our proposed method using two pre-trained face recognition model backbones, i.e. Resnet-101 [9] and MobileFaceNet [4]. These backbones are chosen due to their proven performance in facial recognition applications. Resnet-101 has been used by ArcFace [7], VGGFace2 [3]. MobileFaceNet is a lightweight model capable of running on devices with low computational power at an incredibly fast speed without severely affecting performance. Resnet-101 produces 512-dimensional embeddings, whereas MobileFaceNet produces 128-dimensional embeddings. Both these backbones are pre-trained on the same dataset, i.e. MS1MV2 [7].

4.2 Simulated Mask Generation Technique

As discussed earlier in the literature review section, there is no sizeable masked face recognition dataset; therefore, different mask augmentation techniques have been developed [1,17] to produce a large mask dataset. We have used the mask augmentation technique proposed by Anwar et al. [1] for generating masks of different colours and patterns for the training dataset and the method by Ngan et al. [17] for generating masks of different colours and shapes for the test dataset. The reason for different augmentation techniques for training and testing datasets is to check the generalization capability of the trained model. We shall discuss the details of these datasets in the dataset section.

4.3 Datasets

Training Dataset. We have generated a simulated mask version of the MS1MV2 dataset [7] using the approach by [1]. This dataset is used for training the DRU. The dataset has 5.8 million images of 85 thousand different persons. The Mask augmentation approach is based on the python Dlib library, and it failed to extract facial landmarks of almost 429,000 images. Due to this limitation, mask augmentation could not be done on these images, and we neglected these images.

Test Dataset. We have used one simulated mask dataset, i.e. Simulated Mask LFW and one real-world mask face dataset, i.e. MFR2, to test our approach's verification performance. Simulated Mask LFW is generated using the method proposed by Ngan et al. [17]. LFW [11] contains 13,233 images of 5,749 identities. We have randomly generated 3000 pairs for genuine and 3000 pairs for imposter comparison. MFR2 is a real-world masked face dataset and contains 269 images of 53 identities [1]. We have tested the verification performance on the MFR2 dataset by *N:N* comparison between all unmasked and masked images.

4.4 Experimental Settings

We have used seven different experimental settings for each evaluated dataset. The first one is Unmasked Reference and Unmasked Probe (UMR-UMP), the baseline experiment. The second is Unmasked Reference and Masked Probe (UMR-MP). Here the unmasked reference is compared to a masked probe without being processed additionally by the DRU. It is the baseline for comparing DRU performance on unmasked reference and masked probe scenarios. The third is Unmasked Reference and Masked Probe (UMR-MP(T)). Here masked probe is processed additionally with the DRU trained with Triplet loss before being compared with the unmasked reference. Fourth is Unmasked Reference and Masked Probe (UMR-MP(Q)). Here, the masked probe is processed additionally with DRU trained with Quadruplet loss before being compared with the unmasked reference. Fifth is Masked Reference and Masked Probe (MR-MP). It is the baseline for comparing the DRU performance. Sixth is Masked Reference and Masked Probe (MR-MP(T)). Here masked probe and masked reference are processed additionally with the DRU trained with Triplet loss before comparison. Seventh is Masked Reference and Masked Probe (MR-MP(Q)). In this experimental setting, Masked Reference and Masked Probe are processed additionally with the DRU trained with Quadruplet loss. For ease of reading, all seven experimental settings with their elaborations are listed in Table 1.

4.5 Model Training Setup

We have used the python Pytorch library for the implementation. All the layer weights in DRU are randomly initialized with small but distinct numbers sampled from zero mean and unit standard deviation for symmetry breaking. We have trained the DRU with the help of already extracted M-dimensional

Table 1. Experimental setup summary

Abbreviation	Elaboration
UMR-UMP	Unmasked Reference and Unmasked Probe, both not processed additionally with DRU
UMR - MP	Unmasked Reference and Masked Probe, with Masked Probe not processed additionally with DRU
UMR - MP (T)	Unmasked Reference and Masked Probe, with Masked Probe processed additionally with DRU trained with Triplet loss
UMR - MP (Q)	Unmasked Reference and Masked Probe, with Masked Probe processed additionally with DRU trained with Quadruplet loss
MR - MP	Masked Reference and Masked Probe, both not processed additionally with DRU
MR - MP (T)	Masked Reference and Masked Probe, both processed additionally with DRU trained with Triplet loss
MR - MP (Q)	Masked Reference and Masked Probe, both processed additionally with DRU trained with Quadruplet loss

MS1MV2 dataset facial embeddings from pre-trained backbones (as discussed in Sect. 4.1). DRU is trained with the help of Triplet loss and Quadruplet loss to compare the loss function's performance. For the triplet loss, the margin m is kept equal to 1.0; for the quadruplet loss, the value of margin $m1$ is kept equal to 2.0, and margin $m2$ is kept equal to 1.9. With each loss, DRU is trained for 20 epochs with an initial learning rate of 0.1, learning rate decay step size of 9, and learning rate decay factor of 0.1. The optimizer is set to Stochastic Gradient Descent (SGD). We have trained the model on NVIDIA GeForce RTX 2080 Ti with 12 GB of RAM.

Quadruplet and Triplet Selection Method. We need to select quadruplets and triplets to train the model with Quadruplet and Triplet losses. For each quadruplet selection, we randomly select an anchor image b_k^a. This anchor image is a masked face image of an identity. Secondly, an unmasked face image of the same identity is selected as a positive sample, i.e. b_k^p. The anchor image is not necessarily the masked augmented version of the same identity, and it could be any masked image of the same identity. Then two negative images, i.e. b_k^{n1} and b_k^{n2}, are selected randomly. These are unmasked face images of two different identities other than the identity in the anchor image. For the triplet selection, the process is similar to the quadruplet selection, but the difference is the absence of a second negative example, i.e. b_k^{n2}.

4.6 Evaluation Metrics

We have evaluated the verification performance of our approach using False Match Rate (FMR100 and FMR1000), False Non-Match Rate (FNMR), and

Equal Error Rate (EER). FMR100 is the lowest FNMR when $FMR \leq 1.0\%$ and FMR1000 is the lowest FNMR when $FMR \leq 0.1\%$. The EER is the point at which the False Match Rate is equal to False Reject Rate. Along with it, we have also reported FMR100 and FMR1000 operation thresholds, i.e. $FMR100\_Th^{UMR-UMP}$ and $FMR1000\_Th^{UMR-UMP}$. We give corresponding FMR, FNMR, and average of FMR and FNMR at these thresholds for each of the seven experimental setups based on these operation thresholds. The idea behind these metrics is to simulate a situation where the threshold is set based on UMR-UMP performance. We have also calculated the Fisher Discriminant Ratio (FDR) for studying the separability of imposter and genuine scores. A greater value of FDR means better separation between genuine and imposter scores.

5 Experimental Evaluations

This section presents the experimental results achieved for our designed experiments on the proposed architecture. Experimental results for LFW and MFR2 are shown in Table 2 and 3, respectively. Both tables contain results of Resnet-101 and MobileFaceNet backbones against seven experimental settings.

Table 2. The achieved verification performance of Resnet-101 and MobileFaceNet backbones on LFW dataset with and without additional processing by Dense Residual Unit (DRU) trained with Triplet loss (T) and Quadruplet loss (Q). The best evaluation metrics i.e. lowest EER, lowest average error of FMR100 and FMR1000 at pre-defined threshold and highest FDR are written in bold for each evaluation experiment i.e. UMR-UMP (Unmasked Reference Unmasked Probe), UMR-MP (Unmasked Reference Masked Probe), MR-MP (Masked Reference Masked Probe). The $FMR100\_Th^{UMR-UMP}$ is 0.175 and 0.251 for Resnet-101 and MobileFacenet respectively. Whereas $FMR1000\_Th^{UMR-UMP}$ is 0.243 and 0.349 for MobileFacenet respectively. Significant improvement in verification performance can be observed clearly by our proposed DRU unit trained with Quadruplet loss approach.

LFW/ backbone	Experiments	EER%	FMR 100%	FMR 1000%	$FMR100\_Th^{UMR-UMP}$			$FMR1000\_Th^{UMR-UMP}$			G-mean	I-mean	FDR
					FMR%	FNMR%	Avg%	FMR%	FNMR%	Avg%			
R-101	UMR-UMP	0.3	0.3	0.3	1.0	0.3	0.65	0.1	0.3	0.2	0.7192	−0.0004	34.5764
	UMR-MP(T)	1.6	1.8	3.8	0.7	2.4	1.55	0.2	3.6	1.9	0.3958	0.0004	10.751
	UMR-MP	0.8	0.8	1.0	0.5	0.8	0.65	0.1	1.1	**0.6**	0.5247	−0.0003	13.9739
	UMR-MP(Q)	**0.7**	**0.7**	**0.7**	0.5	0.7	**0.6**	0.6	0.7	0.65	0.6078	0.0002	**21.8056**
	MR-MP(T)	1.9	3.2	9.1	27.7	0.4	14.05	21.3	0.4	10.85	0.6314	0.1026	9.8238
	MR-MP	0.7	0.7	1.0	1.8	0.6	1.2	0.7	0.7	0.7	0.606	0.0089	15.5591
	MR-MP(Q)	**0.5**	**0.5**	**0.7**	1.8	0.5	**1.15**	0.8	0.5	**0.65**	0.6471	0.0142	**19.8189**
MFN	UMR-UMP	0.7	0.7	1.0	1.0	0.7	0.85	0.1	1.0	0.55	0.6838	0.0062	19.218
	UMR-MP(T)	6.1	19.0	36.0	1.0	19.5	10.25	0.0	36.7	18.35	0.2989	−0.0016	4.929
	UMR-MP	3.8	6.4	15.1	1.1	6.4	3.75	0.3	11.7	6.0	0.4464	0.0035	6.4567
	UMR-MP(Q)	**2.0**	**2.8**	**4.4**	1.5	2.1	**1.8**	0.2	4.3	**2.25**	0.522	0.0047	**10.4526**
	MR-MP(T)	7.6	28.9	66.6	92.5	0.1	46.3	84.0	0.2	42.1	0.7253	0.3937	4.1969
	MR-MP	3.6	6.6	17.6	7.6	1.8	4.7	3.1	4.7	3.9	0.5595	0.0508	7.3776
	MR-MP(Q)	**2.1**	**3.2**	**4.5**	3.5	1.4	**2.45**	1.3	2.6	**1.95**	0.5953	0.0195	**9.8732**

Table 3. The achieved verification performance of Resnet-101 and MobileFaceNet backbones on MFR2 dataset with and without additional processing by Dense Residual Unit (DRU) trained with Triplet loss (T) and Quadruplet loss (Q). The best evaluation metrics i.e. lowest EER, lowest average error of FMR100 and FMR1000 at pre-defined threshold and highest FDR are written in bold for each evaluation experiment i.e. UMR-UMP (Unmasked Reference Unmasked Probe), UMR-MP (Unmasked Reference Masked Probe), MR-MP (Masked Reference Masked Probe). The $FMR100\_Th^{UMR-UMP}$ is 0.172 and 0.234 for Resnet-101 and MobileFacenet respectively. Whereas $FMR1000\_Th^{UMR-UMP}$ is 0.233 and 0.312 for MobileFacenet respectively. Significant improvement in verification performance can be observed clearly by our proposed DRU unit trained with Quadruplet loss approach.

MFR2/ backbone	Experiments	EER%	FMR 100%	FMR 1000%	$FMR100\_Th^{UMR-UMP}$			$FMR1000\_Th^{UMR-UMP}$			G-mean	I-mean	FDR
					FMR%	FNMR%	Avg%	FMR%	FNMR%	Avg%			
R-101	UMR-UMP	0.0	0.0	0.0	0.9983	0.0	0.4992	0.1062	0.0	0.0531	0.7585	0.0018	45.0473
	UMR-MP(T)	5.9137	9.5975	11.7647	0.8945	9.5975	5.246	0.1156	10.8359	5.4758	0.3571	0.0031	6.3172
	UMR-MP	4.6434	7.1207	8.6687	1.0284	7.1207	4.0745	0.1339	8.0495	4.0917	0.4415	0.0005	7.4826
	UMR-MP(Q)	**4.0235**	**4.644**	**6.192**	1.3143	4.644	**2.9792**	0.1825	6.192	**3.1872**	0.4746	0.0026	**8.5917**
	MR-MP(T)	9.963	17.7122	23.2472	25.9412	6.2731	16.1072	13.4754	8.4871	10.9813	0.5697	0.1023	4.2514
	MR-MP	9.2224	12.5461	15.8672	4.5152	9.5941	**7.0546**	1.4303	11.8081	6.6192	0.546	0.0177	5.0717
	MR-MP(Q)	**8.115**	**12.1771**	**15.4982**	5.4617	8.8561	7.1589	1.6546	10.3321	**5.9934**	0.545	0.0313	**5.1486**
MFN	UMR-UMP	0.0	0.0	0.0	0.9983	0.0	0.4992	0.1062	0.0	0.0531	0.7512	0.0057	30.2234
	UMR-MP(T)	8.9768	25.0774	48.9164	0.4442	34.3653	17.4048	0.0061	65.9443	32.9752	0.274	0.0022	3.3391
	UMR-MP	7.730	15.7895	29.7214	0.6876	17.3375	9.0125	0.0791	30.031	15.055	0.3788	-0.0029	4.5255
	UMR-MP(Q)	**7.721**	**12.3839**	**21.0526**	0.8397	13.0031	**6.9214**	0.0974	21.3622	**10.7298**	0.4104	0.0003	**4.9714**
	MR-MP(T)	15.1962	55.7196	90.7749	90.8294	1.476	46.1527	75.2296	2.952	39.0908	0.6777	0.413	1.9174
	MR-MP	9.5927	21.4022	**31.7343**	6.268	11.8081	9.038	1.4934	17.7122	9.6028	0.5005	0.0597	3.7232
	MR-MP(Q)	**9.0168**	**16.9742**	**31.7343**	4.4871	11.4391	**7.9631**	0.9816	17.3432	**9.1624**	0.499	0.0386	**3.7944**

5.1 Degradation of Existing Face Recognition Model Performance Due to Mask

There is a degradation in masked face recognition performance for the existing face recognition model backbone. This degradation can be observed if we compare the UMR-UMP and UMR-MP experiments of Resnet-101 and MobileFaceNet backbones on LFW and the MFR2 datasets as shown in Table 2 and 3. We can observe an increase in EER, FMR, and FNMR and a decrease in FDR, which indicates the degradation of the verification performance of the existing face recognition model.

5.2 Improvement in Verification Performance with the Addition of DRU Trained with Quadruplet Loss

It is evident from comparing the UMR-MP and UMR-MP(Q) as well as MR-MP and MR-MP(Q) experiments for both LFW and MFR2 datasets as shown in Table 2 and 3 that there is a reduction in EER, FMR, FNMR, and an increase in FDR, which shows improvement in verification performance of model due to addition of DRU trained with Quadruplet loss.

5.3 Comparison of Triplet and Quadruplet Loss Performance

It is evident from UMR-MP(T) and UMR-MP(Q), MR-MP(T), and MR-MP(Q) experiments for both LFW and MFR2 datasets as shown in Table 2 and 3 that DRU trained with Quadruplet loss has improved the verification performance (i.e. reduced the EER, FMR, FNMR and increased FDR) more as compared to Triplet loss.

5.4 Comparison of Resnet-101 and MobileFaceNet Backbone

Overall, Resnet-101 has produced better results than MobileFaceNet in almost all the experimental settings.

5.5 Comparison of Each Backbones Performance on LFW and MFR2

It can be seen from Table 2 and 3 that Resnet-101 and MobileFaceNet have produced slightly better results on Simulated LFW as compared to the real-world masked face dataset (MFR2).

5.6 Correct vs Incorrect Results

Some of the correctly predicted and incorrectly predicted verification results by our proposed DRU trained with Quadruplet loss are shown in Fig. 2 and Fig. 3 respectively. We shall discuss these incorrectly predicted examples in the discussion section.

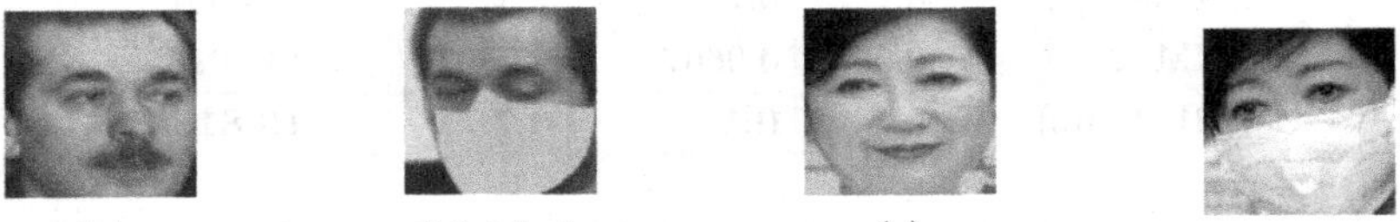

(a) Original Image (b) Mask Augmented (c) Original Image (d) Real Masked

Fig. 2. Correctly predicted verification results. 2(a) and 2(b) are images of same person and are predicted as same. Similarly 2(c) and 2(d) are images of same person and are predicted as same

5.7 Comparison with the State-of-the-Art (SoTA)

We have compared our approach performance with the available state-of-the-art models on LFW and MFR2 datasets. We have found that in the majority of experimental settings our model is performing better. The brief comparison is shown in Table 4, 5, 6 and 7.

(a) Original Image (b) Mask Augmented (c) Original Image (d) Real Masked

Fig. 3. Incorrectly predicted verification results. 3(a) and 3(b) are images of same person but are predicted different. Similarly 3(c) and 3(d) are images of same person but are predicted different.

Table 4. Comparison of our approach with SoTA approaches for Unmasked Reference and Masked Probe experimental setting on LFW dataset. We can clearly observe our proposed MFUM based approaches are performing better than existing SoTA approaches.

UMR-MP (LFW)	EER	FMR100%	FMR1000%	FDR
ArcFace [7]	0.8	0.8	1.0	13.9739
EUM [2]	0.8667	0.8667	1.6	15.0505
DRU (ours)	**0.7**	**0.7**	**0.7**	**21.8056**

Table 5. Comparison of our approach with SoTA approaches for Masked Reference and Masked Probe experimental setting on LFW dataset. We can clearly observe our proposed MFUM based approaches are performing better than existing SoTA approaches.

MR-MP (LFW)	EER	FMR100%	FMR1000%	FDR
ArcFace [7]	0.7	0.7	1.0	15.5591
EUM [2]	0.9667	0.9667	2.0667	14.6018
DRU (ours)	**0.5**	**0.5**	**0.7**	**19.8189**

Table 6. Comparison of our approach with SoTA approaches for Unmasked Reference and Masked Probe experimental setting on MFR2 dataset. We can clearly observe our proposed MFUM based approaches are performing better than existing SoTA approaches.

UMR-MP (MFR2)	EER	FMR100%	FMR1000%	FDR
ArcFace [7]	6.6434	7.1207	8.6687	7.4826
EUM [2]	5.0507	6.5015	8.9783	7.2588
DRU (ours)	**4.0235**	**4.644**	**6.192**	**8.5917**

Table 7. Comparison of our approach with SoTA approaches for Masked Reference and Masked Probe experimental setting on MFR2 dataset. We can clearly observe our proposed MFUM based approaches are performing better than existing SoTA approaches.

MR-MP (MFR2)	EER	FMR100%	FMR1000%	FDR
ArcFace [7]	9.2224	12.5461	15.8672	5.0717
EUM [2]	9.296	13.2841	**14.7601**	4.9276
DRU (ours)	**8.115**	12.1771	15.4982	**5.1486**

6 Discussion

In this section, we discuss the experimental results. The degradation of face recognition performance on masked faces is due to the corruption in facial embeddings due to the presence of face masks which cover essential facial features like nose and mouth. Further, the Resnet-101 backbone verification performance is better than the MobileFaceNet backbone. The primary reason is the number of trainable parameters of these networks. Resnet-101 has almost 66 million parameters compared to MobileFacenet, which has just 1 million parameters. In addition, Resnet-101 produces 512-dimensional embeddings, whereas MobileFaceNet produces 128-dimensional embeddings. The smaller embedding size for MobileFaceNet might be one of the reasons behind the degradation of verification performance. Further, Quadruplet loss produces better results than Triplet loss due to the better generalization capability of Quadruplet loss compared to Triplet loss. Triplet loss ability is limited to good accuracy on the train set as its focus is on obtaining correct training set results. It suffers from weak generalization capability from the training set to the test set. On the other hand, Quadruplet loss generates output with smaller intra-class variation and higher inter-class variation. Now, look into why both backbones have produced better results on LFW than the MFR2 dataset. The reason behind better results on LFW is likely because DRU is trained using masked augmented MS1MV2 dataset, therefore, has better performance on augmented mask dataset, i.e. Simulated Mask LFW. This limitation does not mean our approach does not apply to real-world masked datasets because, despite the degradation of performance, the performance is still better than when DRU is not used; unprocessed base embeddings are used, for example, in experimental cases of UMR-MP and MR-MP. In addition, despite training on simulated mask datasets, the DRU trained with Quadruplet loss has improved the verification performance of existing face recognition models on real-world mask face datasets. Now, discussing the salient feature of the DRU architecture. The DRU architecture performs better with the addition of a residual connection after each Dense Unit Block. This observation comes from the experimentation on the DRU architecture with and without the addition of the residual connection. Basically, the residual connection improves the network performance by making the gradient flow easily in the network. In addition, it makes the context of input to the shallow layer available to the deeper layer. Due to this, the deeper layer learns better, resulting in better training of the model. Further, the value of N is kept equal to 3 in the DRU. It is because increasing the N above three did not improve the performance. So if we increase N above 3, the number of model parameter increase but performance does not. Also, by reducing the N below 3, the performance degraded. Therefore, N optimal value was found to be 3. Similarly, L is kept equal to 2 in Dense Unit Block because it is found to be the most optimal value after the experimentation. Figure 3 contains some incorrectly predicted result pairs. Figure 3(a) and (b) are mispredicted differently because of the considerable variation of the pose in Fig. 3(b). We can see that the face in this image is in the side pose instead of the frontal pose. Also, Fig. 3(c) and (d) are incorrectly predicted as different because the person in Fig. 3(d) has an eye region covered with black sunglasses in addition to the nose and mouth region, which are covered with a mask. Due to this, there are not enough discriminatory facial features left for accurate facial verification.

7 Conclusion

In this paper, we have proposed DRU trained with Quadruplet loss to improve the verification performance of existing face recognition models without the need for retraining. The proposed DRU works upon the masked face embeddings produced by existing face recognition models and tries to produce unveiled face embeddings. The output of the DRU is unveiled face embedding, which is similar to unmasked face embedding of the same person and different from unmasked face embedding of a different person. We have trained the DRU with the Simulated Mask MS1MV2 dataset and tested the performance on two datasets. One is the simulated mask face dataset, i.e. Simulated Masked LFW, and the other is the real-world mask face dataset, i.e. MFR2. We were able to achieve improvement in verification performance in terms of EER, FMR, FNMR, and FDR for both the simulated mask dataset as well as the real-world mask dataset. In addition to this, Quadruplet loss has produced better results than Triplet loss due to its better generalization capability. This is because Quadruplet loss focuses on increasing the inter-class distance and reducing the intra-class distance. The future works include a collection of large real-world masked face recognition datasets for better training and evaluation purposes.

References

1. Anwar, A., Raychowdhury, A.: Masked face recognition for secure authentication. arXiv preprint arXiv:2008.11104 (2020)
2. Boutros, F., Damer, N., Kirchbuchner, F., Kuijper, A.: Self-restrained triplet loss for accurate masked face recognition. Pattern Recogn. **124**, 108473 (2022)
3. Cao, Q., Shen, L., Xie, W., Parkhi, O.M., Zisserman, A.: VGGFace2: a dataset for recognising faces across pose and age. In: 2018 13th IEEE International Conference on Automatic Face & Gesture Recognition (FG 2018), pp. 67–74. IEEE (2018)

4. Chen, S., Liu, Y., Gao, X., Han, Z.: MobileFaceNets: efficient CNNs for accurate real-time face verification on mobile devices. In: Zhou, J., et al. (eds.) CCBR 2018. LNCS, vol. 10996, pp. 428–438. Springer, Cham (2018). https://doi.org/10.1007/978-3-319-97909-0_46
5. Damer, N., Boutros, F., Süßmilch, M., Fang, M., Kirchbuchner, F., Kuijper, A.: Masked face recognition: human vs. machine. arXiv preprint arXiv:2103.01924 (2021)
6. Damer, N., Grebe, J.H., Chen, C., Boutros, F., Kirchbuchner, F., Kuijper, A.: The effect of wearing a mask on face recognition performance: an exploratory study. In: 2020 International Conference of the Biometrics Special Interest Group (BIOSIG), pp. 1–6. IEEE (2020)
7. Deng, J., Guo, J., Xue, N., Zafeiriou, S.: ArcFace: additive angular margin loss for deep face recognition. In: Proceedings of the IEEE/CVF Conference on Computer Vision and Pattern Recognition, pp. 4690–4699 (2019)
8. Fu, B., Kirchbuchner, F., Damer, N.: The effect of wearing a face mask on face image quality. In: 2021 16th IEEE International Conference on Automatic Face and Gesture Recognition (FG 2021), pp. 1–8. IEEE (2021)
9. He, K., Zhang, X., Ren, S., Sun, J.: Deep residual learning for image recognition. In: Proceedings of the IEEE Conference on Computer Vision and Pattern Recognition, pp. 770–778 (2016)
10. Huang, B., et al.: When face recognition meets occlusion: a new benchmark. In: ICASSP 2021–2021 IEEE International Conference on Acoustics, Speech and Signal Processing (ICASSP), pp. 4240–4244. IEEE (2021)
11. Huang, G.B., Mattar, M., Berg, T., Learned-Miller, E.: Labeled faces in the wild: a database for studying face recognition in unconstrained environments. In: Workshop on Faces in 'Real-Life' Images: Detection, Alignment, and Recognition (2008)
12. Li, Y., Guo, K., Lu, Y., Liu, L.: Cropping and attention based approach for masked face recognition. Appl. Intell. **51**(5), 3012–3025 (2021). https://doi.org/10.1007/s10489-020-02100-9
13. Maharani, D.A., Machbub, C., Rusmin, P.H., Yulianti, L.: Improving the capability of real-time face masked recognition using cosine distance. In: 2020 6th International Conference on Interactive Digital Media (ICIDM), pp. 1–6. IEEE (2020)
14. Malakar, S., Chiracharit, W., Chamnongthai, K., Charoenpong, T.: Masked face recognition using principal component analysis and deep learning. In: 2021 18th International Conference on Electrical Engineering/Electronics, Computer, Telecommunications and Information Technology (ECTI-CON), pp. 785–788. IEEE (2021)
15. Muhi, O.A., Farhat, M., Frikha, M.: Transfer learning for robust masked face recognition. In: 2022 6th International Conference on Advanced Technologies for Signal and Image Processing (ATSIP), pp. 1–5 (2022). https://doi.org/10.1109/ATSIP55956.2022.9805960
16. Nawshad, M.A., Zafar, Z., Fraz, M.M.: Recognition of faces wearing masks using skip connection based dense units augmented with self restrained triplet loss. In: 24th IEEE International Multi Topic Conference 2022 (INMIC 2022). IEEE (2022)
17. Ngan, M.L., Grother, P.J., Hanaoka, K.K.: Ongoing face recognition vendor test (FRVT) part 6A: face recognition accuracy with masks using pre-COVID-19 algorithms (2020)
18. Ngan, M.L., Grother, P.J., Hanaoka, K.K.: Ongoing face recognition vendor test (FRVT) part 6B: face recognition accuracy with face masks using post-COVID-19 algorithms (2020)

19. Song, L., Gong, D., Li, Z., Liu, C., Liu, W.: Occlusion robust face recognition based on mask learning with pairwise differential siamese network. In: Proceedings of the IEEE/CVF International Conference on Computer Vision, pp. 773–782 (2019)
20. Wan, W., Chen, J.: Occlusion robust face recognition based on mask learning. In: 2017 IEEE International Conference on Image Processing (ICIP), pp. 3795–3799. IEEE (2017)
21. Wang, Z., et al.: Masked face recognition dataset and application. arXiv preprint arXiv:2003.09093 (2020)

MobileACNet: ACNet-Based Lightweight Model for Image Classification

Tao Jiang[1], Ming Zong[2](✉), Yujun Ma[1], Feng Hou[1](✉), and Ruili Wang[1]

[1] School of Mathematical and Computational Sciences, Massey University, Auckland, New Zealand
{T.Jiang,yma1,F.Hou,Ruili.Wang}@massey.ac.nz
[2] School of Computer Science, Peking University, Beijing, China
zongming@pku.edu.cn

Abstract. Lightweight CNN models aim to extend the application of deep learning from conventional image classification to mobile edge device-based image classification. However, the accuracy of lightweight CNN models currently is not as comparable as traditional large CNN models. To improve the accuracy of mobile platform-based image classification, we propose MobileACNet, a novel ACNet-based lightweight model based on MobileNetV3 (a popular lightweight CNN for image classification on mobile platforms). Our model adopts a similar idea to ACNet: consider global inference and local inference adaptively to improve the classification accuracy. We improve the MobileNetV3 by replacing the inverted residual block with our proposed adaptive inverted residual module (AIR). Experimental results show that our proposed MobileACNet can effectively improve the image classification accuracy by providing additional adaptive global inference on three public datasets, i.e., Cifar-100 dataset, Tiny ImageNet dataset, and a large-scale dataset ImageNet, for mobile-platform-based image classification.

Keywords: Lightweight CNN models · MobileACNet · Mobile-platform-based image classification · Adaptive global inference

1 Introduction

Convolutional neural networks have revolutionized many computer vision tasks (e.g., image classification, object detection) [1–4] since the success of AlexNet [5]. The general trend has been to design deeper and more sophisticated networks, even pre-train some CNN models (such as ResNet [6] and VGG [7]) to achieve higher accuracy. Such large-scale CNN networks need high computational resources beyond the capacities of low-powered edge computing devices [8,9], such as robotics, IoT devices, wearable computing devices and mobile or embedded devices [10].

Lightweight CNN models are proposed for image classification in special environments with limited memory and computational resources [11–13]. Approaches

W. Q. Yan et al. (Eds.): IVCNZ 2022, LNCS 13836, pp. 361–372, 2023.
https://doi.org/10.1007/978-3-031-25825-1_26

for lightweight CNN models can be generally categorized into either pre-trained networks compressing approaches or small networks training approaches. The former approaches usually prune the unimportant weights in a pre-trained model [14], or use a larger model to teach a smaller model by knowledge distillation [15]. The latter approaches usually design efficient small neural architectures with fewer parameters and calculations, and train the networks directly. For example, MobileNet [8] of Google significantly diminishes the number of model parameters using Depthwise Separable Convolution [10,16,17]. The accuracy of MobileNetV3 [18] is superior to that of MobileNetV2 [19] by 3.2% on ImageNet [20] dataset as it uses an inverted residual of a linear bottleneck and an extrusive excitation structure.

However, it is challenging for lightweight CNN models to improve image classification accuracy using limited memory and computing resources. For the aforementioned two categories of approaches, training small efficient neural architectures achieves higher accuracy than compressing large models. Nonetheless, lightweight CNN models, such as GhostNet [21], cannot achieve the same accuracy level as other large-scale networks [22], such as ResNet [6] and VGG [7]. Improving the accuracy of lightweight models with limited memory and computing power has become crucial for effectively applying deep learning on mobile platforms.

Moreover, the lightweight CNN model has the same limitations as the large-scale CNN model. For example, experimental results [23] show that one of the common limitations of CNN is that current CNN training approaches are insufficient to generalize the concepts. In other words, the accuracy of a model trained on regular images (i.e., original image) is considerably worse when tested on negative images (i.e., referred to an image with reversed brightness) than when evaluated on regular images [23]. Since CNN only extracts information from local nearby pixels, the global inference ability is insufficient [24]. Thus, the CNN model possesses a higher inductive bias (i.e., Locality and Translation equivariance) due to the characteristics of the local receptive field [25,25]. Due to CNN lacking the ability of global inference, the convolution procedure struggles to distinguish two similar objects [26]. It has become a crucial bottleneck in enhancing CNN model's accuracy.

To improve the traditional CNN models, ACNet (Adaptively Connected Neural Networks) [27] proposes to combine global and local inference by adaptively learning the correlating weights among the pixels of the feature map. Combining global and local inference has been adopted in traditional large CNN models. However, it has not been extensively used in lightweight models.

In this paper, inspired by ACNet, we propose MobileACNet, a lightweight CNN model with global inference for mobile-platform-based image classification. We take advantage of the adaptive combination of global inference and local inference to improve the accuracy of a lightweight CNN model (MobileNet). We propose a variant ACNet to optimize and reconstruct the Depthwise Separable Convolution and effectively avoid CNN focusing too much on local inference to improve the accuracy. Our proposed model contains an adaptive inverted residual

(AIR) module, and the AIR module possesses global inference capability compared to the inverted residuals module of MobileNet. We test our MobileACNet on three image classification datasets to verify its effectiveness.

2 Related Work

Deep neural networks (DNNs) have formed the foundation of today's computer vision systems [28–37]. Various neural networks have typically been used for wide-field tasks [31,38–45]. Several outstanding models have been proposed for image classification, and they can achieve high accuracy in the classification of image tasks. Even though Deep CNNs (such as VGG16 [7], GoogLeNet [46], ResNet50 [6]) are now used in most classic high-precision image classification models, deep CNNs have certain drawbacks such as difficulty in training and low efficiency. Increasing the overall complexity of the network causes the model to become too large to apply to production environments such as actual edge devices, resulting in most models remaining in the laboratory stage [47].

The lightweight CNN models efficiently address the difficulty of deploying CNNs on resource-constrained [9,48] devices while providing acceptable accuracy. Therefore, research on developing lightweight hardware-efficient convolutional neural networks for mobile vision applications has grown substantially in recent years. For example, SqueezeNet [9] employs a high number of 1×1 convolution kernels instead of 3×3 convolutions and 1×1 convolution to vary the number of channels in the large-scale convolutional layer input feature map to minimise the amount of calculation. This innovative concept is essential, and many works on lightweight CNN models follow the pattern.

Google made significant contributions to light-weight CNN models with MnasNet [49] and MobileNet [8]. MnasNet is based on reinforcement learning and uses a resource-constrained terminal CNN model artificial neural structure search technique, utilising ShuffleNet [48] as a reference. This model achieves high image classification and identification accuracy while addressing the CPU operation latency barrier. MobileNet employs technologies such as an inverted residual network and Depthwise Separable Convolution [10] to significantly reduce the number of model parameters and increase the receptive field simultaneously. MobileNet, based on Depthwise Separable Convolution, is essentially the extreme form of grouped convolution in ResNeXt [50]. To be specific, Depthwise Separable Convolution considers each channel as a group. As Chen et al. [51] noted, although MobileNet theoretically requires only one-eighth of the computational cost by using Depthwise Separable Convolution compared to not using traditional convolution, the model has a small drop in a precision trade-off between computational cost, model parameter size, and accuracy. Although MobileNetV3 [18] provides a 3.2% increase in accuracy with the inverted residual with linear bottleneck and the squeeze and excitation structure [52] compared to MobileNetV2 [19] in the ImageNet classification, the accuracy is not comparable to other large network models such as ResNet and VGG. The model is so small that it can be used in mobile devices, resulting in reduced model parameters with the side effect of reduced accuracy. GhostNet [21] of Huawei uses the Ghost Module to optimize the redundant feature map, thus obtaining a lightweight image classification model. This optimization helps GhostNet achieve better classification accuracy than MobileNetV3. However, the accuracy is still not at the same level as the traditional CNN networks.

In addition, both MobileNetV3 and GhostNet show another CNN model's flaw: convolutional operation collects only information from local pixels, which leads to the lack of global inference capacity in each layer inside the convolutional network [46]. Thus, for lightweight CNN models, it is hard to distinguish two similar objects in the convolution procedure [53]. We hypothesize that the problem can be effectively addressed by equipping the model with both global and local inference abilities, this can improve the robustness of the lightweight CNN models for distinguishing similar objects. Inspired by ACNet [27], we propose a module (Adaptive inverted residual module) that provides adaptive global inference capabilities to lightweight models and applies it to MobileNetV3. This module can enable lightweight CNN models to recognize similar objects, resulting in higher accuracy.

3 MobileACNet

3.1 Adaptively Connected Neural Network (ACNet)

As represented in Eq. (1), ACNet [27] can adaptively combine global and local inference for image classification by using β and γ to adjust the weight of the local CNN operation and global MLP operation (α controls the weight self-transformation). In Eq. (1), x denotes the input node (e.g., images), and y_i implies the $i-$th pixel of the feature map output node. j is the index of some possible nodes related to the $i-$th node and is in the neighbourhood $N(i)$ range. u_{ij}, v_{ij} and w_{ij} represent the learnable weights between the i-th and j-th nodes for the three different operations (self-transformation, CNN, MLP), respectively. Three scalar variables could be concluded from the equation, i.e. when the weights of α and γ are 0, the equation represents the traditional CNN; when the weights of α and β are 0, the equation represents an MLP operation; when the weights of β and γ are 0, the equation represents a 1×1 convolution operation. It contains both global inference and local inference in a single model layer. These three weights are automatically updated by backpropagation, making the equation self-adaptive to adjust the weights of local inference and global inference.

$$y_i = \alpha_i \sum_{j=i} \mathbf{x}_j \mathbf{u}_{ij} + \beta_i \sum_{j \subseteq N(i)} \mathbf{x}_j \mathbf{v}_{ij} + \gamma_i \sum_{\forall j} \mathbf{x}_j \mathbf{w}_{ij} \tag{1}$$

3.2 Our Proposed Variant of ACNet

Inspired by ACNet, we present a variant of ACNet comprising two components of self-transformation and global MLP. Our variant of ACNet can be represented in Eq. (2). The new variance of this equation is that the controlled variables are α and γ, respectively. Note that, α and γ can be simple scalar variables, which are shared across all channels [27]. Therefore, we set $\alpha + \gamma = 1$ forcibly, and $\alpha, \gamma \in [0, 1]$. As shown in Eq. (3), α is computed using a softmax function with λ_α as the control parameter. λ_α and λ_γ can be learned by the standard back-propagation (BP) in the experiments. The unique aspect of Eq. (2) is that it adaptively adjusts the weights of γ and α to balance the importance between self-transformation and global inference. By using gradient learning with two parameters (γ and α), y is the result containing global inference. To further clearly explain the difference between ACNet and our proposed the variant of ACNet, we presents the proposed variant structure of ACNet in Fig. 1. We remove the local inference part in Eq. (1) because the CNN models of MobileNet already have a solid local inference ability. Thus, we assume that the local inference in Eq. (1) has little effect on the CNN models. We also proved this conjecture through subsequent ablation experiments.

$$y_i = \alpha_i \sum_{j=i} \mathbf{x}_j \mathbf{u}_{ij} + \gamma_i \sum_{\forall j} \mathbf{x}_j \mathbf{w}_{ij} \tag{2}$$

$$\alpha = \frac{e^{\lambda_\alpha}}{e^{\lambda_\alpha} + e^{\lambda_\gamma}} \tag{3}$$

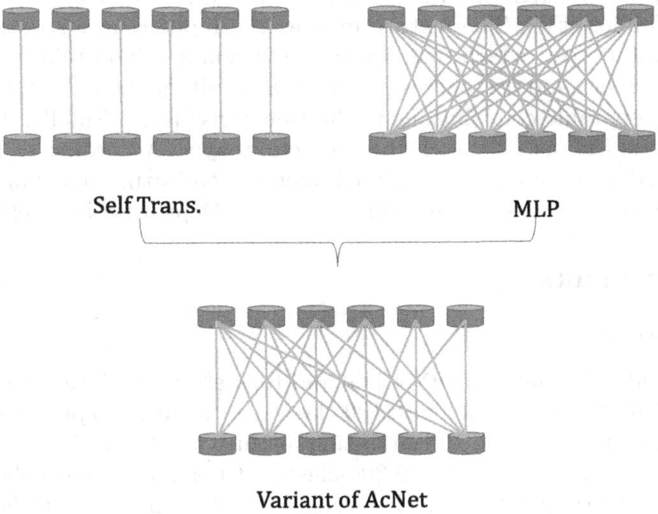

Fig. 1. Equation (2) - The variant structure of ACNet

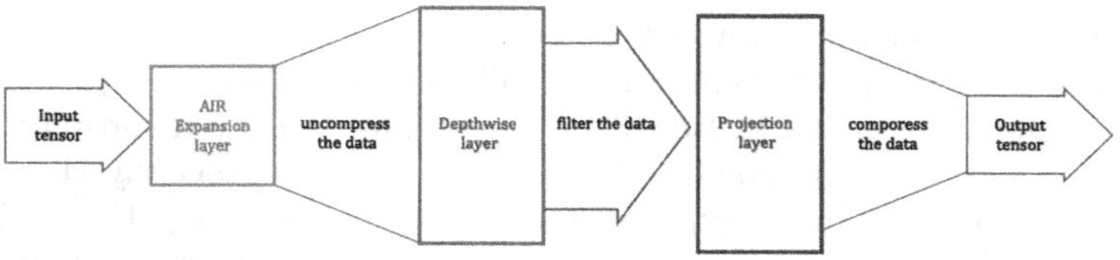

Fig. 2. Adaptive inverted residual module (AIR)

3.3 Architecture of the Proposed MobileACNet

The network structure of MobileNetV3 [18] uses the unique design method of the inverted residuals module, which uses a 1×1 convolution operation to change the inputs' dimensions. The dimensional change will lose significant information. It is the most crucial reason that MobileNet is less accurate than other large networks. In the MobileACNet, we propose an adaptive inverted residual module (AIR) in Fig. 2 which uses Eq. (2) to optimise the 1×1 residual module convolution process, and it equips the reverse module with a specific global capacity to reason. We use the global inference and the self-transformation to adaptively balance weight through two scale variables α and γ. Due to the adaptive updating weights via backpropagation, the model can preserve some useful information to the greatest extent by global inference.

We propose the AIR module to replace the inverted residual module proposed in MobileNetV2. The feature map as the input tensor goes through the AIR Expansion layer, which makes the weight adaptively obtained information by global inference and decompresses the data simultaneously. And then, it filters features by separable convolutions in the Depthwise layer. Finally, dimensionality compresses the data through the projection layer. According to Eq. (2), the proposed AIR module will add a global inference to feature maps in the Expansion layer to pass more input information to the Depthwise layer efficiently.

4 Experiments

4.1 Datasets

The Cifar-100 [54], Tiny ImageNet [55] and ImageNet [20] datasets were used in this study. Cifar-100 is divided into 100 classes, each with 600 pictures. There are 50,000 training pictures and 10,000 testing pictures among the 60,000 pictures. In Tiny ImageNet, there are 100, 000 200-class pictures (500 in each class) reduced to 64×64 pictures in colour. Each class has 500 training pictures, 50 validation pictures, and 50 test pictures. The ImageNet-1K dataset has more categories and images (10,000 categories and 1.2 million images) than Cifar-100 and Tiny-ImageNet. Two error rates were supplied according to the convention: Top 1 accuracy and Top 5 accuracy. The data photographs are uniformly cut to a predetermined size of 224×224 in the central position at the data processing stage. To increase the amount of the dataset, we randomly mirrored and flipped the images. All the implementation details and experiment settings refer to [18].

4.2 Experiments of MobileACNet

We use the same structure and parameters of MobileNetV3 (SMALL) [18] to train and test MobileACNet on Cifar-100, Tiny ImageNet and ImageNet, respectively. Table 1 compares the accuracy of MobileNetV3 with MobileACNet after the 100 epochs iterations. The MobileACNet has achieved higher accuracy compared with MobileNetV3 (SMALL) on all three datasets. On Cifar100, it gained 1.77%, while on Tiny ImageNet, it gained 0.84%. The score on ImageNet is raised by 0.99%. It is due to using the AIR module to optimize the MobileNetV3. The AIR model is adaptive with the global inference ability, which can adapt to more complex image classification problems such as distinguishing two similar objects.

Table 1. MobileACNet results compared with MobileNetV3 (SMALL)

	Datasets	Top-1 acc (%)	Top-5 acc (%)	Training time
MobileNetV3 (SMALL)	Cifar-100	65.9	88.07	2 h 43 min
MobileACNet	Cifar-100	**67.67**	**90.17**	2 h 47 min
MobileNetV3 (SMALL)	Tiny ImageNet	53.3	77.26	14 h 34 min
MobileACNet	Tiny ImageNet	**54.14**	**78.20**	14 h 45 min
MobileNetV3 (SMALL)	ImageNet	52.95	77.26	24 day
MobileACNet	ImageNet	**53.94**	**77.86**	24 day

4.3 Analysis and Discussion

Here we only discuss the training outcomes of Cifar-100 deeply. Although the AIR module can improve accuracy, it has a few side effects. The training time of the whole model is slightly prolonged because more parameters are introduced for global inference. According to Table 1, the training time of MobileACNet is 4 min longer than MobileNetV3 running on a single GPU. As shown in Table 3, MobileACNet has slightly more parameters and Multiply-Accumulate Operations (MACs) than MobileNetV3. Thus, the time and computational complexity are still within an acceptable range. For the Top-1 accuracy in Fig. 3, we can conclude that MobileACNet has a low accuracy rate initially. This is due to the large and complex redundant information to be processed in global inference. However, with the number of epochs increasing, the models of MobileNetV3 and MobileACNet converge at 31 and 32 iterations, respectively. It means the model has learned efficient variables of γ and α. The experimental results show that the AIR module effectively helps the model improve its accuracy, despite a few side effects.

We also conduct experiments with other Eq. (2) variants, including local, global inference, self-transition, and global inference. As shown in Table 2, variations of the AIR can also obtain an approximate accuracy compared to the AIR. One of the most significant improvements is using AIR in the Expansion layer, which is adaptively mixed with global inference and self-transformation.

We conclude that the Expansion layer's 1 × 1 operation filters provide vast and significantly valuable information. The disadvantage of the inverted residuals module can be effectively mitigated by adding an adaptive global inference.

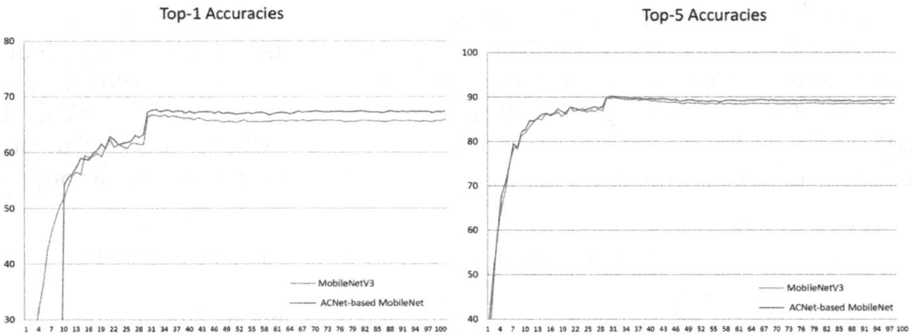

Fig. 3. The Top-1 accuracy and Top-5 accuracy of MobileACNet vs MobileNetV3 during the training process

Table 2. Variants of AIR comparison on Cifar100

Models	Top-1 accuracy (%)	Top-5 accuracy (%)
MobileACNet (self+local)	67.27	89.45
MobileACNet (local+global)	66.91	88.83
MobileACNet (local+global+self)	67.63	88.57
MobileACNet (self+global)	**67.67**	**90.17**

Table 3. Computing performance

Models	MACs	Parameters
MobileNetV3 (SMALL)	65.472M	1.843M
MobileACNet	**65.725M**	**2.007M**

5 Conclusions

In this paper, MobileACNet, a lightweight CNN model with global inference for image classification. In addition, we proposed an AIR module, which play a good role in reducing information loss caused by the change in the dimension of the inverse residual module. As a result, the proposed MobileACNet achieved higher accuracy compared with MobileNetV3 (SMALL) in image classification tasks. Although the training time of the model also increased a little, this consumption is still within the acceptable range. Thus, we believe it is acceptable to utilise some physical memory in exchange for higher image classification accuracy with rapidly developing mobile platform storage devices moving forward.

In the future, we will attempt different hyperparameters and grid searches based on MobileACNet to generate a more accurate lightweight model.

References

1. Tian, Y., et al.: Global context assisted structure-aware vehicle retrieval. IEEE Trans. Intell. Transp. Syst. (2020)
2. Tian, Y., Cheng, G., Gelernter, J., Shihao, Yu., Song, C., Yang, B.: Joint temporal context exploitation and active learning for video segmentation. Pattern Recogn. **100**, 107158 (2020)
3. Tian, Y., Zhang, Y., Zhou, D., Cheng, G., Chen, W.-G., Wang, R.: Triple attention network for video segmentation. Neurocomputing **417**, 202–211 (2020)
4. Jiang, L., et al.: Underwater species detection using channel sharpening attention. In: Proceedings of the 29th ACM International Conference on Multimedia, pp. 4259–4267 (2021)
5. Krizhevsky, A., Sutskever, I., Hinton, G.E.: Imagenet classification with deep convolutional neural networks. Commun. ACM **60**(6), 84–90 (2017)
6. He, K., Zhang, X., Ren, S., Sun, J.: Deep residual learning for image recognition. In: Proceedings of the IEEE Conference on Computer Vision and Pattern Recognition, pp. 770–778 (2016)
7. Simonyan, K., Zisserman, A.: Very deep convolutional networks for large-scale image recognition. arXiv preprint arXiv:1409.1556 (2014)
8. Howard, A.G., et al.: Mobilenets: efficient convolutional neural networks for mobile vision applications. arXiv preprint arXiv:1704.04861 (2017)
9. Iandola, F.N., Han, S., Moskewicz, M.W., Ashraf, K., Dally, W.J., Keutzer, K.: Squeezenet: Alexnet-level accuracy with 50x fewer parameters and <0.5 MB model size. arXiv preprint arXiv:1602.07360 (2016)
10. Zhou, J., Dai, H.-N., Wang, H.: Lightweight convolution neural networks for mobile edge computing in transportation cyber physical systems. ACM Trans. Intell. Syst. Technol. (TIST) **10**(6), 1–20 (2019)
11. Haque, W.A., Arefin, S., Shihavuddin, A.S.M., Hasan, M.A.: Deepthin: a novel lightweight CNN architecture for traffic sign recognition without GPU requirements. Expert Syst. Appl. **168**, 114481 (2021)
12. Valueva, M.V., Nagornov, N.N., Lyakhov, P.A., Valuev, G.V., Chervyakov, N.I.: Application of the residue number system to reduce hardware costs of the convolutional neural network implementation. Math. Comput. Simul. **177**, 232–243 (2020)
13. He, Y., Li, T.: A lightweight CNN model and its application in intelligent practical teaching evaluation. In: MATEC Web of Conferences, vol. 309, p. 05016. EDP Sciences (2020)
14. Luo, J.-H., Wu, J., Lin, W.: Thinet: a filter level pruning method for deep neural network compression. In: Proceedings of the IEEE International Conference on Computer Vision, pp. 5058–5066 (2017)
15. Shipeng, F., Li, Z., Liu, Z., Yang, X.: Interactive knowledge distillation for image classification. Neurocomputing **449**, 411–421 (2021)
16. Kaiser, L., Gomez, A.N., Chollet, F.: Depthwise separable convolutions for neural machine translation. arXiv preprint arXiv:1706.03059 (2017)

17. Chollet, F.: Xception: deep learning with depthwise separable convolutions. In: Proceedings of the IEEE Conference on Computer Vision and Pattern Recognition, pp. 1251–1258 (2017)
18. Howard, A., et al.: Searching for MobileNetV3. In: Proceedings of the IEEE/CVF International Conference on Computer Vision, pp. 1314–1324 (2019)
19. Sandler, M., Howard, A., Zhu, M., Zhmoginov, A., Chen, L.-C.: Mobilenetv 2: inverted residuals and linear bottlenecks. In: Proceedings of the IEEE Conference on Computer Vision And Pattern Recognition, pp. 4510–4520 (2018)
20. Russakovsky, O., et al.: Imagenet large scale visual recognition challenge. Int. J. Comput. Vision **115**(3), 211–252 (2015)
21. Han, K., Wang, Y., Tian, Q., Guo, J., Xu, C., Xu, C.: Ghostnet: more features from cheap operations. In: Proceedings of the IEEE/CVF Conference on Computer Vision and Pattern Recognition, pp. 1580–1589 (2020)
22. Mao, G., Anderson, B.D.O.: Towards a better understanding of large-scale network models. IEEE/ACM Trans. Netw. **20**(2), 408–421 (2011)
23. Hosseini, H., Xiao, B., Jaiswal, M., Poovendran, R.: On the limitation of convolutional neural networks in recognizing negative images. In: 2017 16th IEEE International Conference On Machine Learning And Applications (ICMLA), pp. 352–358. IEEE (2017)
24. Dua, A., Li, Y., Ren, F.: Systolic-CNN: an OpenCL-defined scalable run-time-flexible FPGA accelerator architecture for accelerating convolutional neural network inference in cloud/edge computing. In: 2020 IEEE 28th Annual International Symposium on Field-programmable Custom Computing Machines (FCCM), p. 231. IEEE (2020)
25. Battaglia, P.W., et al.: Relational inductive biases, deep learning, and graph networks. arXiv preprint arXiv:1806.01261 (2018)
26. Zou, S., Chen, W., Chen, H.: Image classification model based on deep learning in internet of things. Wirel. Commun. Mob. Comput. **2020** (2020)
27. Wang, G., Wang, K., Lin, L.: Adaptively connected neural networks. In: Proceedings of the IEEE/CVF Conference on Computer Vision and Pattern Recognition, pp. 1781–1790 (2019)
28. Liu, T., Ma, Y., Yang, W., Ji, W., Wang, R., Jiang, P.: Spatial-temporal interaction learning based two-stream network for action recognition. Inf. Sci. (2022)
29. Zong, M., Wang, R., Chen, Z., Wang, M., Wang, X., Potgieter, J.: Multi-cue based 3D residual network for action recognition. Neural Comput. Appl. **33**(10), 5167–5181 (2021)
30. Ji, W., Wang, R., Tian, Y., Wang, X.: An attention based dual learning approach for video captioning. Appl. Soft Comput. **117**, 108332 (2022)
31. Ji, W., Wang, R.: A multi-instance multi-label dual learning approach for video captioning. ACM Trans. Multimidia Comput. Commun. Appl. **17**(2s), 1–18 (2021)
32. Zong, M., Wang, R., Chen, X., Chen, Z., Gong, Y.: Motion saliency based multi-stream multiplier resnets for action recognition. Image Vision Comput. **107**, 104108 (2021)
33. Chen, Z., Wang, R., Zhang, Z., Wang, H., Lizhong, X.: Background-foreground interaction for moving object detection in dynamic scenes. Inf. Sci. **483**, 65–81 (2019)
34. Jing, C., Potgieter, J., Noble, F., Wang, R.: A comparison and analysis of RGB-D cameras' depth performance for robotics application. In: 2017 24th International Conference on Mechatronics and Machine Vision in Practice (M2VIP), pp. 1–6. IEEE (2017)

35. Wang, L., et al.: Multi-cue based four-stream 3D resnets for video-based action recognition. Inf. Sci. **575**, 654–665 (2021)
36. Liu, Z., Li, Z., Wang, R., Zong, M., Ji, W.: Spatiotemporal saliency-based multi-stream networks with attention-aware LSTM for action recognition. Neural Comput. Appl. **32**(18), 14593–14602 (2020)
37. Shamsolmoali, P., et al.: Image synthesis with adversarial networks: a comprehensive survey and case studies. In. Fusion **72**, 126–146 (2021)
38. Hou, F., Wang, R., He, J., Zhou, Y.: Improving entity linking through semantic reinforced entity embeddings. In: Proceedings of the 58th Annual Meeting of the Association for Computational Linguistics, pp. 6843–6848. Association for Computational Linguistics (2020)
39. Hou, F., Wang, R., Zhou, Y.: Transfer learning for fine-grained entity typing. Knowl. Inf. Syst. **63**(4), 845–866 (2021). https://doi.org/10.1007/s10115-021-01549-5
40. Ma, Z., et al.: Automatic speech-based smoking status identification. In: Arai, K. (ed.) Science and Information Conference, pp. 193–203. Springer, Cham (2022). https://doi.org/10.1007/978-3-031-10467-1_11
41. Ma, Z., Qiu, Y., Hou, F., Wang, R., Chu, J.T.W., Bullen, C.: Determining the best acoustic features for smoker identification. In: ICASSP 2022-2022 IEEE International Conference on Acoustics, Speech and Signal Processing (ICASSP), pp. 8177–8181. IEEE (2022)
42. Qiu, Y., Wang, R., Hou, F., Singh, S., Ma, Z., Jia, X.: Adversarial multi-task learning with inverse mapping for speech enhancement. Appl. Soft Comput. **120**, 108568 (2022)
43. Hou, F., Wang, R., He, J., Zhou, Y.: Improving entity linking through semantic reinforced entity embeddings. arXiv preprint arXiv:2106.08495 (2021)
44. Tian, Y., et al.: 3D tooth instance segmentation learning objectness and affinity in point cloud. ACM Trans. Multimedia Comput. Commun. Appl. (TOMM) **18**(4), 1–16 (2022)
45. Liu, D., Tian, Y., Zhang, Y., Gelernter, J., Wang, X.: Heterogeneous data fusion and loss function design for tooth point cloud segmentation. Neural Comput. Appl. 1–10 (2022)
46. Szegedy, C., et al.: Going deeper with convolutions. In: Proceedings of the IEEE Conference on Computer Vision and Pattern Recognition, pp. 1–9 (2015)
47. Orhan, A.E.: Robustness properties of Facebook's ResNeXt WSL models. arXiv preprint arXiv:1907.07640 (2019)
48. Ma, N., Zhang, X., Zheng, H.-T., Sun, J.: ShuffleNet V2: practical guidelines for efficient CNN architecture design. In: Proceedings of the European Conference on Computer Vision (ECCV), pp. 116–131 (2018)
49. Tan, M., et al.: MnasNet: platform-aware neural architecture search for mobile. In: Proceedings of the IEEE/CVF Conference on Computer Vision and Pattern Recognition, pp. 2820–2828 (2019)
50. Xie, S., Girshick, R., Dollár, P., Tu, Z., He, K.: Aggregated residual transformations for deep neural networks. In: Proceedings of the IEEE Conference on Computer Vision and Pattern Recognition, pp. 1492–1500 (2017)
51. Chen, H.-Y., Su, C.-Y.: An enhanced hybrid mobilenet. In: 2018 9th International Conference on Awareness Science and Technology (iCAST), pp. 308–312. IEEE (2018)
52. Hu, J., Shen, L., Sun, G.: Squeeze-and-excitation networks. In: Proceedings of the IEEE Conference on Computer Vision and Pattern Recognition, pp. 7132–7141 (2018)

53. Chandrarathne, G., Thanikasalam, K., Pinidiyaarachchi, A.: A comprehensive study on deep image classification with small datasets. In: Zakaria, Z., Ahmad, R. (eds.) Advances in Electronics Engineering. LNEE, vol. 619, pp. 93–106. Springer, Singapore (2020). https://doi.org/10.1007/978-981-15-1289-6_9
54. Krizhevsky, A., Hinton, G., et al.: Learning multiple layers of features from tiny images (2009)
55. Wu, J., Zhang, Q., Xu, G.: Tiny imagenet challenge. Technical report (2017)

GRETINA: A Large-Scale High-Quality Generated Retinal Image Dataset for Security and Privacy Assessment

Mahshid Sadeghpour(✉), Arathi Arakala, Stephen A. Davis, and Kathy J. Horadam

Mathematical Sciences, STEM College, RMIT University, Melbourne, Australia
{mahshid.sadeghpour,arathi.arakala,stephen.davis, kathy.horadam}@rmit.edu.au

Abstract. We present a generated dataset that is the largest and the first publicly shared high-quality synthetic retinal dataset. It is known that retinal patterns captured from humans are individual, even between identical twins. Despite the high accuracy and spoof resistance of retinal recognition systems, they have not reached the same level of maturity as the more popular face, fingerprint and iris. One cause is the lack of sufficient data for training and testing these systems. This paper reviews existing publicly available datasets of both real and generated retina images and identifies a lack of a large-scale high-quality retinal image dataset that can be used for security and privacy assessment. We fill this gap by using StyleGAN2-ADA to generate a synthetic dataset of five million high-quality retinal images from the limited available data.

Keywords: Synthetic retina · Retinal recognition · Generative modelling

1 Introduction

Vascular biometric traits are those biometric characteristics that contain vein patterns. These vein patterns represent the structure of blood vessels in the human body. For instance, the vascular pattern in finger vein, dorsal hand vein, palm vein, wrist vein, and retinal images could be used for the purpose of human recognition. This type of biometrics was introduced in 1983 by Joe Rice after he finished examining vascular biometric characteristics of a set of identical twins [55]. Vascular biometrics are far less deployed than the other biometric characteristics such as face, iris, and fingerprint. However, after decades of study in vascular biometrics, researchers have concluded that "vascular patterns are far better and have more entropy than fingerprints" [55].

Retinal vasculature is a light-sensitive type of vascular biometric which is located at the back of the eyeball and is not observable by the naked eye [51].

The first author was supported by an RMIT University RD Gibson Grant and an RMIT University fee-waiver scholarship.

W. Q. Yan et al. (Eds.): IVCNZ 2022, LNCS 13836, pp. 373–387, 2023.
https://doi.org/10.1007/978-3-031-25825-1_27

This modality is unique to individuals even for identical twins [51]. The vessel pattern in the images taken from healthy retinae does not change over time which makes retina suitable for human recognition [36]. There are various advantages of using vascular biometrics over conventional biometric traits, especially in highly-secured systems. These advantages differ based on the type of applied vascular biometric trait [54]. Here, we will review the advantages of applying retina.

In [54], Uhl listed the advantage of using retinal images compared to the three most popular biometric traits in deployed biometrics systems (face, fingerprint, and iris) [26]. To capture retinal images, there is no need to use NIR (Near Infrared) illumination and imaging. We can simply use VIS (Visible Light) imaging [54]. Compared to face images, retinal images are less influenced by intended changes such as make-up and occlusions (masks) or unintended changes such as aging. Compared to fingerprint images, retinal images have more entropy [55]. Vascular biometric samples in general are hard to spoof as the veins are internal to the body. In addition, scanners that have haemoglobin-based liveness detection are used, make spoofing even harder. Because the retina is placed at the back of our eyeballs, it cannot be easily damaged by accidents, unlike fingerprint and face. If compromised by reconstruction attacks, reconstructed face images reveal an individual's identity more easily than a reconstructed retina. Also, retina is considered a less invasive biometric modality. Because of the requirements on the capturing equipment and the capturing distance for retinal image acquisition, it is almost impossible to capture a retina image without the user's consent (unlike e.g., faces). Hence, many individuals prefer this biometric modality. Last but not least, it is possible to capture a retina image from up to one metre [36]. This is significant as the world battles Covid-19 and users opt in for touch-less biometrics systems. This makes retina a suitable candidate during pandemics.

The above reasons justify the development of retina as a mature biometric system. To enable this development, retinal systems must be tested on accuracy, privacy and security at scale. There is good evidence that retina is highly accurate [6,36], but these results are based on datasets of the order of hundreds and there are currently almost no privacy evaluations of such systems. The current available retinal image datasets contain at most thousands of retinal samples. Security and privacy testing against the state-of-the-art security/privacy attacks (which use deep learning techniques), require datasets of the order of millions of samples (as exist for face [42], and iris [16]). It is impractical to collect millions of real retina samples for a publicly accessible dataset as the modality is not yet widely used. This means there is a need for a large dataset of high-quality synthetic retinal images.

There have been some attempts in the literature, from medical image processing researchers in particular, to generate high-quality retinal images using generative models. However, their generated images do not have high quality (fidelity)[1]. Among existent generative modelling techniques, Generative Adversarial Networks (GANs) have superior performance in generating high-quality

[1] Quality and fidelity are used interchangeably to show the resolution and clarity of images. Quality of the images is different from their diversity (distinctiveness).

data. Thus, in this work we use a recent type of GAN which is suitable for generating high-quality retinal images with a limited number of training examples.

In the rest of this paper, publicly available retinal datasets are outlined and compared in Sect. 2. Related work is discussed in Sect. 3. Section 4 reviews the GAN used in this paper to generate the GRETINA dataset and the measures used to evaluate images generated by GANs. Then, it presents the experimental setup and experimental results. The conclusion appears in Sect. 5.

2 Publicly Available Retinal Datasets

There are quite a few public retinal image datasets available. These datasets are mostly collected for medical purposes. Hence, many of them are not suitable for security and privacy assessment. In most medical cases, the images are de-identified. So, it is not clear how many data subjects exist and how many samples each data subject has. It is possible that every sample is collected from a different subject. Figure 1 shows one random sample from each of the available retinal image datasets. Table 1 compares the datasets.

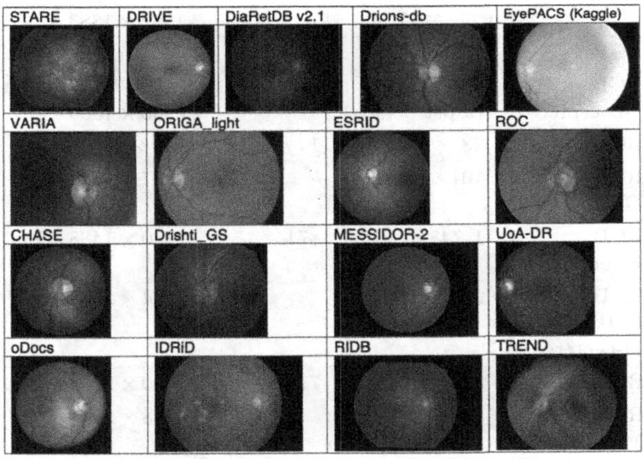

Fig. 1. One example from each of the publicly available retinal image datasets. The dataset name is noted above each image.

Capturing retinal images and collecting a large retinal dataset (with millions of samples) is not as efficient in terms of time and cost as generating a synthetic dataset using a GAN. Also, generating biometric data does not raise privacy concerns related to sensitive human data, as opposed to collecting them from individuals. However, GANs need to be trained on at least hundreds of thousands of training samples.

Table 1. Publicly available retinal datasets. In the second column of the table (Features), ONH, OD, F and M stand for the optic nerve head, optic disc, fovea, and macula, respectively. In the second last column (Purpose), VS, DR, IS, HR, GA, Opht, and MIP stand for Vessel Segmentation, Diabetic Retinopathy, Image Segmentation, Human Recognition, Glaucoma Analysis, Ophthalmology, and Medical Image Processing, respectively.

Name	Features	No. of samples	No. of data subjects	Image size (pixels)	Purpose	Ref, Year
STARE	RGB	400	-	700 × 605	VS	[24], 2000
DRIVE	RGB	40	40	584 × 565	VS	[53], 2004
DiaRetDB v2.1	RGB	89	-	1140 × 960 2240 × 1488 2304 × 1536	DR, IS	[33], 2007
Drions-db	RGB ONH-centered	110	-	600 × 400	IS	[10], 2008
EyePACS (Kaggle)	RGB	88,526	-	Various sizes	DR	[15], 2009
VARIA	Greyscale OD-centered	233	139	768 × 584	HR	[44,45] 2009
ROC	RGB	100	-	768 × 576 1058 × 1061 1389 × 1383	DR	[43] 2009
ORIGA-light	RGB	650	-	-	GA	[57], 2010
ESRID	RGB OD-centered	414 (9 per person)	46	750 × 750 2300 × 1500	Opht	[22], 2012
CHASE	RGB	28	14	999 × 960	VS	[19], 2012
Drishti-GS	RGB OD-centered	101	-	2047 × 1760	IS	[52], 2014
MESSIDOR-2	RGB F-centered	1,748	874	2240 × 1488	DR	[2,17] 2014
UoA-DR	RGB OD-centered M-centered	200	-	2124 × 2056	DR	[11], 2017
oDocs	RGB	142	71	1600 × 1200	DR	[1,34], 2018
IDRiD	RGB M-centered	597	-	4288 × 2848	DR	[47], 2018
RIDB	RGB M-centered	100	20	1504 × 1000	HR	[3], 2020
TREND	RGB M-centered	72	-	2560 × 1920	MIP	[46], 2021

3 Related Work

The medical image processing literature reports research to generate retinal images. Papers can be classified into two categories: 1) those which apply a single GAN, and 2) those which incorporate two types of generative models.

In [8, 9, 28], the authors applied a single GAN to generate their images. Beers *et al.* [8] applied ProGAN [29] to generate retinal images trained on their medical image dataset (i-ROP) with 5, 550 samples. Biswas *et al.* [9] trained a simple GAN on the MESSIDOR-1 dataset [17]. In [27, 28], the authors generated 715+89 retinal images training their GAN on the 35, 126 images in the training set of the EyePACS dataset. The vascular structure in their retina images is not clearly generated. Only the global structure of the retina, such as the circular shape, optic disc, fovea, and macula are generated. In all the above work, the generated images are not convincing, due either to the limited number of training samples, e.g. [8] (where no discriminator augmentation pipeline was used), or to the lack of progressive growing in the training process of the applied GAN, e.g. [9, 27].

In [5, 14, 21], the authors applied two types of GANs or incorporated an autoencoder with the GAN to generate higher quality retinal images. In [13, 14], an adversarial autoencoder [39] is combined with a conditional GAN [41] to generate retinal images. The adversarial autoencoder is used to generate the vessel trees. Then, a GAN is applied to map the generated vessel trees to retinal images. The authors used 946 images from MESSIDOR-1 [17] to train their model. The resolution of their generated images is 256×256 pixels. In [21], the authors used the DCGAN to generate the retinal blood vessel trees. Then, they used the Pix2PixGAN to translate the vessels into photorealistic retinal images. Recently, Anderini *et al.* [5] used a two stage GAN to generate retinal images. They used ProGAN [29] to generate vessel trees and added the color phantom to the vessels using Pix2PixGAN [25]. They used two different training sets, the DRIVE and the CHASE-DB1 datasets, for their GAN. They generated 10, 000 retinal images for each of these training sets.

An overview of the related work is available in Table 2. For easy reference and comparison purposes we included GRETINA in the last row of Table 2. Examples of the resulting generated retinal images are displayed in Fig. 2.

In addition to the above mentioned work, Zhao *et al.* [58], and Yu *et al.* [56] generated the color phantom for existing binary vascular patterns (segmented vessel trees). Hence, the vascular structure is not generated by their GAN. They only generated colorful phantoms for the vascular patterns in a way that the generated color retinal image resembles the existing vascular pattern.

In the next section, we overview a type of GAN that incorporates ProGAN and an adaptive pipeline for augmenting the images that the discriminator is trained on. We will be using this GAN for generating GRETINA. Prior to our work, it was believed that it is not possible to generate high-quality retinal images with a single GAN. "Due to the extreme variation of medical imaging data (various illuminations, noise, patterns, etc.), a single GAN is unable to produce a convincing image. The GAN is unable to determine complex structures, and is only able to identify simple features such as general color, shape, and lighting." [21] However, in this work we show that StyleGAN2-ADA is capable of generating high resolution, high fidelity retinal images that contain clear and complex vascular patterns.

Table 2. Summary of results on synthesising retinal images. NA means that the data is Not Available. All of these datasets except GRETINA are generated for medical purposes.

Reference	Training set name	Training set size	Type of GAN	No. of generated images	Resolution in pixels
Costa *et al.* 2018, [13]	MESSIDOR-1	614	Adv AE [39]+ Pix2PixGAN [25]	614 capable of generating more	256×256
Guibas *et al.* 2018, [21]	DRIVE MESSIDOR-1	40 1,200	DCGAN [48]+ Pix2PixGAN [25]	NA	NA
Beers *et al.* 2018, [8]	i-ROP	5,550	ProGAN	NA	512×512
Biswas *et al.* 2019, [9]	MESSIDOR-1	1,200	Unconditional GAN	1,000	256×256
Kaplan *et al.* 2020, [27]	EyePACS DiaRetDB	35,126 89	DCGAN [48]	715 89	64×64
Anderini *et al.* 2021, [5]	DRIVE CHASE-DB1	40 28	ProGAN [29]+ Pix2PixGAN [25]	10,000 10,000	512×512 1024×1024
GRETINA	MESSIDOR-2	832	StyleGAN2-ADA [30]	5,000,000	1024×1024

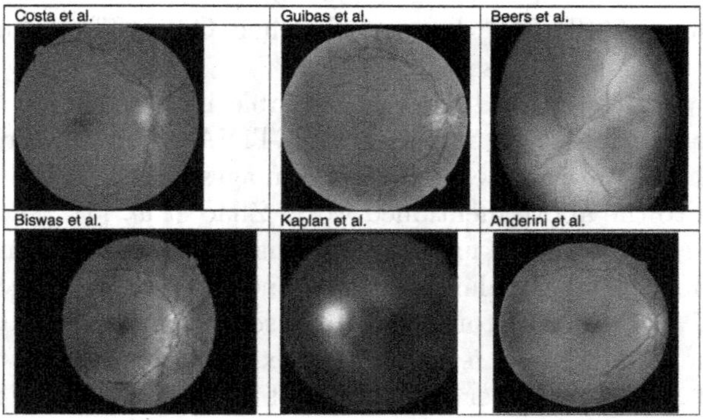

Fig. 2. One sample of the generated retinal images in the literature. The images are adopted from their published work.

4 Synthetic Retinal Data Generation Approach

4.1 Training GANs with Limited Data

Training neural networks requires a large number of training samples. In the absence of such a large training set, the training process of a GAN will diverge. Divergence results from the discriminator overfitting to the training set. When the number of training runs becomes significantly greater than the number of samples in the training set (or in the training batches), the training samples go through the discriminator multiple times. As the number of times that the discriminator sees the training set samples increases, the discriminator starts to memorise the training set (the discriminator overfits) instead of learning the patterns in the training images. Hence, after training reaches the point where discriminator overfitting occurs, the feedback of the discriminator becomes meaningless (becomes 100% fake to all generated images) to the generator and training starts to diverge.

A solution to the problem of training GANs with limited data was introduced in [30] by Karras *et al.*. In their work, StyleGAN2-ADA (SG2-ADA) was introduced, employing an augmentation strategy for expanding the training set, which prevents the augmentations from leaking to the discriminator. What makes this GAN special is that it can be trained on small training sets. This architecture is desirable for our purpose because there exists no large public retinal image dataset. It is composed of StyleGAN2 [32] which contains progressive growing (hence, suitable for training images with high quality) and ADA (adaptive discriminator augmentation) which is a technique to augment small-sized training sets without leaking augmented data to generated images.

The main idea of the SG2-ADA architecture is designing augmentations that do not leak to the generated images. Since this type of GAN is designed to be trained on comparatively smaller datasets, it has to augment the training samples. Augmentations are normally performed using transformations such as flipping, rotations, scaling, color transformations, and image space filtering. We want to avoid leaking these transformations to the generated images. This means that we want to generate fake samples that look extremely similar to our real training samples (not to their transformed/augmented versions).

The authors of [30] showed that if the augmentation transformation is invertible, it will not leak to the generated images. So, invertibility is the key criteria for each augmentation transformation. Through exhaustive experiments, they found that as long as the probability of augmentations remains below a certain point, $P < 0.85$, leaks are not likely to happen. They proposed a heuristic method for finding the best P value. The adaptive part of the name in the SG2-ADA stems from the fact that the value of P is adapted based on the output of the discriminator using a heuristic method. In [30], P was tuned for different datasets of various sizes from less than $2K$ to $140K$ images. Their experiments showed that the same probability value, $P = 0.6$, works for all the training sets with different sizes. Thus, $P = 0.6$ can be used as the default augmentation probability to train this architecture.

4.2 Evaluating the Generated Images

In training generative adversarial networks, it is important to know when to stop the training process to avoid the discriminator from overfitting. When overfitting occurs, any sample that is not from the training set is predicted to be fake by the discriminator, even if the sample looks extremely real. This is undesirable because our aim is to generate realistic looking images that can deceive the discriminator. In addition, we need to stop training the GAN as soon as the images have fidelity and diversity enough for our goal. Training the GAN more than we require will incur extra cost. By checking the diversity of the generated images, we ensure that "mode collapse" [37] has not affected the generator. If mode collapse occurs, the generator will memorise specific modes (classes) of data that can fool the discriminator, and will only generate samples from those modes. This is not desired when training GANs. The trade-off between quality and diversity of images can be controlled by the "truncation trick" [40].

In conventional deep learning tasks such as classification, there are techniques such as early stopping [20] that can help us to determine when to stop the training. In early stopping, we stop the training process as soon as the validation error begins to rise. However, when training a GAN we do not have a validation set. We require a measure that helps us to determine when the fidelity and diversity of images are good enough to stop the training process. That is, the measure specifies that the generated images have similar quality to real images (fidelity), and the generated images look different enough from each other (diversity).

There are metrics such as Fréchet Inception Distance (FID) [23], Inception Score (IS) [7], and Precision and Recall [35,50] to evaluate the fidelity and diversity of generated images. These scores have been used in the literature to evaluate the outputs of different types of GANs. FID is commonly used and hence proper to compare the performance of different GANs on the same datasets or the same GAN on different datasets. Thus, we will use FID as the core metric to realise when to stop the training process.

4.3 Experimental Setup

The largest public retinal image dataset is the Diabetic Retinopathy dataset (called the Kaggle dataset by medical image processing researchers) which contains 88,526 images. The images in this dataset come from different models and types of cameras and contain a great number of low-quality images. Overall, the quality and size of images vary significantly among samples in this dataset. Some images contain artifacts, some are out of focus, underexposed, or overexposed. Thus, we chose the second largest dataset, MESSIDOR-2 [2,17], which contains 1,748 macula-centered retinal images. It includes samples from 874 data subjects. Each data subject has one sample from each of their left and right eyes.

We used left eye images from MESSIDOR-2 in our experiments. Since this dataset is collected from Diabetic Retinopathy (DR) examinations, it contains samples with degenerated ODs or samples with damaged retinal blood vessels. We excluded 42 examples of those samples from the left eye samples that we used. We have shared this list on our GitHub repository[2] The images in the MESSIDOR-2 dataset are centered on their macula. Hence, no registration was needed to center them on the same frame of reference. The images in MESSIDOR-2 had black borders around the retinal image area. We cropped the images to reduce the extra black border. Figure 3 presents a random sample from the MESSIDOR-2 images.

The StyleGAN2 is a modification of the StyleGAN architecture [31] designed by researchers from Nvidia. This architecture is capable of generating high resolution realistic fake images, such as face images[3]. The StyleGAN [31] includes three main components, 1) a progressive growing GAN (ProGAN) [29], 2) a noise mapping network, and 3) an adaptive instance normalisation (AdaIN). In the StyleGAN2, AdaIN is replaced with the weight demodulation technique [32].

The progressive part (ProGAN) of the StyleGAN2 requires its input images to be square with height and width being equal to powers of 2, e.g., size should be $2^n \times 2^n$ pixels, where $n = 8, \ldots, 11$. We chose $n = 10$ to generate 1024×1024 pixel images. Thus, the training set images were resized to be of size 1024×1024 pixels. The SG2-ADA require images to be of PNG format. We converted the MESSIDOR-2 images to PNG format before applying them to the SG2-ADA.

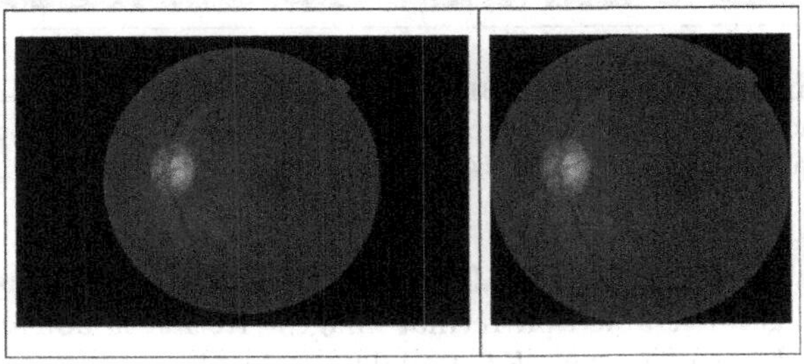

Fig. 3. Preprocessing MESSIDOR-2 images. The left side image is an example of a raw MESSIDOR-2 image. The preprocessed MESSIDOR-2 image to be used by the SG2-ADA is shown on the right side of the figure. The images are resized to 1024×1024 pixels and PNG format before being used by the SG2-ADA.

4.4 Experimental Results

We applied the SG2-ADA to generate retinal images using the 832 preprocessed MESSIDOR-2 left eye images (The same could be done for the right eye samples in MESSIDOR-2 to generate more retinal images). Training was performed for 90 hours on Google Collaborate using a 16GB Tesla P100-PCIE GPU. We generated the GRETINA dataset with 5,000,000 synthetic retinal images using the SG2-ADA with the default augmentation parameter $P = 0.6$. The FID score between

[2] https://github.com/mahshidsa/SG2-ADA-TheseRetinaeDoNotExist to support Reproducible Research (RR).

[3] https://thispersondoesnotexist.com/.

the real images and our generated images is 10.3. The truncation trick [40] with value $\psi = 0.7$ can be used to generate images with higher quality (lower diversity) in case the quality of images is more crucial than their diversity. The generated images are 1024×1024 pixel images. Figure 4 compares real MESSIDOR-2 images with our generated images from GRETINA. More examples of the generated images are available on our GitHub repository.

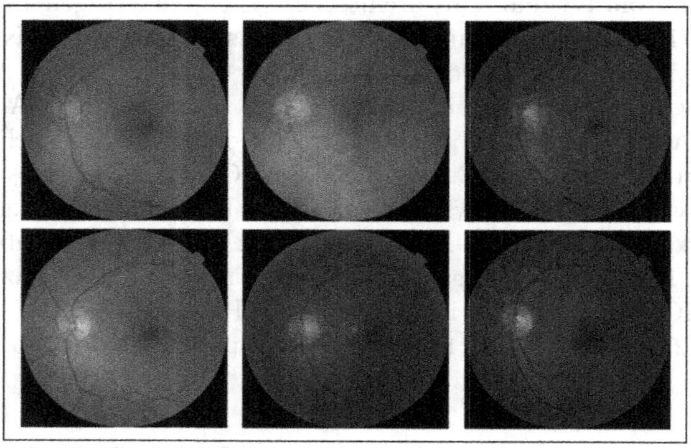

Fig. 4. Comparison of the MESSIDOR-2 with GRETINA samples. The top row shows 3 random samples from preprocessed MESSIDOR-2. The bottom row images are 3 random samples from GRETINA. The size of images is 1024 × 1024 pixels.

We are aware that because the Inception network uses the ImageNet dataset for training, it is not trained on retinal images (Retina is not among the ImageNet classes). For this reason, FID might not be the perfect measure for evaluating our generated retinal images. However, the main reason that we applied FID to evaluate the quality of our generated retinal images is that it is currently the most widely used measure for evaluating the performance of GANs. This would allow for a fair comparison of performance of different GANs on the same datasets. This is also consistent with how the developers of SG2-ADA evaluated the performance of adaptive discriminator augmentation on the BreCaHAD (Breast Cancer Histopathological Annotation and Diagnosis) dataset [4] (Cancer tissues are not among the ImageNet dataset's classes). They trained the SG2-ADA on 1944 partially overlapping crops of size 512 × 512 pixels from the BreCaHAD dataset. To evaluate the quality of their generated images they used FID. Their results shows an FID of 15.7 for their generated breast cancer histopathology images of size 512 × 512 pixels. FID is a distance measure that calculates the distance between the distributions of scores for real and generated images. Hence, a lower FID is better. Table 3 compares our results in generating retinal images training the SG2-ADA on the MESSIDOR-2 dataset with the results of [30] in generating Breast cancer images trained on the BReCaHAD dataset and dog images trained on the AFHQ-Dog dataset [12]. Dog images are included among the ImageNet classes. The FID value for our generated retina images is comparable to those of AFHQ-Dog and BreCaHAD, which shows that our generated retinal images have high quality and diversity.

It is worth mentioning that FID is biased to the number of images sampled from the real and fake distributions. The common practice is to choose $50,000$ images from reals and $50,000$ images from fakes to calculate the FID between real and fake images. We followed this standard to be consistent with [30].

We were not able to calculate the FID score for the generated retinal images in the literature since their images were not publicly available. Only Kaplan *et al.* calculated the FID of their generated images. Their reported FID value is 161.202 [27]. Kaplan *et al.* generated at most 715 retinal images and hence could not follow the common practice in calculating the FID.

Table 3. A comparison of the results of [30] in generating dog images and breast cancer tissue images with our results in generating retinal images when the augmentation probability $P = 0.6$ is used.

Image class	Training set name	Training set size	Image size	FID
Dog	AFHQ-Dog	$4,739$ samples	512×512 pixels	7.4
Breast Cancer	BreCaHAD	$1,944$ samples	512×512 pixels	15.7
Retina	MESSIDOR-2	832 samples	1024×1024 pixels	10.3

A link to the code, data, model, and the generated images (GRETINA), is shared on our GitHub repository[4]. This generated dataset can be used to assess privacy and security of retinal recognition systems: e.g., as for evaluating reversibility of unprotected deep face templates in [18,38], or as for evaluating privacy like our work conducted in [49], or as for a brute-force attack on a retinal recognition system.

5 Conclusion

In this work, we applied StyleGAN2-ADA as a single GAN trained on a limited number (832) of retinal images. This is, to the best of our knowledge, the first high-quality synthetic retinal dataset that is generated by applying a single generative model. We have generated $5,000,000$ retinal images and made our generated images and the model publicly accessible. This synthetic retinal image dataset is the largest synthetic dataset generated so far. The resolution of our generated images is 1024×1024 pixels, and the vascular structure is generated clearly. This allows feature extraction modules to extract the vasculature pattern in them effectively, as we have shown elsewhere. Our aim in making the

[4] https://github.com/mahshidsa/SG2-ADA-TheseRetinaeDoNotExist to support RR.

model and the dataset publicly available, is to draw the attention of biometrics researchers to developing privacy-preserving retinal recognition systems, since this characteristic has promising performance in recognition. Generating this dataset costs less than collecting retinal images from individuals. More importantly, by training our models using synthetic data, we can avoid the privacy concerns related to the use of individuals' personal information.

Availability of large training sets in training deep learning models is crucial. This will allow researchers to evaluate the security and privacy of retinal recognition schemes and to develop image segmentation models using deep learning techniques.

GRETINA is generated for assessing the privacy and security of a developed retinal recognition system. It can be used for presentation attack detection, too. However, a dataset suitable for developing biometric recognition systems is one that has multiple samples of each data subject to be used for training, testing, and validation. For future work, our aim is to explore the possibility of combining conditional GANs (CGANs) with SG2-ADA, and Pix2PixGAN to explore the possibility of generating a large retinal dataset that contains multiple samples of the same class which are not identical but look similar enough to be classified as being from the same fake identity. This would allow us to train deep neural networks for retinal recognition.

References

1. oDocs retinal database. https://www.odocs-tech.com/database. Accessed 12 Nov 2022
2. Abràmoff, M.D., et al.: Automated analysis of retinal images for detection of referable diabetic retinopathy. JAMA Ophthalmol. **131**(3), 351–357 (2013)
3. Akram, M.U., Salam, A.A., Khawaja, S.G., Naqvi, S.G.H., Khan, S.A.: RIDB: a dataset of fundus images for retina based person identification. Data Brief **33**, 106433 (2020)
4. Aksac, A., Demetrick, D.J., Ozyer, T., Alhajj, R.: BreCaHAD: a dataset for breast cancer histopathological annotation and diagnosis. BMC Res. Notes **12**(1), 1–3 (2019)
5. Andreini, P., et al.: A two-stage GAN for high-resolution retinal image generation and segmentation. Electronics **11**(1), 60 (2021)
6. Arakala, A., Davis, S., Horadam, K.J.: Vascular biometric graph comparison: theory and performance. In: Uhl, A., Busch, C., Marcel, S., Veldhuis, R. (eds.) Handbook of Vascular Biometrics. ACVPR, pp. 355–393. Springer, Cham (2020). https://doi.org/10.1007/978-3-030-27731-4_12
7. Barratt, S., Sharma, R.: A note on the inception score. arXiv preprint arXiv:1801.01973 (2018)
8. Beers, A., et al.: High-resolution medical image synthesis using progressively grown generative adversarial networks. arXiv preprint arXiv:1805.03144 (2018)
9. Biswas, S., Rohdin, J., Drahanský, M.: Synthetic retinal images from unconditional GANs. In: 2019 41st Annual International Conference of the IEEE Engineering in Medicine and Biology Society (EMBC), pp. 2736–2739. IEEE (2019)

10. Carmona, E.J., Rincón, M., Garcí a Feijoó, J., Martínez-de-la Casa, J.M.: Identification of the optic nerve head with genetic algorithms. Artif. Intell. Med. **43**(3), 243–259 (2008). https://doi.org/10.1016/j.artmed.2008.04.005
11. Chalakkal, R.J., Abdulla, W.H., Sinumol, S.: Comparative analysis of university of Auckland diabetic retinopathy database. In: Proceedings of the 9th International Conference on Signal Processing Systems, pp. 235–239 (2017)
12. Choi, Y., Uh, Y., Yoo, J., Ha, J.W.: StarGAN v2: diverse image synthesis for multiple domains. In: Proceedings of the IEEE/CVF Conference on Computer Vision and Pattern Recognition, pp. 8188–8197 (2020)
13. Costa, P., et al.: Towards adversarial retinal image synthesis. arXiv preprint arXiv:1701.08974 (2017)
14. Costa, P., et al.: End-to-end adversarial retinal image synthesis. IEEE Trans. Med. Imaging **37**(3), 781–791 (2018)
15. Cuadros, J., Bresnick, G.: EyePACS: an adaptable telemedicine system for diabetic retinopathy screening. J. Diabetes Sci. Technol. **3**(3), 509–516 (2009)
16. Daugman, J.: 600 million citizens of India are now enrolled with biometric ID. SPIE Newsroom, vol. 7 (2014)
17. Decencière, E., et al.: Feedback on a publicly distributed image database: the messidor database. Image Anal. Stereol. **33**(3), 231–234 (2014)
18. Dong, X., Jin, Z., Guo, Z., Teoh, A.B.J.: Towards generating high definition face images from deep templates. In: 2021 International Conference of the Biometrics Special Interest Group (BIOSIG), pp. 1–11. IEEE (2021)
19. Fraz, M.M., et al.: An ensemble classification-based approach applied to retinal blood vessel segmentation. IEEE Trans. Biomed. Eng. **59**(9), 2538–2548 (2012). https://doi.org/10.1109/TBME.2012.2205687
20. Goodfellow, I., Bengio, Y., Courville, A.: Deep Learning. MIT Press, Cambridge (2016). http://www.deeplearningbook.org
21. Guibas, J.T., Virdi, T.S., Li, P.S.: Synthetic medical images from dual generative adversarial networks. arXiv preprint arXiv:1709.01872 (2017)
22. Hao, H., et al.: Does retinal vascular geometry vary with cardiac cycle? Investig. Ophthalmol. Vis. Sci. **53**(9), 5799–5805 (2012)
23. Heusel, M., Ramsauer, H., Unterthiner, T., Nessler, B., Hochreiter, S.: GANs trained by a two time-scale update rule converge to a local nash equilibrium. In: Advances in Neural Information Processing Systems, vol. 30 (2017)
24. Hoover, A.D., Kouznetsova, V., Goldbaum, M.: Locating blood vessels in retinal images by piecewise threshold probing of a matched filter response. IEEE Trans. Med. Imaging **19**(3), 203–210 (2000). https://doi.org/10.1109/42.845178
25. Isola, P., Zhu, J.Y., Zhou, T., Efros, A.A.: Image-to-image translation with conditional adversarial networks. In: Proceedings of the IEEE Conference on Computer Vision and Pattern Recognition, pp. 1125–1134 (2017)
26. Jain, A.K., Nandakumar, K., Ross, A.: 50 years of biometric research: accomplishments, challenges, and opportunities. Pattern Recogn. Lett. **79**, 80–105 (2016)
27. Kaplan, S., Lensu, L., Laaksonen, L., Uusitalo, H.: Evaluation of unconditioned deep generative synthesis of retinal images. In: Blanc-Talon, J., Delmas, P., Philips, W., Popescu, D., Scheunders, P. (eds.) ACIVS 2020. LNCS, vol. 12002, pp. 262–273. Springer, Cham (2020). https://doi.org/10.1007/978-3-030-40605-9_23
28. Kaplan, S., et al.: Deep generative models for synthetic retinal image generation (2017)
29. Karras, T., Aila, T., Laine, S., Lehtinen, J.: Progressive growing of GANs for improved quality, stability, and variation. arXiv preprint arXiv:1710.10196 (2017)

30. Karras, T., Aittala, M., Hellsten, J., Laine, S., Lehtinen, J., Aila, T.: Training generative adversarial networks with limited data. arXiv preprint arXiv:2006.06676 (2020)
31. Karras, T., Laine, S., Aila, T.: A style-based generator architecture for generative adversarial networks. In: Proceedings of the IEEE/CVF Conference on Computer Vision and Pattern Recognition, pp. 4401–4410 (2019)
32. Karras, T., Laine, S., Aittala, M., Hellsten, J., Lehtinen, J., Aila, T.: Analyzing and improving the image quality of StyleGAN. In: Proceedings of the IEEE/CVF Conference on Computer Vision and Pattern Recognition, pp. 8110–8119 (2020)
33. Kauppi, T., et al.: The DIARETDB1 diabetic retinopathy database and evaluation protocol. In: BMVC, vol. 1, pp. 1–10 (2007)
34. Kim, T.N., et al.: A smartphone-based tool for rapid, portable, and automated wide-field retinal imaging. Transl. Vis. Sci. Technol. **7**(5), 21–21 (2018)
35. Kynkäänniemi, T., Karras, T., Laine, S., Lehtinen, J., Aila, T.: Improved precision and recall metric for assessing generative models. arXiv preprint arXiv:1904.06991 (2019)
36. Lajevardi, S.M., Arakala, A., Davis, S.A., Horadam, K.J.: Retina verification system based on biometric graph matching. IEEE Trans. Image Process. **22**(9), 3625–3635 (2013)
37. Lin, Z., Khetan, A., Fanti, G., Oh, S.: PacGAN: the power of two samples in generative adversarial networks. In: Advances in Neural Information Processing Systems, vol. 31 (2018)
38. Mai, G., Cao, K., Yuen, P.C., Jain, A.K.: On the reconstruction of face images from deep face templates. IEEE Trans. Pattern Anal. Mach. Intell. **41**(5), 1188–1202 (2018)
39. Makhzani, A., Shlens, J., Jaitly, N., Goodfellow, I., Frey, B.: Adversarial autoencoders. arXiv preprint arXiv:1511.05644 (2015)
40. Marchesi, M.: Megapixel size image creation using generative adversarial networks. arXiv preprint arXiv:1706.00082 (2017)
41. Mirza, M., Osindero, S.: Conditional generative adversarial nets. arXiv preprint arXiv:1411.1784 (2014)
42. Ngan, M.L., Grother, P.J., Hanaoka, K.K., et al.: Ongoing face recognition vendor test (FRVT) part 6B: face recognition accuracy with face masks using post-COVID-19 algorithms (2020)
43. Niemeijer, M., et al.: Retinopathy online challenge: automatic detection of microaneurysms in digital color fundus photographs. IEEE Trans. Med. Imaging **29**(1), 185–195 (2009)
44. Ortega, M., Penedo, M.G., Rouco, J., Barreira, N., Carreira, M.J.: Personal verification based on extraction and characterisation of retinal feature points. J. Vis. Lang. Comput. **20**(2), 80–90 (2009)
45. Ortega, M., Penedo, M.G., Rouco, J., Barreira, N., Carreira, M.J.: Retinal verification using a feature points-based biometric pattern. EURASIP J. Adv. Signal Process. **2009**, 1–13 (2009)
46. Popovic, N., Vujosevic, S., Radunović, M., Radunović, M., Popovic, T.: Trend database: retinal images of healthy young subjects visualized by a portable digital non-mydriatic fundus camera. PLoS ONE **16**(7), e0254918 (2021)
47. Porwal, P., et al.: IDRiD: diabetic retinopathy-segmentation and grading challenge. Med. Image Anal. **59**, 101561 (2020)
48. Radford, A., Metz, L., Chintala, S.: Unsupervised representation learning with deep convolutional generative adversarial networks. arXiv preprint arXiv:1511.06434 (2015)

49. Sadeghpour, M., Arakala, A., Davis, S., Horadam, K.: Protection of sparse retinal templates using cohort-based dissimilarity vectors. TechrXiv preprint techrxiv.20278923.v1 (2022)
50. Sajjadi, M.S., Bachem, O., Lucic, M., Bousquet, O., Gelly, S.: Assessing generative models via precision and recall. arXiv preprint arXiv:1806.00035 (2018)
51. Semerád, L., Drahanský, M.: Retinal vascular characteristics. In: Uhl, A., Busch, C., Marcel, S., Veldhuis, R. (eds.) Handbook of Vascular Biometrics. ACVPR, pp. 309–354. Springer, Cham (2020). https://doi.org/10.1007/978-3-030-27731-4_11
52. Sivaswamy, J., Krishnadas, S., Joshi, G.D., Jain, M., Tabish, A.U.S.: Drishti-GS: Retinal image dataset for optic nerve head (ONH) segmentation. In: 2014 IEEE 11th International Symposium on Biomedical Imaging (ISBI), pp. 53–56. IEEE (2014)
53. Staal, J., Abramoff, M., Niemeijer, M., Viergever, M., van Ginneken, B.: Ridge based vessel segmentation in color images of the retina. IEEE Trans. Med. Imaging **23**(4), 501–509 (2004)
54. Uhl, A.: State of the art in vascular biometrics. In: Uhl, A., Busch, C., Marcel, S., Veldhuis, R. (eds.) Handbook of Vascular Biometrics. ACVPR, pp. 3–61. Springer, Cham (2020). https://doi.org/10.1007/978-3-030-27731-4_1
55. Uhl, A., Busch, C., Marcel, S., Veldhuis, R.: Handbook of Vascular Biometrics. Springer, Cham (2020). https://doi.org/10.1007/978-3-030-27731-4
56. Yu, Z., Xiang, Q., Meng, J., Kou, C., Ren, Q., Lu, Y.: Retinal image synthesis from multiple-landmarks input with generative adversarial networks. Biomed. Eng. Online **18**(1), 1–15 (2019)
57. Zhang, Z., et al.: ORIGA-light: an online retinal fundus image database for glaucoma analysis and research. In: 2010 Annual International Conference of the IEEE Engineering in Medicine and Biology, pp. 3065–3068. IEEE (2010)
58. Zhao, H., Li, H., Maurer-Stroh, S., Cheng, L.: Synthesizing retinal and neuronal images with generative adversarial nets. Med. Image Anal. **49**, 14–26 (2018)

Texture Generation Using a Graph Generative Adversarial Network and Differentiable Rendering

K. C. Dharma[1](✉), Clayton T. Morrison[1], and Bradley Walls[2]

[1] The University of Arizona, Tucson, AZ 85721, USA
{kcdharma,claytonm}@arizona.edu
[2] Areté Associates, Tucson, AZ 85712, USA

Abstract. Novel photo-realistic texture synthesis is an important task for generating novel scenes, including asset generation for 3D simulations. However, to date, these methods predominantly generate textured objects in 2D space. If we rely on 2D object generation, then we need to make a computationally expensive forward pass each time we change the camera viewpoint or lighting. Recent work that can generate textures in 3D requires 3D component segmentation that is expensive to acquire. In this work, we present a novel conditional generative architecture that we call a *graph generative adversarial network (GGAN)* that can generate textures in 3D by learning object component information in an unsupervised way. In this framework, we do not need an expensive forward pass whenever the camera viewpoint or lighting changes, and we do not need expensive 3D part information for training, yet the model can generalize to unseen 3D meshes and generate appropriate novel 3D textures. We compare this approach against state-of-the-art texture generation methods and demonstrate that the GGAN obtains significantly better texture generation quality (according to Fréchet inception distance). We release our model source code as open source (https://github.com/ml4ai/ggan).

Keywords: 3D texture synthesis · Graph neural networks · Differentiable rendering

1 Introduction

Synthesizing novel photorealistic textures for 3D mesh models is an important task for the generation of novel scenes in static images or realistic 3D simulations. Such generated textures can be applied to 3D mesh models and rendered with different lighting conditions and camera angles quite easily. The generative adversarial network (GAN) framework [11] is a promising approach to training models capable of novel image generation. However, extending the GAN framework to support texture generation in 3D that can generalize to novel, previously unseen 3D models poses interesting challenges. Generating textures in 2D space (u, v coordinate system) and then wrapping to 3D mesh models won't generalize

W. Q. Yan et al. (Eds.): IVCNZ 2022, LNCS 13836, pp. 388–401, 2023.
https://doi.org/10.1007/978-3-031-25825-1_28

to unseen meshes because the UV mapping function is different for different 3D meshes. However, humans are able to identify the components of a 3D object and could texture them consistently. This raises an interesting research question: can we design an algorithm that can generate realistic textures for unseen 3D mesh models by identifying object components as humans do? We present here a system that addresses this challenge. Our model can learn to distinguish 3D part information shared across instances of an object class (*e.g.*, wheels, doors, hood, windows, tail, headlights, etc. of a car) in an unsupervised way that supports generating specific texture features for these components. Recent work presented a new model called TM-NET [9] that can generate textures in 3D but requires prior supervised segmentation of 3D parts, while our model can identify 3D part information in an unsupervised way. Another closely related work to ours is [34], but they work on 2.5D space rather than on the original 3D space, forcing us to make an expensive forward pass each time the viewpoint changes. We use PyTorch [30] and PyTorch3D [35] for the implementation of our system and use the ShapeNet dataset [3] to train and evaluate our framework. We adopt the commonly used Fréchet Inception Distance (FID) [15] to assess the quality of textures generated by model. FID is typically applied to 2D images. To extend the FID measure to texture map generation for 3D models, we apply the generated texture map that is to be evaluated to the given 3D model and render it from multiple views, producing multiple 2D images, and the FID scores across these images are aggregated to produce a summary FID score. The major contributions of our paper are as follows:

- We present a simple solution to the challenging problem of 3D texture generation for 3D mesh models rather than 2D or 2.5D images.
- Our framework is capable of unsupervised learning of part information shared across a class of objects, therefore avoiding a costly, separate supervised learning task in order to learn textures appropriate to object parts.
- We present a thorough review, analysis, and evaluation of various techniques that can be used for texture generation for 3D meshes and demonstrate the advantages and disadvantages of each.

2 Related Work

In this section, we describe five threads of work related to our problem and proposed framework.

2.1 Generative Adversarial Networks

Generative adversarial networks (GANs) [11] are known for their ability to generate photorealistic images with very high resolution [20]. The GAN framework consists of a generator $\mathcal{G}$ and discriminator $\mathcal{D}$. The generator $\mathcal{G}$ attempts to generate realistic images that can fool the discriminator. At the same time, the discriminator $\mathcal{D}$ tries to predict whether the image is "real" (comes from the real data distribution) or "fake" (generated by the generator). This constitutes a two-player minimax game with the following value function:

$$\mathcal{V}(\mathcal{G},\mathcal{D}) = \mathbb{E}_{\boldsymbol{x}\sim p_{data}(\boldsymbol{x})}[log D(\boldsymbol{x})] + \mathbb{E}_{z\sim p_z(z)}[log(1-\mathcal{D}(\mathcal{G}(\boldsymbol{z})))] \tag{1}$$

Here, $\boldsymbol{x}$ denotes a sample from a real distribution, p_{data}, and $\boldsymbol{z}$ denotes a "noise" vector from distribution p_z. $D(\boldsymbol{x})$ denotes the probability that $\boldsymbol{x}$ comes from the real distribution. Multiple applications have adapted GANs for the task of generating realistic images. Specifically, Radford *et al.* [33] propose deep convolutional GAN (DCGAN), which uses convolutional neural networks (CNNs) to generate low-resolution photorealistic images. Recently, Karras *et al.* [18–20] developed a novel architecture for the generator. Arjovsky *et al.* [1] propose the Wasserstein GAN (WGAN) to improve the stability of training. Xian *et al.* [39] propose image synthesis with texture, but this only works on 2D images. Mirza *et al.* [29] propose a conditional GAN framework that can generate samples from a specified class. In this framework, the noise vector is combined with a class label to generate a sample image from that class. This work is closely related to our work as we seek to generate a texture conditioned on an input mesh. This work is a simplified version of our problem as they work on 2D images and the combination of the noise vector with the class label (a one-hot vector) can be easily achieved with a simple concatenation while a simple concatenation of the noise vector with a 3D mesh model is not possible. This makes our problem challenging and requires a new architecture.

2.2 Differentiable Rendering

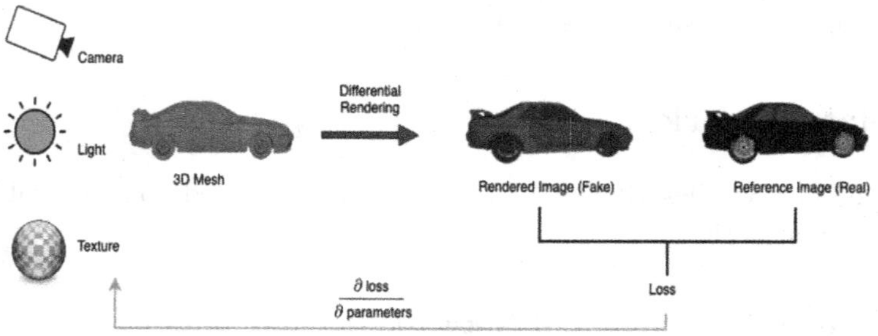

Fig. 1. Differentiable rendering

The second line of work related to ours is *differentiable rendering*. Rendering in computer graphics is a process of generating a 2D image from a 3D mesh, light source, camera properties, texture properties, and other scene properties. Classical rendering using rasterization, or ray tracing, is not differentiable. This means we cannot propagate the gradients of the loss in image space (2D space) with respect to mesh properties such as vertices and textures. Given that we want to generate a realistic texture for a given 3D mesh model with supervision from 2D images, we need a way to propagate the gradients of the loss from these projected (rendered) 2D images back to the 3D scene properties. Differentiable rendering is a process that enables backpropagating these gradients from the 2D image loss back into the 3D scene properties. Figure 1 illustrates the differentiable rendering part of our architecture. Recent methods propose approximate solutions for making the rendering process differentiable [21,25–27,35]. We use PyTorch3D [35] for differentiable rendering.

2.3 Texturing

Texturing is the process of applying a texture to a given 3D mesh model. There are multiple ways to apply a texture to a 3D mesh. Given that we want to generate textures for 3D polygonal meshes that can be applied directly to 3D shapes, we have the following options for texturing [35]:

UV Textures: A *UV texture* is a 2D image that can be mapped to 3D mesh model. For UV textures to work, we need a 2D UV-coordinate image and a mapping function that maps every vertex in the 3D object space to a (u, v) coordinate in a 2D UV image. The advantage of this method is that it can represent high-resolution textures, but the mapping function is different for different meshes, making the texturing process hard to generalize across varieties of 3D models. We refer to this method as TextureUV in the following sections.

Vertex Textures: A *vertex texture* defines a texture per vertex (e.g. r, g, b color). If the mesh has V vertices, and the dimension of texture per vertex is D, the texture can be represented by a tensor of shape (V, D) for a given 3D mesh. In this approach, the texture within faces between vertices has to be interpolated from vertex textures. This makes the vertex texture suitable only for low-resolution textures. We refer to this method as TextureVertex in the following sections.

Face Textures: The *face texture* method defines a separate texture per face. The texture per face can be an RxR dimensional texture, where R is the resolution of a texture image of a single face. This can be modeled using a (F, R, R, D) tensor where F is the number of faces, R is the resolution of a texture and D is the dimension of a texture (*e.g.*, $D = 3$: r, g, b colors). This allows us to learn very high-resolution textures. We refer to it as TextureFace in the following sections.

2.4 Deformable Models

This family of work learns to generate the 3D mesh along with textures from 2D images. The method generally starts with a fixed geometry (e.g. sphere) and a fixed UV mapping. Given input images, the model extracts information from these images, represented as a latent vector. This vector is then used to predict the deformation of the vertices of the sphere template to approximate the 3D mesh and the texture image. Then the estimated 3D shape, estimated texture, and fixed UV map go through a differentiable rendering step to generate a 2D image. The main idea then is to make these generated images similar to the original images, which can be achieved using reconstruction loss and adversarial loss. Figure 2 shows the architecture.

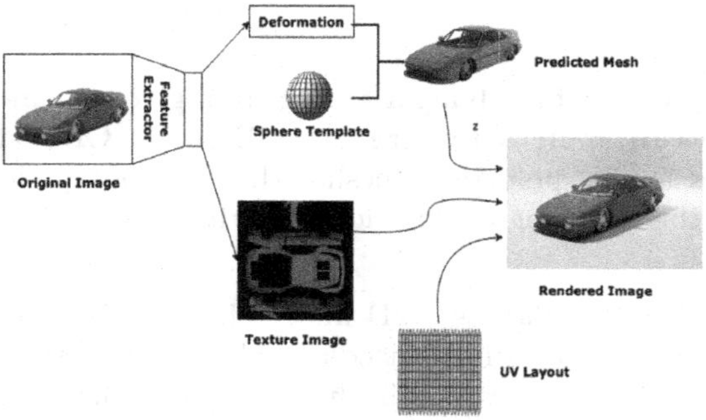

Fig. 2. General architecture of deformable models

Recent work has explored variations of this idea and has achieved good results [5,10,14,17,31,32,32,41]. This is a really good approach when we don't have ground truth 3D meshes. But, when the 3D meshes are available, as in our case, the major disadvantage of this method is that the predicted mesh tends to be relatively poor quality compared to the ground-truth mesh.

2.5 Graph Neural Networks

Graph neural networks (GNNs) are powerful models for learning from graph-structured data. They work on the theory of message passing, where a node gets some information from its neighbors and updates its state. Consider a graph $\mathcal{G} = (\mathcal{V}, \mathcal{E})$, where $\mathcal{V}$ is the set of nodes and $\mathcal{E}$ is the set of edges. Let, $\mathcal{X} \in \mathbb{R}^{|v|*d}$ be the set of node features where each node $v \in \mathcal{V}$ has a d dimensional feature. The kth message passing iteration of a GNN can be modeled as a variation of the following equation [12]:

$$\begin{aligned} h_v^{(k+1)} &= \texttt{update}^{(k)}(h_v^{(k)}, \texttt{aggregate}^{(k)}(h_u^{(k)}), \\ &= \texttt{update}^{(k)}(h_v^{(k)}, \boldsymbol{m}_{\mathcal{N}(v)}^{(k)}) \quad \forall u \in \mathcal{N}(v)) \end{aligned} \tag{2}$$

Here, $\mathcal{N}(v)$ denotes the neighbors of node v. At any iteration of the GNN, the `aggregate` function takes the embedding of the neighbors of node v and combines them into one embedding vector. The `update` function takes the embedding of the node v at the previous time step and the output embedding vector of the `aggregate` function to give us the new embedding for the node v. Here, `update` and `aggregate` can be any differentiable functions. In our work here, we convert an input 3D mesh model to a graph and use the power of GNNs to learn latent part information of a class of objects.

3 Models

In this section, we describe multiple different approaches to addressing the problem of how to generate novel but realistic textures for variant 3D meshes of an object class. We discuss the advantages and disadvantages of these methods. In summary:

- MODEL-BASELINE: This model utilizes simple UV mapping. We found that this architecture doesn't generalize to unseen 3D meshes.
- MODEL-UV: This model takes UV layout as extra input information but suffers similar problems to MODEL-BASELINE.
- MODEL-DEFORMABLE: This method is based on the idea of deformable models. The disadvantage of this method is that the approximated mesh is of low quality compared to the original 3D mesh.
- MODEL-GRAPH: This model consists of two variants, MODEL-GCN, and MODEL-GGAN. These models transform an input 3D mesh into a graph and generate a texture conditioned on the graph. The final variant of this model, called MODEL-GGAN, produced higher quality and more diverse results than the other previous methods.

In the following section, we describe each model in detail.

3.1 Model-Baseline

In this first model, the texturing is performed using the TEXTUREUV mechanism. The framework is summarized in Fig. 3. For this model, we adapted a deep convolutional generative adversarial network (DCGAN) architecture [33] for the generator and the discriminator. We modified the DCGAN architecture to support the generation of higher-resolution textures. For applying the texture map to these diverse 3D models, we need a way to map the 3D vertices in these models into a 2D texture map (a procedure called UV mapping). We use the *smart UV project* feature of the Blender Python API to automatically generate these mappings for a given 3D model. Recall that our challenge is to adapt the GAN framework so that we can take advantage of training on multiple real-world examples and have the learned generation capability transfer to new 3D objects. However, the UV image for each 3D model has a different coordinate system. This means that, for example, the features of the given 3D model, such as the tires or windshields of different cars, project to different regions in the UV space for each 3D model.

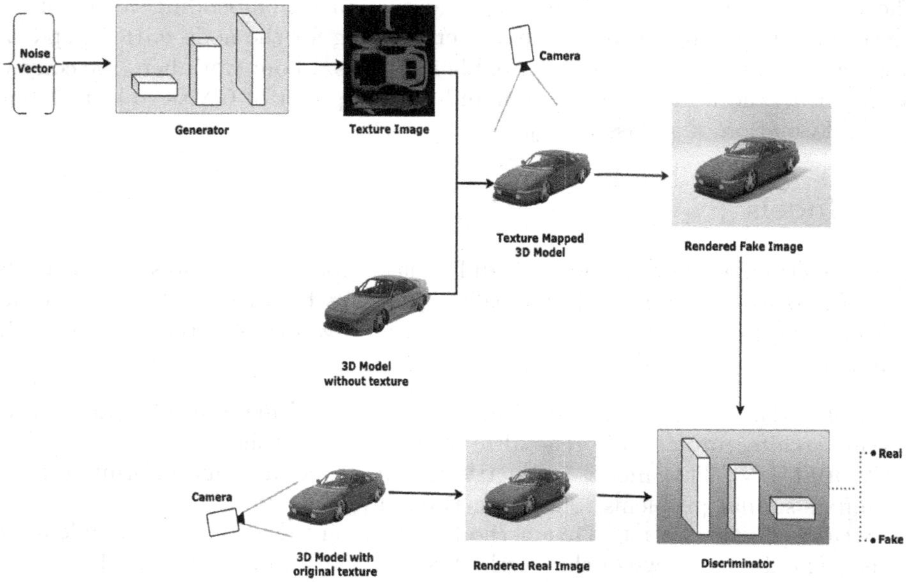

Fig. 3. Initial architecture for texture synthesis

To establish a baseline, we directly applied the UV mapping and found that this led the generator to converge to a mean texture across examples, rather than learning how to produce varied textures constrained by the input 3D mesh. This happens because the generator has no information about the UV mapping function and 3D mesh model to which the texture will be applied.

3.2 Model-UV

To address the above issue, we next explored the idea of injecting the UV map layout into the generator with the hypothesis that the generator might adapt during training in order to learn where the different parts of the given 3D model project in the 2D texture image. We used the same texturing mechanism, TextureUV, as described in the above Model-Baseline. The architecture produced results similar to Model-Baseline. The reason was the model couldn't learn complicated UV mapping information just from the UV layout image.

3.3 Model-Deformable

In this model, we use the idea of deformable models similar to Fig. 2, but with some modifications. We use the same TextureUV method described above.

The main problem with the above two models is that the UV mapping function is different for different 3D mesh models, making it harder for the generator to learn features that can work across different 3D mesh models. To mitigate this problem, we explored the idea of starting from a common mesh model with a fixed UV mapping. We used a sphere template 3D model as the starting point and used azimuth and elevation as the UV map function. We then used 3D chamfer loss [7] to predict the deformations of the sphere template to approximate the 3D mesh. This model more directly addresses our overall challenge by learning a generalized mapping from the space of textures to different 3D meshes enabling us to swap the texture learned from one model to another. However, the generated textures must still be applied to the approximate model, which reduces the quality of the 3D mesh model and the texture. The distortions in the model shape mesh are significant as demonstrated in the example in Fig. 6. Another disadvantage of this method is that we need to approximate the deformation for every new 3D model, creating extra computational overhead for training and inference.

3.4 Model-Graph

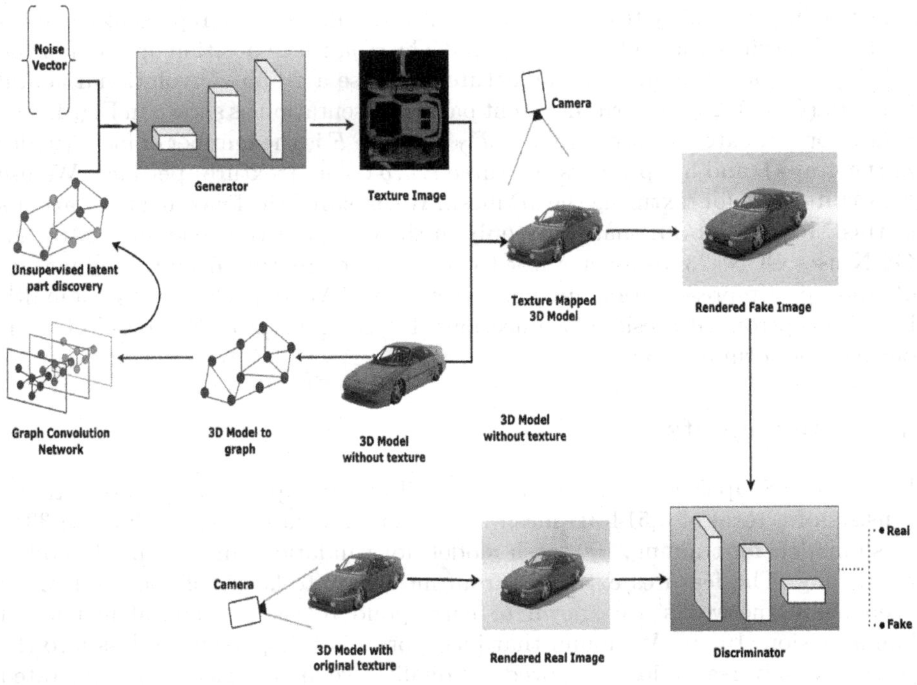

Fig. 4. Graph-based methods for texture synthesis

In this section, we describe our architecture that incorporates information from the 3D mesh model to guide the generator. We first convert the input 3D mesh model into a graph by taking each face as a node in the graph and connecting neighboring faces using graph edges. The graph neural network is then used to learn a latent representation of the structural components of the given 3D model. In the latent representation, the topological features of the 3D mesh graph can be clustered, so as to learn features that could correspond to structural components that tend to share texture properties, such as wheels, windows, lights, and hood of a car. In turn, this latent component representation can then provide an inductive bias for the generator to produce a texture for the given 3D mesh model. The architecture is shown in Fig. 4. An interesting aspect of this design is that the generator can take the unsupervised latent part representation as node features and combine it with the input noise vector to generate a texture for the particular 3D mesh. Node features is a 2D tensor of shape v, f where v is the number of nodes and f is the dimension of the node feature. The noise is a 1D vector of shape d. We sample a noise vector $z \in \mathbb{R}^d$ from a multivariate normal distribution. This d dimensional noise vector is then replicated to have a shape of $v * d$. This allows our model to process 3D mesh models with a different number of nodes. This noise tensor is then concatenated with the node feature tensor $v * f$. The concatenated tensor is then input to the generator (e.g. MLP) that, in turn, generates the textures for the given 3D mesh model. We use TEXTUREFACE for texturing as it enables us to generate higher quality textures than TEXTUREVERTEX. We represent the faces of the 3D mesh as the nodes in the graph. The x, y, z face position and its normal (n_x, n_y, n_z) form the initial node features. We use a graph convolutional neural network (GCN) [23] to learn the latent part representation as shown in Fig. 4. The generator generates a tensor of shape $F * 3$, where F is the number of faces (nodes in the graph), and 3 represents the three r, g, b colors (texture) per face. We use TEXTUREFACE for texturing the 3D mesh. We explored the following two variants of this MODEL-GRAPH that differ only in the design of the generator. MODEL-GCN uses GCN [23] as a generator to generate the texture from a combination of latent part representation and noise vector. And MODEL-GGAN uses a multi-layer perceptron with residual connections [13] as a generator. MODEL-GGAN is our best-performing model.

4 Experiments

We use the ShapeNet [3] car data set for all of our experiments. This data set consists of a total of 3,514 3D mesh models of cars with textures. We use 3314 mesh models for training, 100 mesh models for validation, and 100 mesh models for testing. The features extracted from intermediate layers of the pre-trained deep neural networks are known to correspond to the perceptual metrics of human vision [16,40]. We found that incorporating this perceptual loss into the generative adversarial loss improved the qualitative appearance of the generated textures. Thus our overall loss function is as follows:

$$\texttt{Loss(L)} = \texttt{gan\_loss} + \lambda * \texttt{perceptual\_loss} \tag{3}$$

We use the validation dataset to select the best value of λ. We use the library of Zhang *et al.* [40] to extract features from the intermediate layer of pre-trained AlexNet architecture [24] that are in turn used to calculate the perceptual loss. All of the models are trained with the above loss function. At each minibatch iteration, we render a 3D mesh with real and synthetic textures from eight different viewpoints. The loss is calculated from these real and synthetic ("fake") images. We use a learning rate of 0.0001 for both the generator and the discriminator. We use a hidden size of 64 for both GNN and the MLP generator. Here, hidden size is the dimension that's being used to project the node features of the graph. We render images of size 512 × 512 from the differentiable renderer. We use the Adam optimizer [22] for training all of our models. We used $d = 16$ for the random noise vector. For MODEL-GRAPH variants, we use 3 graph convolution layers [23] for learning the latent part representation. Each convolutional layer has a hidden size of 64. The noise vector is concatenated to the output of the last graph convolutional layer. We use TEXTUREFACE to texture the 3D mesh model. For the MODEL-GCN, we use a 7-layered GCN [23] as a generator with a hidden size of 64. The generator does not use residual connections [13]. For the MODEL-GGAN architecture, we use a 7-layered MLP as a generator with a hidden size of 64. The generator uses residual connections [13].

5 Results

The MODEL-BASELINE and MODEL-UV were only able to learn to generate textures for a single mesh model and were not able to generalize across unseen 3D mesh models. The MODEL-DEFORMABLE was able to generate a texture for unseen 3D mesh models, but the quality of approximated mesh and texture was not good, as demonstrated in contrast between Figs. 5 and 6. Moreover, the approximation of the 3D mesh model created extra computational overhead. Thus we didn't move forward with this approach for the full ShapeNet car experiment. The graph-based models based on the MODEL-GRAPH architecture were able to learn textures across different 3D mesh models. This general approach has multiple advantages compared to existing solutions that generate textures in 2D. First, the model is able to learn about the parts of the given 3D mesh model in an unsupervised way. This removes the effort and cost required for manual labeling of the 3D part segmentation. Second, the approach generates textures in 3D, so that a texture can be applied once and the 3D model can be viewed from multiple directions and under multiple light conditions without a need to generate texture each time we change these parameters. We evaluated

Fig. 5. Original mesh with original texture

Fig. 6. Approximate mesh with learned texture using MODEL-DEFORMABLE

Table 1. FID scores on the test dataset

Model	FID
MODEL-GCN	0.75
MODEL-NERF	0.93
Model-GGAN	**0.70**

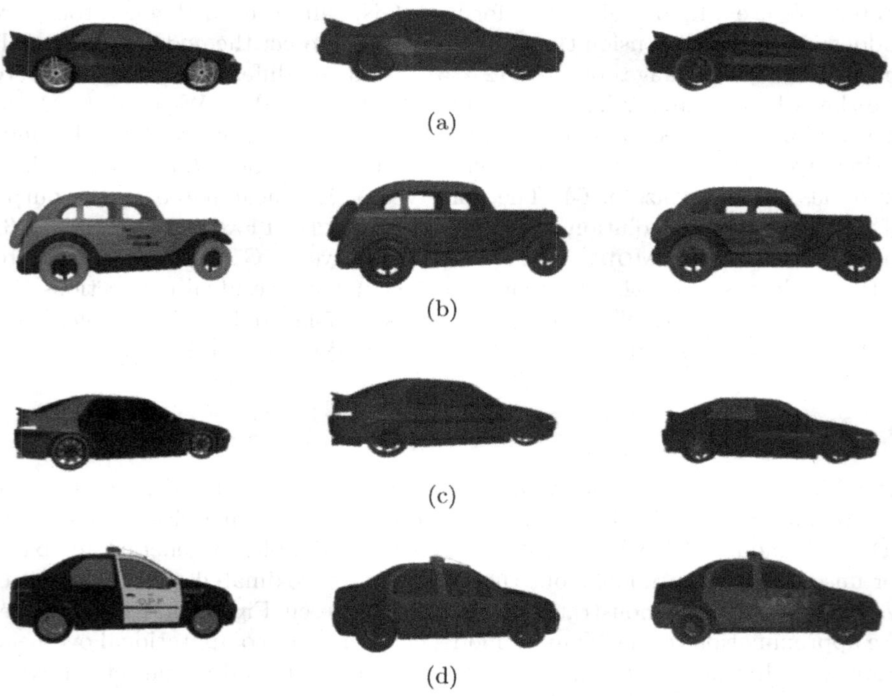

Fig. 7. First column: original images, second column: MODEL-GGAN, third column: MODEL-GCN

these models by applying the synthetic texture generated from respective models and rendering them from multiple viewpoints. We then calculated the FID score based on these projected images and actual original images rendered from the same views with the original texture. The MODEL-BASELINE, MODEL-UV, and MODEL-DEFORMABLE architectures were not suitable for learning textures across different 3D mesh models, so we did not compute FID scores for these. Table 1 shows the average FID values (lower is better) for different models per 3D mesh model. For further comparison, we also experimented with a simple variant of the NeRF [28] model as a generator (MODEL-NERF), but it did not produce results as good as the graph neural network approaches. The low quality of results is reasonable because it doesn't have a way for learning part information like our MODEL-GRAPH.

Figure 7 shows a set of selected examples of rendered images generated from different models with a fixed viewpoint. Textures are applied to the 3D mesh models and rendered as projected 2D images for visualization. The first column shows the images rendered with original textures, the second column shows the images rendered with textures generated from the MODEL-GGAN model, and the third column shows the images rendered with textures generated from the MODEL-GCN. In Fig. 7, we observe that the MODEL-GCN lacks diversity in the generated images: it produces images with the same texture for every random noise input. We hypothesize that this is due to the over-smoothing problem observed in GNNs [2,4] as the model uses GNN-only layers for the generator. Finally, the model MODEL-GGAN (GGAN) produces images (second column, Fig. 7) that respect the object boundaries, are visually better than other models and are diverse (the model produces new textures on each run with different random noise input). Some of the images generated from our final model MODEL-GGAN (third row, second column) look even better than the original image itself (third row, first column).

6 Conclusion

In this work, we have presented and evaluated the graph generative adversarial network (GGAN), a new architecture that can learn to generate a texture for a given 3D mesh with high fidelity and that can learn 3D part information in an unsupervised way. We think GGAN will be useful in various domains to generate graph-structured representation. However, there are multiple directions for improvement. The first important research direction for future work is to introduce symmetry constraints on the system such that all components with symmetrical structures will generate textures that respect symmetries. Second, we want to increase the diversity of the generated textures. Third, we want to improve the controlled synthesis of part-specific textures. Another important research direction would be to incorporate encoder-decoder graph architectures [8] within our framework. Another important direction would be to couple with a semi-supervised labeling approach. Finally, we would like to explore the use of flow-based models [36,37] and diffusion models [6,38] for the generation of texture within our current framework.

References

1. Arjovsky, M., Chintala, S., Bottou, L.: Wasserstein GAN. arXiv 2017. arXiv preprint arXiv:1701.07875 (2017)
2. Cai, C., Wang, Y.: A note on over-smoothing for graph neural networks. arXiv preprint arXiv:2006.13318 (2020)
3. Chang, A.X., et al.: Shapenet: an information-rich 3D model repository. arXiv preprint arXiv:1512.03012 (2015)
4. Chen, D., Lin, Y., Li, W., Li, P., Zhou, J., Sun, X.: Measuring and relieving the over-smoothing problem for graph neural networks from the topological view. In: Proceedings of the AAAI Conference on Artificial Intelligence, vol. 34, pp. 3438–3445 (2020)

5. Chen, W., et al.: Learning to predict 3D objects with an interpolation-based differentiable renderer. In: Advances in Neural Information Processing Systems, vol. 32 (2019)
6. Dhariwal, P., Nichol, A.: Diffusion models beat GANs on image synthesis. Adv. Neural. Inf. Process. Syst. **34**, 8780–8794 (2021)
7. Fan, H., Su, H., Guibas, L.J.: A point set generation network for 3D object reconstruction from a single image. In: Proceedings of the IEEE Conference on Computer Vision and Pattern Recognition, pp. 605–613 (2017)
8. Gao, H., Ji, S.: Graph U-Nets. In: International Conference on Machine Learning, pp. 2083–2092. PMLR (2019)
9. Gao, L., Wu, T., Yuan, Y.J., Lin, M.X., Lai, Y.K., Zhang, H.: TM-NET: deep generative networks for textured meshes. ACM Trans. Graph. (TOG) **40**(6), 1–15 (2021)
10. Goel, S., Kanazawa, A., Malik, J.: Shape and viewpoint without keypoints. In: Vedaldi, A., Bischof, H., Brox, T., Frahm, J.-M. (eds.) ECCV 2020. LNCS, vol. 12360, pp. 88–104. Springer, Cham (2020). https://doi.org/10.1007/978-3-030-58555-6_6
11. Goodfellow, I., et al.: Generative adversarial nets. In: Advances in Neural Information Processing Systems, vol. 27 (2014)
12. Hamilton, W.L.: Graph representation learning. Synth. Lect. Artif. Intell. Mach. Learn. **14**(3), 1–159 (2020)
13. He, K., Zhang, X., Ren, S., Sun, J.: Deep residual learning for image recognition. In: Proceedings of the IEEE Conference on Computer Vision and Pattern Recognition, pp. 770–778 (2016)
14. Henderson, P., Tsiminaki, V., Lampert, C.H.: Leveraging 2D data to learn textured 3D mesh generation. In: Proceedings of the IEEE/CVF Conference on Computer Vision and Pattern Recognition, pp. 7498–7507 (2020)
15. Heusel, M., Ramsauer, H., Unterthiner, T., Nessler, B., Hochreiter, S.: GANs trained by a two time-scale update rule converge to a local nash equilibrium. In: Advances in Neural Information Processing Systems, vol. 30 (2017)
16. Johnson, J., Alahi, A., Fei-Fei, L.: Perceptual losses for real-time style transfer and super-resolution. In: Leibe, B., Matas, J., Sebe, N., Welling, M. (eds.) ECCV 2016. LNCS, vol. 9906, pp. 694–711. Springer, Cham (2016). https://doi.org/10.1007/978-3-319-46475-6_43
17. Kanazawa, A., Tulsiani, S., Efros, A.A., Malik, J.: Learning category-specific mesh reconstruction from image collections. In: Proceedings of the European Conference on Computer Vision (ECCV), pp. 371–386 (2018)
18. Karras, T., et al.: Alias-free generative adversarial networks. In: Advances in Neural Information Processing Systems, vol. 34 (2021)
19. Karras, T., Laine, S., Aila, T.: A style-based generator architecture for generative adversarial networks. In: Proceedings of the IEEE/CVF Conference on Computer Vision and Pattern Recognition, pp. 4401–4410 (2019)
20. Karras, T., Laine, S., Aittala, M., Hellsten, J., Lehtinen, J., Aila, T.: Analyzing and improving the image quality of StyleGAN. In: Proceedings of the IEEE/CVF Conference on Computer Vision and Pattern Recognition, pp. 8110–8119 (2020)
21. Kato, H., Ushiku, Y., Harada, T.: Neural 3D mesh renderer. In: Proceedings of the IEEE Conference on Computer Vision and Pattern Recognition, pp. 3907–3916 (2018)
22. Kingma, D.P., Ba, J.: Adam: a method for stochastic optimization. arXiv preprint arXiv:1412.6980 (2014)

23. Kipf, T.N., Welling, M.: Semi-supervised classification with graph convolutional networks. arXiv preprint arXiv:1609.02907 (2016)
24. Krizhevsky, A., Sutskever, I., Hinton, G.E.: Imagenet classification with deep convolutional neural networks. In: Advances in Neural Information Processing Systems, vol. 25 (2012)
25. Li, T.M., Aittala, M., Durand, F., Lehtinen, J.: Differentiable Monte Carlo ray tracing through edge sampling. ACM Trans. Graph. (TOG) **37**(6), 1–11 (2018)
26. Liu, S., Li, T., Chen, W., Li, H.: Soft rasterizer: a differentiable renderer for image-based 3D reasoning. In: Proceedings of the IEEE/CVF International Conference on Computer Vision, pp. 7708–7717 (2019)
27. Loper, M.M., Black, M.J.: OpenDR: an approximate differentiable renderer. In: Fleet, D., Pajdla, T., Schiele, B., Tuytelaars, T. (eds.) ECCV 2014. LNCS, vol. 8695, pp. 154–169. Springer, Cham (2014). https://doi.org/10.1007/978-3-319-10584-0_11
28. Mildenhall, B., Srinivasan, P.P., Tancik, M., Barron, J.T., Ramamoorthi, R., Ng, R.: NeRF: representing scenes as neural radiance fields for view synthesis. In: Vedaldi, A., Bischof, H., Brox, T., Frahm, J.-M. (eds.) ECCV 2020. LNCS, vol. 12346, pp. 405–421. Springer, Cham (2020). https://doi.org/10.1007/978-3-030-58452-8_24
29. Mirza, M., Osindero, S.: Conditional generative adversarial nets. arXiv preprint arXiv:1411.1784 (2014)
30. Paszke, A., et al.: Pytorch: an imperative style, high-performance deep learning library. In: Advances in Neural Information Processing Systems, vol. 32 (2019)
31. Pavllo, D., Kohler, J., Hofmann, T., Lucchi, A.: Learning generative models of textured 3D meshes from real-world images. In: Proceedings of the IEEE/CVF International Conference on Computer Vision, pp. 13879–13889 (2021)
32. Pavllo, D., Spinks, G., Hofmann, T., Moens, M.F., Lucchi, A.: Convolutional generation of textured 3D meshes. Adv. Neural. Inf. Process. Syst. **33**, 870–882 (2020)
33. Radford, A., Metz, L., Chintala, S.: Unsupervised representation learning with deep convolutional generative adversarial networks. arXiv preprint arXiv:1511.06434 (2015)
34. Raj, A., Ham, C., Barnes, C., Kim, V., Lu, J., Hays, J.: Learning to generate textures on 3D meshes. In: Proceedings of the IEEE/CVF Conference on Computer Vision and Pattern Recognition Workshops, pp. 32–38 (2019)
35. Ravi, N., et al.: Accelerating 3D deep learning with PyTorch3D. arXiv preprint arXiv:2007.08501 (2020)
36. Rezende, D., Mohamed, S.: Variational inference with normalizing flows. In: International Conference on Machine Learning, pp. 1530–1538. PMLR (2015)
37. Weng, L.: Flow-based deep generative models. lilianweng.github.io (2018). https://lilianweng.github.io/posts/2018-10-13-flow-models/
38. Weng, L.: What are diffusion models? lilianweng.github.io (2021). https://lilianweng.github.io/posts/2021-07-11-diffusion-models/
39. Xian, W., et al.: Texturegan: controlling deep image synthesis with texture patches. In: Proceedings of the IEEE Conference on Computer Vision and Pattern Recognition, pp. 8456–8465 (2018)
40. Zhang, R., Isola, P., Efros, A.A., Shechtman, E., Wang, O.: The unreasonable effectiveness of deep features as a perceptual metric. In: Proceedings of the IEEE Conference on Computer Vision and Pattern Recognition, pp. 586–595 (2018)
41. Zhang, Y., et al.: Image GANs meet differentiable rendering for inverse graphics and interpretable 3D neural rendering. arXiv preprint arXiv:2010.09125 (2020)

Medical VQA: MixUp Helps Keeping it Simple

Jitender Singh[1], Dwarikanath Mahapatra[2(✉)], and Deepti R. Bathula[1(✉)]

[1] Indian Institute of Technology Ropar, Punjab 140001, India
bathula@iitrpr.ac.in
[2] Inception Institute of Artificial Intelligence, Abu Dhabi, UAE
dwarikanath.mahapatra@inceptioniai.org
https://iitrpr.ac.in/, http://www.inceptioniai.org/en

Abstract. Recently, Medical Visual Question Answering (VQA) became an active area of research with the induction of several publicly available benchmark datasets and the organization of challenges. Like many competitions, the quest for success has driven the use of increasingly complex neural networks. Winning strategies generally leverage multi-scale architectures and model ensembling to achieve state-of-the-art performance. However, several studies have established the capability of simpler architectures in learning more meaningful features and avoiding over-parameterization. Specifically, the use of MixUp based image augmentation with a simple VGG16 network helped achieve significant improvement in performance for medical VQA. Inspired by this finding, we propose a modified version, VQAMixUp, that leverages both images and questions for augmenting VQA datasets. VQAMixUp combined with a few enhanced training strategies help simple models (with $\approx 65\%$ reduced parameters) achieve state-of-the-performance on benchmark ImageCLEF-VQA-MED validation datasets.

Keywords: Medical VQA · Deep learning · MixUp · ImageCLEF

1 Introduction

Visual Question Answering (VQA) is a semantic task that involves answering questions about the content of the image. Machine learning models require an understanding of both vision and language to effectively answer a question regarding an image. Recently, Medical VQA became an active area of research with the induction of several publicly available benchmark datasets and the organization of challenges. A sophisticated Medical VQA system can provide invaluable assistance to overburdened and under-resourced healthcare systems worldwide. However, training a reliable VQA model for the medical domain is difficult due to limited annotated data and domain-specific characteristics. In 2018, ImageCLEF released a radiological dataset (VQA-MED) and organized the first VQA grand challenge in medicine. Figure 1 depicts a VQA sample from the ImageCLEF dataset. With increasing interest in utilizing medical artificial

W. Q. Yan et al. (Eds.): IVCNZ 2022, LNCS 13836, pp. 402–414, 2023.
https://doi.org/10.1007/978-3-031-25825-1_29

intelligence for improved patient care, several editions of ImageCLEF-VQA-MED along with other datasets have been introduced. Concurrently, several researchers employed state-of-the-art machine learning algorithms to address this challenge.

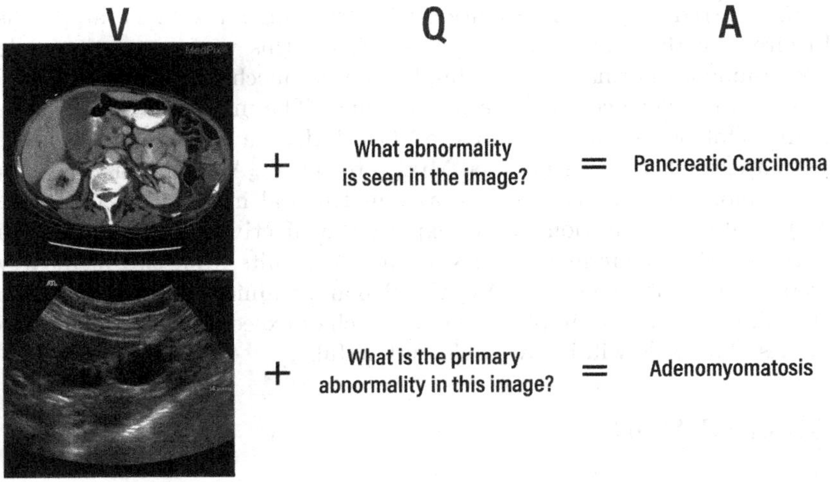

Fig. 1. Two VQA Samples - CT Images with the corresponding free-form question and answer pairs.

For a better understanding of the inherent challenges, we investigated the approaches adopted by some of the top-ranking participants on the leader-boards. For ImageCLEF-MED-VQA 2020, six of the top ten teams published their methods. While the top-ranking team leveraged an ensemble of multiple network architectures, most others used either VGG16 [16] or variants of ResNet [9] as baseline architectures for extracting image features. Three of these teams considered the VQA task as a multiclass image classification problem without modelling the QA strings. The other groups used either BioBERT [13] or GRU [6] to extract QA-related features. With ImageCLEF-MED-VQA 2021, seven of the top ten teams published their techniques. Continuing the trend from the previous year's challenge, three teams simplified the VQA task into image classification without any text models. The remaining groups used either BioBERT or LSTM [10] for QA feature extraction. While four participants used VGG16 as a backbone, others used ResNet, ResNest [22], DenseNet121 [11] or Bilateral-Branch Networks (BBNs) [23] for image-based features. Most of the high-ranking participants used multi-scale architectures and models ensemble to achieve their best performance.

Models used for VQA tasks have also grown increasingly complex to achieve state-of-the-art performance. Specifically for Medical VQA that suffers from low sample space, training such large models is very difficult. Consequently, several recent studies have questioned the use of such complex architectures and established the effectiveness of simpler models when trained systematically. For instance, SYSU-HCP [8] used a simple VGG16 network with MixUp [21] based data augmentation and some simple training mechanisms to achieve one of the best performances in ImageCLEF-VQA-MED-2021. However, as *MixUp* can only be used with images, they posed the VQA task as a simple image classification task and omitted question information while treating answers as target classes.

Inspired by these findings and observations, this work attempts to leverage multi-modal information and simple training mechanisms that enable simple models to achieve competitive performance. The main contributions of this work are as follows: Firstly, we propose VQAMixUp - a data augmentation technique that extends the conventional MixUp to the VQA task. It uses text-based questions along with images to leverage multimodal information for augmenting VQA datasets. Additionally, we explore the effectiveness of several training schemes and their combinations. Experimental results from two medical VQA datasets establish the efficacy of VQAMixUp and training mechanisms in boosting the performance of a simple model to match or exceed that of complex model ensembles. The code will be available on GitHub.

2 Related Work

In ImageCLEF 2020 [1], there are six participants under the top ten who published their papers. Their positions are 1, 2, 3, 4, 5 and 7th. Except for the first-place team, who uses multiple network architectures, three out of six teams used VGG16 and the other three used ResNet variants for image features. Three of the six participants considered the VQA task as multiclass image classification by using image-only features. Two out of the remaining three used BioBERT and one used GRU to extract question features. The 1st place participant SYSU-HCP [8] team considered the VQA task as image classification. They used VGG16 (without batch norm) architecture for image feature extraction with Hierarchical Adaptive Global Average Pooling (HAGAP) (1984 dimensional latent feature vector) pooling. The training process includes MixUp for data augmentation, label smoothing and SuperLoss [4] curriculum learning. Validation and prediction include test time augmentation by flipping the image and taking the average of the predictions. The second position participant YNU team [18] uses VGG16 with HAGAP (1472 dimensional latent feature vector) for images and BioBERT for questions. They use Co-attention and Multi-modal Factorized High-order (MFH) [24] pooling to fuse the image and text features. They include 2019 abnormality subset data in training. The third-place participant, TeamS [7], proposed a BBN-Orchestra model based on Bilateral-Branch Networks (BBN) and solved the task as a multiclass classification. The backbone architecture is ResNeSt50. The rest of the participants use various model architectures such as modified VGG16, BioBERT, DenseNet121, modified ResNet34, and LSTM along with MFH, co-attention mechanism, and simple concatenation. Some of the top methods used by the participants are explained in more details in *Experimental Results* section.

In ImageCLEF 2021 [2], seven of the top ten Med-VQA 2021 participants published their papers. Their leaderboard position on 2021's test set is 1, 2, 3, 6, 7, 8, and 10th. Out of these seven, three of them considered this VQA task as image classification by omitting questions and using images and answers as classes. Four out of seven used VGG16 as a backbone for image feature extraction. Others used ResNet, ResNeSt, DenseNet121, and Bilateral-Branch Networks (BBN). Out of the four participants considering both image and questions, two of them used BioBERT for question features extraction and the other two used LSTM. Most of the participants used an ensemble of multiple models with different architectures and image resolutions. The first-place team AIML [14] used a large multi-scale multi-model ensemble of VGGs [16], ResNets [9], ResNexts [19], DenseNets [11], and MobileNets [15]. They used a method called Skeleton-based Sentence Mapping (SSM) in which they extract information, such as type of modality and organ, from the questions. The ensemble is trained as a multi-task classification. They also include the 2019 dataset. The second-place Inception team [3] also considered the VQA task as image classification and omitted questions. They used VGG16 to extract image features. The third position team *bumjun_jung* [12] used VGG16 with GAP (1472 dimensional latent feature vector) for image features and BioBERT for text feature extraction. The image and text models are fused using MFH and co-attention. They used the abnormality subset from 2019's dataset along with 2020's dataset. HCP-MIC team [5] used BBN with ResNet34 and ResNeSt50 for image and BioBERT for question feature extraction. They used KL Divergence for training set expansion along with retrieval-based candidate answer selection.

3 Datasets

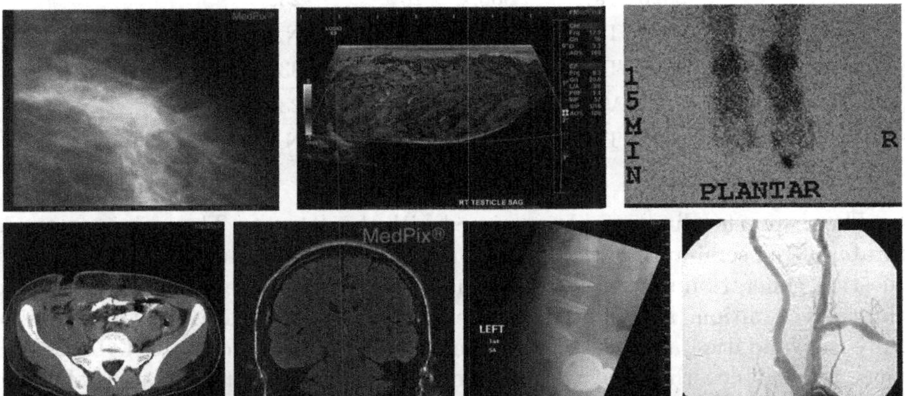

Fig. 2. Sample images of different modalities present in the ImageCLEF datasets.

There are four ImageCLEF-VQA-MED competition datasets (2018–2021) in which 2018's dataset is the first publicly accessible medical VQA dataset. However, 2018's dataset is very different from 2019 onwards. Additionally, as our main focus is on the abnormality subset, we used 2019-2021's datasets for our experiments and comparison. Specifically, the 2020 and 2021 versions of the challenge focused on abnormality-related questions. The ImageCLEF dataset sample images consist of seven different modalities including MRI, X-Ray, CT, Angiogram, Mammogram, Ultrasound, and PET scans. Figure 2 depicts sample images of different modalities.

Following the official suggestion and for consistency with other methods, a subset of VQA-Med 2019 related to abnormality is used to extend the VQA-Med 2020 training dataset. Hence, the training set contains a total of 6583 images with 57 unique questions and 332 unique answers. Furthermore, for a fair comparison, we used the SYSU-HCP team's publicly available curated training dataset for 2021 experiments. In addition to the VQA-MED 2021 training set, it includes the abnormality subset of the VQA-MED 2019 dataset and the test set of the VQA-MED 2020 dataset. The yes-no-type QA samples are removed from the training set as they are not part of validation or test sets. This generates a training set with 5683 images, 24 unique questions and 330 unique answers. Dataset details from different editions are given in Table 1.

Table 1. ImageCLEF-VQA-MED datasets details

Year	Set type	Images	Questions	Answers
2019	Train	3200	247	1552
2019	Val	500	186	470
2019	Test	500	138	166
2020	Train	4000	38	332
2020	Val	500	26	232
2020	Test	500	40	NA
2021	Train*	4500	40	332
2021	Val	500	16	236
2021	Test	500	24	NA

There are a few limitations associated with the datasets. The training sets are curated using semi-automatic methods to construct QA pairings. An automated question generation system used a combination of sentence reduction, response phrase recognition, and candidate question rating to create viable QA pairs. Only the validation and test sets are manually verified and corrected. Therefore, there is a chance that the training annotations contain noisy labels. Due to the low number of samples and a large variety of answers, the datasets are highly imbalanced on both image modalities and ground truths. For example, there are 833 CT samples in the 2020's training set and only 4 samples of PET scans. Also, there are 453 MRI samples and only 15 mammograph samples.

Apart from the modalities, there are 2047 samples with unknown modalities. All other datasets have similar statistics. Additionally, the images are noisy. For example, the X-Ray scan (sixth image) depicted in Fig. 2 is not aligned and barely visible and the PET scan (third image) contains salt-pepper noise. As there are many modalities and each one requires a different type of preprocessing, the overall preprocessing pipeline will be highly complex. The questions asked for each image are limited. Two questions are most frequent which span 48.8% of the dataset. These questions are *What abnormality is seen in the image?* and *What is the primary abnormality in this image?* spanning 25% and 23.8% samples of the total questions. Even these two questions are very similar to each other. To address the noisy images and automated label limitations, we use label smoothing (discussed in Sect. 4.6) to soften the confidence of the model. Whereas, our proposed data augmentation method is helpful to overcome the limited samples issue and train a more generalized model which also avoids overfitting. Using combinations of these methods, the effect of these limitations is reduced. The results of the corresponding experiments are depicted in Table 3 and 4.

4 Methodology

To develop a simple yet powerful VQA model, we evaluated top-performing models to help select the best baseline models. Consequently, we chose VGG16 (pre-trained on ImageNet) as the visual/image model (IM) and a single-layer GRU with word embedding as the question/text model (TM). Furthermore, based on preliminary experiments, simple multiplication-based fusion is utilized to combine the image and text features. Subsequently, we explored the effectiveness of several strategies to boost the performance of our baseline model. These strategies include:

4.1 MixUp (MX)

MixUp Generates a new image V_{mix} with ground truth labels A_{mix} as a weighted combination of original images V_x and V_y and their corresponding ground truth labels A_x and A_y using the following equations:

$$V_{mix} = \lambda V_x + (1 - \lambda) V_y \tag{1}$$

$$A_{mix} = \lambda A_x + (1 - \lambda) A_y \tag{2}$$

where the value of mixing parameter $\lambda \in [0, 1]$ is chosen randomly from a *Beta* distribution with α as the hyper-parameter. While Eq. 2 works for one-hot encoded labels, a Loss MixUp is used for other types of targets as shown below:

$$loss = \lambda \mathcal{L}(\hat{P}_x, A_x) + (1 - \lambda) \mathcal{L}(\hat{P}_y, A_y) \tag{3}$$

where $\mathcal{L}$ can be any loss function that takes predictions ($\hat{P}_x$ and $\hat{P}_y$) and target values or ground truth labels (A_x and A_y).

4.2 VQAMixUp (VMX)

Algorithm 1: *VQAMixUp* algorithm.

Data: VQA samples - (V_x, Q_x, A_x), (V_y, Q_y, A_y)
Model: VQA model - $\mathcal{M}$
Loss: Loss function ($\mathcal{L}$) - cross-entropy in our case.

1 **Function** `VQAMixup`(V_x,Q_x,A_x,V_y,Q_y,A_y,$\mathcal{M}$,α):
2 $\lambda \leftarrow Beta(\alpha, \alpha)$ where $\alpha \in (0, \infty)$
3 $\hat{V} \leftarrow \lambda V_x + (1 - \lambda) V_y$ `/*mixed image*/`
4 $\hat{P}_x \leftarrow \mathcal{M}(\hat{V}, Q_x)$
5 $\hat{P}_y \leftarrow \mathcal{M}(\hat{V}, Q_y)$
6 $\hat{P}_{vqa} \leftarrow \lambda \hat{P}_x + (1 - \lambda) \hat{P}_y$ `/*predictions*/`
7 $loss \leftarrow \lambda \mathcal{L}(\hat{P}_x, A_x) + (1 - \lambda) \mathcal{L}(\hat{P}_y, A_y)$
8 **return** $loss$, $\hat{P}_{vqa}$

We propose a modified version of MixUp (MX) data augmentation technique for the VQA task by leveraging both images and questions. The technique is described in Algorithm 1. An illustration of the algorithm is depicted in Fig. 3. Similar to conventional MX, a new image sample ($\hat{V}$) is generated as a weighted combination of sample images as shown in Step 3. However, generating a combination of two questions is not as straightforward as questions are represented as vectors of tokens. Hence, we avoid generating a mixed question and use LossMixUp to generate the target answer for the mixed image. To this end, the influence of questions is accounted for by generating two target answer predictions for the new mixed image linked with two sample questions. This mechanism is depicted in Steps 4 and 5, where the VQA model ($\mathcal{M}$) is used to generate predictions $\hat{P}_x$ and $\hat{P}_y$ for a new image ($\hat{V}$) corresponding to questions Q_x and Q_y, respectively. These individual QA-specific predictions are further combined using the mixing parameter (λ) to generate the final prediction as shown in Step 6. Finally, LossMixUp is used to calculate the task-specific loss using predicted and ground-truth answers as in Step 7.

4.3 MixPool (MP)

The Mixpool [17] uses a weighted combination of average and max pooling, as shown below, to boost invariance to data transformations and provide rich latent features

$$MixPool(x) = \frac{1}{\theta}\, AvgPool(x) + \frac{1}{\phi}\, MaxPool(x) \tag{4}$$

where θ and ϕ are trainable parameters both initialized with value 2.

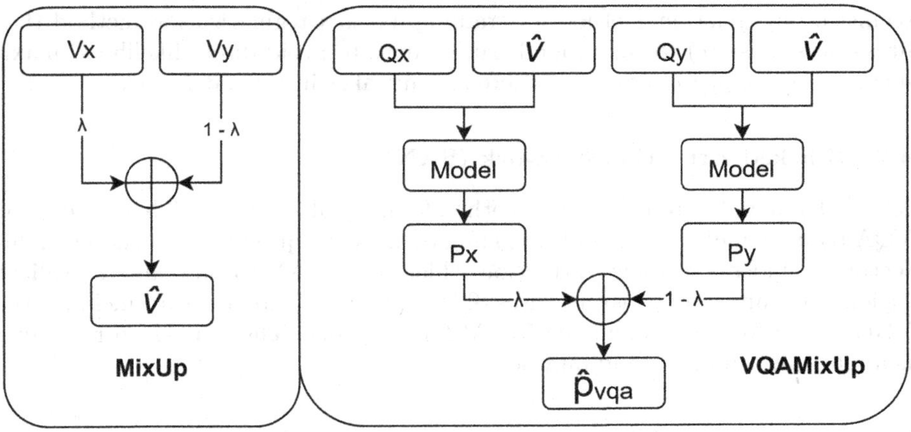

Fig. 3. MixUp and VQAMixUp Algorithm illustration. V_x and V_y represent two sample images. MixUp output image, $\hat{V}$, represents the weighted (λ represents the image weight) sum of V_x and V_y sample images. Q_x and Q_y are the two sample questions. *Model* represents the VQA model ($\mathcal{M}$) which outputs P_x and P_y predictions for each sample pair of an image and a question. $\hat{P}_vqa$ is the weighted sum, similar to the *MixUp*, of P_x and P_y weighted using λ parameter.

4.4 Question Generation (QG)

QG is used for manually augmenting the dataset with slightly different versions of original questions associated with images, keeping the answers constant. For example, given an original question *What abnormality is seen in the image?*, sample generated questions include *What is most alarming about this image?*, *What is the primary abnormality in this image?*, etc.

4.5 SuperLoss (SL)

SL is a curriculum learning approach where a model is presented with easy samples before difficult ones during training. It is used in conjunction with an existing task-based loss function. We use SL with cross-entropy loss. As described in [4], any loss function can be transformed to confidence-aware loss as depicted in Eq. 5. Similarly, Eq. 6 depicts the SuperLoss.

$$loss(x_i, y_i) \rightarrow L(loss(x_i, y_i), \sigma_i) \tag{5}$$

$$SuperLoss(loss) = min_\sigma L(loss, \sigma) \tag{6}$$

4.6 Label Smoothing (LS)

ImageCLEF's training datasets contain noise labels due to automatic dataset construction from MedPix images and their corresponding annotations. Consequently, label smoothing that converts 'hard' class labels to 'soft' score labels seems to be a relevant method for training. LS is a regularization method that introduces noise (ϵ) to the ground truth values to avoid direct likelihood maximization of $log\ p(y|x)$ assuming there are mistakes in the data labels.

4.7 Stacked Attention Network (SAN)

SAN [20] is a multi-step reasoning method for natural VQA task. They argue that VQA requires multiple steps of reasoning to answer a question. The model architecture includes multi-layer attention which helps SAN to pinpoint the salient regions and filter out noisy features. The base model architecture includes the VGG model for images and the LSTM for questions. The multi-step reasoning is applied to the image and question's features.

5 Experimental Results

As ImageCLEF-VQA-MED challenges for 2020 and 2021 are no longer accepting submissions, performance comparison on test sets is not possible. However, as several top-ranking teams have published their results on validation sets, we use them for evaluation. For 2020, teams AIML [14] and HCP-MIC [5] ranked 1st and 4th respectively. Similarly, for 2021, SYSU-HCP [8] ranked 1st and TeamS [7] was placed 3rd. Some of the features used by these winning teams include multi-scale networks, multi-model ensembles, Skeleton-based Sentence Mapping (SSM) for extracting relevant information from questions using knowledge inference methodology which transforms the VQA task into multiple image classification tasks, Bilateral Branch Network (BBN) that can perform both representation learning and classifier learning simultaneously where each branch can work independently, Hierarchical Adaptive Global Average Pooling (GAP) which takes output of every convolution block followed by a global average pooling layer and concatenates the features before classifier layer, MixUp data augmentation, Label smoothing (LS) and SuperLoss curriculum learning (SL). It was not practically feasible to try all possible combinations of strategies due to limited computational resources. Nevertheless, we tried numerous combinations and report the most significant ones due to space limitations.

5.1 Model Training

All models were trained using the PyTorch framework with cross-entropy loss and SGD optimizer, 0.001 learning rate, 0.0005 weight decay, 0.9 momentum, 32 batch size, and 60 epochs. Step-wise learning rate decay was used with a factor of 0.6 for every 20 epochs. Both α and ϵ were set to 0.1 for VQAMixUp and Label Smoothing, respectively. SL hyper-parameters are the same as SYSU-HCP. The SAN fusion method uses single-layer attention because of the simplicity of the questions. QG generates an average of 7 questions per sample.

Table 2. Models' size comparison with 2020 and 2021's top teams. Value 1 in the *number of models in ensemble* column means a single model. The number of parameters for each ensemble is the sum of the image model and text model's parameters multiplied by the number of models in the ensemble. Note that the number of parameters calculated here are approximate because the authors used modified model architectures.

Year	Author	No. of parameters	No. of models in ensemble	Accuracy (%)
2020	HCP-MIC	275.0M	2	57.2
2020	AIML	630.3+M	30+	59.6
2020	Ours	169.9M	1	**60.4**
2021	TeamS	220.0M	4	61.3
2021	SYSU-HCP	422.8M	8	69.2
2021	Ours	169.9M	1	**70.2**

5.2 Model Comparison

The winning participants used multiple model ensembles to boost their accuracy. We compared the models with the VQA-MED competition's top teams in 2020 and 2021. Table 2 depicts the comparison of parameters of the models used in recent studies to our method. As shown, we are getting better results as compared to winners by using a single VGG (for images) and GRU (for text) model with VQAMixUp and a few training tricks. Given the results in Table 2, it is clear that VQAMixUp can help a single model to be as generalized as an ensemble of several models. Note that we are using the same model for both 2020 and 2021's datasets. The only difference is that 2021's model is trained on additional images from 2020's training and validation datasets. The results show that the model and training methods are well scalable. Hence, more data can improve the results further using the same architecture and training methods.

5.3 Results

We compare the performance of our proposed approach with the winning groups on both the 2020 and 2021 ImageCLEF Med-VQA competitions' official validation sets. These results are shown in Tables 3 and 4 respectively. Specific strategies employed by each method are indicated with a (✓) in the corresponding column. For ensembles, the number of models used is specified. Rows 5–7 of Table 4 clearly show that VQAMixUp helps boost the performance by 1.6% on baseline methods. Different rows of Table 3 and 4 help identify the combination of analysis strategies that complement VQAMixUp and constitute the ablation studies of this work. Results are sorted using the best accuracy to identify the combination of strategies that provides the best performance. It can be observed that our simple model can achieve state-of-the-art (SOTA) performance with the proposed *VQAMixUp* and a few simple strategies. To prove that our results are significant, we used the unpaired t-test for hypothesis testing. Improvement of

Table 3. Results from ImageCLEF-VQA-MED 2020 Validation Set: (✓) indicates the techniques used in a particular method. Performance is reported as best accuracy followed by mean (μ) and standard deviation (σ) across 4 runs. ($\star$) indicates that μ and σ were not reported.

Method	IM	TM	QG	SAN	BBN	SSM	KA	LS	SL	MP	VMX	Acc ($\mu \pm \sigma$) (%)
Ours	VGG16	GRU		✓								53.0 (52.55 ± 0.35)
Ours	VGG16	GRU	✓									55.2 (54.20 ± 0.78)
AIML [14]	3xR50, R101, R152	-				✓						55.2*
Ours	VGG16	GRU	✓					✓	✓	✓	✓	56.2 (55.40 ± 0.51)
Ours	VGG16	GRU		✓				✓		✓	✓	56.6 (55.65 ± 0.59)
Ours	VGG16	GRU	✓					✓			✓	57.2 (57.10 ± 0.10)
HCP-MIC [5]	RS50	BB			✓		✓					57.2*
Ours	VGG16	GRU	✓	✓				✓			✓	58.4 (57.85 ± 0.43)
Ours	VGG16	GRU							✓		✓	58.6 (58.35 ± 0.17)
Ours	VGG16	GRU										59.6 (58.80 ± 0.58)
AIML [14]	3xR, 2xRX, 2xD, M	-				✓						59.6*
Ours	VGG16	GRU							✓	✓	✓	**60.4 (60.05 ± 0.38)**

ACRONYMS: IM – Image Model, TM – Text Model, QG – Question Generation, SAN - Stacked Attention Network, BBN – Bilateral Branch Network, SSM – Skeleton-based Sentence Mapping, KA – KL Divergence with Answer Selection, LS – Label Smoothing, SL – Super-Loss, MP – Mixed Pooling, VMX – VQAMixUp

Table 4. Results from ImageCLEF-VQA-MED 2021 Validation Set: (✓) indicates the techniques used in a particular method. Performance is reported as best accuracy followed by mean (μ) and standard deviation (σ) across 4 runs. ($\star$) indicates that μ and σ were not reported.

S. No.	Method	IM	TM	Q	BBN	GAP	SAN	LS	SL	QG	MP	MX	VMX	Acc ($\mu \pm \sigma$) (%)
1	TeamS [7]	RS	-		✓									61.3*
2	Ours	VGG16	GRU	✓			✓							63.2 (63.00 ± 0.14)
3	Ours	VGG16	GRU	✓			✓	✓		✓	✓		✓	63.8 (63.34 ± 0.45)
4	Ours	VGG16	GRU	✓			✓						✓	65.2 (64.95 ± 0.32)
5	Ours	VGG16	GRU	✓										66.2 (65.85 ± 0.29)
6	SYSU-HCP [8]	VGG16	-											66.6 (66.25 ± 0.60)
7	Ours	VGG16	GRU	✓									✓	68.2 (68.00 ± 0.14)
8	Ours	VGG16	GRU	✓				✓			✓		✓	68.4 (67.95 ± 0.29)
9	Ours	VGG16	GRU	✓				✓	✓		✓		✓	68.6 (67.95 ± 0.43)
10	SYSU-HCP [8]	VGG16	-			✓		✓	✓			✓		69.2 (68.35 ± 0.85)
11	Ours	VGG16	GRU					✓	✓				✓	69.4 (68.35 ± 0.60)
12	Ours	VGG16	GRU	✓			✓	✓		✓			✓	69.4 (69.10 ± 0.22)
13	Ours	VGG16	GRU	✓				✓					✓	**70.2 (69.70 ± 0.30)**

ACRONYMS: IM – Image Model, TM – Text Model, Q - Used questions, BBN – Bilateral Branch Network, GAP – Global Average Pooling, SAN – Stacked Attention Network, LS – Label Smoothing, SL – SuperLoss, QG – Question Generation, MP – Mixed Pooling, MX – MixUp, VMX – VQAMixUp

our model is statistically significant as $p < 0.02$ and $p < 0.05$ compared to second-best methods, SYSU-HCP (2021) and AIML (2020) respectively, on validation sets.

5.4 VQAMixUp Discussion

Our proposed method is a modification of the popular MixUp data augmentation technique. MixUp is limited to images only. VQAMixUp overcomes the limitation of images only and extends the MixUp to include text information as well. Using our method, MixUp applies to all such tasks which include neural networks that require both images and text as input. These tasks are Visual Question Answering (VQA), image captioning, multi-class classification, multi-label classification, etc. Being an augmentation method, it helps the network to be more generalized and reduces overfitting.

6 Conclusion

In the context of Medical VQA, we demonstrate the effectiveness of simple training mechanisms in boosting the performance of a simple model to match or exceed that of large and complex models like ensembles. While most winning strategies treated VQA as only an image classification problem by omitting the questions, our work demonstrates the significance of including textual information, even if some of the questions are redundant. Remarkably, a simple VGG16 (for images) combined with a single layer GRU (for text) aided by our proposed VQAMixUp and generic training methods achieved state-of-the-art performance on benchmark ImageCLEF-VQA-MED validation sets. Compared to the winning teams of 2020 and 2021, our model achieves better performance with 73% and 60% reduction in the number of parameters, respectively. In future, a systematic and comprehensive ablation study that determines the relative contribution from each training strategy can further help establish their efficacy for VQA in general.

References

1. Abacha, A.B., Datla, V.V., Hasan, S.A., Demner-Fushman, D., Muller, H.: Overview of the VQA-med task at ImageCLEF 2020: visual question answering and generation in the medical domain. In: CEUR Workshop Proceedings (2020)
2. Abacha, A.B., Sarrouti, M., Demner-Fushman, D., Hasan, S.A., Müller, H.: Overview of the VQA-med task at ImageCLEF 2021: visual question answering and generation in the medical domain. In: CEUR Workshop Proceedings (2021)
3. Al-Sadi, A., Al-Theiabat, H.A., Al-Ayyoub, M.: The inception team at VQA-med 2020: pretrained VGG with data augmentation for medical VQA and VQG. In: CEUR Workshop Proceedings, vol. 2696 (2020)
4. Castells, T., Weinzaepfel, P., Revaud, J.: Superloss: a generic loss for robust curriculum learning, vol. 33, pp. 4308–4319. Curran Associates, Inc. (2020)
5. Chen, G., Gong, H., Li, G.: HCP-mic at VQA-med 2020: effective visual representation for medical visual question answering, vol. 2696. CEUR (2020)
6. Chung, J., Gulcehre, C., Cho, K., Bengio, Y.: Empirical evaluation of gated recurrent neural networks on sequence modeling. CoRR, abs/1412.3555 (2014)

7. Eslami, S., de Melo, G., Meinel, C.: Teams at VQA-med 2021: BBN-orchestra for long-tailed medical visual question answering. In: CEUR, vol. 2936, pp. 1211–1217 (2021)
8. Gong, H., Huang, R., Chen, G., Li, G.: SYSU-HCP at VQA-med 2021: a data-centric model with efficient training methodology for medical visual question answering. In: CLEF (2021)
9. He, K., Zhang, X., Ren, S., Sun, J.: Deep residual learning for image recognition. CoRR, abs/1512.03385 (2015)
10. Hochreiter, S., Schmidhuber, J.: Long short-term memory. Neural Comput. **9**(8), 1735–1780 (1997)
11. Huang, G., Liu, Z., Maaten, L., Weinberger, K.Q.: Densely connected convolutional networks. CoRR, abs/1608.06993 (2016)
12. Jung, B., Gu, L., Harada, T.: bumjun jung at VQA-med 2020: VQA model based on feature extraction and multi-modal feature fusion. In: CEUR Workshop Proceedings, vol. 2696 (2020)
13. Lee, J., et al.: BioBERT: a pre-trained biomedical language representation model for biomedical text mining. CoRR, abs/1901.08746 (2019)
14. Liao, Z., Wu, Q., Shen, C., Hengel, A.V., Verjans, J.W.: AIML at VQA-med 2020: knowledge inference via a skeleton-based sentence mapping approach for medical domain visual question answering, vol. 2696, pp. 1–14. CEUR (2020)
15. Sandler, M., Howard, A., Zhu, M., Zhmoginov, A., Chen, L.C.: Inverted residuals and linear bottlenecks: mobile networks for classification, detection and segmentation. CoRR, abs/1801.04381 (2018)
16. Simonyan, K., Zisserman, A.: Very deep convolutional networks for large-scale image recognition. CoRR, arXiv:1409.1556 (2014)
17. Virk, J.S., Bathula, D.R.: Domain-specific, semi-supervised transfer learning for medical imaging. In: CODS COMAD, pp. 145–153 (2021)
18. Xiao, Q., Zhou, X., Xiao, Y., Zhao, K.: Yunnan university at VQA-med 2021: pretrained BioBERT for medical domain visual question answering. In: CEUR Workshop Proceedings, vol. 2936, pp. 1405–1411 (2021)
19. Xie, S., Girshick, R.B., Dollár, P., Tu, Z., He, K.: Aggregated residual transformations for deep neural networks. CoRR, abs/1611.05431 (2016)
20. Yang, Z., He, X., Gao, J., Deng, L., Smola, A.J.: Stacked attention networks for image question answering. CoRR, abs/1511.02274 (2015)
21. Zhang, H., Cisse, M., Dauphin, Y.N., Lopez-Paz, D.: mixup: beyond empirical risk minimization. CoRR, abs/1710.09412 (2018)
22. Zhang, H., et al.: Resnest: split-attention networks. CoRR, abs/2004.08955 (2020)
23. Zhou, B., Cui, Q., Wei, X.S., Chen, Z.: BBN: bilateral-branch network with cumulative learning for long-tailed visual recognition. CoRR, arXiv:1912.02413 (2019)
24. Zhou, Y., Yu, J., Xiang, C., Fan, J., Tao, D.: Beyond bilinear: generalized multimodal factorized high-order pooling for visual question answering. IEEE Trans. Neural Netw. Learn. Syst. **29**, 5947–5959 (2018)

TTF-ST: Diversified Text to Face Image Generation Using Best-Match Search and Latent Vector Transformation

Srinidhi Temkar(✉), Amrutha Ukkalam, Sanket Donty, Mahesh Dorsala, and S. S. Shylaja

Department of Computer Science and Engineering, PES University, Bengaluru, India
{srinidhiptemkar,sanketnd}@pesu.pes.edu, shylaja.sharath@pes.edu

Abstract. Text to face generation is a multimodal task which involves generation of facial images of people from the textual description of their faces. It requires joint modeling of the language space and the image space. This task is less studied due to numerous problems, some of which are - high diversity and abstractness in descriptions of human facial features, lack of datasets, and difficulty in independent manipulation of discrete facial features in the image space to produce the desired, high quality human face images. In our work, we attempt to resolve some of the aforementioned challenges that exist in the context of the task. This paper presents the Text-to-Face Search and Transform model, TTF-ST for short, which introduces a novel technique of searching and selecting images from a custom dataset that best match the encoded textual description and subsequently transforming their latent vectors to enhance the accuracy of the generated images. The methodology proposed in this paper also allows for the usage of complex and diversified textual descriptions, while generating images of high resolution of 1024 $\times$ 1024 pixels. A new metric is introduced to allow for a more accurate evaluation of the model's performance by accounting for only those facial attributes specified in the description. Extensive experiments and comparisons with the current models demonstrate that the proposed method achieves significant improvements in accuracy while generating diverse high-quality photo-realistic images.

Keywords: Text-to-face synthesis · Face description · High resolution · Face image generation · Multi-label · Diversity

1 Introduction

Text-to-Face (TTF) image synthesis is a research topic that has gained increasing interest in recent years. The task involves generating a photorealistic human face image conditioned on an input textual description. This topic is a subset of the Text-to-Image (TTI) problem. Most of the existing TTI models are limited for simple tasks like generating images of flowers, birds, etc. TTF synthesis, however, is a more challenging subdomain because of a slew of ways in which the features of a human face can be described. The number of attributes required to describe a flower or a bird is much lower than what is

W. Q. Yan et al. (Eds.): IVCNZ 2022, LNCS 13836, pp. 415–429, 2023.
https://doi.org/10.1007/978-3-031-25825-1_30

needed to describe a human face [1]. Other factors that make TTF synthesis challenging include lack of benchmark datasets, absence of standardized metrics to evaluate the performance of a TTF model, and poor overlapping between the image distribution of an attributed facial dataset and the output distribution of an image generating model impacting the accuracy of image encoders.

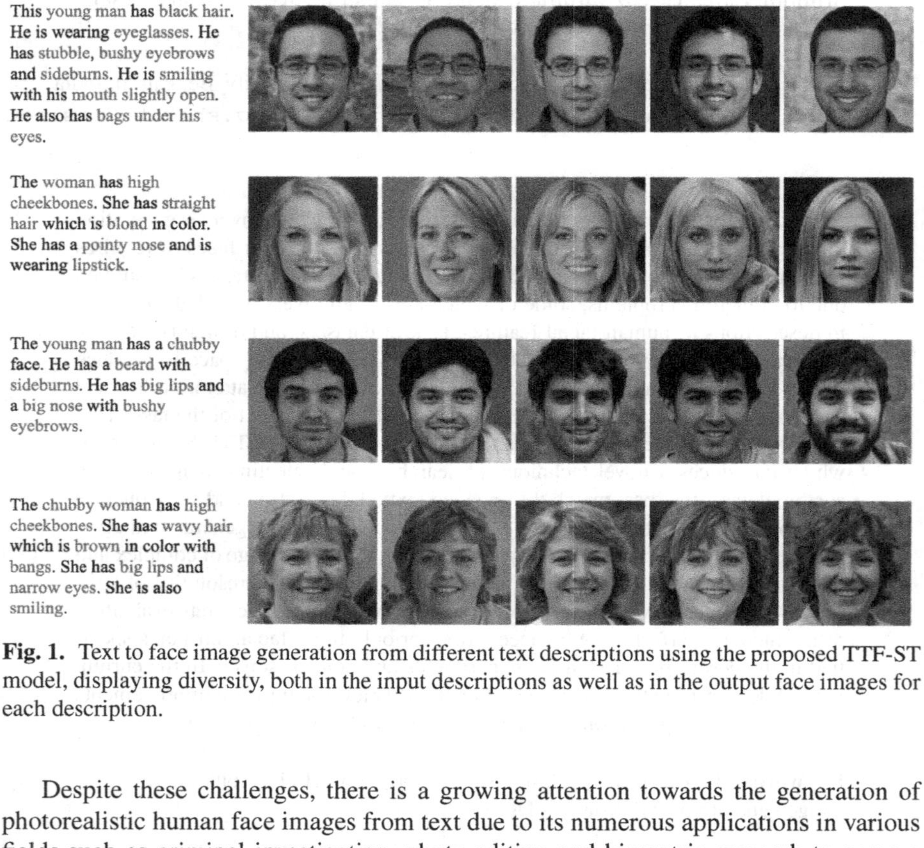
This young man has black hair. He is wearing eyeglasses. He has stubble, bushy eyebrows and sideburns. He is smiling with his mouth slightly open. He also has bags under his eyes.
The woman has high cheekbones. She has straight hair which is blond in color. She has a pointy nose and is wearing lipstick.
The young man has a chubby face. He has a beard with sideburns. He has big lips and a big nose with bushy eyebrows.
The chubby woman has high cheekbones. She has wavy hair which is brown in color with bangs. She has big lips and narrow eyes. She is also smiling.

Fig. 1. Text to face image generation from different text descriptions using the proposed TTF-ST model, displaying diversity, both in the input descriptions as well as in the output face images for each description.

Despite these challenges, there is a growing attention towards the generation of photorealistic human face images from text due to its numerous applications in various fields such as criminal investigation, photo editing and biometric research to name a few. In criminal investigations, a TTF model can be used to mitigate the requirement of a composite artist and generate faces with high accuracy and speed from eyewitnesses' descriptions. Photo-editing apps can let users tweak certain attributes of a base picture iteratively just by taking input textual descriptions of the desired image. In addition to being able to edit existing attributes like skin complexion and hair color, one could also add or remove cosmetic attributes such as eyeglasses, necklace, etc. In cosmetic surgeries, the surgeons can provide patients with viable restructuring options based on the patient's requirements and description. In each of these use cases, the input facial descriptions can be first taken in the form of speech and converted to text using a speech-to-text synthesizer before feeding it to a TTF model, which would make such a model even more powerful in its applications.

The existing TTF models [1–3] have been fairly successful in synthesizing images conditioned on text. They do, however, have a few shortcomings. The state-of-the-art work done in [1] handles only simple captions that consist of attributes from the dataset and doesn't account for any synonyms that the user might use to describe facial features. Due to insufficient disentanglement of features in the latent space, only few of the generated images are highly consistent with the textual descriptions. Thus, the model doesn't produce images of high accuracy which is reflected in the model's cosine scores. Nasir et al. in their paper [2] generate images of resolution 64 × 64 which lack clarity and details. These shortcomings necessitated a model that could process complex captions and synthesize images of high accuracy, diversity and resolution.

The methodology proposed in this paper attempts to solve the issue of poor accuracy of images generated by transforming the latent vectors of images that best match the textual descriptions, instead of using a randomly generated face as the base image as done in [1]. The principle behind using this technique is that two images that are visually similar have similar positioning of attribute vectors in higher dimensional latent space. The proposed model also handles complex text descriptions by modifying the caption generation algorithm proposed in [2] and pre-processing the text by incorporating widely used synonyms of the existing facial attributes. Varying length captions are handled by fine-tuning a pretrained BERT [4] model. Other major contributions include the addition of a novel best-match search algorithm by creation and utilization of a custom dataset consisting of facial images and their encodings, and the introduction of a new metric which we call as the Cosine Similarity for Specified Attributes, or CSSA for short. This metric is a modified version of a simple cosine similarity score, and is used for the selection of the best-match image as well as for better evaluation of the TTF model itself.

2 Related Work

2.1 Text Classification and Caption Generation

The first step in any TTF synthesis pipeline would be to extract the facial features specified in the input text describing the desired facial image. A text classification or text encoder model is used to accomplish this by mapping the open-ended text descriptions to binary attributed vectors. Many of the existing works around facial attribute recognition, including some of the TTF models [1–3] utilize the celebA dataset [5] which consists of over 200,000 images of celebrities and corresponding attribute vectors describing the presence or absence of 40 specific features. These TTF models leverage the attribute vectors for identification of facial features specified in the input text description and mapping them to the desired face image.

As the celebA dataset only consists of binary attributed vectors and not textual descriptions of the images, there is a need for a corpus of natural language text on which the text encoder can be trained. Such a corpus of text can be generated from the attribute vectors present in the celebA dataset. However, manual generation of captions for each image is non-viable and may introduce human bias. Nasir et al., in [2], address this problem by presenting an algorithm that automates the process of generating meaningful, well-structured captions from a list of attributes. The algorithm categorizes the celebA attributes into 6 groups and generates captions which collectively and iteratively define the face structure via a top-down approach, starting with the outline of the face and then including finer inner details.

As there are numerous ways in which a facial feature can be described, the synthesized captions can be diverse. Additionally, the captions can be of varying numbers of words and sentences. The text encoder should be able to handle such variability. The work done in TTF-HD [1], which is the previous state-of-the-art TTF model, proposes to use BERT [4], an advanced natural language processing model, as the text encoder because of its high performance on high dimensional training data and its ability to handle varying length texts. This provides users with a greater leeway of describing a face by not limiting the description to a fixed size. However, the authors of TTF-HD do not discuss the dataset on which it was trained and leave the fine-tuning of the text encoder unexplored.

BERT is a context aware, transformer-based language representation model. It leverages the encoder network of transformers, attention mechanism and transfer learning to learn deep bidirectional representation of words. It uses two mechanisms, namely masked language modeling (MLM) and next sentence prediction (NSP) to learn the bidirectional context of words. The fine-tuning phase of BERT facilitates its usage for multiple downstream tasks. One such task, in the context of the present problem, is to obtain a binary attribute vector from natural language text. Chi Sun et al., in [6], explore various fine-tuning strategies of BERT. They present different methods to handle the problem of overfitting in BERT.

2.2 Image Generation

Generative Adversarial Networks (GAN) [7] have been used extensively to accomplish the task of image generation. The major issue that accompanies the usage of GANs for generation of photorealistic images conditioned on text is that a traditional GAN does not provide the ability to control the image generated as it is generated from random noise. Thus, there is a need for a better GAN model.

StyleGAN [8] introduced an extension to the typical GAN architecture and was able to produce high-quality facial images of large 1024×1024 resolution. More importantly, it offered higher control over the style of the image being generated at different levels of detail. This was made possible with five modifications in the progressive GAN architecture. This model achieved impressive Fréchet inception distance (FID) scores. However, StyleGAN suffered from the 'water droplet' effect [9], where images produced often suffered from the presence of artifacts such as water droplets in them, which made them easily distinguishable from real images.

In order to deal with this problem and other issues associated with the original StyleGAN model, StyleGAN2 [10] was introduced. StyleGAN2 used the original StyleGAN model as a base model and introduced five modifications to the architecture. With the modifications in effect, the quality of the images was found to be better with lower FID scores. There was also a considerable reduction in the training time. This model is now the current state-of-the-art model for generating human face images, and hence is used in the proposed model.

2.3 Text-To-Face Image Synthesis

The work in [3] proposed using two models - a bidirectional LSTM as the text encoder and a conditional GAN as the image decoder. The authors proposed training them together as opposed to training them separately. The model outperformed several state-of-the-art models such as AttnGAN [11], StackGAN [12] and FTGAN [13] when compared using Face Semantic Distance (FSD) and FID scores. However, this model could only produce low resolution images of size 256 × 256.

Wang et al., presented TTF-HD, a TTF model that produced high quality images of resolution 1024 × 1024 with good accuracy and diversity. This model modifies the latent vector values of a randomly generated image to maximize the similarity between the text encoding and the image encoding to generate diverse images with desired features that accurately match with the description. This model outperformed AttnGAN [11] in all the 3 metrics - Inception Score (IS) used to judge the quality of images, Cosine Similarity (CS) used to evaluate similarity of text corpus and learned perceptual image patch similarity (LPIPS) used to assess diversity of images were used. This model has produced one of the best results so far in TTF work.

3 Proposed Methodology

In this section, we discuss TTF-ST methodology in detail. Figure 2 illustrates the architecture of the proposed methodology.

3.1 Overview

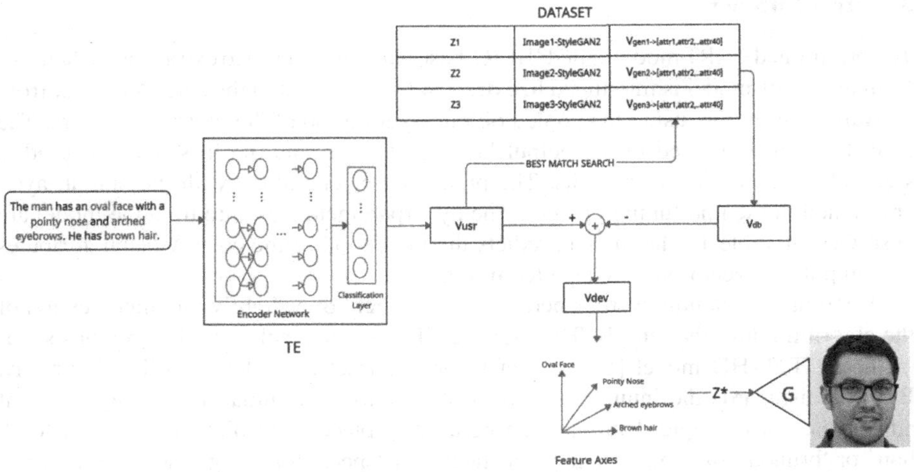

Fig. 2. TTF-ST architecture

Let the user description of a face be represented by t_0. Facial attributes from celebA and any of their synonyms are extracted from t_0 using a text encoder TE, which produces a text encoding v_{usr}. A static custom dataset is created which contains facial images generated by StyleGAN2, the latent vectors of these images (z_n), and their image encodings (v_{gen}) generated by the image encoder IE. The text encoding, v_{usr}, is compared with all the image encodings, $v_{gen\ i}$, present in the dataset based on the metric of Cosine Similarity for the Specified Attributes (CSSA), and these image encodings are rank ordered from the highest to the least value of CSSA. Top five image encodings that are most similar to v_{usr} and their corresponding latent vectors are selected. Of these, the image encoding having the highest CSSA is chosen as the best-match image encoding, represented by v_{db}, and the corresponding image is the best-match image, thus completing the best-match search. The rest four are used to account for diversity as they can have any values for the attributes not specified in the text description.

A difference vector, v_{dev}, is computed by differencing v_{usr} and v_{db}, in that order. This vector represents the deviation of the image from the text description. It is used to represent the attributes specified by the user that are absent in the best-match image. It is then used as a reference to guide the transformation of the latent vector, z_i, of the best-match image. This transformation happens by iteratively modifying the latent vector for each missing attribute. At each step, the modifier component attempts to add a missing attribute, either until the attribute is successfully added or a heuristically predefined maximum number of attempts is reached. The resultant latent vector after transformation, represented by z^*, is finally passed through a generator network, G, that synthesizes the desired output image. This is repeated for the rest of the four other image encodings to synthesize the diversity images. Figure 1 shows the results of the architecture on different text descriptions, where the first image is the best-match image and the remaining four are the diversity images.

3.2 Text Encoder

The pre-trained BERT model called 'BERT base uncased' is used to extract facial features from text. This model is fine-tuned to extract attributes listed in the celebA dataset from a textual description, using the pooled output layer on top of the pretrained layers. The activation function used in the output layer is sigmoid and the loss function used is sigmoid-cross-entropy with logits. The pretrained layers, along with the output layer, are trained in the fine-tuning process. The hyperparameters used to train the model are described in Table 1. The model predicts the likelihood of the celebA facial attributes and outputs the vector v_{usr} as depicted in Fig. 2.

Captions for training were generated with the help of a slightly modified version of the algorithm described in [2]. The existing TTF models, including the previous state-of-the-art TTF-HD model [1], attempt to only extract the celebA attributes directly. But this disregards the innumerous ways of describing a human face using different adjectives. For example, 'stubble' can be used in place of '5 o'clock shadow'; 'dark hair' or 'brunette' for 'black hair'; 'spectacles' or 'specs' for 'eyeglasses'; 'stout face', 'plump face' or 'round face' for 'chubby', etc. To address this problem, TTF-ST pipeline also extracts numerous synonyms of the facial features from the input text while encoding it.

Table 1. Hyperparameters for the BERT model

Hyperparameter	Value
Batch size	32
Learning rate	2e − 5
Training epochs	2.0
Warmup proportion	0.1
Output layer dropout	0.1

3.3 Image Generator

The image generator used in the architecture is a pre-trained StyleGAN2 [10] model due to its ability to generate high-quality facial images of large resolution similar to the FFHQ [14] dataset on which it was trained. It works by decoding a latent vector of 512 dimensions and generating a 1024 × 1024 resolution image. It is used in two places in the architecture - during the generation of the custom dataset as shown in Fig. 3, and during the final step of the pipeline for the synthesis of the desired face image(s) from the best-match image's latent vector as shown in Fig. 2.

3.4 Image Encoder

MobileNetV2 [15] model is chosen to be used as the image encoder IE, given its light-weighted nature, lesser parameters to train and good performance in terms of speed and accuracy. This is inspired by the work done in [16]. This work dropped three celebA attributes during training, namely 'pale skin', 'attractive' and 'blurry'. For its usage as IE, the pre-trained MobileNetV2 is fine-tuned by addition of a fully connected layer, a batch normalization layer and finally a multi-label dense layer whose output is the facial attributes of the given sample. Cosine proximity metric is used as the loss function during fine-tuning.

It is used in two places in the architecture - during encoding of the images present in the custom dataset as shown in Fig. 3, and during the iterative process of latent vector transformation where the current image is encoded at each step of the process to detect the addition of a specified facial attribute.

3.5 Dataset

A custom dataset is generated as shown by Fig. 3 consisting of 10,000 images of resolution of 1024 × 1024 using the image generator G of the StyleGAN2 model, their corresponding latent vectors, z_n, and their image encodings, v_{gen}. This dataset is generated for the purpose of performing the best-match search.

First, the 10,000 images and their latent vectors, z_n, are generated and stored. Then, the image encoder IE is used for encoding all the generated images into image encodings, v_{gen}. To perform the task of encoding, the IE is trained on the celebA dataset.

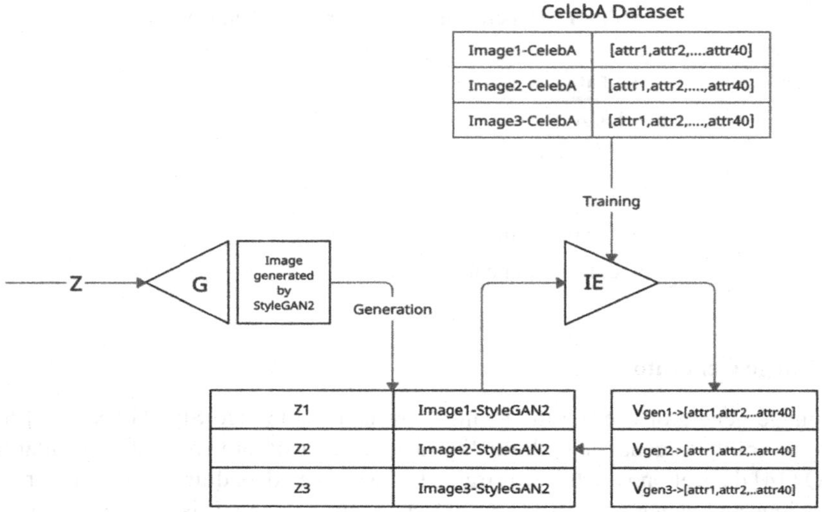

Fig. 3. Custom dataset generation

3.6 Best-Match Search

Upon obtaining the text encoding (v_{usr}) of the input text, best-match search is performed on all the image encodings in the custom dataset and this text encoding. The best-match search routine returns an image from the dataset that best meets the input text description. To perform this, a new metric is introduced for computing the similarity between a text encoding and an image encoding in this problem statement. This metric is referred to as Cosine Similarity for the Specified Attributes, or CSSA for short. It is calculated as the cosine similarity between the encodings for only those attributes which are specified in the input text and thus captured in the text encoding. Apart from being used for the best-match search in our work, CSSA is also helpful for evaluating the accuracy of any TTF model as it can be used as a means for comparing the input text and output image of such a model. Unlike a simple cosine similarity between the encodings, CSSA helps in accounting for diversity in the output images by unconstraining the presence of those facial features that are not mentioned in the input text, thus providing a more accurate representation of the performance of the model.

The face image corresponding to the image encoding (v_{db}) with the highest CSSA is chosen as the best-match image. However, not all features described in the text may have been present in the selected best-match image due to a large number of possible combinations in the description. So, to improve the accuracy of the desired output image of the model, the facial features in the best-match image that are different from those in the text encoding are modified via latent vector transformation. The deviation of the best-match image from the expected facial image is computed as the deviation vector v_{dev} which is obtained by differencing v_{usr} and v_{db} as given in (1).

$$v_{dev} = v_{usr} - v_{db} \tag{1}$$

3.7 Latent Vector Transformation

Latent Vector Transformation (LVT) was inspired by the work done in [1]. This is performed in order to guide the latent vector of the best-match image in the direction of attributes specified in the text encoding, but absent in the best-match image encoding.

First, linear regression is performed to establish a mapping between latent space represented by matrix $L \in R^{512}$ and the attribute space represented by matrix $A \in R^{37}$ to obtain a matrix M of dimension 512×37 as given in (2). Then, orthonormalization on the matrix M is performed in order to disentangle the feature axes.

$$A = LM \quad (2)$$

We introduce a few changes to the procedure described in the work done in [1]. Firstly, instead of locking features, we orthogonalize the one feature column being modified at any given time with respect to all the other features. This helps in iteratively attempting to add a given feature to the face image with as little modification to the other attributes as possible. Secondly, we eliminate the process of nonlinear reweighting because only one feature is being added at a time instead of multiple features. Thirdly, the attributes are changed to indicate only their presence and not absence. Thus, LVT only works on addition of features and not removal of features. This is done in order to improve diversity in the images generated, with variability being present for attributes not specified by the user. In other words, if a feature is present in the best-match image, but is not mentioned in the text description, we do not attempt to remove it, but we keep it to account for diversity.

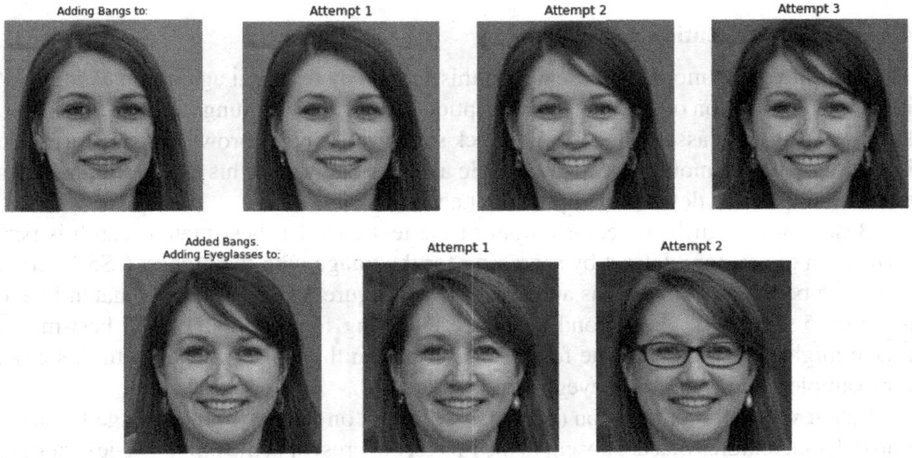

Fig. 4. Adding bangs and eyeglasses to the base image using latent vector transformation with a scaling factor of 2

This process is done iteratively for all the features mentioned in the text description that are absent in the best-match image. At each step of the iteration, a column vector, c, is added to the latent vector, z, as given by (3) in order to attempt to add a missing facial feature.

$$z* = z + c \tag{3}$$

The column vector, c, controls the direction and speed of the movement along the feature axes. It is calculated by multiplying the matrix M with the vector v_{dev}, and by multiplying this product with a scaling factor, s, as given in (4).

$$c = (Mv_{dev})s \tag{4}$$

We introduce the scaling factor, s, in order to control how big the jumps are between each attempt in the transformation. Having a high value for s could disrupt features other than the one that is being added at a given moment because of imperfect disentanglement [1] in the latent space; however, having a low value for s could lead to higher number of attempts for adding a feature. Heuristically, s is set at a default value of 2, but can be modified based on the use case.

Figure 4 shows the LVT process for addition of two facial attributes of 'bangs' and 'eyeglasses' to a randomly chosen starting image. At the end of each step in the iteration, the transformed image is encoded using IE and the addition of the facial attribute is checked in the encoding. Once the attribute is successfully added, the iterative process stops and LVT is performed again on the end image to attempt to add the next missing attribute. During the iterative process, if a heuristically predefined maximum number of attempts is reached, then the process is stopped despite the current attribute not being added in order to prevent introduction of random artifacts in the image.

4 Results

4.1 Pipeline Execution Example

The working of the model is presented in this section by testing it against an example of input text description of a face. The description used is, "This young man has black hair. He is wearing eyeglasses. He has a 5 o'clock shadow, bushy eyebrows and sideburns. He is smiling with his mouth slightly open. He also has bags under his eyes." Fig. 5 shows the encoding of the description by the text encoder.

Based on the attribute vector output of the text encoder, best-match search is performed on the custom dataset by searching for the image with the highest CSSA score. Here, the best-match image has a CSSA of 0.84. Figure 6 shows the best-match image and Fig. 5 shows the corresponding image encoding for this image. The best-match image might not contain all the features described in the text description. In this case, for example, it does not have eyeglasses.

Latent vector transformation (LVT) is performed on the best-match image for addition of facial features that are absent in the image, but present in the input text description. This is an iterative process, and the model checks for the addition of the facial feature at each step. This process is shown in Fig. 7. The end result of LVT is shown in Fig. 8, where the best-match image and the final output image are compared. When the CSSA values for the image encodings of the best-match and the latent vector transformed images are compared, a significant increase from 0.91 to 0.97 is observed, thus demonstrating that the final image is much closer to the desired image.

Attribute	Text encoding	Image Encoding	Attribute	Text Encoding	Image Encoding
5_o_Clock_Shadow	1	0.98	Mouth_Slightly_Open	1	0.99
Arched_Eyebrows	0	0.05	Moustache	0	0.13
Bags_Under_Eyes	1	0.91	Narrow_Eyes	0	0.19
Bald	0	0.0	Beard	0	0.01*
Bangs	0	0.0	Oval_Face	0	0.08
Big_Lips	0	0.46	Pointy_Nose	0	0.2
Big_Nose	0	0.8	Receding_Hairline	0	0.02
Black_Hair	1	0.21	Rosy_Cheeks	0	0.01
Blond_Hair	0	0.01	Sideburns	1	0.75
Brown_Hair	0	0.06	Smiling	1	0.93
Bushy_Eyebrows	1	0.99	Straight_Hair	0	0.46
Chubby	0	0.04	Wavy_Hair	0	0.46
Double_Chin	0	0.04	Wearing_Earrings	0	0.02
Eyeglasses	1	0.0	Wearing_Hat	0	0.02
Goatee	0	0.19	Wearing_Lipstick	0	0.03
Gray_Hair	0	0.0	Wearing_Necklace	0	0.01
Heavy_Makeup	0	0.01	Wearing_Necktie	0	0.05
High_Cheekbones	0	0.12	Young	1	0.99
Male	1	1			

Fig. 5. Output of the text and image encoders for an input text description and the best-match search image respectively. All the attributes in the input text description captured by the text encoder are highlighted.

Fig. 6. Best-match image corresponding to the description

4.2 Quantitative Evaluation

The text encoder was evaluated using the sigmoid cross entropy with logits and hamming score metrics on a dataset generated using a slightly modified version of the algorithm described in [2]. Table 2 consists of values of training and evaluation loss values computed using sigmoid cross entropy metric. The hamming score metric measures the fraction of labels correctly predicted to the total labels. The text encoder scored an average hamming score of 0.999 on the evaluation dataset of over 48k rows. The image encoder was evaluated using binary accuracy metric on celebA dataset. It scored an accuracy of 90.95% on the test dataset.

We use two metrics to compare the performance of our model with other state-of-the-art models. These metrics are Cosine Similarity (CS) and Learned Perceptual Image

Fig. 7. Latent vector transformation of best-match image

Fig. 8. Final Image after latent vector transformation of the best-match image

Table 2. Measure of loss during training and evaluation of text encoder on celebA

	Training	Evaluation
Samples	144215 (75%)	48018 (25%)
Loss	0.010	0.008

Patch Similarity (LPIPS). We only compare CS with other models, and not the newly introduced CSSA metric as the existing models have not been scored using CSSA. CS is used to evaluate the accuracy of a model, while LPIPS is used to evaluate the level of diversity present in the images generated by a model for a given text description.

To evaluate the pipeline, we generate 100 random descriptions using our caption generation algorithm, and generate 10 images for each description - the same as what is done in the previous state-of-the-art TTF-HD model in [1]. We measure the three metrics on these images. Table 3 shows the comparison of TTF-ST with TTF-HD [1] and AttnGAN [11]. TTF-ST was able to significantly outperform the previous models in terms of cosine similarity. Both maximum and average CS are markedly higher than the maximum CS of TTH-HD.

Table 3. Evaluation results of different models

Model	Maximum CS	Average CS	Average LPIPS
TTF-ST (ours)	**0.98**	**0.785**	0.518
TTF-HD	0.664	----	0.583
AttnGAN	0.511	----	----

The average CSSA score of TTF-ST was found to be **0.932**, with the maximum CSSA being **1.0**. A score of 1.0 signifies that all the attributes in the text encoding of the input description were captured in the image encoding of the final transformed image.

The decrease in LPIPS score is likely due to the nature of the proposed pipeline itself. Unlike TTF-HD which modifies the latent vectors of randomly generated images, the TTF-ST pipeline starts LVT with images which are much closer to the desired image obtained using best-match search, thereby significantly boosting the accuracy at the cost of some diversity.

5 Conclusion and Future Work

In this paper, we present the Text-To-Face Search and Transform model, TTF-ST for short, which introduces a novel technique to accomplish the task of text-to-face image synthesis by searching and selecting images from a custom dataset that best match the encoded input textual description, and subsequently transforming their latent vectors to enhance the accuracy of the generated images. As there are numerous ways in which a particular facial feature can be described, the model accounts for diversity in the input descriptions by incorporating several synonyms of the facial features and using a highly accurate text classifier. The model also accounts for diversity in the output face images by generating multiple images for each description, with each image potentially having different facial features not specified in the input text. The model consists of image generator, text encoder and image encoder models working harmoniously together in a pipeline. We also introduce a new metric of Cosine Similarity for the Specified Attributes (CSSA) to help better evaluate any TTF model which uses a fixed set of facial attributes. This metric is also used during the best-match search in the custom dataset. The proposed model outperforms the relevant state-of-the-art models presented in literature in terms of the accuracy and quality of images generated.

In future, the individual components of image encoder and image generator in the pipeline can be further enhanced to improve the overall performance. The current pipeline is trained to work with the attributes in the celebA dataset, but it can be extended by training the modules on a dataset with more facial attributes. More techniques for feature disentanglement in latent space can be explored to further improve the accuracy of images generated.

References

1. Wang, T., Zhang, T., Lovell, B.: Faces à la carte: text-to-face generation via attribute disentanglement. In: 2021 IEEE Winter Conference on Applications of Computer Vision (WACV), pp. 3379–3387. IEEE (2021). https://doi.org/10.1109/WACV48630.2021.00342
2. Nasir, O., Jha, S., Grover, M., Yu, Y., Kumar, A., Shah, R.: Text2FaceGAN: face generation from fine grained textual descriptions. In: 2019 IEEE Fifth International conference on Multimedia Big Data (BigMM), pp. 58–67. IEEE (2019). https://doi.org/10.1109/BigMM.2019.00-42
3. Khan, M., et al.: A realistic image generation of face from text description using the fully trained generative adversarial networks. In: IEEE Access **9**, 1250–1260 (2021). https://doi.org/10.1109/ACCESS.2020.3015656
4. Devlin, J., Chang, M., Lee, K., Toutanova, K.: BERT: Pre-training of deep bidirectional transformers for language understanding. In: arXiv preprint arXiv: 1810.04805 (2018)
5. Liu, Z., Luo, P., Wang, X., Tang, X.: Deep learning face attributes in the wild. In: 2015 IEEE International Conference on Computer Vision (ICCV), pp. 3730–3738. IEEE (2015). https://doi.org/10.1109/ICCV.2015.425
6. Sun, C., Qiu, X., Xu, Y., Huang, X.: How to fine-tune BERT for text classification? In: Sun, M., Huang, X., Ji, H., Liu, Z., Liu, Y. (eds.) CCL 2019. LNCS (LNAI), vol. 11856, pp. 194–206. Springer, Cham (2019). https://doi.org/10.1007/978-3-030-32381-3_16
7. Goodfellow, I., et al.: Generative adversarial nets. In: Advances in Neural Information Processing Systems 2014, pp. 2672–2680 (2014)
8. Karras, T., Laine, S., Aila, T.: A style-based generator architecture for generative adversarial networks. In: 2019 IEEE/CVF Conference on Computer Vision and Pattern Recognition (CVPR), pp. 4396–4405. IEEE (2019). https://doi.org/10.1109/CVPR.2019.00453
9. Shorten, C.: StyleGAN2. https://towardsdatascience.com/stylegan2-ace6d3da405d. Accessed 2 Apr 2021
10. Karras, T., Laine, S., Aittala, M., Hellsten, J., Lehtinen, J., Aila, T.: Analyzing and improving the image quality of StyleGAN. In: 2020 IEEE/CVF Conference on Computer Vision and Pattern Recognition (CVPR), pp. 8107–8116. IEEE (2020). https://doi.org/10.1109/CVPR42600.2020.00813
11. Xu, T., et al.: AttnGAN: fine-grained text to image generation with attentional generative adversarial networks. In: 2018 IEEE/CVF Conference on Computer Vision and Pattern Recognition, pp. 1316–1324. IEEE (2018). https://doi.org/10.1109/CVPR.2018.00143
12. Zhang, H., et al.: StackGAN: text to photo-realistic image synthesis with stacked generative adversarial networks. In: 2017 IEEE International Conference on Computer Vision (ICCV), pp. 5908–5916. IEEE (2017). https://doi.org/10.1109/ICCV.2017.629
13. Chen, X., Qing, L., He, X., Luo, X., Xu, Y.: FTGAN: a fully trained generative adversarial network for text to face generation. arXiv:1904.05729. http://arxiv.org/abs/1904.05729 (2019)
14. Karras, T., Aittala, M., Hellsten, J., Laine, S., Lehtinen, J., Aila, T.: Flickr-Faces-HQ Dataset (FFHQ) (2019). https://github.com/NVlabs/ffhq-dataset

15. Sandler, M., Howard, A., Zhu, M., Zhmoginov, A., Chen, L.: MobileNetV2: inverted residuals and linear bottlenecks. In: 2018 IEEE/CVF Conference on Computer Vision and Pattern Recognition, pp. 4510–4520. IEEE (2018). https://doi.org/10.1109/CVPR.2018.00474
16. Anzalone, L., Barra, P., Barra, S., Narducci, F., Nappi, M.: Transfer learning for facial attributes prediction and clustering. In: Wang, G., El Saddik, A., Lai, X., Martinez Perez, G., Choo, K.-K. (eds.) iSCI 2019. CCIS, vol. 1122, pp. 105–117. Springer, Singapore (2019). https://doi.org/10.1007/978-981-15-1301-5_9

Lensless Image Reconstruction with an Untrained Neural Network

Abeer Banerjee[1,2](✉), Himanshu Kumar[1,2], Sumeet Saurav[1,2], and Sanjay Singh[1,2]

[1] Academy of Scientific and Innovative Research (AcSIR), CSIR-Human Resource Development Centre, (CSIR-HRDC) Campus, Postal Staff College Area, Sector 19, Kamla Nehru Nagar, Ghaziabad 201 002, Uttar Pradesh, India
{abeer.ceeri20a,himanshu.ceeri20a}@acsir.res.in
[2] CSIR-Central Electronics Engineering Research Institute (CSIR-CEERI), Pilani 333 031, Rajasthan, India
{sumeet,sanjay}@ceeri.res.in
https://www.acsir.res.in, https://www.ceeri.res.in

Abstract. Lensless image reconstruction is an ill-posed inverse problem in computational imaging, having several applications in machine vision. Existing approaches rely on large datasets for learning to perform deconvolution and are often specific to the point spread function of a particular lensless imager. Generating pairs of lensless images and their corresponding ground truths requires a specialized laboratory setup, thus making the dataset collection procedure challenging. We propose a reconstruction method using untrained neural networks that relies on the underlying physics of lensless image generation. We use an encoder-decoder network for reconstructing the lensless image for a known PSF. The same network can predict the PSF when supplied with a single example of input and ground-truth pair, thus acting as a one-time calibration step for any lensless imager. We used a physics-guided consistency loss function to optimize our model to perform reconstruction and PSF estimation. Our model generates accurate non-blind reconstructions with a PSNR of 24.55 dB.

Keywords: Lensless image reconstruction · Untrained neural networks · Computational imaging

1 Introduction

Cameras have evolved into an important part of human life, acting as external sensors for observing the physical environment. Cameras of with various functional attributes are used for several purposes, ranging from macro photography to astronomical photography. The recent advancements in technologies like computer vision-based wearables, augmented reality, and microrobotics were essential factor that demanded the miniaturizing of imaging systems. The volume and size of cameras have decreased over time for various applications, but the

W. Q. Yan et al. (Eds.): IVCNZ 2022, LNCS 13836, pp. 430–441, 2023.
https://doi.org/10.1007/978-3-031-25825-1_31

dependence on lenses prevents further reductions in size because the conventional lens-based optical elements contribute to more than 90% of the volume of the imager. Computational imaging research is attempting to eliminate the need for lenses in the camera arrangement to drastically reduce camera size [1–4]. Removing lenses reduces the size of the camera nearly to the flat form-factor, and various sensor shapes are possible to use.

The lensless imaging setup captures the object information by multiplexing light rays onto the sensors, and the image is then reconstructed by solving an inverse problem. Different optical elements are used for multiplexing, such as diffusers, coded apertures, or diffraction gratings. Existing research in lensless image reconstruction work has produced results comparable to lensed images utilizing two different approaches: optimization-based and learning-based [5]. Lensless cameras have potential use-cases in 3D microscopy [6,7], monocular depth estimation, and other such fields [8]. They are being employed for privacy-protecting applications [9,10] because, without the knowledge of the point spread function, the mother image can not be robustly reconstructed with the currently available methods.

Image reconstruction using a lensless setup can be understood as the inverse computational imaging problem. Traditionally, domain-specific recovery algorithms have been utilized to create hand-crafted mathematical models that draw conclusions from their understanding of the basic forward model related to the measurement. These techniques often do not rely on a dataset to learn the mapping, and because the problem is ill-posed, the models frequently show poor discriminative performance [11]. On the other side, deep learning-based methods provide a breakthrough compared to the hand-crafted methods to solve the inverse computational imaging problem [12]. The most recent advancement in Generative Adversarial Networks(GANs) has demonstrated its capability to recreate high-resolution pictures with lesser information from the sample data. With hand-crafted methods, it was impossible to achieve this high compression ratio. However, the success of the deep-learning-based approaches majorly rely on large labelled datasets, and medical imaging and microscopic imaging are a few of the applications in which the enriched labelled data is not available. In contrast to the earlier training-based deep learning approaches, untrained neural networks are able to estimate the faithful reconstruction using the corrupted measurement as an input, without having any prior exposure to the ground truth.

Various optimization algorithms are used to solve inverse problems to recover the original image. The alternating direction method of multipliers (ADMM) [13], regularised L1 [14], or total variation regularisation [15] are common variations that have been modified for lensless imaging [1,2]. The main advantage of this method is that they are data-agnostic, but there is a trade-off between computational complexity and reconstruction quality. The deep-learning based

ResNet architecture is utilized with raw sensor measurements to create the image [16]. However, the reconstruction of the image is only possible if the acquired image was taken up close since this makes it easier to project high-intensity features onto the sensor. In continuation, deep learning for lensless optics has been coupled with spatial light modulators, scattering media [17], and glass diffusers [18] to produce reconstructions. With the use of lensless images and object-detectors based on Convolutional Neural Networks, FlatCam [19] was able to recognize faces using deep learning. Khan et al. [20] proposed that Flatcam reconstruction was performed using GANs without the need for a point-spread PSF during testing. Another proposed method optimizes unrolled ADMM using a learnable parameter that could be integrated with U-Net and tested on diffuser images [21]. The existing work either relies on the large labelled dataset to perform the reconstruction task or the iterative optimization method which is computationally complex. In this paper, we reconstruct lensless images with untrained neural networks. No training data is used in the reconstruction process that is guided by a physics-informed consistency loss. We verified our approach with images captured using multiple random-diffusers, each with a unique random point spread function. We performed a detailed performance evaluation of our method against the traditional optimization-based and deep-learning-based methods with evaluation metrics like Peak Signal-to-Noise Ratio (PSNR) and Structural SIMilarity index (SSIM). We have also provided visual comparison results that indicate our method was able to outperform the existing methods in the majority of test cases.

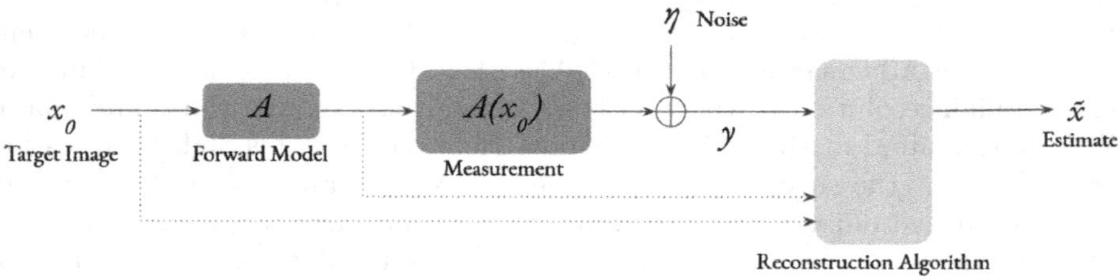

Fig. 1. A typical pipeline for solving inverse problems in computational imaging. The forward model A generates the measurement. The reconstruction algorithm could require priors about the target image, the forward measurement procedure, etc. to reconstruct the estimate $\tilde{x}$.

2 Theoretical Basis

Lensless image reconstruction is a classic example of an inverse problem in computational imaging. Typically, in every inverse problem, the main task is the reconstruction of a signal for the available observed measurement. In our case, the observed measurement is the lensless image captured using the bare image sensor with a random diffuser protecting it. A general approach toward solving an inverse problem is to formulate the forward problem as:

$$y = A(x_0) + \eta, \tag{1}$$

where y is the measurement obtained via a forward operation A that signifies the physical measurement process that acts on the input x_0, and η represents the noise process. In mathematical terms, the inverse problem refers to the faithful reconstruction of the target image x_0, given the available measurement y. There are a wide variety of inverse problems such as denoising, super-resolution, phase-retrieval, inpainting, and deconvolution that are fundamentally differentiated by the forward operator used for generating the measurement. In this paper, we concern ourselves with lensless image reconstruction, which is essentially a deconvolution problem. The forward operator for a deconvolution problem is formulated as:

$$A(x) = k * x, \tag{2}$$

where k is called the point spread function (PSF), and $*$ denotes the convolution operation. Reconstruction algorithms that utilize the prior knowledge of the PSF that characterizes the whole lensless image formation process, are called non-blind deconvolution techniques. Algorithms that do not require explicit knowledge about the PSF are called blind deconvolution techniques, but they are ill-posed and can result in multi-modal solutions.

Deep Learning based approaches are being extensively used for solving inverse problems for their ability to leverage large datasets for learning a mapping from y to x. Generative adversarial training of U-Nets has resulted in excellent lensless image reconstruction models [22,23], especially for the non-blind deconvolution task. However, the task-specificity of these discriminative approaches reduces the generalizability of the model and these approaches are yet not well equipped at handling even subtle changes in the forward measurement process. Furthermore, the generation and recording process of datasets related to inverse problems is often complicated. The huge computational cost involved in the retraining and reconfiguration of learning-based methods subject to the availability of datasets with acceptable quality is major setback that need to be addressed.

Physics-informed neural networks are being used for handling ill-posed inverse problems due to their capability of integrating mathematical physics and data. Incorporating the physics of operation of a particular problem simplifies the task and helps in faster convergence of the model, and also attempts to address the problem of big data requirements [24]. Untrained Neural Networks (UNN) completely solve the problem of training data requirements making them perfect for problems where paired data collection is challenging or cumbersome. An approach for phase imaging using untrained neural networks has been explored by Wang et al. [25]. There, they use an iterative approach for optimizing a physics-enhanced deep neural network to produce the object phase. Our approach is

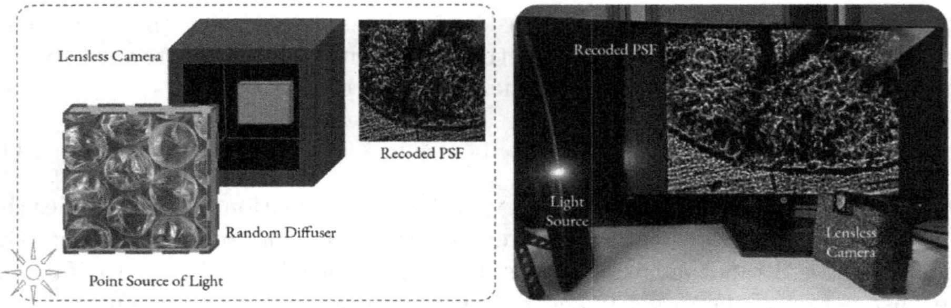

Fig. 2. Experimental setup for capturing the point spread function obtained for a random diffuser. The same setup was used for capturing lensless images corresponding to the lensed images displayed on the monitor.

inspired by their concept of incorporating a physical model into a neural network, which we use to solve the deconvolution problem in lensless computational imaging.

In this paper, we attempt to address the inverse computational imaging problem using a UNN to get reconstruction of the lensless image. We iteratively optimize our model with a physics-informed consistency loss, thereby achieving faster convergence and better performance compared to traditional optimization-based techniques. Our approach leverages random diffusers prepared using inexpensive materials like bubble wraps to capture lensless images making the whole setup compact and easily reproducible.

3 Experimental Setup

Figure 2 illustrates the experimental setup used for capturing the PSF with a random diffuser. We used a bright light source as a point source to illuminate the lensless camera system. We used a random diffuser, i.e., a bubble wrapping plastic sheath, to protect the camera sensor. The resulting PSF recorded using this setup was convolved with a lensed image to generate a lensless image. The lensless image, thus generated, was compared with the lensless image captured using the same setup, and we found that they were very similar. Ideally, they should have matched to each pixel, but practically, the lensless image capturing process introduces a slight amount of noise into the captured image.

4 Methodology

This section discusses the network architecture designed for the reconstruction pipeline. We have not used any training dataset for reconstruction. The images used for explaining the pipeline are obtained from the lensless image test set of nine images captured with the DiffuserCam, provided by Monakhova et al. [21].

4.1 Network Architecture

For a reconstruction task, it is very common to use an encoder-decoder architecture. U-Nets are the most commonly used networks for image reconstruction since they have skip connections that help in modelling identity transformations. However, the lensless images are heavily multiplexed, leading to an image that is incomprehensible to humans and that shares almost no structural attributes with their lensed counterparts. Therefore, the presence of identity connections need not imply a faster convergence, since an identity transformation does not help.

We have used an encoder-decoder framework with the encoder network being popular convolutional architectures like a ResNet or a DenseNet. The CNN architectures available off-the-shelf are truncated to exclude the fully-connected layers, such that the resulting encoder structure is fully convolutional. The structure of the decoder network is mostly fixed with a varying number of input channels according to the feature maps produced by the truncated encoder.

4.2 Pipeline

Figure 3 illustrates the PSF estimation and the reconstruction process. The lensless image reconstruction pipeline follows an untrained iterative optimization that uses a physics-based consistency loss for optimizing the encoder-decoder framework. In the forward path, the lensless image is set as the input to the neural network that produces an intermediate reconstruction y. The output of the neural network is passed through the mathematical process of lensless image generation H, which refers to the convolution of the intermediate reconstruction with the PSF to produce an intermediate lensless image. Theoretically, the neural network achieves perfect reconstruction if the generated intermediate lensless image via process H is the same as the input lensless image. Therefore, we backpropagate the Mean-Squared Error loss between the generated intermediate lensless image $\tilde{I}$ via process H, and the original lensless image x, i.e., the physics-based consistency loss.

The same framework can be repurposed to estimate the PSF, subject to the condition that we already have a pair lensed-lensless pair. This is known as the calibration step that is specific to every lensless camera system. Here, we provide the network with the lensless image and expect the PSF as the output. The model can be forced to predict the PSF by changing the mathematical process H, i.e., by convolving the intermediate output with the available lensed image corresponding to the lensless image. The remaining process of calculating the consistency loss is the same.

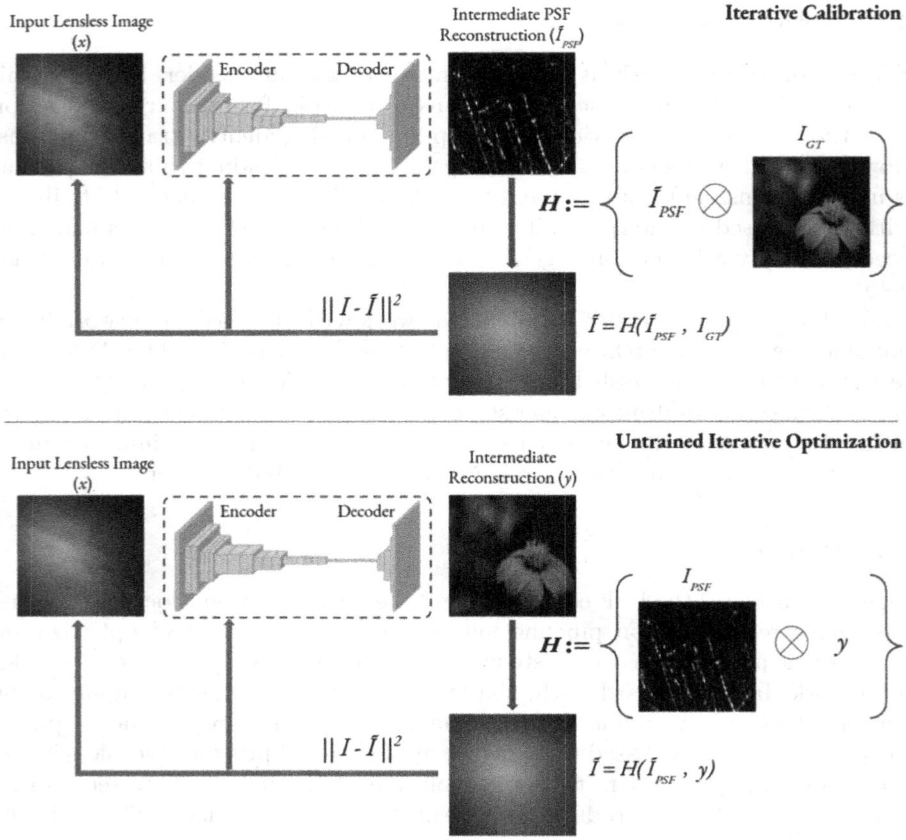

Fig. 3. The complete pipeline for lensless image reconstruction. The top section of the image illustrates the calibration step for PSF estimation using a single pair of lensless and ground truth images. The bottom section illustrates the iterative reconstruction approach. It is to be noted that the same network is being utilized for calibration and reconstruction.

5 Results and Analysis

This section presents the results obtained using the testing set of the DiffuserCam dataset [21]. The images were created with a random diffuser, making them suitable for the evaluation of our framework. We have selected five images from the test set to provide our results using two commonly used metrics in the image reconstruction domain, namely, peak signal-to-noise ratio PSNR and structural similarity index SSIM. Table 1 shows the PSNR obtained by our model corresponding to different images, and Table 2 shows the SSIM results.

Table 1. Image-specific PSNR comparison of our method with the existing popular reconstruction methods.

PSNR					
Method	Flower	Face	Butterfly	Flowers	Tokens
ADMM [13]	14.40	11.10	13.02	12.00	13.36
U-Net [26]	16.57	18.51	14.42	17.43	9.83
GAN [22]	19.32	17.48	14.96	17.08	11.21
Ours	**24.55**	**22.95**	**17.72**	16.59	**13.09**

Table 2. Image-specific SSIM comparison of our method with the existing popular reconstruction methods.

SSIM					
Method	Flower	Face	Butterfly	Flowers	Tokens
ADMM [13]	0.50	0.38	0.44	0.39	0.55
U-Net [26]	0.62	0.61	0.69	0.60	0.43
GAN [22]	0.67	0.73	0.71	0.75	0.42
Ours	**0.84**	**0.80**	**0.73**	0.70	**0.57**

5.1 Visual Comparison

Figure 4 compares the visual performance of our method against the existing methods. Metrics like PSNR and SSIM are helpful in determining the reconstruction performance, but they often fail to match the human visual perception, as pointed out by Rego et al. [22]. It can be evidently observed that the visual performance of our model is superior to the existing methods in most scenarios. A noteworthy observation was that ADMM performs exceptionally well compared to the other recent models when it comes to the reconstruction of fine features, as seen in the "Token" image. Otherwise, in most cases, it produces dark and grainy reconstructions.

5.2 Ablation Study

We performed an ablation study of the encoder architecture to determine the variation in reconstruction performance. We used different versions of the DenseNet and the ResNet architectures to obtain truncated encoders corresponding to each architecture. These encoders extracted the features from the lensless image, which were decoded to form the reconstructed image. In the increasing order of parameter count, we used DenseNet-121, DenseNet-201, ResNet-50, and ResNet-101 for the reconstruction task. The results are displayed in Fig. 5.

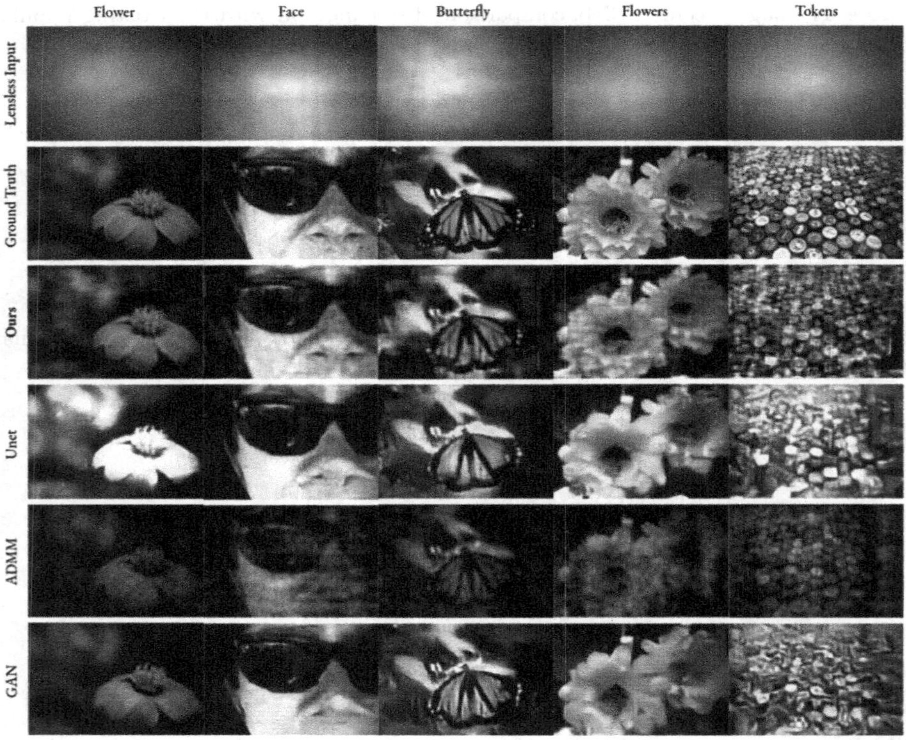

Fig. 4. A visual comparative study of our approach against the existing popular reconstruction approaches.

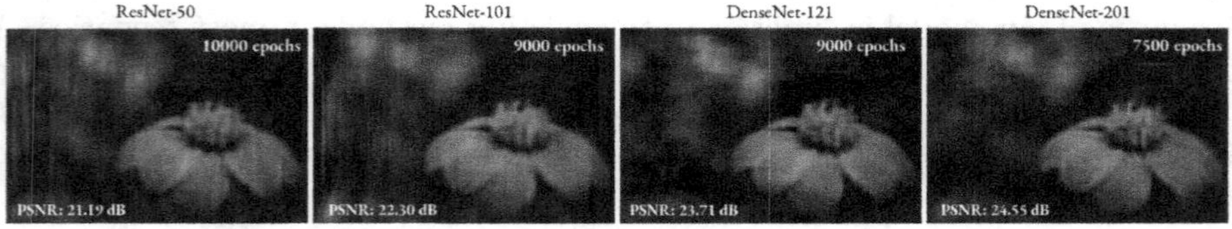

Fig. 5. Ablation study of the encoder backbone used for reconstructing the *Flower* image.

The results obtained by DenseNet-201 appear to be of the best visual quality, but since there is a trade-off between the model performance and the parameter size, DenseNet-121 should be declared the best performer since it achieves a reconstruction performance comparable to the DenseNet-201 with drastically reduced parameter size. The number of epochs for which the iterative optimization had to be performed is plotted on the top right section of each image in Fig. 5. The convergence was significantly faster for the DenseNet framework compared to the ResNet framework, which might be because dense connections strengthen feature propagation, thereby facilitating feature reuse.

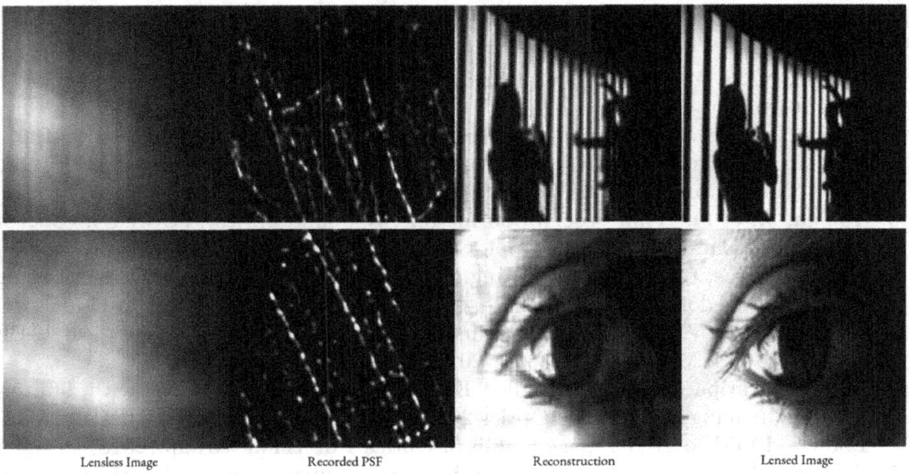

Fig. 6. Reconstructions achieved using PSFs captured by our setup.

5.3 Discussion

Lensless images captured using our own setup took less than 9000 epochs to produce a faithful reconstruction. Figure 6 shows some of the images from the DiffuserCam Training set that were captured using our lensless imager. The PSF was captured using the setup shown in Fig. 2. Our main observation is that the reconstruction of finer features presents inside an image is difficult to reconstruct with the current resolution of PSF convolution. If the PSF convolution is performed at a higher resolution during the physics-based consistency loss calculation, the network might be able to reserve the details, although the optimization time would severely suffer.

6 Conclusion

We have achieved a faithful reconstruction of a lensless image captured using a random diffuser without any training data. We compared the resulting reconstructions with the existing ADMM-based, U-Net-based, and GAN-based approaches and evaluated our performance with metrics like PSNR and SSIM. In almost all cases, our method was able to outperform the existing methods by a significant margin. To support our claims, we present a visual comparison report using the lensless test images provided in the DiffuserCam dataset. The optimization pipeline with the untrained neural networks is a general pipeline that can be repurposed to any inverse imaging task, provided that we have the correct physics-based consistency loss to model the system.

References

1. Antipa, N., et al.: Diffusercam: lensless single-exposure 3d imaging. Optica **5**(1), 1–9 (2018)
2. Salman Asif, M., Ayremlou, A., Sankaranarayanan, S.C., Veeraraghavan, A., Baraniuk, R.: FlatCam: Thin, lensless cameras using coded aperture and computation. IEEE Trans. Comput. Imag. **3**(3), 384–397 (2017)
3. Tanida, J., et al.: Thin observation module by bound optics (TOMBO): concept and experimental verification. Appl. Opt. **40**(11), 1806–1813 (2001)
4. Gill, P.R., Lee, C., Lee, D.-G., Wang, A., Molnar, A.: A microscale camera using direct Fourier-domain scene capture. Opt. Lett. **36**(15), 2949–2951 (2011)
5. Boominathan, V., et al.: Lensless imaging: a computational renaissance. IEEE Signal Process. Mag. **33**(5), 23–35 (2016)
6. Kuo, G., Antipa, N., Ng, R., Waller, L.: 3D fluorescence microscopy with diffusercam. In: Imaging and Applied Optics 2018 (3D, AO, AIO, COSI, DH, IS, LACSEA, LS&C, MATH, pcAOP), page CM3E.3. Optica Publishing Group (2018)
7. Adams, J.K., et al.: Single-frame 3D fluorescence microscopy with ultraminiature lensless flatscope. Sci. Adv. **3**(12) (2017)
8. Zheng, Y., Salman Asif, M.: Joint image and depth estimation with mask-based lensless cameras. In: 2019 IEEE 8th International Workshop on Computational Advances in Multi-Sensor Adaptive Processing (CAMSAP) (2019)
9. Wang, Z.W., Vineet, V., Pittaluga, F., Sinha, S.N., Cossairt, O., Kang, S.B.: Privacy-preserving action recognition using coded aperture videos. In: 2019 IEEE/CVF Conference on Computer Vision and Pattern Recognition Workshops (CVPRW), pp. 1–10 (2019)
10. Canh, T.N., Nagahara, H.: Deep compressive sensing for visual privacy protection in flatcam imaging. In: 2019 IEEE/CVF International Conference on Computer Vision Workshop (ICCVW), pp. 3978–3986 (2019)
11. Hegde, C.: Algorithmic Aspects of Inverse Problems Using Generative Models. IEEE Press (2018)
12. Ongie, G., Jalal, A., Metzler, C.A., Baraniuk, R.G., Dimakis, A.G., Willett, R.: Deep learning techniques for inverse problems in imaging. IEEE J. Select. Areas Inf. Theory **1**(1), 39–56 (2020)
13. Boyd, S., Parikh, N., Chu, E., Peleato, B., Eckstein, J.: Distributed optimization and statistical learning via the alternating direction method of multipliers. Found. Trends Mach. Learn. **3**(1), 1–122 (2011)
14. Beck, A., Teboulle, M.: A fast iterative shrinkage-thresholding algorithm for linear inverse problems. SIAM J. Imaging Sci. **2**, 183–202 (2009)
15. Rudin, L.I., Osher, S., Fatemi, E.: Nonlinear total variation based noise removal algorithms. Phy. D Nonlinear Phenomena **60**, 259–268 (1992)
16. Sinha, A., Lee, J., Li, S., Barbastathis, G.: Lensless computational imaging through deep learning. Optica **4**(9), 1117–1125 (2017)
17. Li, Y., Xue, Y., Tian, L.: Deep speckle correlation: a deep learning approach toward scalable imaging through scattering media. Optica **5**(10), 1181–1190 (2018)
18. Li, S., Deng, M., Lee, J., Sinha, A., Barbastathis, G.: Imaging through glass diffusers using densely connected convolutional networks. Optica **5**(7), 803–813 (2018)
19. Tan, J., et al.: Face detection and verification using lensless cameras. IEEE Trans. Comput. Imaging **5**(2), 180–194 (2018)

20. Khan, S.S., Adarsh, V.R., Boominathan, V., Tan, J., Veeraraghavan, A., Mitra, K.: Towards photorealistic reconstruction of highly multiplexed lensless images. In: Proceedings of the IEEE/CVF International Conference on Computer Vision, pp. 7860–7869 (2019)
21. Monakhova, K., Yurtsever, J., Kuo, G., Antipa, N., Yanny, K., Waller, L.: Learned reconstructions for practical mask-based lensless imaging. Opt. Express **27**(20), 28075–28090 (2019)
22. Rego, J.D., Kulkarni, K., Jayasuriya, S.: Robust lensless image reconstruction via PSF estimation. In: Proceedings of the IEEE/CVF Winter Conference on Applications of Computer Vision, pp. 403–412 (2021)
23. Nelson, S., Menon, R.: Bijective-constrained cycle-consistent deep learning for optics-free imaging and classification. Optica **9**(1), 26–31 (2022)
24. George Em Karniadakis: Ioannis G Kevrekidis, Lu Lu, Paris Perdikaris, Sifan Wang, and Liu Yang. Phys. Inf. Mach. Learn. Nat. Rev. Phys. **3**(6), 422–440 (2021)
25. Wang, F., et al.: Phase imaging with an untrained neural network. Light: Sci. Appl. **9**(1), 1–7 (2020)
26. Ronneberger, O., Fischer, P., Brox, T.: U-Net: convolutional networks for biomedical image segmentation. In: Navab, N., Hornegger, J., Wells, W.M., Frangi, A.F. (eds.) MICCAI 2015. LNCS, vol. 9351, pp. 234–241. Springer, Cham (2015). https://doi.org/10.1007/978-3-319-24574-4_28

A Lexicon and Depth-Wise Separable Convolution Based Handwritten Text Recognition System

Lalita Kumari[1], Sukhdeep Singh[2], V. V. S. Rathore[3], and Anuj Sharma[1](✉)

[1] Department of Computer Science and Applications, Panjab University, Chandigarh, India
{lalita,anujs}@pu.ac.in
[2] D.M. College (Aff. to Panjab University, Chandigarh), Moga, India
[3] Physical Research Laboratory, Ahmedabad, India
vaibhav@prl.res.in
https://anuj-sharma.in

Abstract. Cursive handwritten text recognition is a challenging research problem in pattern recognition. The current state-of-the-art approaches include models based on convolutional recurrent neural networks and multi-dimensional long short-term memory recurrent neural network techniques. These methods are highly computationally extensive as well model is complex at the design level. In recent studies, a combination of convolutional neural networks and gated convolutional neural networks based models demonstrated less number of parameters in comparison to convolutional recurrent neural networks based models. In the direction to reduced the total number of parameters to be trained, in this work, we have used depthwise separable convolution in place of standard convolutions with a combination of gated-convolutional neural network and bidirectional gated recurrent unit to reduce the total number of parameters to be trained. Additionally, we have also included a lexicon-based word beam search decoder at the testing step. It also helps in improving the overall accuracy of the model. We have obtained 3.84% character error rate and 9.40% word error rate on IAM dataset, 3.15% character error rate and 11.8% word error rate on RIMES dataset and 4.88% character error rate and 14.56% word error rate in George Washington dataset respectively.

Keywords: Depthwise separable convolution · Cursive handwritten text line recognition · Word beam search · Deep learning

1 Introduction

Handwritten Text Recognition (HTR) is a complex and widely studied computer vision problem in the research community. In HTR, cursive strokes of handwritten text needs to be recognized. Available text can be either in online or offline form [1]. In online HTR, the time-ordered sequence of pen tip is captured. While

W. Q. Yan et al. (Eds.): IVCNZ 2022, LNCS 13836, pp. 442–456, 2023.
https://doi.org/10.1007/978-3-031-25825-1_32

in offline HTR, static images of handwritten text are available. In this work, we have focused on offline HTR. Handwriting recognition, especially offline HTR systems poses challenges, such as variability of strokes varying in single and multi writers environment, poor and degraded quality historical document images, slop and slant present in the text, variable space present in between lines and characters of handwritten text and limited availability of labelled dataset needed for training of the HTR model. Modern deep learning based techniques are used to solve this complex task.

Initially, Hidden Markov Models (HMM) based techniques are used to solve HTR problems. In this technique, a text image is pre-processed using various computer vision techniques and hand-crafted features as aspect ratios of individual characters are manually extracted from an image and fed to HMM-based classifiers for recognition. Due to HMM's limited capacity for extracting contextual information and manual feature selection, recognition results are poor. In the last few decades, deep learning based methods are primarily used for this task. Convolutional Neural Networks (CNN), Recurrent Neural Networks (RNN), Convolutional Recurrent Neural Networks (CRNN), Gated CNN (GCNN), Multi-Dimensional Long Short-Term Memory (MDLSTM) are some state-of-the-art machine learning techniques used to solve HTR problem. The convolution based techniques are used to extract fine features of the input image and RNN based techniques provide memory to remember long character sequences. Connectionist Temporal Classification (CTC) is widely used to train and test these Neural Networks (NNs) based systems in an end-to-end manner. Since the emergence of artificial intelligence and machine learning based devices in day-to-day life, the latest research trends in the HTR domain are favoured for a robust, less complex system with fewer trainable parameters and acceptable accuracy. In this work, we have proposed an HTR model favoured in this direction. We have applied depth-wise separable convolutional operation in place of standard convolution to reduce the number of trainable parameters. A combination of convolutions, depth-wise convolutions and fully gated convolutions along with Bidirectional Gated Recurrent Units (BGRUs) are used to recognize text lines of benchmarked datasets in an end-to-end manner. Following are the key contributions of our present study,

- A novel end-to-end HTR system is given to recognize text lines.
- Proposed model is able to improve recognition accuracy with less number of trainable parameters (Number of training parameters 820,778).
- Recognition text is lexically confined with the help of Word Beam Search (WBS) decoder [2].
- Overall pipeline of HTR system including essential steps are presented for better understanding of HTR system.
- We have achieved 3.84% Character Error Rate (CER) and 9.40% Word Error Rate (WER) on the IAM dataset, 3.15% CER and 11.8% WER on the RIMES dataset and 4.88% CER and 14.56% WER on the George Washington (GW) dataset, respectively. These comparable state-of-the-art results were achieved using fewer training parameters.

Extensive experiments are performed on benchmarked datasets such as IAM, RIMES and GW. The rest of the paper is organized as follows. Section 2 includes the key contributions in text line recognition. Section 3 demonstrates the proposed architecture. Extensive experiments are presented in Sect. 4. Results obtained reported in Sect. 5 and Sect. 6 include the conclusion of the present study.

2 Related Work

In this section, previous works in the domain of HTR have been discussed. We have focused on the text line HTR recognition task. This section presents the genesis of HTR that helps readers to understand the evaluation of significant HTR techniques. At the start, researchers tackle the HTR task by using Dynamic Programming (DP) based approaches on word-level images, where an optimum path finding based algorithm is used. In DP based approaches, character accuracy improves but this does not guarantee the improvement in the overall word accuracy [3,4]. Later, HMM based techniques were used in HTR [5]. The combination of N-Gram language models and optical recognition using HMM with Gaussian mixture emission probability (HMM-GMM) is one of the primarily studied HTR techniques [6,7]. Optical models of HMMs are further improved by using Multi-Layer Perceptron (MLP) as emission probability [8,9] and discriminative training techniques in HMM-GMMs [10]. The MLP is constrained by fixed length input, thus Long Short-Term Memory Recurrent Neural Networks (LSTM)-HMM based techniques are used in HTR system [11,12]. In the HMM the probability of each observation depending only on the current state; therefore, contextual information is less utilized [13].

As an alternative to the HMM, the RNN do not suffer from these limitations [14]. The CTC is used in sequence labelling task [15]. Models trained with the help of CTC do not require pre-segmented data and are able to provide probability distribution of label sequences. The variation of LSTM such as BLSTM and MDLSTM are used with CTC to give state-of-the-art recognition results [13,16–19]. The LSTM and BLSTM work in one-dimensional sequences and MDLSTM captures long-term dependencies across both the directions. The MDLSTM is highly computationally expensive. Similar recognition result can be acquired by stacking the BLSTM layers [20]. The CRNN is first introduced in the scene-text recognition task [21]. It consists of a stack of convolutional layers followed by one or more layers of BLSTM and softmax output layers, that provides occurrence probability of N (numbers of characters in the dataset) +1 (CTC blank) tokens [22]. A similar model was proposed using GCNN architecture [23]. Later the benefits of [23] and [20] is combined in [24]. Page level HTR systems are compiled as a line level HTR systems using various internal and external segmentation techniques to convert page into lines [25,26]. Table 1 summarizes the key contributions and their techniques.

Table 1. Genesis of key methods used in the HTR system

S. No.	Techniques
[14] (1994)	Pre-segmented Data + RNN at character level
[3] (1999)	A dynamic programming based approach to find best path
[6] (2004)	Stochastic finite-state transducers + HMM
[15] (2006)	CTC techniques which are widely in HTR introduced in the speech recognition task
[9] (2011)	Multi-layer perceptron + HMM
[8] (2011)	Artificial Neural Network + HMM
[13] (2012)	BLSTM + CTC based recognition and end-to-end training
[16] (2013)	MDLSTM + Convolution Layers + Fully Connected Network (FCN) + CTC
[12] (2013)	LSTM-RNN tandem HMM system
[11] (2014)	LSTM-RNN and accelerating backpropagation through time using mini batching
[17] (2014)	MDLSTM + CTC + Dropout. It helps to reduce overfitting and is part of every model usually
[18] (2016)	MDLSTM+ Vertical Attention to recognize text at line level and paragraph level
[19] (2016)	Large MDLSTM + CTC
[27] (2016)	CNN + BLSTM + CTC
[20] (2017)	CNN+ BLSTM + CTC. Stacking of BLSTM layers gives MDLSTM equivalent performance
[23] (2017)	GCNN + BLSTM
[28] (2018)	MDLSTM + HMM to boost recognition speed and improve accuracy
[29] (2018)	CNN-RNN hybrid model + pre-training with synthetic data and transformations
[30] (2018)	CNN + Map2Seq + BLSTM Encoder + Attention + LSTM Decoder
[31] (2019)	Seq2Seq Encoder and Decoder model with Attention
[32] (2020)	Transformers (CNN feature extractor + multi-headed self-attention layers). Non recurrent model
[24] (2020)	CNN + GCNN + BLSTM + CTC on text line images of benchmarked dataset
[26] (2022)	FCN+LSTM+Attention of paragraph text recognition with internal line segmentation

3 System Design

In this section, we have discussed the basic building blocks of the system proposed. The Fig. 1 shows the detailed view of proposed architecture. Here, a text line image I is given as input. The pre-processing is performed to reduce the noise in the image as well as to improve accuracy. This preprocessed image is further convolve by a series of convolutional, gated convolutional and Depthwise Separable Convolutional (DSC) layers to extract features from it. A series of gated recurrent layers propagate the extracted features. A dense layer processes the output of the recurrent unit and character probability is obtained by applying the softmax as the last layer. The CTC is used to calculate the loss during the training phase. The WBS decoder is used in the testing phase of the system. This system replicates [24] design with fewer number of trainable parameters using the DSC layers. Overview of the sub components of the system used in this work has been discussed in the following subsections.

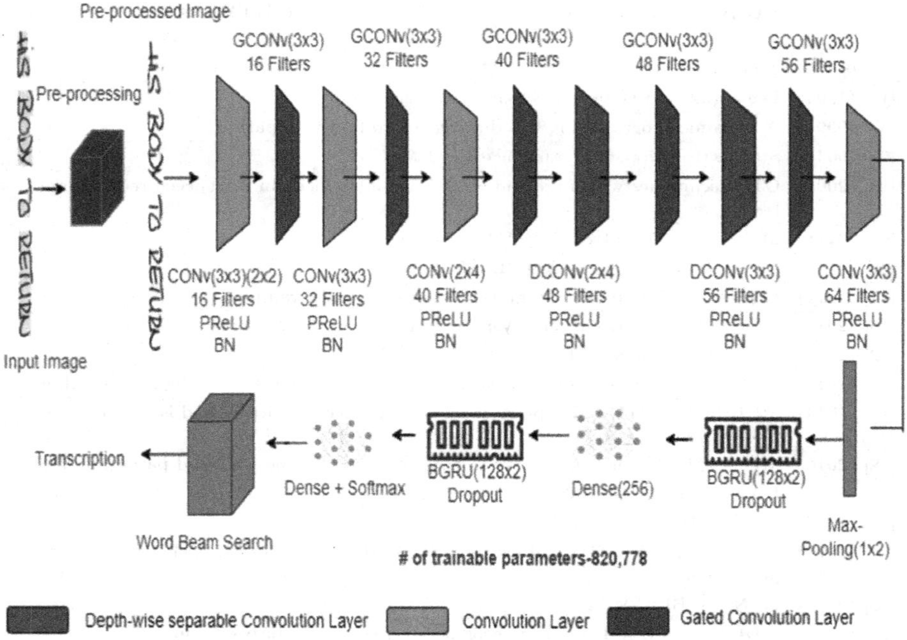

Fig. 1. Proposed architecture

3.1 Convolutional Layer

More abstract features can be identified, as the network becomes deeper and such systems are able to extract features regardless of their position in the image using convolutional layers [33]. The pooling is used alongside with convolution to downsample the image, thus, having fewer parameters to be taken care of at the next layer. Further, gates are introduced in convolutional layers to extract a larger context [23]. So, as deep learning techniques progress, the research community focuses on identifying the model that helps reduce the training time of the HTR model with fewer parameters.

The DSC layer is a variation of the convolutional layer in which convolution operation is performed to a single channel at a time. In Fig. 2, input data of size, ($\mathrm{P}_f \times \mathrm{P}_f \times \mathrm{M}$) where $\mathrm{P}_f \times \mathrm{P}_f$ is image size and M is the number of channels. Assume, we have N filters of size ($\mathrm{P}_k \times \mathrm{P}_k \times \mathrm{M}$). The standard convolutional operation produces the output size of ($\mathrm{P}_p \times \mathrm{P}_p \times \mathrm{N}$) with total number of operations as ($N \times {\mathrm{P}_p}^2 \times {\mathrm{P}_p}^2 \times \mathrm{M}$). While in the case of the DSC with input of ($\mathrm{P}_f \times \mathrm{P}_f \times \mathrm{M}$) and kernal size of ($\mathrm{P}_k \times \mathrm{P}_k \times 1$) the total number of multiplicative operations are $M \times Pp^2 \times \left(Pk^2 + N\right)$ and where $M \times Pk^2 \times Pp^2$ operations contributed by depthwise convolution and $M \times Pp^2 \times N$ added by pointwise convolution.

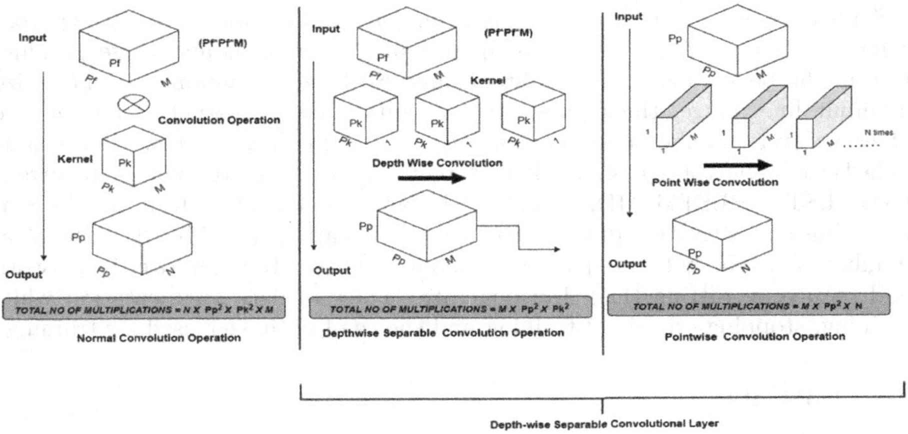

Fig. 2. Comparison between standard convolution and depthwise separable convolution operation

3.2 Recurrent Layer

Recurrent layers are used to remember part of the sequence in a sequence learning task such as HTR. It can utilize the context information of input under processing. In this study, we use BGRU since it has fewer parameters than BLSTM. It consists of two GRUs, one for taking the input in the forwarding direction and the other for taking the input in the backward direction [34].

3.3 Decoder

In the present work, we have used a CTC based decoder [2]. The softmaxed RNN's output is fed into WBS decoder for further processing. It makes a prefix tree of words from the corpus given to it as input. Based upon the prefix tree it creates an empty beam. For each iteration of the decoding process, this beam is extended along the path of the prefix tree. Each beam is scored on the basis of probability of characters along a certain path. At each time step, only the number of beams equal to the beam width is considered for the next time step. At the end of the last time step the beam with the highest probability considered as the final recognized text for given input image.

4 Experimental Setup and Results

In this section, the experimental setup of the present study has been discussed. The recognition model and the WBS decoder have been taken from [24][1] and

[1] https://github.com/arthurflor23/handwritten-text-recognition.

[2][2] respectively. An HTR system is a unique combination of many hyperparameters that needs to be looked upon while designing it and these parameters might be unique to that system only. Such as the minimum image size of input line image, the number of convolution layers, kernel and filter size of each layer, number and position of the max-pooling layer with the suitable kernel, the total number of recurrent layers, choice of type of recurrent layer (LSTM/BLSTM/MDLSTM/BGRU) and number of units it should contain, choice of activation function to have non-linearity, size of batches, the total number of epochs, rate and place of dropout (it is used to regularize the network and reduce overfitting [17]), learning rate, choice of data augmentation while training, stopping criteria of training and choice of optimizer used for training.

4.1 Datasets

Benchmarked datasets such as IAM [35], RIMES [36] and George Washington (GW) [37] have been used in this study to evaluate the proposed architecture.

IAM Dataset. The IAM dataset contains English handwritten forms that are used to train and text HTR models. It is obtained from the LOB corpus. It was first published in ICDAR 1999 and currently, 3.0 version is for public access including 657 writers and 1539 scanned pages. It has 13353 isolated and labelled text lines. Table 2 shows the train, test and validation split used in this study.

RIMES Dataset. The RIMES dataset contains handwritten letters in the French Language. It has more than 12,000 pages written by 1300 different writers. However, recognition is considered comparatively easy due to the good writing of the text but faced the challenge of accented letters. Table 2 shows the train, test and validation split used in this study.

GW Dataset. This dataset is created from George Washington Papers in English at the Library of Congress. It consists of 20 pages, 656 text lines, 4894 word instances and 82 unique characters. The availability of fewer data makes this dataset challenging for the HTR task. Table 2 shows the train, test and validation split used in this study.

Table 2. Train, validation and test splits of benchmarked datsets

S. No.	Dataset	Train set	Validation set	Test set
1	IAM Dataset (# of characters = 79)	6,161	900	1,861
2	RIMES Dataset (# of characters = 100)	10,193	1,133	778
3	GW Dataset (# of characters = 82)	325	168	163

[2] https://github.com/githubharald/CTCWordBeamSearch.

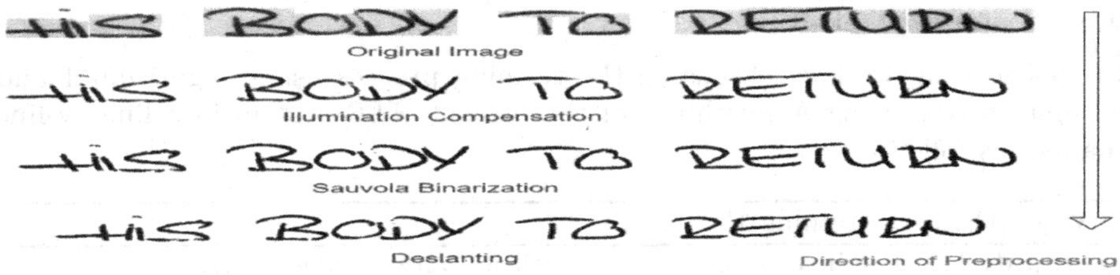

Fig. 3. Pre-processing steps

4.2 Preprocessing

Preprocessing steps are applied to reduce the noise in the raw data and make it more convenient for NN models to learn. The handwritten text documents, especially historical ones, are of poor quality and in the degraded stage. Pre processing techniques such as binarization and normalization have been applied to remove noise in text images. Figure 3 shows the part of the text line image of the IAM dataset before and after applying pre processing methods.

Illumination Compensation. With the help of the illumination compensation technique, the uneven light distribution of the documents has been balanced [38]. The light-balanced image is produced by performing contrast enhancement, edge detection, text location and light distribution on the input image.

Binarization. Binarization is applied as a part of preprocessing in most HTR systems. A bi-level image is obtained in this process. It reduces the computational load of the system as compared to 256 levels of grey scale or colour image. In this study, Sauvola binarization is used [39]. It uses a hybrid approach for deciding threshold values and considering the region properties of document classes while binarizing.

Deslanting. It is a normalization technique. These techniques are used to address the variability among different authors and same author's writing style by making each slanted handwritten text resemble other handwritten text. In this study, a slant and slope removal technique is used to normalize text data by utilizing probability density distributions graphs [40].

4.3 Evaluation Metric

We have used evaluation metrics CER and WER to compare this system with other systems. The CER is the number of operations required at the character level to transform the recognised text into the ground truth. As shown in Eq. 1 where S_{char} is the number of substitutions, D_{char} is the number of deletions and I_{char} is the number of insertions in the recognised text required at the character level and N_{char} is the total number of characters in ground truth word.

$$CER = \frac{S_{char} + D_{char} + I_{char}}{N_{char}} \tag{1}$$

4.4 Training Details

In this section, we have discussed the training process using Algorithm 1 and testing process using Algorithm 2 of the present HTR system in a line-by-line manner as follows.

Algorithm 1: Training Process

Input: Text line images $I_1, I_2, ...I_n$ and their transcriptions $y_1, y_2, ...y_n$
Result: Trained model weights of minimizing the validation loss
1 *epochs*=1000, *batch*=16, *lr*=0.001, *stop_tolerence*=20, *reduce_tolerence*=15 ;
`// initiliaze the training parameters`
2 init model() ; `// initialize the model framework`
3 **for** *i=1* **to** *batch* **do**
4 | augmentImage(I_i); `// Augment text line images`
5 | $\hat{y_i}$=model(I_i) ; `// Image prediction from model`
6 | δ_{ctc}+=$L_{ctc}(y_i,\hat{y_i})$; `// compute CTC loss`
7 **end**
8 Backward(δ_{ctc}) `// Updated model weights using back propagation`

Explanation. We will discuss the training process in line by line manner as follows,

Line 1: Initialize the model parameters required for training, such as the total number of epochs, batch size learning rate, early stopping criteria for training and reduced learning rate on the plateau (factor 0.2).
Line 2: Preparation of the model architecture.
Line 3–4: For each image in the training batch, random data augmentation is performed such as random morphological and displacement transformations that include rotation, resizing, erosion and dilation.
Line 5: Augmented image, obtained in earlier step is fed into NN the model.
Line 6–7: For each text line image CTC loss is calculated and combined.
Line 8: The model weights are updated based on loss value using backpropagation through time algorithm. This training is continued until the maximum number of epochs reached or stopping criteria have been met whichever is earlier.

Algorithm 2: Prediction Process

Input: Text line image I, D_{corpus}, D_{chars}, $D_{wordchars}$
Result: Transcription of the image along with CER and WER
BW=50, *mode*='NGrams', *smooth*=0.01 ; `// initialize the WBS decoding parameters`
initModel() ; `// Loading of the trained model`
out=predict(I) ; `// Predict the text of image`
swapaxis(out) ; `// Swapping of the output axis to make it suitable for WBS decoder`
$\hat{y}$= WBS(BW,*mode*, *smooth*=0.01,D_{corpus}, D_{chars}, $D_{wordchars}$) ;
`// Applying WBS decoding algorithm`
CER,WER= accuracy(y,$\hat{y}$) ; `// compute accuracy`

Explanation. We will discuss the prediction process in line by line manner as follows,

Line 1: Input of the model is text line image I. First, we initialize the WBS decoding parameters such as beam width to 50, mode of the algorithm to 'NGrams' and smoothing factor as 0.01.
Line 2: Preparation of the model.
Line 3: The model takes input text line image I and produces probabilities of each character at each time step.
Line 4: The output matrix dimensions are swapped as per predefined input accepted by the WBS decoder.
Line 5: Computation of the text using WBS decoding algorithm.
Line 6: Calculation of CER and WER.

4.5 Results and Comparison

As discussed, we have used benchmarked datasets and compared this with other state-of-the-art techniques. For recognition, we have used the base model of [24] and for WBS decoding we have used the implementation of [2]. The total number of trainable parameters in flor et al. [24] are 822,770, while, in this study total number of trainable parameters are 820,778 due to use of the DSC layers in Table 3) instead of standard convolutional layers. Table 3, Table 4 and Table 5 summarizes our findings. We are able to achieve 3.84% CER and 9.40% WER on IAM dataset, 4.88% CER and 14.56% WER on GW dataset and 3.15% CER and 11.8% WER on RIMES dataset.

Table 3. Comparision of present work with other state-of-the-art works on IAM Dataset

S. No.	Reference	Method/Technique	CER	WER
1	Puigcerver et al. [20]	CNN + LSTM + CTC	4.4	12.2
2	Chowdhury et al. [30]	CNN + BLSTM + LSTM	8.1	16.7
3	Michael et al. [31]	CNN + LSTM+ Attention	4.87	-
4	Kang et al. [32]	Transformer	4.67	15.45
5	Yousef et al. [41]	CNN + CTC	4.9	-
6	Flor et al. [24]	CNN + BGRU + CTC	3.72	11.18
7	Present Work	CNN + DSC + BGRU+ CTC	**3.84**	**9.40**

Table 4. Comparision of present work with other state-of-the-art works on GW Dataset

S. No.	Reference	Method/Technique	CER	WER
1	Toledo et al. [46]	CNN + BLSTM + CTC	7.32	-
2	Almazan et al. [47]	Word Embedding	17.40	-
3	Fischer et al. [48]	HMM + RNN	20	-
4	Present Work	CNN + DSC + BGRU + CTC	**4.88**	**14.56**

Table 5. Comparision of present work with other state-of-the-art works on RIMES Dataset

S. No.	Reference	Method/Technique	CER	WER
1	Chen et al. [42]	MDLSTM + CTC	8.29	30.54
2	Bluche [43]	MDLSTM + Attention	3.2	13.6
3	Huang et al. [44]	CNN + CTC	3.4	-
4	Flor et al. [24]	CNN + BGRU + CTC	3.27	11.14
5	Poulos et al. [45]	CNN + BLSTM + Attention	12.1	-
6	Present Work	CNN + DSC + BGRU+ CTC	**3.15**	**11.8**

5 Discussion

Here, we have done changes in the position of the layers while studying this system. While doing variations, we did not change the settings of the gated convolutional layer and recurrent layer configuration. We only varied the positions of the convolutional and the DSC layer. We have studied 6 different variants of the proposed architecture. In the Table 6, C represents standard convolution and D represents depth-wise separable convolution. The number of trainable parameters are given in front of each system architecture. The comparitive study of training and validation loss curves of different variants of proposed model shown

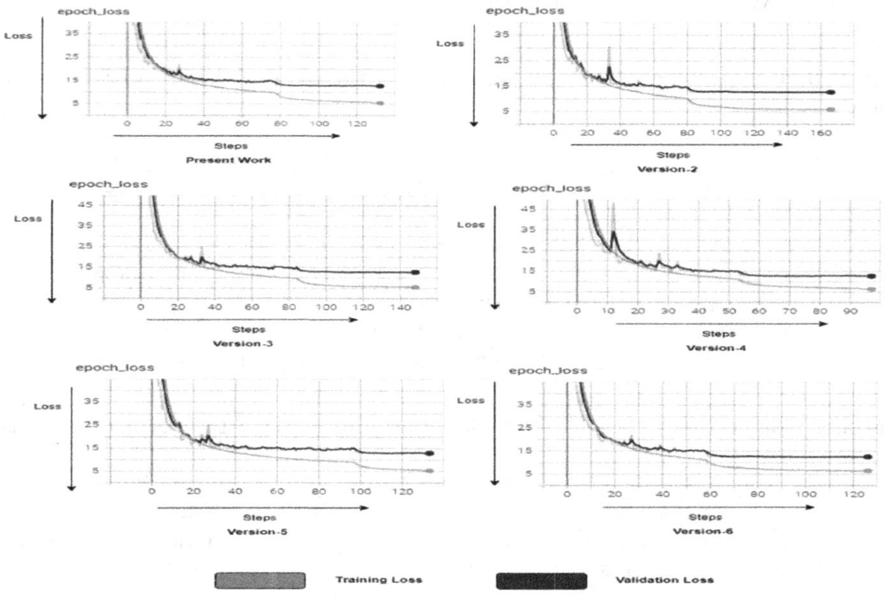

Fig. 4. Traning and Validation loss curves of proposed model and its variants

Table 6. Variations of layers in proposed model

Name	Model	# of trainable parameters
Present work	C–C–C–D–D–C	820,778
Version-2	D–D–D–D–D–D	818,492
Version-3	C–C–C–D–D–D	821,122
Version-4	C–D–C–D–C–D	819,682
Version-5	C–C–D–D–D–D	820,386
Version-6	C–D–D–D–D–D	818,610

in Fig. 4. From the Fig. 4 it is observed that variation in the number and place of the DSC layers does not affect much in the model's performance.

For statistical analysis, since each iteration of NN training is independent of each other, we have executed thirty training iterations of each variation of Table 6 on the IAM dataset and used one-way Analysis Of Variance (ANOVA) test [49] with 5% significance. As null hypothesis, we considered $H_0 : \mu_1 = \mu_2 = \mu_3 = \mu_4 = \mu_5 = \mu_6$ and alternative hypothesis H_1 : at least one of the μ_i (where $i = 1$ to 6) is different. In this, $\mu_1, \mu_2, \mu_3, \mu_4, \mu_5, \mu_6$ are the different rows of Table 6. We analyze for both CER and WER. The obtained p-value is 0.68614, which is greater than 0.05; thus, we failed to reject the null hypothesis.

6 Conclusion

We have presented a text line handwritten text recognition model using state-of-the-art approaches in each module of the HTR system. We have used the DSC layers to reduce the number of parameters for training and obtained results similar to state-of-the-art techniques. Different model architectures are discussed by changing the number and position of depth-wise separable convolution layers. The ANOVA statistical test has been performed on these models to show their performance similarity irrespective of the model architecture. We have also implemented the WBS algorithm at the decoding step while testing, which improves the obtained results.

Acknowledgements. This research is funded by the Government of India, University Grant Commission, under the Senior Research Fellowship scheme.

References

1. Kumari, L., Sharma, A.: A review of deep learning techniques in document image word spotting. Arch. Comput. Methods Eng. **29**(2), 1085–1106 (2022)
2. Scheidl, H., Fiel, S., Sablatnig, R.: Word beam search: a connectionist temporal classification decoding algorithm. In: 2018 16th International Conference on Frontiers in Handwriting Recognition (ICFHR), pp. 253–258 (2018)

3. Chen, W.T., Gader, P., Shi, H.: Lexicon-driven handwritten word recognition using optimal linear combinations of order statistics. IEEE Trans. Pattern Anal. Mach. Intell. **21**(1), 77–82 (1999)
4. Bellman, R.E., Dreyfus, S.E.: Applied Dynamic Programming, vol. 2050. Princeton University Press, Princeton (2015)
5. Vinciarelli, A.: A survey on off-line cursive word recognition. Pattern Recogn. **35**(7), 1433–1446 (2002)
6. Toselli, A., et al.: Integrated handwriting recognition and interpretation using finite-state models. Int. J. Pattern Recognit. Artif. Intell. **18**(4), 519–539 (2004)
7. Sánchez, J.A., Romero, V., Toselli, A.H., Villegas, M., Vidal, E.: A set of benchmarks for handwritten text recognition on historical documents. Pattern Recogn. **94**, 122–134 (2019)
8. Espana-Boquera, S., Castro-Bleda, M., Gorbe-Moya, J., Zamora-Martinez, F.: Improving offline handwritten text recognition with hybrid HMM/ANN models. IEEE Trans. Pattern Anal. Mach. Intell. **33**(4), 767–779 (2011)
9. Dreuw, P., Doetsch, P., Plahl, C., Ney, H.: Hierarchical hybrid MLP/HMM or rather MLP features for a discriminatively trained gaussian HMM: a comparison for offline handwriting recognition. In: 2011 18th IEEE International Conference on Image Processing, pp. 3541–3544 (2011)
10. Toselli, A.H., Vidal, E.: Handwritten text recognition results on the bentham collection with improved classical N-gram-hmm methods. In: Proceedings of the 3rd International Workshop on Historical Document Imaging and Processing, pp. 15–22 (2015)
11. Doetsch, P., Kozielski, M., Ney, H.: Fast and robust training of recurrent neural networks for offline handwriting recognition. In: 2014 14th International Conference on Frontiers in Handwriting Recognition, pp. 279–284. IEEE (2014)
12. Kozielski, M., Doetsch, P., Ney, H., et al.: Improvements in RWTH's system for offline handwriting recognition. In: 2013 12th International Conference on Document Analysis and Recognition, pp. 935–939. IEEE (2013)
13. Liwicki, M., Graves, A., Bunke, H.: Neural networks for handwriting recognition. In: Ogiela, M., Jain, L. (eds.) Computational Intelligence Paradigms in Advanced Pattern Classification, pp. 5–24. Springer, Heidelberg (2012). https://doi.org/10.1007/978-3-642-24049-2_2
14. Bourbakis, N.G., Koutsougeras, C., Jameel, A.: Handwriting recognition using a reduced character method and neural nets. In: Nonlinear Image Processing VI, vol. 2424, pp. 592–601. SPIE (1995)
15. Graves, A., Fernández, S., Gomez, F., Schmidhuber, J.: Connectionist temporal classification: labelling unsegmented sequence data with recurrent neural networks, vol. 2006, pp. 369–376 (2006)
16. Louradour, J., Kermorvant, C.: Curriculum learning for handwritten text line recognition (2014)
17. Pham, V., Bluche, T., Kermorvant, C., Louradour, J.: Dropout improves recurrent neural networks for handwriting recognition. In: 2014 14th International Conference on Frontiers in Handwriting Recognition, pp. 285–290 (2014)
18. Bluche, T., Louradour, J., Messina, R.: Scan, attend and read: end-to-end handwritten paragraph recognition with MDLSTM attention. In: 2017 14th IAPR International Conference on Document Analysis and Recognition (ICDAR), vol. 1, pp. 1050–1055 (2017)

19. Voigtlaender, P., Doetsch, P., Ney, H.: Handwriting recognition with large multidimensional long short-term memory recurrent neural networks. In: 2016 15th International Conference on Frontiers in Handwriting Recognition (ICFHR), pp. 228–233 (2016)
20. Puigcerver, J.: Are multidimensional recurrent layers really necessary for handwritten text recognition? In: 2017 14th IAPR International Conference on Document Analysis and Recognition (ICDAR), vol. 1, pp. 67–72 (2017)
21. Shi, B., Bai, X., Yao, C.: An end-to-end trainable neural network for image-based sequence recognition and its application to scene text recognition. IEEE Trans. Pattern Anal. Mach. Intell. **39**(11), 2298–2304 (2017)
22. Scheidl, H.: Handwritten text recognition in historical document. Diplom-Ingenieur in Visual Computing, Master's thesis, Technische Universität Wien, Vienna (2018)
23. Bluche, T., Messina, R.: Gated convolutional recurrent neural networks for multilingual handwriting recognition. In: 2017 14th IAPR International Conference on Document Analysis and Recognition (ICDAR), vol. 1, pp. 646–651 (2017)
24. de Sousa Neto, A.F., Bezerra, B.L.D., Toselli, A.H., Lima, E.B.: HTR-flor: a deep learning system for offline handwritten text recognition. In: 2020 33rd SIBGRAPI Conference on Graphics, Patterns and Images (SIBGRAPI), pp. 54–61 (2020)
25. Kumari, L., Singh, S., Sharma, A.: Page level input for handwritten text recognition in document images. In: Kim, J.H., Deep, K., Geem, Z.W., Sadollah, A., Yadav, A. (eds.) Proceedings of 7th International Conference on Harmony Search, Soft Computing and Applications, pp. 171–183. Springer, Singapore (2022)
26. Coquenet, D., Chatelain, C., Paquet, T.: End-to-end handwritten paragraph text recognition using a vertical attention network. IEEE Trans. Pattern Anal. Mach. Intell. (2022)
27. Doetsch, P., Zeyer, A., Ney, H.: Bidirectional decoder networks for attention-based end-to-end offline handwriting recognition. In: 2016 15th International Conference on Frontiers in Handwriting Recognition (ICFHR), pp. 361–366 (2016)
28. Castro, D., L. D. Bezerra, B., Valença, M.: Boosting the deep multidimensional long-short-term memory network for handwritten recognition systems. In: 2018 16th International Conference on Frontiers in Handwriting Recognition (ICFHR), pp. 127–132 (2018)
29. Dutta, K., Krishnan, P., Mathew, M., Jawahar, C.: Improving CNN-RNN hybrid networks for handwriting recognition. In: 2018 16th International Conference on Frontiers in Handwriting Recognition (ICFHR), pp. 80–85 (2018)
30. Chowdhury, A., Vig, L.: An efficient end-to-end neural model for handwritten text recognition (2018). https://arxiv.org/abs/1807.07965
31. Michael, J., Labahn, R., Gruning, T., Zollner, J.: Evaluating sequence-to-sequence models for handwritten text recognition, pp. 1286–1293 (2019)
32. Kang, L., Riba, P., Rusiñol, M., Fornés, A., Villegas, M.: Pay attention to what you read: non-recurrent handwritten text-line recognition (2020). https://arxiv.org/abs/2005.13044
33. Albawi, S., Mohammed, T.A., Al-Zawi, S.: Understanding of a convolutional neural network. In: 2017 International Conference on Engineering and Technology (ICET), pp. 1–6 (2017)
34. Cho, K., van Merrienboer, B., Bahdanau, D., Bengio, Y.: On the properties of neural machine translation: encoder-decoder approaches (2014). https://arxiv.org/abs/1409.1259

35. Marti, U.V., Bunke, H.: A full english sentence database for off-line handwriting recognition. In: Proceedings of the Fifth International Conference on Document Analysis and Recognition, ICDAR 1999, p. 705. IEEE Computer Society, USA (1999)
36. Grosicki, E., El-Abed, H.: ICDAR 2011 - French handwriting recognition competition. In: 2011 International Conference on Document Analysis and Recognition, pp. 1459–1463 (2011)
37. Fischer, A., Keller, A., Frinken, V., Bunke, H.: Lexicon-free handwritten word spotting using character HMMs. Pattern Recognit. Lett. **33**(7), 934–942 (2012)
38. Chen, K.N., Chen, C.H., Chang, C.C.: Efficient illumination compensation techniques for text images. Digit. Signal Process. **22**(5), 726–733 (2012)
39. Sauvola, J., Pietikäinen, M.: Adaptive document image binarization. Pattern Recogn. **33**(2), 225–236 (2000)
40. Vinciarelli, A., Luettin, J.: A new normalization technique for cursive handwritten words. Pattern Recogn. Lett. **22**(9), 1043–1050 (2001)
41. Yousef, M., Hussain, K.F., Mohammed, U.S.: Accurate, data-efficient, unconstrained text recognition with convolutional neural networks. Pattern Recognit. **108**, 107482 (2020)
42. Chen, Z., Wu, Y., Yin, F., Liu, C.L.: Simultaneous script identification and handwriting recognition via multi-task learning of recurrent neural networks. In: 2017 14th IAPR International Conference on Document Analysis and Recognition (ICDAR), vol. 1, pp. 525–530 (2017)
43. Bluche, T.: Joint line segmentation and transcription for end-to-end handwritten paragraph recognition. In: Proceedings of the 30th International Conference on Neural Information Processing Systems, NIPS 2016, pp. 838–846. Curran Associates Inc., Red Hook (2016)
44. Huang, X., Qiao, L., Yu, W., Li, J., Ma, Y.: End-to-end sequence labeling via convolutional recurrent neural network with a connectionist temporal classification layer. Int. J. Comput. Intell. Syst. **13**, 341–351 (2020)
45. Poulos, J., Valle, R.: Character-based handwritten text transcription with attention networks. Neural Comput. Appl. **33**(16), 10563–10573 (2021)
46. Toledo, J.I., Dey, S., Fornes, A., Llados, J.: Handwriting recognition by attribute embedding and recurrent neural networks. In: 2017 14th IAPR International Conference on Document Analysis and Recognition (ICDAR), vol. 1, pp. 1038–1043 (2017)
47. Almazan, J., Gordo, A., Fornes, A., Valveny, E.: Word spotting and recognition with embedded attributes. IEEE Trans. Pattern Anal. Mach. Intell. **36**(12), 2552–2566 (2014)
48. Fischer, A.: Handwriting recognition in historical documents. Ph.D. thesis, Verlag nicht ermittelbar (2012)
49. Scheffe, H.: The Analysis of Variance, vol. 72. Wiley, Hoboken (1999)

Face Recognition System Using Multicolor Image Analysis and Template Protection with BioCryptosystem

Alamgir Sardar[1](✉), Saiyed Umer[1], and Ranjeet Kumar Rout[2]

[1] Department of Computer Science and Engineering, Aliah University, Kolkata, India
alamgir.india@gmail.com, saiyed.umer@aliah.ac.in

[2] Department of Computer Science and Engineering, National Institute of Technology, Srinagar, India
ranjeetkumarrout@nitsri.net

Abstract. This paper presents a multicolor image analysis-based face recognition system in an encrypted domain using the Tree Structure Part Model (TSPM) for face detection and a statistical feature extraction technique. The features are extracted from the RGB and YCbCr components independently of the input images. Then, an enhanced Face-Hashing technique has been employed on the extracted feature vectors to generate cancelable biometrics (CB), i.e., cancelable biometrics for red (CB_R), green (CB_G), and blue (CB_B) of an RGB image, and similarly green (CB_Y), blue (CB_{Cb}), and red (CB_{Cr}) for an YCbCr image. Further, we encrypted these cancelable biometrics using a novel BioCryptosystem technique and then stored them in a template database. This BioCryptosystem generates a security code called "CryptoCode" during the encryption process, which is stored on a smart card and in the template database as well. This "CryptoCode" is used as a decryption key and smart card verification. Then, a multi-class linear SVM classifier has been employed on the decrypted and query-cancelable feature vectors for both RGB and YCbCr biometrics individually. Finally, we employed the score fusion technique between the best score obtained from the classification of RGB biometrics and the best score obtained from YCbCr biometrics. For the performance evaluation, two benchmark facial databases, CVL and FEI, have been used. The system provides 100% identification accuracy even after reducing the dimension from 500 to 100. The optimal security, reliability, efficiency, and speed of the proposed BioCryptosystem provide the novelty of this system.

Keywords: Face recognition · FaceHashing · BioCryptosystem · Template protection scheme

1 Introduction

Accuracy and security are important aspects of biometric systems. Only optimal accuracy is not the necessary and sufficient criterion for biometric authentication systems, and hence the system must be secured enough so that any third party

W. Q. Yan et al. (Eds.): IVCNZ 2022, LNCS 13836, pp. 457–473, 2023.
https://doi.org/10.1007/978-3-031-25825-1_33

cannot control the system or misuse biometric information. The face is one of the most widely used and universally accepted biometric traits among several biometric traits because of its acquisition simplicity. Apart from any injury or deformation, it remains the same throughout life. On the other hand, there is more risk of attack (such as a replay attack, a print attack, etc.) on the face. Face biometrics can be captured without the user's consent, like gait and voice. Basically, there are two template protection approaches at the feature level: biometric cryptosystems and cancelable biometrics. Cancelable biometrics [21] use a one-way function to distort the raw biometric so that an attacker cannot reverse it to its original form. Query cancelable biometrics are compared with database templates in the transformed domain during authentication, and a match or non-match determination is made. However, there is still a chance that cancelable biometrics could fail [17,23] against database attacks, channel attacks, Trojan horses, brute-force attacks, pre-image attacks [9], spoofing attack against [8,11], and record multiplicity attack [10]. The privacy of the parameter (token) used for the distortion process determines how secure the cancelable biometric approach is. An attacker can reverse the initial biometric feature if this key is lost. Cancelable biometrics are therefore insufficient to prevent biometric attacks. The biometric cryptosystem encrypts data before storing it. The secured template is then decrypted and used for comparison with query data for user authentication. A cryptographic key or decryption key is released during the encryption process. As a result, the confidentiality of this key is necessary for the system to be secure. On the other hand, real-time identification systems require faster comparisons with more secure and time-saving factors, whereas classic cryptographic algorithms take a lot of time.

In the last few years, researchers have made massive contributions to face recognition systems (FRS) and their template protection schemes. Bah et al. [2] proposed an attendance management system using an advanced face recognition method where a few modern image processing methods such as image blending, histogram equalization, bilateral filtering, and contrast adjustment are integrated with the "local binary pattern" (LBP) to improve its performance. Sardar et al. [21] proposed a face template protection scheme using the RSA algorithm to protect cancelable biometric features from several attacks, such as channel attacks, brute force attacks, database attacks, etc. Winarno et al. [28] proposed a face recognition-based attendance system using a real-time camera and the CNN-PCA algorithm. Rajput et al. [18] proposed a high-resolution algorithm to handle the low-resolution face images captured by surveillance systems. Sardar et al. [22] proposed an IoT-enabled healthcare system using face biometrics with BioCrypto-Circuit and BioCrypto-Protection methods to monitor the patients. Umer et al. [27] proposed a deep learning feature-based facial expression recognition system. Umer et al. [26] proposed a feature learning technique-based FRS using the tree-structured part model for frontal and profile face preprocessing. Wu et al. [29] proposed a face detection and recognition system using FACENET and MTCNN for access control applications. Sardar et al. [20] proposed an FRS with a template protection scheme using the Huffman algorithm as the BioCryptosystem.

Objectives: The objectives of this paper include (1) the implementation of efficient feature extraction for face recognition in an encrypted domain, (2) the implementation of a score fusion technique to improve the performance of the FRS, (3) the implementation of a faster BioCryptosystem, and (4) the implementation of long biometric code as the security key from user biometric.

Contributions: To fulfill the above objectives we have four major contributions in this work: (1) we have implemented a face recognition system using an efficient classical feature extraction technique on multicolor images, (2) we have implemented a score fusion technique to improve the performance of the system where we fused scores obtained after classification tasks on individual image components of RGB and YCbCr images, (3) we have introduced a novel two-step BioCryptosystem to preserve biometric templates where encryption and decryption time complexity is $O(n^2)$, and (4) we have introduced a novel security code (called "CryptoCode") generation technique from user biometrics during the encryption process which is used in the decryption process.

2 Proposed Methodology

The proposed methodology consists of two main components: (1) face recognition, and (2) BioCryptosystem for biometric template protection. Face recognition consists of four sub-components: (a) face preprocessing; (b) feature extraction; (c) FaceHashing; and (d) classification. The biocryptosystem consists of two sub-components: encryption and decryption. The block diagram of the proposed methodology has been shown in Fig. 1. To access the system, the user need to insert the smart card into the system; if the "CryptoCode" recorded on the smart card and in the database are matched, access is granted; otherwise, access is denied. After successfully completing smart card verification, the system will employ the decryption algorithm to decrypt the encrypted cancelable feature vectors using the "CryptoCode".

2.1 Face Recognition

(a) **Face Preprocessing:** The image samples in both the CVL and FEI databases are challenging due to several facial expressions on the frontal and profile faces. To handle these challenges, we employed an efficient face detection method called Tree-Structure Part Model (TSPM) [19]. This model detects 68 landmark points for the profile faces and 39 landmark points for the frontal faces. From these pixel positions of computed landmark points, four corner points (x_{min}, x_{max}, y_{min}, y_{max}) are detected from the x abscissa and y ordinate, respectively. Then the detected faces are employed in the feature extraction method in the form of RGB, YCbCr, and grayscale separately. The preprocessing steps are shown in Fig. 2.

(b) **Feature Extraction:** In feature extraction, preprocessed images are used to extract local distinct, and discriminant features. And hence we considered a distinct patch of size 25 × 25 ($p^{25\times25}$) for the image samples. For each 200×200 image sample, we obtain $(200 \times 200)/(25 \times 25) = 64$ distinct patches. Sliding patches are not considered due to the very slow process.

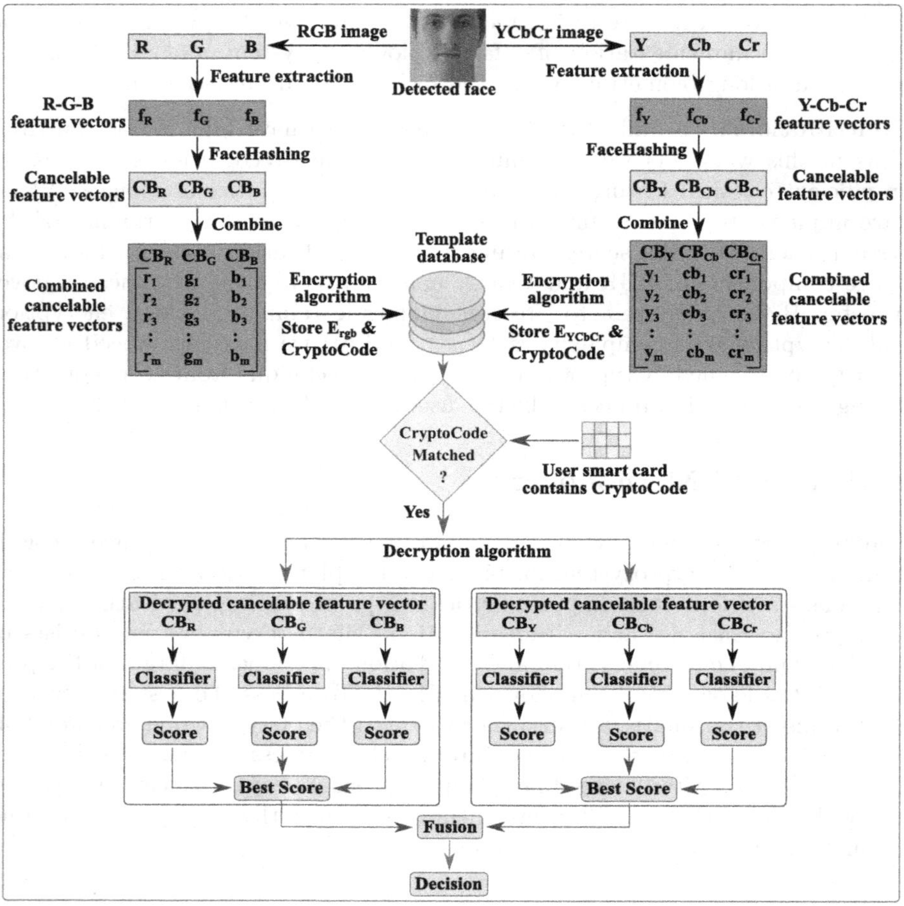

Fig. 1. The block diagram of the proposed methodology. Here, (R, G, B) and (Y, Cb, Cr) refers to the R-G-B components of the RGB image and Y-Cb-Cr components of the YCbCr image. E_{rgb} and E_{YCbCr} represents encrypted cancelable RGB and YCbCr feature vectors, respectively.

Here, the feature extraction technique consists of two steps. In the first step, we generated a dictionary using one image sample of some subjects. Distinct patches of these image samples are employed in the K-means clustering algorithm to find K-centres (here K = 250). In the second step, we extracted sliding patches of size 25×25 from all the input image samples, which generate n patches for each image sample. Then we computed the largest Euclidean distance to obtain local discriminant features. We extracted features from various image forms, such as feature extraction from (i) R-G-B components from RGB image, (ii) R-G-B components from RGB image samples by partitioning each image into two halves vertically, (iii) Y-Cb-Cr components from YCbCr image, and (iv) Y-Cb-Cr components from YCbCr

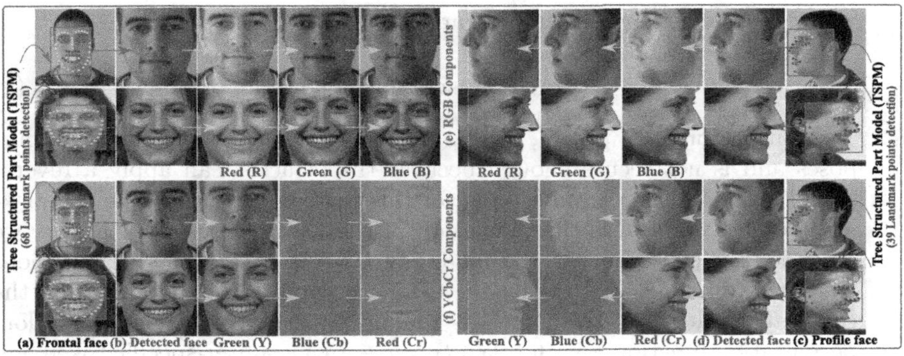

Fig. 2. The steps of face preprocessing for the frontal and profile faces.

Algorithm 1. Feature extraction (x) from $\mathscr{F}$

Input: Preprocessed face image $\mathscr{F}$, dictionary $\mathscr{D}$
Output: Feature vector x

Partition $\mathscr{F}$ into p_i using vertical partition.
for $p_i \in \mathscr{F}$ **do**
  Set $U = \phi$, ϕ is null set.
  Sliding over $\mathscr{F}$ horizontally find m distinct patch $p_i^{25\times25}$.
  Find normalised vectors u_i for each p_i.
  Combine normalised feature vectors: $U = U \parallel u_i$, '$\parallel$' refers to concatenation operator
end for
Initialize all x's by zeros for K texels ($K \in \mathscr{D}$) i.e. $x(1 \cdots K) \leftarrow 0$.
for $u_i \in U$ **do**
  Compute V=$\{v_1, v_2, ...v_n\}$ such that $v_j = dist\,(u_i, \mathscr{D}_j)$ for $1 \leq j \leq n$ 'dist' refers to Euclidean distance.
  Select $v_j = m^{th}$ least element of V and then update $v_{ij} = \begin{cases} v_{ij}, & \text{if } v_{ij} < v_j \\ \infty, & \text{else} \end{cases}$
  Update $x(j) = x(j) + exp\{-\frac{a_j}{(n)^2}\}$, n=patch size.
end for
Combine feature vectors of two vertical segments: $x_v = [x_1, x_2]$. /* x_v is the feature vector obtained from vertically segmented image partitions.

---

image samples by partitioning each image into two halves vertically. The feature extraction processes are shown in Algorithm 1.

(c) **FaceHashing:** Here, we employed a three-level FaceHashing method wherein, at the first level, a user token-based randomly generated matrix R is normalized by the Gram-Schmidt orthogonalization method, and then a projection operation has been applied between the original face feature vector and the normalized random matrix $\mathscr{R}$. This projection generates a FaceHashed feature vector x' which is further scaled to [0, 1] i.e., $x \odot \mathscr{R} \xrightarrow{\rho} x' \xrightarrow{\{0,1\}} b_{x'}$.

The steps of existing FaceHashing are shown in Eq. (1).

$$x \odot \mathscr{R} \xrightarrow{\rho} x' \xrightarrow{\{0,1\}} b_{x'} \tag{1}$$

This bit vector $b_{x'} = [x_1, x_2, ...x_n] \in \{0, 1\}$ is used only for verification purposes and is not secure enough because the attacker can apply a reverse process to get back x from $b_{x'}$ and for this reverse operation time complexity is $O(d)$ (where n is the dimension of the feature vector x). Hence, we extended Eq. (1) to Eq. (2) by applying two successive random permutations using two different subject-specific tokens, t_1 and t_2. In the first step, the cancelable feature vector $x'^{(1\times d)}$ is randomly permuted using the random permutation function π with token $t_1 = \rho + t$ (ρ is a user-specific token and t is a system-specific token), which generates x''. In the second step, the cancelable feature vector $x''^{(1\times d)}$ is randomly permuted using the random permutation function π with token $t_2 = \rho + t'$ (t' is another system-specific token) which generates x'''. This x''' is further scaled to an integer of n-bit binary, which generates the final cancelable vector $CB_x \in \mathbb{R}^{(1\times d)}$ and is used for both identification and verification purposes. The reason behind scaling the feature vector x'''. This $CB_x \in \mathbb{R}^{(1\times d)}$ is more secured than $b_{x'}$ and gives better performance. If CB_x is compromised, then the predictions of x from CB_x via x''', x'' and x' are almost impossible. Applying the π function on x' and x'' successively increases performance as well as security. Moreover, CB_x is more discriminant than x'''. The proposed enhanced FaceHashing method has been represented in Eq. (2).

$$x \odot \mathscr{R} \xrightarrow{\rho} x' \xrightarrow{\pi_{t_1}(x')} x'' \xrightarrow{\pi_{t_2}(x'')} x''' \xrightarrow{\text{integer transformation}} CB_x \tag{2}$$

This CB_x refers to the cancelable biometric feature vector where x is the image components name i.e. R, G, B, Y, Cb, and Cr.

(d) Classification: A multi-class linear SVM classifier with a k-fold cross-validation technique has been used for user classification. We employed a classifier on the cancelable RGB feature vector (CB_R, CB_G, CB_B) and query RGB feature vectors (CB_R^Q, CB_G^Q, CB_B^Q) separately, i.e. $<CB_R, CB_R^Q>$, $<CB_G, CB_G^Q>$, $<CB_B, CB_B^Q>$ where CB indicates a cancelable biometric feature vector of the enrolled image's (i.e., training sample) RGB components and CB^Q indicates a cancelable biometric feature vector of the query image's (i.e., testing sample) RGB components. Similarly, the SVM classifier has been employed on the YCbCr components of the enrolled cancelable biometric feature vector and with the corresponding YCbCr components of the query cancelable biometric feature vector, i.e. $<CB_Y, CB_Y^Q>$, $<CB_{Cb}, CB_{Cb}^Q>$, $<CB_{Cr}, CB_{Cr}^Q>$ respectively. In the case of the CVL database, six image samples are considered training images, and one image sample has been used as a testing sample. Similarly, for the FEI database, 13 image samples are employed as training samples and one image sample as a testing image. We employed the score fusion technique to maximize the performance of the system. The best rank-1 score of each classification of the RGB components has been fused with the best rank-1 score of each rank-1 classification of the YCbCr components. This fusion technique has been shown in Fig. 1.

2.2 BioCryptosystem

The proposed biocryptosystem consists of two components: encryption and decryption. These are discussed below:

(a) **Encryption:** The cancelable biometric feature vectors (CB's) of RGB components are combined to form a single matrix (say, RGB) where CB_R ($= [r_1, r_2, \cdots, r_m]$) in the first column, CB_G ($= [g_1, g_2, \cdots, g_m]$) in the second column, and CB_B ($= [b_1, b_2, \cdots, b_m]$) is in the third column (shown in Fig. 1). The proposed encryption scheme consists of two steps. In the first step, each element of the RGB matrix is converted to n-bit binary form. Then, all first bits of r_i, g_i, and b_i are combined and converted in a decimal form, and then all second bits of r_i, g_i, and b_i are combined and converted in a decimal form, and this process will continue for all n-bits of this feature elements. This transformation of 3-bit binary to integer gives an integer matrix (say rgb) with values in the range $(0 \cdots 7)$. In the second step, we computed an encryption key λ_1 where $\lambda_1 = \sum_{i=1,j=1}^{m,n} rgb(i)$. Finally, each column value of the rgb matrix is subtracted from the corresponding column-sum values to obtain the required encrypted feature matrix. This computations process is shown in Eq. (3). Similarly, the cancelable biometric feature vectors (CB's) of YCbCr components are combined to form a single matrix (say, $YCBCR$) where CB_Y ($= [y_1, y_2, \cdots, y_m]$) in the first column, CB_{Cb} ($= [cb_1, cb_2, \cdots, cb_m]$) in the second column, and CB_{Cr} ($= [cr_1, cr_2, \cdots, cr_m]$) is in the third column (shown in Fig. 1). Then the same processes are employed for y_i, cb_i, and cr_i to compute the first step encrypted feature matrix (say $ycbcr$) from $YCBCR$. Then we computed λ_2, where $\lambda_2 = \sum_{i=1,j=1}^{m,n} ycbcr(i)$. Finally, the encrypted matrix E_{ycbcr} is computed by the Eq. (4). The keys λ_1 and λ_2 are used as encryption keys. We computed decryption keys λ_3 and λ_4 from the encrypted matrices E_{rgb} and E_{ycbcr}, where $\lambda_3 = \sum_{i=1,j=1}^{m,n} E_{rgb}(i))/(m-1)$ and $\lambda_4 = \sum_{i=1,j=1}^{m,n} E_{ycbcr}(i))/(m-1)$, m refers to the column size. The combined form of λ_1, λ_2, λ_3, and λ_4 (each of containing n elements) forms a key vector λ which is called here as "CryptoCode" (λ) i.e. $\lambda = [\lambda_1, \lambda_2, \lambda_3, \lambda_4]$. This "CryptoCode" is stored in a smart card and in a database. This "CryptoCode" will be used for smart card verification as well as decryption. The entire encryption process is summarised in Algorithm 2.

$$E_{rgb}(i,j) = \sum_{i=1,j=1}^{m,n} [\lambda_1(j) - rgb(i,j)] \tag{3}$$

$$E_{ycbcr}(i,j) = \sum_{i=1,j=1}^{m,n} \left[\lambda_2(j) - ycbcr(i,j)\right] \quad (4)$$

Algorithm 2. Encryption algorithm

Input: Cancelable RGB (CB_{RGB}) and YCbCr(CB_{YCbCr}) feature vector.
Output: Encrypted cancelable RGB (E_{rgb}) and YCBCR (E_{ycbcr}) feature vector.

```
Convert RGB (CB_RGB) and YCbCr(CB_YCbCr) feature vectors to n-bit binary form.
r=(CB_R)_2, g=(CB_G)_2, b=(CB_B)_2, y=(CB_Y)_2, cb=(CB_Cb)_2, cr=(CB_Cr)_2.
Combine first bits of r, g, b and covert to decimal, combine second bits of r, g, b
  and covert to decimal, and so on. Apply same process for y, cb, cr.
for i ← 1 to m do                /*m=elements of corresponding feature vector*/
    for j ← 1 to n do            /*n=number of bits in binary form of each feature elements*/
        α = [r(i, j),  g(i, j),  b(i, j)]
        β = [y(i, j),  cb(i, j),  cr(i, j)]
        rgb(i, j) = (α)_10,   ycbcr(i, j) = (β)_10
    end for
end for
Compute sum of all column values of rgb and ycbcr separately.
i.e. λ1 = Σ_{i=1,j=1}^{m,n} rgb(i) and λ2 = Σ_{i=1,j=1}^{m,n} ycbcr(i).
Subtract each element of rgb from its corresponding column sum values to obtain
  the encrypted feature matrix E_rgb. Similarly, subtract each element of ycbcr from
  the corresponding column sum values to obtain the encrypted feature matrix E_ycbcr.
for i ← 1 to m do                /*m=elements of corresponding feature vector*/
    for j ← 1 to n do            /*n=number of bits in binary form of each feature elements*/
        E_rgb(i, j) = λ1(j) − rgb(i, j)
        E_ycbcr(i, j) = λ2(j) − ycbcr(i, j)
    end for
end for
Compute λ3 = Σ_{i=1,j=1}^{m,n} E_rgb(i)/(m − 1) and λ4 = Σ_{i=1,j=1}^{m,n} E_ycbcr(i)/(m − 1). This
  λ3 and λ4 will be used to decrypt E_rgb and E_ycbcr respectively.
Finally, compute a security code called 'CryptoCode' (λ) combining λ1, λ2, λ3,
  and λ4 i.e. λ = [λ1, λ2, λ3, λ4].
```

(b) **Decryption:** The decryption process includes two steps: first, each column elements of the encrypted matrix (E_{rgb}) is subtracted from corresponding λ_3 values, i.e. $\gamma(i,j) = \lambda_3(j) - E_{rgb}(i,j)$, and then each column elements of the encrypted matrix (E_{ycbcr}) is subtracted from corresponding λ_4 values i.e. $\delta(i,j) = \lambda_4(j) - E_{ycbcr}(i,j)$. These subtraction processes are shown in Eq. (5) and Eq. (6). Then each γ_j and δ_j are transformed into 3-bit binary form. The collection of all bits from γ_j yields $n-$ bit red (R'), green (G'), blue (B') components for RGB and the collection of all bits from δ_j yields green(Y'), blue(Cb'), and red(Cr') components for YCbCr. Then convert n-bit R', G', and B' to decimal form to obtain decrypted cancelable biometrics of R-G-B components, i.e. $D_R = (R')_{10}$, $D_G = (G')_{10}$, and $D_B = (B')_{10}$. Similarly, convert n-bit Y', Cb', and Cr' to decimal form to obtain decrypted cancelable biometrics of Y-Cb-Cr components, i.e. $D_Y = (Y')_{10}$, $D_{Cb} = (Cb')_{10}$, and $D_{Cr} = (Cr')_{10}$. Finally, combine D_R, D_G, and D_B to obtain decrypted cancelable RGB feature vector $D_{rgb} = [D_R, D_G, D_B]$ and combine D_Y, D_{Cb}, D_{Cr} to obtain a decrypted cancelable $YCbCr$ feature vector. The entire decryption process is summarised in Algorithm 3.

$$D_{rgb}(i,j) = \sum_{i=1,j=1}^{m,n} \left[\lambda_3(j) - E_{rgb}(i,j)\right], \tag{5}$$

$$D_{ycbcr}(i,j) = \sum_{i=1,j=1}^{m,n} \left[\lambda_4(j) - E_{ycbcr}(i,j)\right] \tag{6}$$

3 Experimental Results

3.1 Databases Used

In this work, two benchmark facial databases, FEI [14] and CVL [16] have been employed for the experiments. The FEI database contains $200 \times 14 = 2800$ colour image samples of 200 subjects, each of which has 14 samples with various poses, expressions, and illumination. Out of 14 image samples, one is captured in a fully dark mode, which makes it very challenging to detect the face region. The CVL database contains $114 \times 7 = 798$ color image samples of 114 subjects, each with seven image samples of frontal and profile faces in various poses and expressions. Figure 3 shows the image samples of CVL (Fig. 3(a)) and FEI (Fig. 3(b)) face databases.

3.2 Results and Discussion

In face preprocessing, we extracted 200×200 color image samples as input face $\mathcal{F}$. Then, from these color images, R, G, and B components and Y, Cb, and Cr components are considered separately for feature extraction and performance comparison. The proposed feature extraction technique is based on a dictionary computed by the K-means clustering algorithm. To generate the dictionary, we considered a single image sample for each subject and computed a patch p of size $25 \times 25 = 625$ pixels sliding over $\mathcal{F}$ horizontally. Then the computed feature vectors from each patch, $p^{25\times25}$ are normalised to a vector $u \in \mathbb{R}^{n^2\times1}$ (n = 25). After

Algorithm 3. Decryption algorithm

Input: Encrypted cancelable RGB (E_{rgb}) and YCbCr(E_{ycbcr}) feature vector, λ_3, λ_4.

Output: Decrypted cancelable RGB (D_{rgb}) and YCbCr(D_{ycbcr}) feature vector.

Subtract the first column elements of E_{rgb} from the first value of λ_3, subtract the second column elements of E_{rgb} from the second value of λ_3, and so on similar subtraction between E_{ycbcr} and λ_4.

for $i \leftarrow 1$ to m **do** /*m=number of rows in E_{rgb} or E_{ycbcr}*/

  for $j \leftarrow 1$ to n **do** /*n=number of bits in binary form of each feature elements*/

    $\gamma(i,j) = \lambda_3(j) - E_{rgb}(i,j)$

    $\delta(i,j) = \lambda_4(j) - E_{ycbcr}(i,j)$

  end for

end for

γ and δ are the decimal values of 3-bits. If γ_i is an element of γ and $(\gamma_i)_2 = [b_3, b_2, b_1]$ then bit b_3 will be the leftmost bit of the R component, bit b_2 will be the leftmost bit of the G component, and bit b_1 will be the leftmost bit of the B component of CB_{RGB}. Similarly, if δ_i is an element of δ and $(\delta_i)_2 = [b'_3, b'_2, b'_1]$ then bit b'_3 will be the leftmost bit of the Y component, bit b'_2 will be the leftmost bit of the Cb component, and bit b'_1 will be the leftmost bit of the Cr component of CB_{YCbCr}.

for $i \leftarrow 1$ to m **do**

  for $j \leftarrow 1$ to n **do**

    $b = [b_3, b_2, b_1] = \left(\gamma(:,j)\right)_2$ /*Convert γ_j to 3 bit binary form */

    $R'(:j) = b(:,1), G'(:j) = b(:,2), B'(:j) = b(:,3)$, /*Distribute bits of γ_j */

    $b' = [b'_3, b'_2, b'_1] = \left(\delta(:,j)\right)_2$ /*Convert δ_j to 3 bit binary form*/

    $Y'(:j) = b'(:,1), Cb'(:j) = b'(:,2), Cr'(:j) = b'(:,3)$, /*Distribute bits of δ_j */

  end for

  Convert n-bit R', G', and B' to decimal form to obtain decrypted cancelable biometrics of R-G-B Components i.e. $D_R = (R')_{10}$, $D_G = (G')_{10}$, and $D_B = (B')_{10}$.

  Convert n-bit Y', Cb', and Cr' to decimal form to obtain decrypted cancelable biometrics of Y-Cb-Cr Components i.e. $D_Y = (Y')_{10}$, $D_{Cb} = (Cb')_{10}$, and $D_{Cr} = (Cr')_{10}$.

end for

Finally, combine D_R, D_G, and D_B to obtain decrypted cancelable RGB feature vector $D_{rgb} = [D_R, D_G, D_B]$ and combine D_Y, D_{Cb}, D_{Cr} to obtain decrypted cancelable $YCbCr$ feature vector $D_{ycbcr} = [D_Y, D_{Cb}, D_{Cr}]$.

that, all normalized feature vectors u_i are concatenated to obtain a combined feature vector called texels U. Then for each $u_i \in U$, we computed the Euclidean distance between u_i and $\mathcal{D}_i$ i.e., $v_i = \text{dist}(u_i, \mathcal{D}_i)$. Finally, we obtain a dictionary of size 625×250. Using this process, we generated individual dictionaries for each of the image components and extracted individual feature vectors for experimen-

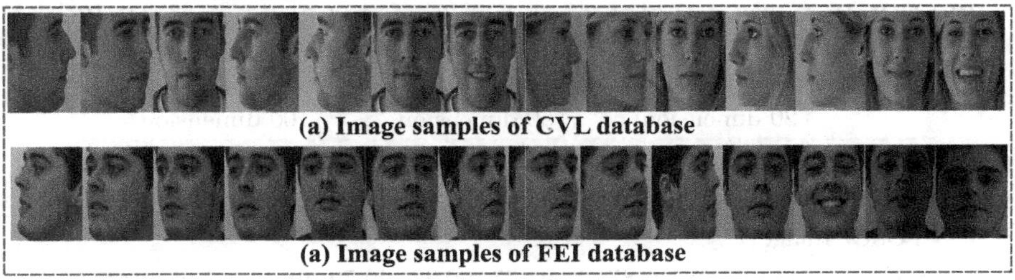

Fig. 3. Image samples of (a) CVL and (b) FEI databases

tation. Here, dictionary elements determine the size of the feature vector. In feature extraction, two approaches have been employed, such as (i) feature extraction from the original image components, i.e., features of R_o, G_o, B_o, Y_o, Cb_o, and Cr_o, and (ii) feature extraction from the vertically segmented two halves of an image component, i.e., features of R_v, G_v, B_v, Y_v, Cb_v, and Cr_v.

For the performance evaluation, we employed a multi-class linear SVM classifier for the feature vectors. From Table 1, it is clear that the performance of vertically segmented images is better than the performance of the original image (i.e., without segmentation) for both RGB and $YCbCr$. Hence, the feature vectors obtained from vertically segmented image components of RGB and $YCbCr$ are considered for the performance evaluation of the system.

Table 1. Performance comparison of the original image and vertically segmented image.

Database	RGB image			YCbCr image		
	R_o	G_o	B_o	Y_o	Cb_o	Cr_o
	R_v	G_v	B_v	Y_v	Cb_v	Cr_v
CVL	79.66	76.44	72.24	80.66	73.19	75.15
	83.85	84.73	80.82	84.49	82.06	81.62
FEI	92.02	92.28	90.24	91.43	91.46	95.38
	94.88	95.56	94.71	92.26	95.54	94.58

To enhance the performance of the original feature vector, we employed the modified FaceHashing technique in two steps such as cancelable FRS level-1 and level-2 i.e. $CFRS_1$ (Eq. (1)) and $CFRS_2$ (Eq. (2)). For the experiment, we used the feature vectors generated from vertically segmented image samples since their performance is better than feature vectors without segmentation. In $CFRS_1$, the original feature vector $x \in \mathbb{R}^{1\times 500}$ has been projected on each column of $\mathscr{R} \in \mathbb{R}^{500\times d}$ (d = dimension of random matrix) to compute $x' \in \mathbb{R}^{1\times d}$ i.e. $\left[x \in \mathbb{R}^{1\times 500} \odot \mathscr{R} \in \mathbb{R}^{500\times d}\right] = x' \in \mathbb{R}^{1\times d}$ ($\odot$ indicates projection operator). Then $x' \in \mathbb{R}^{1\times d}$ is quantized to $b_x \in [0,1]^{1\times d}$ which can be used

Table 2. Performance of level-1 cancelable face recognition system ($CFRS_1$).

Database	RGB image								
	20 dimension			50 dimension			100 dimension		
CVL	87.30	88.17	84.44	90.72	91.26	89.56	97.61	98.54	97.08
FEI	97.40	97.77	97.22	99.63	98.51	99.03	99.79	100	99.63
YCbCr Image									
CVL	87.62	86.90	87.01	91.77	90.23	90.79	98.15	97.65	98.07
FEI	94.27	97.49	97.80	97.20	98.19	99.31	99.87	99.93	100

Table 3. Performance of level-2 cancelable face recognition system ($CFRS_2$).

Database	RGB image								
	20 dimension			50 dimension			100 dimension		
CVL	91.26	92.60	90.49	98.45	99.04	98.67	100	100	100
FEI	99.72	99.22	99.42	99.89	100	99.74	100	100	100
YCbCr Image									
CVL	93.39	92.50	93.47	99.21	98.29	99.11	100	100	100
FEI	98.33	99.14	99.68	99.99	100	100	100	100	100

for the verification tasks. Table 2 shows the performance for $CFRS_1$. Existing cancelable biometric systems use $CFRS_1$ to authenticate subjects based on their biometric features and their assigned tokens. To use the system for identification purposes, the $CFRS_1$ (Eq. (1)) has been extended to $CFRS_2$ (Eq. (2)). The final result obtained from $CFRS_2$ is considered as cancelable face biometrics CB_x and elements of these CB_x are in decimal form. Therefore, the $CFRS_2$ is more efficient and reliable to preserve biometric data from unauthorized usage and several attacks. When a subject's cancelable features are compromised in the $CFRS_2$ system, a new token is assigned to the user, and the subject's original feature vector (which is kept offline) and the newly assigned token are used to replace the compromised subject's cancelable feature vector and restore the subject's original feature vector's functionality. Hence, $CFRS_2$ is highly effective in the recognition systems based on cancelable biometrics. This $CFRS_2$ enhances security as well as provides better performances than the original biometrics. In the proposed FaceHashing, after employing $CFRS_2$ system can identify a user with 100% accuracy even after dimension reduction from 500 to 100. Table 3 shows the performance of the system for $CFRS_2$.

3.3 Security Analysis of the Proposed Template Protection Scheme

The proposed template protection scheme is the integration of two template protection schemes: cancelable biometrics and BioCryptosystem which adopts the advantages of both cancelable biometrics and bio-cryptography. Since cancelable

Table 4. Performance comparison of the existing and proposed system for CVL and FEI database.

Method	Training-testing	CRR (%)	EER
CVL			
Sardar [22]	(90%–10%)	70.15	0.0451
Sardar [21]	(90%–10%)	57.79	0.1002
Umer [25]	(90%–10%)	56.36	0.1034
Goel [6]	(90%–10%)	50.60	0.1961
Gou [7]	(90%–10%)	41.64	0.2112
Proposed	(90%-10%)	**84.73**	0.0256
FEI			
Elmahmudi [5]	(90%–10%)	76.00	0.2157
Milad [12]	(90%–10%)	78.57	0.2103
Belavadi [3]	(90%–10%)	87.43	0.0151
Shnain [24]	(90%–10%)	87.91	0.0148
Cai [4]	(90%–10%)	61.31	0.6081
Abhishree [1]	(90%–10%)	94.75	0.0017
Muqeet [13]	(90%–10%)	91.14	0.0063
Proposed	(90%–10%)	**95.56**	0.0038

biometrics suffer from some security concerns [15] such as pre-image attacks, replay attacks, channel attacks, score matching attacks, and database attacks, on the other hand, BioCryptosystem cannot provide irreversibility, performance preservation, unlinkability, or revocability properties. Hence, to overcome the shortcomings of both schemes, we integrated them. The use of the feature transformation scheme, i.e., cancelable biometrics, in three steps, such as random projections and then two-level permutations based on the user token, generates a distorted feature vector, and it is impossible to revert from this distorted form to its original form. This transformation requires exponential time complexity, i.e. $O(2^d)$ (where d is the dimension of the feature vector). Moreover, this transformation holds the necessary and sufficient criteria, i.e., irreversibility, performance preservation, unlinkability, and revocability properties of cancelable biometrics. Additionally, the proposed biocryptosystem enhances the security level of the system. This biocryptosystem has been employed in such a way that classification operations can be done in an encrypted domain, which prevents pre-image attacks, replay attacks, channel attacks, score-matching attacks, and database attacks at the same time. Since the generated "CryptoCode" is also long and dependent on six individual image components, i.e. CB_R, CB_G, CB_B, CB_Y, CB_{Cb}, and CB_{Cr} so it is very hard to produce and guess. Moreover, the decryption algorithm's multiple computational steps make it more challenging computationally and overcome the chances of dictionary and brute-force attacks.

3.4 Complexity Analysis of the Proposed Template Protection Scheme

The proposed template protection scheme consists of two consecutive schemes: FaceHashing and BioCryptosystem. The FaceHashing method performs in three steps, such as random projections of the column elements between the randomly generated matrix and orthonormalized face biometric feature vector followed by two-level permutations. The BioCryptosystem consists of encryption and decryption operations. Table 5 summarises the time complexity of independent operations of the proposed FaceHashing and Biocryptosystem.

Table 5. Time complexity of the proposed FaceHashing and BioCryptosystem

Operations	Time complexity
FaceHashing	
Projection operation: $\left[x \in \mathbb{R}^{1\times 500} \odot \mathscr{R} \in \mathbb{R}^{500\times d}\right] = x' \in \mathbb{R}^{1\times d}$	$O(1 \times 500 \times d) = O(d)$, d = dimension of random matrix
Level-1 and level-2 permutations	$max\left(O(2^d), O(2^d)\right) = O(2^d)$ d = dimension of feature vector (or random matrix)
Complexity of the FaceHashing	$max(O(d), O(2^d)) = O(2^d)$
BioCryptosystem	
Decimal to binary conversion (Line no. 2 of Algorithm 2)	$O(logn)$ where n is the number of digits in decimal number
Binary to decimal conversion (Line no. 4 to 10 of Algorithm 2)	$O(m.n) = O(n^2)$
Computation of λ_1 and λ_2 (Line no. 12 of Algorithm 2)	$O(n)$
Subtraction from λ_1 and λ_2 (Line no. 14 to 19 of Algorithm 2)	$O(m.n) = O(n^2)$
Computation of λ_3 and λ_4 (Line no. 20 of Algorithm 2)	$O(n)$
Time complexity of the encryption algorithm	$max(O(logn), O(n^2), O(n), O(n^2), O(n)) = O(n^2)$
Subtraction from λ_3 and λ_4 (Line no. 2 to 7 of Algorithm 3)	$O(m.n) = O(n^2)$
Computation of n-bit binary (R', G', B' etc.) (Line no. 9 to 15 of Algorithm 3)	$O(m.n) = O(n^2)$
Binary to decimal conversion (Line no. 16 of Algorithm 3)	$O(logn)$ where n is the number of bits in binary number.
Time complexity of the encryption algorithm	$max(O(n^2), O(n^2), O(logn)) = O(n^2)$
Complexity of the Bio-Cryptosystem	$max(O(nlogn), O(n^2)) = O(n^2)$

3.5 Novelty of the Proposed System

The novelties of this paper are: (1) **efficient feature extraction technique:** in this paper, a novel feature extraction technique has been employed where features are extracted from red, green, and blue components separately from an RGB image sample. Similarly, features are extracted from the Y, Cb, and Cr components separately from a YCbCr image sample. This feature extraction technique provides better performance than the grayscale image format. On the other hand, the feature extraction method from a three-dimensional image provides flexibility to employ BioCryptosystem, (2) **face recognition in an encrypted domain:** the classification of the proposed FRS performs in an encrypted domain, which protects classification using the pre-stored feature of the attacker, (3) **score fusion technique:** this fusion technique improves the performance of the system, (4) **generation of a novel "CryptoCode":** this bio-code enhances security level during smart card verification as well as the decryption process, and (5) **faster BioCryptosystem:** the proposed biocryptosystem is faster, more robust, and more reliable. Compared to some other traditional cryptographic algorithms, this biocryptosystem provides less complexity and requires less computational time for both encryption and decryption.

4 Conclusion and Future Directions

A face recognition system with both cancelable and BioCryptosystem security schemes is presented in this paper. To handle the challenges of various poses and expressions on both frontal and profile faces, the TSPM has been employed as an efficient face detection technique. Instead of feature extraction from the grayscale image of the original color image like traditional methods, we employed the feature extraction technique on the individual image components of the RGB and YCbCr images. This method improves the performance of the system. Moreover, the FaceHashing technique has been employed on these individual image components to generate independent cancelable biometrics of the RGB and YCbCr images. These cancelable biometrics are transformed into an encrypted domain and kept online for user authentication. During system access, encrypted cancelable biometrics are decrypted, and a classification task is performed between similar cancelable biometric components (R-R, G-G, and B-B) of both enrolled and query RGB and YCbCr images. This method improves the performance of the system by introducing the score fusion technique. The novel "CryptoCode" generation technique also enhances the security of the proposed system. Above all, we obtained 100% accuracy after reducing the feature dimension from 500 to 100. In future work, we will implement a deep-learning-based feature extraction technique and also improve BioCryptosystem by introducing some other reliable methods.

References

1. Abhishree, T., Latha, J., Manikantan, K., Ramachandran, S.: Face recognition using Gabor filter based feature extraction with anisotropic diffusion as a pre-processing technique. Procedia Comput. Sci. **45**, 312–321 (2015)
2. Bah, S.M., Ming, F.: An improved face recognition algorithm and its application in attendance management system. Array **5**, 100014 (2020)
3. Belavadi, B., Sanjay, G., Prashanth, K.M., Shruthi, J.: Gabor features for single sample face recognition on multicolor space domain. In: 2017 International Conference on Recent Advances in Electronics and Communication Technology (ICRAECT), pp. 211–215. IEEE (2017)
4. Cai, J., Chen, J., Liang, X.: Single-sample face recognition based on intra-class differences in a variation model. Sensors **15**(1), 1071–1087 (2015)
5. Elmahmudi, A., Ugail, H.: Deep face recognition using imperfect facial data. Futur. Gener. Comput. Syst. **99**, 213–225 (2019)
6. Goel, N., Bebis, G., Nefian, A.: Face recognition experiments with random projection. In: Defense and Security, pp. 426–437. International Society for Optics and Photonics (2005)

7. Gou, G., Huang, D., Wang, Y.: A hybrid local feature for face recognition. In: Anthony, P., Ishizuka, M., Lukose, D. (eds.) PRICAI 2012. LNCS (LNAI), vol. 7458, pp. 64–75. Springer, Heidelberg (2012). https://doi.org/10.1007/978-3-642-32695-0_8
8. Izu, T., Sakemi, Y., Takenaka, M., Torii, N.: A spoofing attack against a cancelable biometric authentication scheme. In: 2014 IEEE 28th International Conference on Advanced Information Networking and Applications, pp. 234–239. IEEE (2014)
9. Lacharme, P., Cherrier, E., Rosenberger, C.: Preimage attack on biohashing. In: 2013 International Conference on Security and Cryptography (SECRYPT), pp. 1–8. IEEE (2013)
10. Li, C., Hu, J.: Attacks via record multiplicity on cancelable biometrics templates. Concurr. Comput. Pract. Exp. **26**(8), 1593–1605 (2014)
11. Li, L., Correia, P.L., Hadid, A.: Face recognition under spoofing attacks: countermeasures and research directions. IET Biometrics **7**(1), 3–14 (2018)
12. Milad, A., Yurtkan, K.: An integrated 3D model based face recognition method using synthesized facial expressions and poses for single image applications. Appl. Nanosci. 1–11 (2022)
13. Muqeet, M.A., Holambe, R.S.: Local binary patterns based on directional wavelet transform for expression and pose-invariant face recognition. Appl. Comput. Inform. **15**(2), 163–171 (2019)
14. de Oliveira Junior, L., Thomaz, C.E.: Captura e alinhamento de imagens: Um banco de faces brasileiro. Department of Electrical Engineering, FEI (2006)
15. Patel, V.M., Ratha, N.K., Chellappa, R.: Cancelable biometrics: a review. IEEE Signal Process. Mag. **32**(5), 54–65 (2015)
16. Peer, P.: CVL face database. Computer Vision Lab., Faculty of Computer and Information Science, University of Ljubljana, Slovenia (2005). http://www.lrv.fri.uni-lj.si/facedb.html
17. Punithavathi, P., Subbiah, G.: Can cancellable biometrics preserve privacy? Biom. Technol. Today **2017**(7), 8–11 (2017)
18. Rajput, S.S., Arya, K.: A robust face super-resolution algorithm and its application in lowresolution face recognition system. Multimedia Tools Appl. **79**(33), 23909–23934 (2020)
19. Ramanan, D., Zhu, X.: Face detection, pose estimation, and landmark localization in the wild. In: Proceedings of the 2012 IEEE Conference on Computer Vision and Pattern Recognition (CVPR), pp. 2879–2886. Citeseer (2012)
20. Sardar, A., Umer, S.: Implementation of face recognition system using BioCryptosystem as template protection scheme. J. Inf. Secur. Appl. **70**, 103317 (2022)
21. Sardar, A., Umer, S., Pero, C., Nappi, M.: A novel cancelable facehashing technique based on non-invertible transformation with encryption and decryption template. IEEE Access **8**, 105263–105277 (2020)
22. Sardar, A., Umer, S., Rout, R.K., Wang, S.H., Tanveer, M.: A secure face recognition for IoT-enabled healthcare system. ACM Trans. Sensor Netw. (TOSN) (2022)
23. Sardar, A., Umer, S., Rout, R.K., Khan, M.K.: A secure and efficient biometric template protection scheme for palmprint recognition system. IEEE Trans. Artif. Intell. (2022)
24. Shnain, N.A., Hussain, Z.M., Lu, S.F.: A feature-based structural measure: an image similarity measure for face recognition. Appl. Sci. **7**(8), 786 (2017)
25. Umer, S., Dhara, B.C., Chanda, B.: Biometric recognition system for challenging faces. In: 2015 Fifth National Conference on Computer Vision, Pattern Recognition, Image Processing and Graphics (NCVPRIPG), pp. 1–4. IEEE (2015)

26. Umer, S., Dhara, B.C., Chanda, B.: Face recognition using fusion of feature learning techniques. Measurement **146**, 43–54 (2019)
27. Umer, S., Rout, R.K., Pero, C., Nappi, M.: Facial expression recognition with trade-offs between data augmentation and deep learning features. J. Ambient. Intell. Humaniz. Comput. **13**(2), 721–735 (2022)
28. Winarno, E., Al Amin, I.H., Februariyanti, H., Adi, P.W., Hadikurniawati, W., Anwar, M.T.: Attendance system based on face recognition system using CNN-PCA method and real-time camera. In: 2019 International Seminar on Research of Information Technology and Intelligent Systems (ISRITI), pp. 301–304. IEEE (2019)
29. Wu, C., Zhang, Y.: MTCNN and FaceNet based access control system for face detection and recognition. Autom. Control Comput. Sci. **55**(1), 102–112 (2021)

Conformer-Based Lip-Reading for Japanese Sentence

Taiki Arakane[1], Takeshi Saitoh[1(✉)], Ryuuichi Chiba[2], Masanori Morise[2], and Yasuo Oda[3]

[1] Kyushu Institute of Technology, 680-4 Kawazu, Iizuka, Fukuoka, Japan
saitoh@ai.kyutech.ac.jp
[2] Meiji University, 4-21-1 Nakano, Nakano-ku, Tokyo, Japan
mmorise@meiji.ac.jp
[3] SSS LLC, 2-1-2 Sanno, Ota Ward, Tokyo, Japan

Abstract. Various applications of lip-reading technology are considered, and it is an important issue. In recent years, lip-reading research on sentences has attracted attention. However, most of the published datasets are English-talking scenes, and there are few datasets other than English. Therefore, in this research, we are researching Japanese sentence-level lip-reading. In this paper, we construct Japanese sentence utterance scene datasets ITA and ROHAN4600 and propose the Conformer-based lip-reading method. Recognition experiments were conducted using the Transformer model as a conventional method. As a result, it was confirmed that the Conformer model obtained high recognition accuracy both at the phoneme and the mora levels.

Keywords: Lip-reading · Sentence-level · Conformer

1 Introduction

Lip-reading is a technique for estimating the utterance content from the talking speaker's mouth. The video's temporal resolution is lower than audio data, which targets voice, and the inside of the mouth cannot be observed from the camera image. Even words such as /ta-ma-go/, /na-ma-ko/, and /ta-ba-ko/ that are easy to recognize in audio-based speech recognition are difficult to recognize in the video because the mouth movements are the same. However, lip-reading technology is expected to be used in conditions where it is difficult to speak, such as in noisy environments and waiting rooms, and to support communication for speech disorders who cannot speak due to laryngectomy. In addition, various applications such as security, entertainment, and human-computer interaction are expected.

However, although audio-based speech recognition technology has been put to practical use, lip-reading technology has not yet been put to practical use. This is due not only to the difficulty of lip-reading but also to the lack of data for training. In lip-reading research, recognition targets are roughly divided into

W. Q. Yan et al. (Eds.): IVCNZ 2022, LNCS 13836, pp. 474–485, 2023.
https://doi.org/10.1007/978-3-031-25825-1_34

three categories: single sounds such as /a/, /i/, /u/, /e/, and /o/, words, and sentences. Single-sound-level lip-reading has been studied since the 1980s and is being researched to support communication for disabilities [10]. Word-level lip-reading has long been and continues to be the focus of research [6,9]. However, recent subjects are shifting from words to sentences [1,2,4,5,14,17].

A well-known sentence-level lip-reading model LipNet was proposed by Assael et al. [4]. This model consists of two stages; (1) three layers of spatiotemporal convolution and spatial pooling layers and (2) two bi-directional GRU layers, a linear layer, and a softmax. The word error rate (WER) of 11.4% was obtained in the unseen speaker recognition task with GRID [7]. Chung et al. built a large-scale sentence dataset LRS2 [5]. They proposed a Watch, Listen, Attend and Spell network. In [5], a character error rate of 39.5% is obtained using only video data. Regarding lip-reading for English sentences, two large-scale datasets of LRS2 and LRS3 have been released. Thus, lip-reading for English sentences has been actively studied in recent years [1,4,5,17]. These datasets will be introduced in Sect. 2.1.

On the other hand, sentence-level lip-reading for Japanese has not progressed because there are not enough large-scale public datasets compared to English. Noda et al. [11] proposed a combined method with CNN and hidden Markov model. They collect speech scenes of ATR phoneme balance words from six speakers, conduct recognition experiments for 40 phonemes, and report an average recognition rate of 58%. Unfortunately, this dataset is private and is not available to us. Our previous research [14] collected utterance scenes of Japanese sentences and worked on lip-reading using Transformer. The error rates of phoneme-level and mora-level in speaker-dependent recognition tasks were 35.5% and 11.4%, respectively. In the speaker-independent recognition task, we obtained error rates of 75.3% and 82.6%, respectively. We showed the possibility of Japanese sentence lip-reading by preparing datasets and applying the Transformer model, but the accuracy needed to be improved.

The contributions of this paper are as follows.

- Construction of Japanese sentence utterance scene dataset
- Proposal of the sentence-level lip-reading method using the Conformer model.

This paper is organized as follows. Section 2 describes the datasets open to the public and the datasets constructed in this research. Section 3 describes the proposed recognition method. Section 4 describes recognition experiments. Finally, Sect. 5 concludes this paper.

2 Datasets

2.1 Existing Public Datasets

Here, we introduce public datasets that can be used for sentence-level lip-reading. The datasets are summarized in Table 1.

Table 1. Datasets for lip-reading.

Name	Year	# of speakers	Language	Content
CUAVE [12]	2002	36	En	Digits sentences
GRID [7]	2006	34	En	Formed sentences
LRS2 [5]	2017	1,000+	En	BBC
LRS3 [2]	2018	10,000+	En	TED
Lombard GRID [3]	2018	54	En	Formed sentences
CENSREC-1-AV [15]	2010	93	Ja	1–7 connected digits
ITA (ours)	2022	4	Ja	424 sentences
ROHAN4600 (ours)	2022	1	Ja	6400 sentences

The GRID [7][1] consists of 1,000 sentence utterance scenes for each of 33 (17M+16F) speakers. A sentence contains a command, a color, a preposition, an alphabet, a digit, and an adverb one by one, such as "put red at G9 now". The constituent words are four commands ("bin," "lay," "place," "set"), four colors ("blue," "green," "red," "white"), four prepositions ("at," "by," "in," "with"), 25 alphabets (from "A" to "Z" without "W"), ten digits (from "zero" to "nine"), and four adverbs ("again," "now," "please," "soon"), all randomly assigned to form sentences. This dataset provides scenes at two different resolutions: 360 × 288 [pixels] and 720 × 576 [pixels]. The frame rate is 25 fps. The duration time of all scenes is three seconds. That is 75 frames per scene. The total duration time of the entire dataset is 27.5 h.

The Lombard GRID [3][2] is an extended dataset of GRID. This dataset includes 54 (24M+30F) speakers, with 100 utterances per speaker (50 Lombard and 50 plain utterances). This dataset follows the same sentence format as the GRID corpus. However, the sentence sets used in the Lombard Grid corpus are unique and have not been utilized by the Grid corpus. The major difference between this dataset and other datasets is that it includes both frontal and profile scenes. The front and profile scene image sizes are 720 × 480 [pixels] and 854 × 480 [pixels], respectively. Both frame rates are 24 fps.

The CUAVE dataset [12] contains speech scenes of 36 (19M+17F) speakers. The target utterances contain connected and isolated ten digits ("zero", "one", "two", "three", "four", "five", "six", "seven", "eight", and "nine") in English. The background of each speaker is green. This dataset consists of two major sessions: one of the individual speakers and one of the speaker pairs. The first session includes a scene in which the speaker stands naturally still and a scene in which the speaker moves side-to-side, back-to-forth, or tilts the head while speaking. The second session includes 20 pairs of speakers (labeled speakers A and B). Speaker A speaks a connected-digit sequence, followed by speaker B and vice-versa a second time, and both speakers A and B overlap each other while

[1] https://spandh.dcs.shef.ac.uk/gridcorpus/.
[2] https://spandh.dcs.shef.ac.uk//avlombard/.

speaking each speaker's separate digit sequence. The image size is 720 × 480 [pixels], and its frame rate is 29.97 fps.

LRS2 [5][3] contains various English speech scenes from BBC. It provides both training and test data. The former consists of 70,783 scenes, and its total duration is more than 100 h. The latter consists of 48,165 scenes, and its total duration is approximately 30 h. It also provides not only video data but also speech text. The face ROI extracted from the video is provided. The image size is 160 × 160 [pixels], and the frame rate is 25 fps. Furthermore, label information on each word's utterance start/end time is also provided.

LRW and LRS2 are based on the BBC, while LRS3 [2][4] is based on TED and TEDx. It consists of thousands of spoken sentences. In this dataset, pre-trained data (118,516 scenes), trainval data (31,982 scenes), and test data (1,321 scenes) are provided. The total duration time is more than 400 h. The image size is 224 × 224 [pixels], the frame rate is 25 fps, and the speech scene of the face area and the content text are provided similarly to LRS2.

CENSREC-1-AV is an audio-visual speech corpus [15]. It consists of 3,234 and 1,963 Japanese connected digits utterances. The numbers of speakers for training and testing data are 42 (22M+20F) and 51 (25M+26F), respectively. Each utterance consists of 1–7 digit sequences, and each digit is pronounced as /i-chi/ (one), /ni/ (two), /sa-N/ (three), /yo-N/ (four), /go/ (five), /ro-ku/ (six), /na-na/ (seven), /ha-chi/ (eight), /kyu/ (nine), /ze-ro/ (zero) or /ma-ru/ (zero) in Japanese, respectively. The image size is 720 × 480 [pixels], and its frame rate is 29.97fps. The speech scenes were taken from the frontal side, and the background color was almost blue.

2.2 Our Datasets

As mentioned above, a large-scale Japanese speech scene dataset has yet to be released. Therefore, to activate this research field, we are promoting the enhancement of datasets. This paper introduces our publicly available datasets of ITA and ROHAN4600.

For the ITA corpus multimodal dataset, utterance scenes of four female voice actors (CV1, CV2, CV3, CV4) were collected. Here, 324 recitation sentences are normal utterances, but 100 emotional sentences are recorded with scenes of multiple emotions such as normal, sweet, pungent, sexy, and whispered. Our group creates the ITA corpus to extract a chain of two consecutive phonemes in Japanese and cover all of them.

On the other hand, ROHAN4600 is another Japanese text consisting of 4,600 sentences that include all reading of common Japanese Kanjis and covers moras that infrequently appear by manually creating sentences that meet the conditions presented by the corpus sentence generation system[5] which is a corpus. For ROHAN4600, voice actor CV2's utterance scene was recorded.

[3] https://www.robots.ox.ac.uk/~vgg/data/lip_reading/lrs2.html.
[4] https://www.robots.ox.ac.uk/~vgg/data/lip_reading/lrs3.html.
[5] https://github.com/mmorise/rohan4600.

Table 2. Collected Japanese speech scenes.

Corpus	ITA				ROHAN4600
Speaker ID	CV1	CV2	CV3	CV4	CV2
Sex	Female				
# sentences	724	724	724	824	6,400
Total length [h:m:s]	1:15:28.14	1:15:38.74	1:19:53.36	1:45:36.31	8:33:49.86
Average length [s]	6.25	6.27	6.62	8.75	6.70

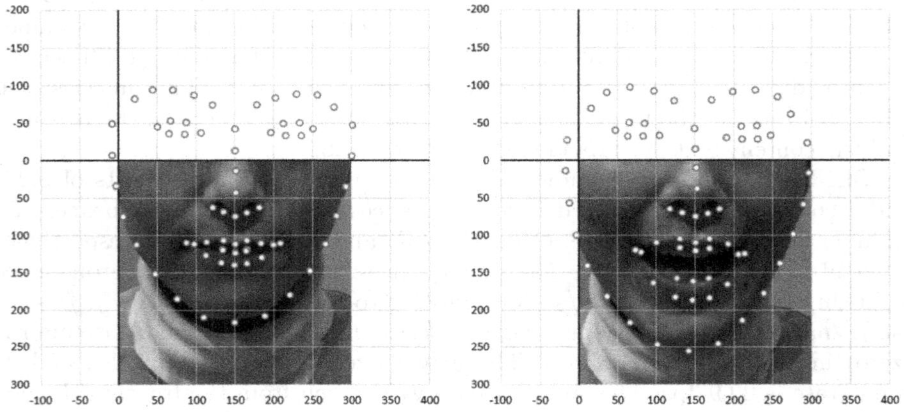

Fig. 1. LFROI and feature points provided by our datasets (ITA-CV1).

We have already constructed an utterance scene dataset SSSD of 25 Japanese words collected with a smart device [13][6]. As with SSSD, preprocessing was applied to the collected scenes, and the lower half of the face ROI image (LFROI) and 68 facial feature points were extracted, as shown in Fig. 1. The size of LFROI is 300 × 300 [pixels]. In addition, we maintain and publish phoneme label data, including speech data and speech boundaries. An overview of both databases is summarized in Table 2. Both datasets are available free on the same site[7].

3 Proposed Method

This Section explains the proposed lip-reading method. Figure 2 shows the processing flow of the proposed method. The figure shows the horizontal dashed line's upper half as datasets. These provide LFROI, facial feature points, and label data. Below the horizontal dashed line is the proposed method. The unit of the provided data is one utterance scene, and the final output is the predicted utterance text. However, the core processing is performed in units of short-term

[6] https://www.saitoh-lab.com/SSSD/index_ja.html.
[7] https://zunko.jp/multimodal_dev/login.php.

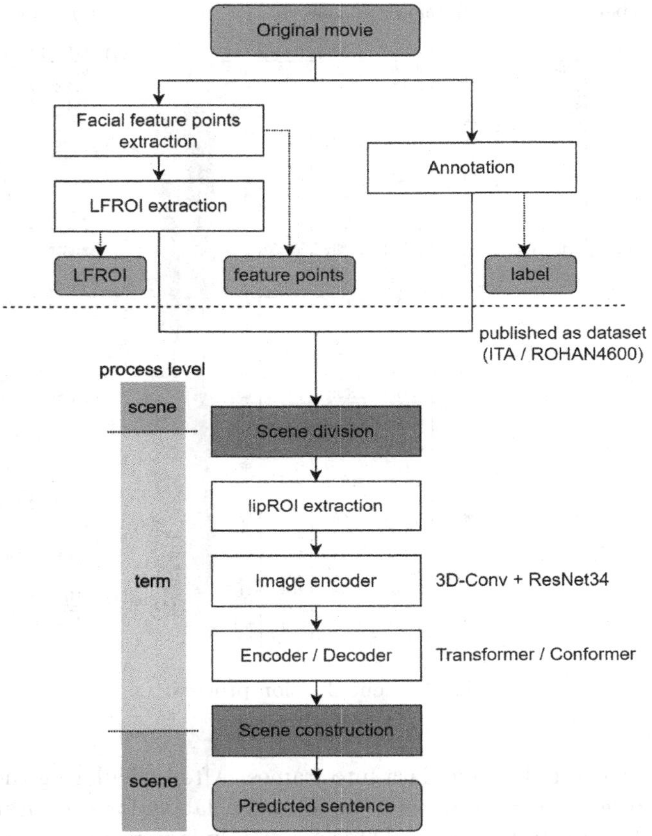

Fig. 2. Process flow of the proposed method.

divisions, similar to speech recognition tasks. Scene division, reconstruction, and recognition processes are described below.

A square region centered on the mouth is cropped as a lipROI for the LFROI. All lipROIs are resized to 96×96 [pixels].

3.1 Scene Division and Construction

It is desirable to be able to divide the scene into meaningful units. However, predicting the utterance content cannot be divided into meaningful units in the inference task. Therefore, the scene is divided by length F frames.

The lipROI can be easily divided here since the number of frames is counted. However, since the label data is in units of time, as shown in Fig. 3(a), it is

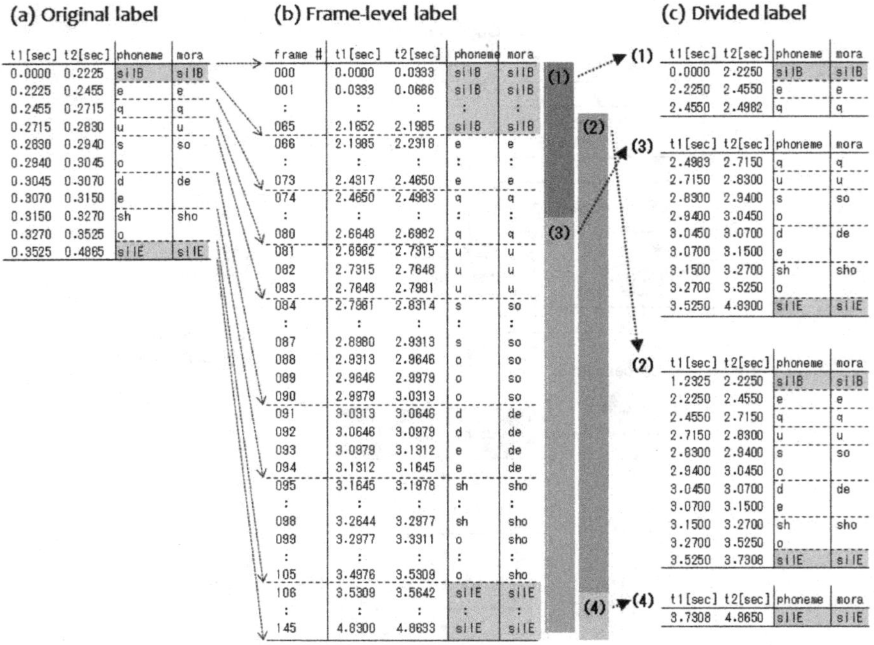

Fig. 3. Scene division process.

necessary to convert the time data into frames. After calculating the time in all frames, the original label data is used to assign labels for each frame. Finally, divide by F frames. In the figure, (1) is a scene divided from the beginning, and (3) is a continued scene. (2) is a scene divided from $F/2$, and (4) is a continuation scene of (2). (2) (4) is $F/2$ shifted from (1) (3).

The reconstruction process concatenates the predicted text term by term.

The number of scenes after segmentation in the utterance scenes of speakers ITA-CV1, ITA-CV2, ITA-CV3, ITA-CV4, and ROHAN4600-CV2 was 3974, 3977, 4173, 6308, and 26768, respectively.

3.2 Conformer-Based Lip-Reading

In recent years, the accuracy of neural network-based speech recognition systems has improved dramatically. RNNs, which can efficiently consider the temporal dependence of speech, are at the forefront, and more recently, Transformer based on self-attention have been proposed [16]. Transformer can capture longer time dependencies and can be trained efficiently. Besides, CNN successfully captures local context through local receptive fields layer by layer.

However, self-attention and CNN each have their own limitations. Transformer is good at considering global context, i.e., long temporal dependencies, but not at extracting local context, i.e., local relationships. Conversely, CNN is

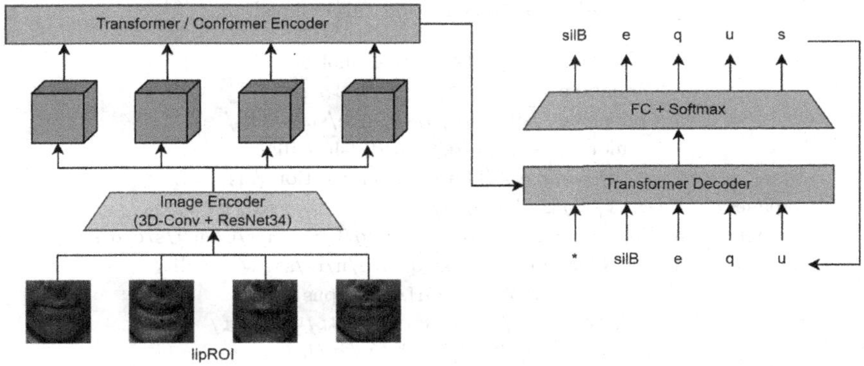

Fig. 4. Proposed model structure.

good at subdividing into blocks and extracting local information, but it requires many layers and parameters to capture connections from a broader perspective. Therefore, in recent research, movements that exceed individual performance are becoming active by combining CNN and self-attention. It uses picking the best of CNN and self-attention and captures both local and global contexts. Conformer [8] is a model that combines CNN and Transformer. It is based on the assumption that capturing local and global information will lead to more accurate parameter determination.

Figure 4 shows the model structure of recognition processing. LipROI is image data converted into 512-dimensional feature values for every input LipROI through an Image Encoder. The Image Encoder consists of a 3D convolutional layer in the front stage and a ResNet34 in the rear stage. After extracting feature values with Image Encoder, phoneme or mora of utterance content are predicted with Conformer or Transformer.

Chung et al. [5] recognize English sentences in alphabetical units rather than word units. Although Japanese is targeted in this paper, the recognition level can be phoneme and mora, as shown in Table 3. The table shows three sentences as examples. Mora is a technical term linguists use to define a phonological unit of a language. It is different from a syllable. In Japanese, each kana corresponds to a mora. The mora has the longest time among these units, and the phoneme has the shortest time. In the table, `q` means double consonant, `silB` means a silent section before the start of an utterance, `silE` means a silent section after the utterance, and `sp` means a short pause. Since `sp` is based on actual utterances, the insertion position differs depending on the speaker.

Table 3. Recognition target level.

(a) Sample 1 (ITA corpus emotion #1)

Japanese	/え/っ/う/そ/で/しょ/
phoneme	/silB/e/q/u/s/o/d/e/sh/o/silE/
mora	/silB/e/q/u/so/de/sho/silE/

(b) Sample 2 (ITA corpus recitation #1)

Japanese	/お/ん/な/の/こ/が/き/っ/き/っ/う/れ/し/そ/う/
phoneme	/silB/o/N/n/a/n/o/k/o/g/a/k/i/q/k/i/q/u/r/e/sh/i/s/o/o/silE/
mora	/silB/o/N/na/no/ko/ga/ki/q/ki/q/u/re/shi/so/o/silE/

(c) Sample 3 (ROHAN corpus #1)

Japanese	/な/が/し/ぎ/り/が/か/ん/ぜ/ん/に/は/い/れ/ば/ で/ば/ふ/の/こ/う/か/が/ふ/よ/さ/れ/る/
phoneme	/silB/n/a/g/a/sh/i/g/i/r/i/g/a/k/a/N/z/e/N/n/i/h/a/i/r/e/b/a/ sp/d/e/b/a/f/u/n/o/k/o/o/k/a/g/a/f/u/y/o/s/a/r/e/r/u/silE/
mora	/silB/na/ga/shi/gi/ri/ga/ka/N/ze/N/ni/ha/i/re/ba/ sp/de/ba/fu/no/ko/o/ka/ga/fu/yo/sa/re/ru/silE/

4 Evaluation Experiment

4.1 Experimental Condition

In this experiment, we use two datasets, ITA and ROHAN4600, to work on predicting sentences from Japanese utterance scenes. Considering the small number of speakers, 4, we conducted eight recognition experiments shown in Table 4. The test data for each experiment is the utterance scene of one ITA speaker. Of the experimental conditions #1 to #8, the test data speakers of #1 and #5, #2 and #6, #3 and #7, and #4 and #8 are CV1, CV2, CV3, and CV4, respectively. Although ROHAN4600 has only one speaker, it has the largest number of scenes, so it is used as training data in all experiments. The training data from #1 to #4 is only ROHAN4600, but the training data from #5 to #8 uses ROHAN4600 and ITA of other speakers that are not used as test data. In other words, #2 and #6 are speaker-dependent (SD) recognition tasks in which the training data includes the test data speaker, and the others are speaker-independent (SI) recognition tasks in which the training data does not include the test data speaker.

Levenshtein distance or edit distance in phonemes or mora is used as an evaluation metric for estimation results. It can quantitatively indicate "how similar a string `S` is to another string `T`." The smaller the distance, the more "similar." This distance is an error rate, so the smaller the value, the higher the accuracy. This distance is the amount of effort required to change to the string `T` by performing the three operations of substitute, insert, and delete for the string `S`. A cost is set for each of the three operations. Let S, I, and D denote the costs of substitute, insert, and delete, respectively, and Nt and Ns denote the string lengths of the two strings `T` and `S` to be compared. Then the distance is $d = (S + I + D)/\max(Nt, Ns)$.

Image Encoder uses a pre-trained model trained by LRW. Two recognition models, Transformer and Conformer, were used. The Transformer is the same as

Table 4. Experimental condition.

Corpus	ITA				ROHAN4600	Task
Speaker	CV1	CV2	CV3	CV4	CV2	
#1	Test	—	—	—	Training	SI
#2	—	Test	—	—	Training	SD
#3	—	—	Test	—	Training	SI
#4	—	—	—	Test	Training	SI
#5	Test	Training	Training	Training	Training	SI
#6	Training	Test	Training	Training	Training	SD
#7	Training	Training	Test	Training	Training	SI
#8	Training	Training	Training	Test	Training	SI

existing research [1,14]. In this experiment, two recognition targets, phoneme, and mora are evaluated. The former has 47 classes, and the latter has 166 classes. The recognition target is one sentence, but the recognition model treats it at the term-level. Therefore, accuracy is sought from two viewpoints: term-level and sentence-level.

4.2 Experimental Result

Table 5(a) (b) show the recognition results of the phoneme-level and the mora-level under the eight experimental conditions, respectively. In the table, the left half indicates the average error rate of the term-level, and the right half indicates the average error rate of the sentence-level. The results of applying the Transformer and Conformer as recognition models are shown, respectively.

Concerning the recognition model, the accuracy of Conformer is higher than the accuracy of Transformer under all conditions. This result confirms that the proposed method is effective.

Comparing the phoneme-level and the mora-level concerning the recognition accuracy for each term, the recognition accuracy at the phoneme-level is high. It is presumed that the number of classes at the phoneme-level is smaller than that at the mora-level.

The accuracy of #5 to #8 is higher than that of #1 to #4. It is assumed that this is because the training data is large.

Observing the recognition accuracy of the SI task (#1, #3, #4, #5, #7, and #8) and the SD task (#2 and #6), the recognition accuracy of the SD task is high, as expected. There is no significant difference in recognition accuracy for the three speakers in the SI task. Although it is an SD task, the error rates of sentence-level at the phoneme-level and the mora-level were 0.384 and 0.373, respectively, indicating the possibility of lip-reading Japanese sentences.

Table 5. Recognition result (error rate).

	Term-level		Sentence-level	
	Transformer [1,14]	Conformer	Transformer [1,14]	Conformer
(a) Phoneme-level				
#1	0.646	0.503	0.797	0.638
#2	0.473	0.399	0.551	0.460
#3	0.595	0.531	0.729	0.652
#4	0.642	0.595	0.755	0.714
#5	0.509	0.459	0.639	0.556
#6	0.430	**0.343**	0.485	**0.384**
#7	0.521	0.489	0.650	0.593
#8	0.606	0.540	0.699	0.633
(b) Mora-level				
#1	0.846	0.831	0.650	0.625
#2	0.730	0.650	0.602	0.537
#3	0.875	0.831	0.674	0.636
#4	0.895	0.857	0.729	0.673
#5	0.745	0.693	0.564	0.519
#6	0.578	**0.446**	0.473	**0.373**
#7	0.744	0.686	0.563	0.523
#8	0.798	0.736	0.648	0.598

5 Conclusion

In this paper, we constructed two Japanese utterance scene databases for lip-reading Japanese sentences, ITA and ROHAN4600, and proposed the method using the Conformer model. Although the scale of our datasets is smaller than the published English datasets, our datasets are characterized by the cooperation of professional voice actors. Regarding the recognition model, it was confirmed that Conformer is superior to Transformer, an existing method, under all experimental conditions. In the SD recognition task, we obtained the highest error rate of 0.384 at the phoneme-level and 0.373 at the mora-level.

The lack of data is obvious; we are collecting utterance scenes and expanding our datasets. In this paper, data augmentation is not applied, so we will work on recognition experiments with data augmentation in the future. There is room for improvement since the scene reconstruction is only concatenated. In the future, we will consider the division considering overlap and the introduction of language models.

Acknowledgments. This work was supported by JSPS KAKENHI Grants Number 19KT0029.

References

1. Afouras, T., Chung, J.S., Zisserman, A.: Deep lip reading: a comparison of models and an online application. In: Interspeech (2018)
2. Afouras, T., Chung, J.S., Zisserman, A.: LRS3-TED: a large-scale dataset for visual speech recognition. arXiv:1809.00496 (2018). https://doi.org/10.48550/arXiv.1809.00496
3. Alghamdi, N., Maddock, S., Marxer, R., Barker, J., Brown, G.J.: A corpus of audio-visual lombard speech with frontal and profile views. J. Acoust. Soc. Am. **143**(6), EL523–EL529 (2018)
4. Assael, Y.M., Shillingford, B., Whiteson, S., de Freitas, N.: LipNet: end-to-end sentence-level lipreading. arXiv:1611.01599 (2016). https://doi.org/10.48550/arXiv.1611.01599
5. Chung, J.S., Senior, A., Vinyals, O., Zisserman, A.: Lip reading sentences in the wild. In: IEEE Conference on Computer Vision and Pattern Recognition (CVPR), pp. 6447–6456 (2017). https://doi.org/10.1109/CVPR.2017.367
6. Chung, J.S., Zisserman, A.: Lip reading in the wild. In: Asian Conference on Computer Vision (ACCV) (2016)
7. Cooke, M., Barker, J., Cunningham, S., Shao, X.: An audio-visual corpus for speech perception and automatic speech recognition. J. Acoust. Soc. Am. **120**(5), 2421–2424 (2006). https://doi.org/10.1121/1.2229005
8. Gulati, A., et al.: Conformer: convolution-augmented transformer for speech recognition. In: Interspeech (2020)
9. Kodama, M., Saitoh, T.: Replacing speaker-independent recognition task with speaker-dependent task for lip-reading using first order motion model paper. In: 13th International Conference on Graphics and Image Processing (ICGIP) (2021). https://doi.org/10.1117/12.2623640
10. Nakamura, Y., Saitoh, T., Itoh, K.: 3DCNN-based mouth shape recognition for patient with intractable neurological diseases. In: 13th International Conference on Graphics and Image Processing (ICGIP) (2021). https://doi.org/10.1117/12.2623642
11. Noda, K., Yamaguchi, Y., Nakadai, K., Okuno, H.G., Ogata, T.: Lipreading using convolutional neural network. In: INTERSPEECH, pp. 1149–1153 (2014)
12. Patterson, E.K., Gurbuz, S., Tufekci, Z., Gowdy, J.N.: Moving-talker, speaker-independent feature study, and baseline results using the CUAVE multimodal speech corpus. EURASIP J. Adv. Signal Process. **2002**(11), 1–13 (2002). https://doi.org/10.1155/S1110865702206101
13. Saitoh, T., Kubokawa, M.: SSSD: speech scene database by smart device for visual speech recognition. In: 24th International Conference on Pattern Recognition (ICPR), pp. 3228–3232 (2018). https://doi.org/10.1109/ICPR.2018.8545664
14. Shirakata, T., Saitoh, T.: Japanese sentence dataset for lip-reading. In: IAPR Conference on Machine Vision Applications (MVA) (2021). https://doi.org/10.23919/MVA51890.2021.9511353
15. Tamura, S., et al.: CENSREC-1-AV: an audio-visual corpus for noisy bimodal speech recognition. In: International Conference on Auditory-Visual Speech Processing (AVSP) (2010)
16. Vaswani, A., et al.: Attention is all you need. arXiv:1706.03762 (2017). https://doi.org/10.48550/arXiv.1706.03762
17. Zhang, X., Cheng, F., Wang, S.: Spatio-temporal fusion based convolutional sequence learning for lip reading. In: International Conference on Computer Vision (ICCV), pp. 713–722 (2019). https://doi.org/10.1109/ICCV.2019.00080

Immuno-Inspired Augmentation of Siamese Neural Network for Multi-class Classification

Suraj Kumar Pandey(✉) and Shivashankar B. Nair

Indian Institute of Technology Guwahati, Assam, India
{suraj18a,sbnair}@iitg.ac.in

Abstract. Siamese Neural Networks (SNN) provide a robust mechanism to learn similarities/dissimilarities between objects of different classes. The distinguishing features learnt by SNNs make them a good candidate for multi-class classification as well. However, the potential of an SNN to create a classification space that has, both higher accuracy and lower inference time, needs to be exploited further. In this paper, we present a novel multi-class classification approach using SNNs by drawing concepts from the Immune Network theory. This bio-inspired strategy aids in injecting class specific characteristics into the SNN architecture, thereby enhancing the classification process. Experimental results conducted on three benchmark datasets indicate that the approach consistently provides higher accuracies and lesser inference times as compared to recent SNN based multi-class classification approaches, indicating its efficacy.

Keywords: Siamese Neural Network · Artificial immune system · Classification · Bio-inspired algorithms

1 Introduction

SNNs have been used in a plethora of applications, ranging from signature verification [1] to several instances of multi-class classification [2–5]. SNNs achieve a similarity metric in the form of a learnt model that allows the discrimination between similar and dissimilar data, resulting in a similarity space. Owing to useful features like low data requirement, SNNs have also been leveraged into the multi-class classification domain by transforming this similarity space into a classification space [2–5]. However, the usage of SNNs for multi-class classification is often done at the cost of losing inter-class relations [3], by using random class representations [4] or by involving further complex training procedures such as plugging additional neural network layers on top of the SNN [2,5], which also tends to limit their scalability. Some other approaches take a heavy toll in terms of high inference times [6–9]. Finding an approach that can enhance both accuracy and the speed of inference, thus, remains a desideratum.

In this paper, we propose a novel immuno-inspired technique that enhances an SNN based multi-class classification approach for images. The technique

W. Q. Yan et al. (Eds.): IVCNZ 2022, LNCS 13836, pp. 486–500, 2023.
https://doi.org/10.1007/978-3-031-25825-1_35

involves distillation of the underlying multi-class dataset to find a better generic representation, for the samples of each class. The representations, referred to as *Gold Standards* (GS), in turn interact with each other to form a network that brings out the similarities and dissimilarities between various classes, in the form of *stimulations* and *suppressions*, much like an Immune Network (IN) [10], resulting in better prediction performance. The main contributions of the work presented herein are enumerated below:

- Rather than making an SNN to merely learn the characteristics of all the classes, this work emphasises its usage towards data distillation by generating a GS per class, based on the information collated from the dataset. These GSs interact with one another, resulting in an IN which inherently contains the inter-class similarities/dissimilarities, thereby contributing towards the improvement of classification accuracy, viz. the ratio of the number of correct predictions to the total number of predictions.
- The proposed approach augments the SNN with the IN, during run time, thereby enhancing the accuracy and decreasing the inference time.

2 Background and Related Work

This section briefly discusses the two main entities - the SNN and the IN, used in the proposed approach along with the related work.

2.1 Siamese Neural Network

First introduced by Bromley et al. [1] for signature verification, SNNs predominantly aid in finding the similarity between inputs. A typical SNN consists of two Artificial Neural Networks (ANN) branches (Fig. 1), fed by an input each, and whose outputs are combined as a single output. Each of the two neural networks transforms the high-dimensional input samples into what are referred to as *embeddings* in generally a lower dimension. The *embeddings* are then blended together via a distance metric. This is followed by a loss function (cross-entropy loss [11], contrastive loss [12], triplet loss [13], etc.) whose minimisation ensures that the similarity of the inputs is reflected in the output as well.

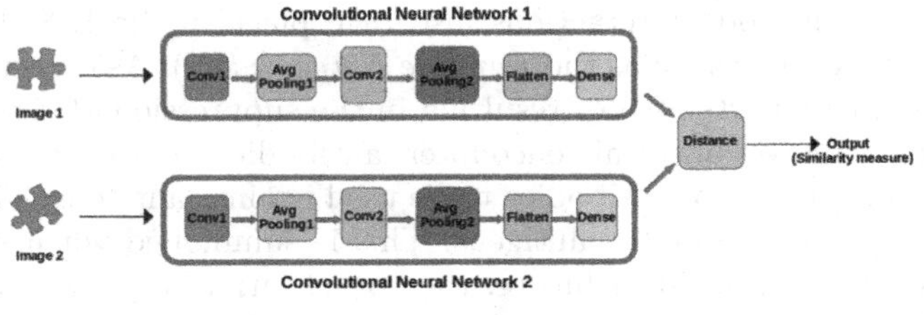

Fig. 1. A typical CNN based Siamese Neural Network

SNNs have been widely used across varied domains including speech processing, robotics, image analysis, etc. [14]. They have also been used for multi-class classification. However, the transformation from the similarity space to the classification space has been carried out in different ways. Swati et al. [2] have used SNNs with a Multi-Layer Perceptron (MLP) for chromosome classification. The usage of MLP adds an intricate training overhead and also limits the scalability. Nanni et al. [3] attempted animal sound classification by combining SNNs with different clustering techniques such as *K-Medoids* and *K-Means*, for speeding up inferencing. However, the cluster centers did not leverage any inter-class relations. Hindy et al. [4] advocated the usage of SNN for intrusion detection since SNN requires less data and has scalability towards new cyber attacks, thereby obviating the need for retraining. However, classification was accomplished by combining the SNN model with randomly chosen class representatives that need not always be the best representatives, thus negatively impacting the accuracy. Jiang and Zhang [5] used SNNs for handwritten numeral recognition by plugging a *softmax* layer at the end. This yielded good recognition performance but limited the scalability. Several researchers have proposed classification systems by combining SNNs and the K-Nearest Neighbours (KNN) algorithm [6–9]. However, as also mentioned in [2], the involvement of KNNs essentially leads to higher inference times and a substantial memory footprint for retention of training data.

2.2 Immune Network

The human Biological Immune System (BIS) is composed of a rich repertoire of cells, aimed at the identification and containment of the invading pathogens. The T- and B-cells together, co-operate to accomplish the task of identifying pathogens. Antibodies, which emanate from the B-cells, have protruding structures on their surface called *paratopes*. Pathogens have antigens on their surface, parts of which could have shapes that can be complementary to some paratopes. These parts are termed as *antigenic epitopes*. When the paratope of an antibody (Ab_1), binds (complementarily) with the epitope of an antigen (Ag), it leads to the recognition of the pathogen. Figure 2 depicts such a scenario. This binding results in the suppression of Ag and stimulation of Ab_1, the latter of which then proliferates to create clones of itself. Antibodies, such as Ab_2 whose paratopes bind less, tend to get suppressed, thereby decreasing in population. Thus, *recognisers* dominate the antibody population, allowing better and early containment of the pathogen. Jerne [10] proposed that apart from antigen-antibody interactions, antibody-antibody interactions also take place in the BIS, leading to a complex network referred to as the *Immune Network* (IN). As depicted in Fig. 2, Ab_1 also recognises parts of Ab_3, resulting in the suppression of Ab_3 and stimulation of Ab_1. After an antigenic encounter, antibodies interact with each other to maintain a good mix of antibodies to be used against antigens. This mix acts as the IN's memory about the antigens. This is summoned when similar antigens are encountered again in future, thereby ensuring a quicker and effective response to the invasion. This theory is often referred to as the Immune Network theory.

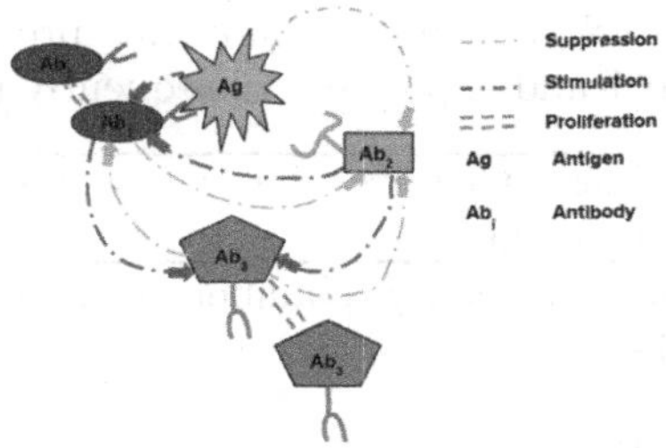

Fig. 2. Schematic of antibody-antigen/antibody-antibody interactions in an immune network

Concepts from BIS have been extensively used in applications in the computational domain that has given rise to an area often referred to as *Artificial Immune Systems* (AIS). In an AIS, the degree of binding among antibodies and antigens is quantified by *affinity* functions. AISs have been used in a plethora of applications including optimisation, anomaly detection, pattern recognition, robotics, etc. [15]. The proposed work described in the subsequent section, combines an SNN and an IN for multi-class classification to obtain better classification accuracy and also lower inference time.

3 Methodology

SNNs create a similarity space to tackle incoming test samples similar to an IN setting in a BIS, wherein antibodies employ their similarity space to identify and contain an incoming antigen. Further, just as a BIS generates *memory cells* which in turn release the right kind of antibodies facilitating a quicker response, an SNN based classification method also requires a mechanism to provide a quick response at the time of inferencing. This stark similarity between the two domains inspired us to draw concepts from the BIS and augment them to SNN based multi-class classification. As suggested in the IN theory, the formation of the IN aids the immune system in two ways:

- The immune system already has the apt collection of antibodies to respond at the time of an antigenic attack, making the response quicker.
- The network formed amongst the antibodies is able to mount a more holistic response as compared to the response of the antibodies in isolation.

The work proposed herein enhances an SNN based multi-class classification approach by exploiting the aforementioned characteristics of the IN. The training samples within the multi-class dataset used, represent a repertoire of antibodies while the test samples constitute the antigens. Instead of a conventionally used affinity function, we have used an SNN to assess the degrees of binding amongst the antibodies as well as between an antibody and an antigen. These bindings yield good antibodies in the form of GSs. The GSs together form an IN, which in turn, coupled with the SNN, aids in the inferencing process. The overall process is provided in Algorithm 1 and explained subsequently in detail.

Algorithm 1: Multi-class Classification using Immune Network augmented Siamese Neural Network

---

Input: *Abs* (Training Dataset), *q* (Number of classes), *Ag* (Test Data)
Output: Class label of *Ag*

Train SNN()
$GoldStandards \leftarrow SNN(Abs)$
$Prediction[q] \leftarrow 0$
for *each GoldStandard* Ab_{GS_i} *in GoldStandards* **do**
  for *each GoldStandard* Ab_{GS_j} *in GoldStandards* **do**
    $\text{IN}(Ab_{GS_i}, Ab_{GS_j}) \leftarrow SNN(Ab_{GS_i}, Ab_{GS_j})$
for *each GoldStandard* Ab_{GS_i} *in GoldStandards* **do**
  $Antigenic\_Contribution_i \leftarrow SNN(\text{Ag}, Ab_{GS_i})$
  for *each GoldStandard* Ab_{GS_j} *in GoldStandards* **do**
    $IN\_Contribution_j \leftarrow IN(Ab_{GS_i}, Ab_{GS_j})$
    $Prediction_j$
    $\leftarrow Prediction_j + Antigenic\_Contribution_i * IN\_Contribution_j$
return argmax(*Prediction*)

3.1 Affinity Function

Due to its notable discriminative ability, an SNN is pre-trained and used as the affinity function to distinguish between samples. The approach begins by training the SNN on a given dataset having q classes. Every sample, S_j^i, is associated with another sample, $S_{k(k \neq j)}^i$, belonging to the same class, i, leading to the first pair referred to as a *positive pair*, $[S_j^i, S_k^i]$. Another pair is generated by pairing S_j^i, with a sample S_n^m of a different class, m. This constitutes a *negative pair*, $[S_j^i, S_n^m]$. The SNN is trained using all possible positive and negative pairs along with distinct associated labels viz. 0 for positive pairs and 1 for negative pairs. The contrastive loss used for training the SNN is calculated as follows:

$$\mathcal{L} = (1 - l) * D^2 + l * max((\tau - D), 0)^2 \quad (1)$$

where l is the label of the pair fed to the SNN and D is the Euclidean distance between the *embeddings* of the pair of samples. τ forms the distance margin beyond which the samples of a pair are considered dissimilar. Once the SNN is trained, similar samples will yield a value close to 0 while dissimilar ones yield value close to 1, in accordance with the labelling convention. We have therefore taken the affinity as the inverse of the output of the SNN.

3.2 Churning Gold Standards

In a BIS, all antibodies stimulate and suppress each other to eventually yield the best set of representatives that can recognise an antigen. In this work, we have used a similar technique to churn out the best set of GSs, S_{GS}, having a representative from each of the classes. S_{GS} acts as a distilled form of the underlying data, which is used during inferencing as a reference for comparison. Each GS belonging to S_{GS} is chosen via the following process:

- For each class i, for every sample S_j^i, a pre-trained SNN is used to find the affinity of S_j^i with every other sample within the class i. All these affinities obtained for S_j^i are summed together. Since the samples are from the same class, and hence similar in characteristics, the sum of their affinities is considered to be the *stimulation*, ST_j^i, received by S_j^i, as in Eq. (2).

$$ST_j^i = \sum_{k=1}^{t} Aff_{jk}^i \tag{2}$$

  where ST_j^i is the *stimulation* for the sample j of class i, Aff_{jk}^i is the affinity of sample k for sample j in class i, having t number of samples. Among all the samples of each class i, the set of top N most stimulated samples, S_i, is chosen.
- With the samples in the set S_i of class i, the affinities of samples in the corresponding sets of other classes is calculated using the SNN. Such affinities are referred as *suppressions*, calculated using Eq. (3).

$$SU_j^i = \sum_{c=1(c \neq i)}^{q} \sum_{k=1}^{N} Aff_{jk}^{ic} \tag{3}$$

  where SU_j^i is the *suppression* received by sample j of class i and Aff_{jk}^{ic} is the affinity of sample k of set S_c of class c for sample j of class i.
- For each set, S_i, of a class i, the most suppressed sample is designated as the GS of that class. The suppressive interactions push the GS representative away from class boundaries, thereby refining the GS representation.

The stimulative interactions yield a medoid sample of the class as a potential representative for the GS. If time and computational resources form a constraint, the GS may be found using only *stimulations* and not *suppressions*. This process is performed offline so as to yield the set, $S_{GS} = [GS_1, GS_2 \ldots GS_q]$.

3.3 Creating a Network of Gold Standards

In a BIS, the IN not only yields the best antibodies but also allows antibodies to relay information about one another across the network [10]. As depicted in Fig. 2, the entire network of antibodies participates in relaying the respective information regarding the compatibility of each of the connected antibodies with the antigen. The GSs, which form the metaphors of the antibodies, mimic such a mechanism as and when a new and unknown test sample (antigen) is presented. Prior to inference, the GSs of different classes form an IN, based on affinity values provided by the pre-trained SNN. Any two GSs interact with one another based on this affinity which represents the weight of an edge between them within the IN.

3.4 Classification of a Test Sample

Classification of a test sample, S_t, involves finding two contributions towards prediction and aggregating them as shown in Fig. 3.

- *Antigenic contribution*: The trained SNN is used to compare S_t individually with the members of the set S_{GS}, to generate a set of antigenic affinities, $Aff^{Ag} = [Aff_1^{Ag}, Aff_2^{Ag} \dots Aff_q^{Ag}]$ constituting the antigenic contributions of each GS.
- *IN contribution*: A $GS_i \in S_{GS}$ is connected to form the IN, based on its affinities, with all other GSs ($GS_j \in S_{GS}$). Hence, the decision of GS_j regarding S_t, is influenced by the decision made by GS_i regarding S_t. This influence is equal to the mutual affinity, Aff_{ij}^{IN}, between GS_i and GS_j, which constitutes the contribution made by the IN.

Both the antigenic and the IN contributions are combined for each GS. Finally, the net contributions by all the GSs are aggregated to yield an affinity distribution across all the classes (as detailed in Algorithm 2). The class having the maximal affinity is chosen as the prediction.

Algorithm 2: Combining Antigenic and IN contributions for prediction

Input: S_{GS} (GS set), IN (Network of GSs), q (Number of classes), S_t (Test Data)
Output: Class label of S_t

Prediction$[q] \leftarrow 0$
for *each* GS_i *in* S_{GS} **do**
  for *each* GS_j *in IN* **do**
    Prediction$_j$ = Prediction$_j$ + $Aff_{ij}^{IN} * Aff_i^{Ag}$
return argmax(*Prediction*)

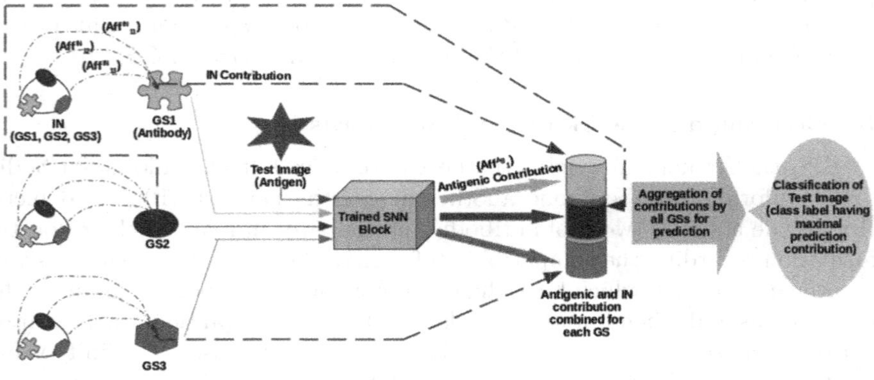

Fig. 3. Classification of a Test sample using both Antigenic & IN contributions

The method used in Algorithm 2 adds a degree of confidence to the contribution made by the IN in proportion to the antigenic contribution. If Aff_i^{Ag} itself is low, the impact made by each of the GSs via the IN (Aff_{ij}^{IN}) over the prediction of the test sample will be curtailed by weighing it down. This interlinking of contributions allows a holistic method of prediction as opposed to prediction by individual GSs in isolation. The overall classification process discussed so far primarily utilises just one set of GSs. However, as also suggested in [4], one may employ multiple sets of GSs to enhance the decision making process. We therefore propose an additional method as briefed in Algorithm 3 to achieve this. Multiple sets of GSs are found by using *stimulations*. This is followed by execution of Algorithm 1 for each set of GSs so as to yield its corresponding inference. These inferences are combined via majority voting to yield final prediction.

Algorithm 3: Multi-class Classification using Immune Network augmented Siamese Neural Network with multiple Gold Standard Sets and Majority Voting

Input: *Abs* (Training Dataset), q (Number of classes), *Ag* (Test Data)
Output: Class label of *Ag*

Train SNN()
$GoldStandardsSet \leftarrow SNN$(Abs)
$CombinedPrediction[q] \leftarrow 0$
for *each* $GoldStandards_i$ *in* $GoldStandardsSet$ **do**
  $ClassIndex \leftarrow$ **Algorithm 1**(Abs, Ag)
  $CombinedPrediction[ClassIndex] \leftarrow$
  $CombinedPrediction[ClassIndex]$ +1
return argmax($CombinedPrediction$)

4 Experimental Analyses and Results

Due to high dimensionality and multiple levels of features, images prove to be an apt candidate for the rigorous testing of the approach discussed herein. We have thus, tested the algorithms by conducting experiments using the MNIST [16], the Fashion-MNIST (FMNIST) [17] and the Kuzushiji-MNIST (KMNIST) datasets [18]. These datasets contain 60,000 training and 10,000 testing samples as images of size 28×28 pixels each. The MNIST dataset consists of images of handwritten digits from 0 to 9. The FMNIST dataset consists of ten different classes of images of clothing items. The KMNIST dataset consists of images of characters from the Japanese syllabary, Hiragana, spread across 10 classes. These three datasets differ in terms of features and extent of overlap among classes, presenting diverse scenarios and making classification a rigorous exercise. The experimental setup was realised on *Jupyter* [19], a web-based computation platform. Architectures involving ANNs, were realised using *Keras* [20] and *Tensorflow* [21] frameworks running on a computer with an *Intel CORE i5* processor with 16GB RAM. We have used a CNN based SNN, as depicted in Fig. 1, to act as the base model. As can be observed, the SNN has two CNN branches connected by a distance layer.

The distance layer is used to find the Euclidean distance between the *embeddings* generated from the two CNNs. This distance was used to train the SNN using contrastive loss minimisation method. The configuration of each CNN branch is given below:

- A *Convolutional* layer having four filters of size 5×5 using *tanh* as the activation function
- An *Average Pooling* layer of size 2×2
- A *Convolutional* layer having sixteen filters of size 5×5 with *tanh* as the activation function
- An *Average Pooling* layer of size 2×2
- A *Flatten* layer
- A *Dense* layer having ten neurons which use *tanh* as the activation function

As explained in Sect. 3.1, we began the implementation of our approach by creating a *positive pair* and a *negative pair* for each sample belonging to each of the ten classes within the MNIST dataset. These pairs were used to train the SNN. We then followed the next step (Sect. 3.2) to obtain the set of GSs, which comprised one representative training sample image per class. This was followed by the creation of a 2-dimensional matrix in the form of a Python list structure that stored the affinities amongst the GSs thereby acting as the IN (Sect. 3.3). This matrix, along with the set of GSs and the SNN was then used to classify images from the test set of the MNIST dataset based on the explanation provided in Sect. 3.4. In order to analyse the significance of the major components of our proposed method, ablation studies were conducted which have been discussed in subsequent sections. After gathering enough empirical justification for the components involved in our approach, we conducted rigorous experimentation to compare our approach with existing methods, viz. the popular KNN based approach [6–9] and the recent approach of random sampling of class representatives [4,9]. In all the experiments, the underlying SNN was trained for 100 epochs. All the methods were tested against 800 samples per class. The details of the experiments and the results obtained are discussed in the following sections.

4.1 Comparison of GS Selection Methods

GSs play a major role in representing the underlying knowledge about the dataset. Hence, the strategy used for selection of a GS directly impacts the classification accuracy. For conducting ablation studies of this impact, experiments were carried out using the following four methods for selection of a GS:

- **RS**: *Random Sampling* as in [4,9]
- **RMV**: Random selection of multiple GS sets, followed by classification based on *Majority Voting* amongst the sets, as suggested in [4]
- **ST**: Using only *Stimulations* (Eq. (2))
- **STSU**: Using *Stimulations* and *Suppressions* (Eqs. (2) and (3))

The multi-class classification of the MNIST dataset was performed with the set(s) of GSs obtained using each of the above methods. In order to closely analyse the impacts of the modes of selection of the GS on classification accuracy, we conducted the experiments by removing the IN and hence its own impact. To achieve this, we ignored the portion in Algorithm 2 where IN is used to update the $Prediction_i$ vector (line 3) and also omitted Aff_{ij}^{IN} from line 4. The SNN was initially trained separately using four discrete training dataset sizes made by using 5, 50, 100 and 400 distinct training samples per class. These four SNNs were then used individually for classification in conjunction with each of the above GS selection methods amounting to sixteen different experiments. In order to take care of any possible stochastic variations, we conducted each of these sixteen experiments, five times (leading to eighty experiments in total) and compared their average testing accuracies. In addition, five GS sets were used for voting in the RMV method. As can be observed in Fig. 4(a), the accuracies in case of ST and STSU methods were found to be consistently more than those of the RS and RMV. This can be attributed to the improved characterisation of features of a class by the corresponding GS selected using ST and STSU methods. This effect becomes more pronounced with increase in the size of the training dataset, thus validating the proposed stimulation-suppression mechanism for GS selection.

4.2 Impact of in During Inferencing

The significance of the involvement of IN for inference was analysed through an ablative procedure. This involved comparing the classification accuracies when both *Stimulations* and the *IN* were used (STIN method) with the case when the *IN* was not used (NOIN method). The procedure discussed in Sect. 4.1 was used to remove the IN. In both the cases, the underlying SNN was initially trained separately using four discrete training dataset sizes made by using 10, 50, 100 and 200 distinct training samples per class. These four SNNs were then used individually for classification both under STIN and NOIN setting, leading to eight experiments. Further, each of these experiments were averaged across five repeated separate executions to yield average testing accuracies. It may be seen from Fig. 4(b) that there is a consistent increase in accuracy across different training sizes in case of STIN against NOIN, highlighting the role of IN.

4.3 Comparison of STIN with KNN Based Approach for Multi-class Classification

Since KNN is often employed with SNN to perform classification tasks [6–9], we have compared a KNN based approach (KNN) with our approach (STIN). We have used the same trained SNN for classifying the MNIST dataset while using KNN and STIN methods. After the SNN was trained, the test sample was compared with all the training samples for clustering under the KNN based approach. As the simplest case, requiring least computations, a cluster of size one was used for KNN. Five experiments were performed for both KNN and STIN approaches using a training set size of fifty samples per class. Table 1

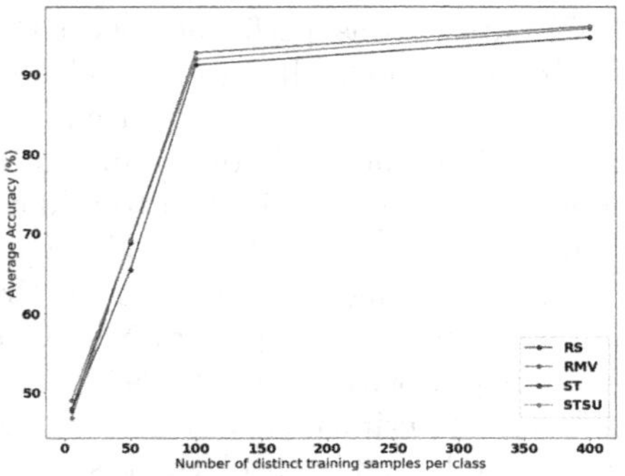

(a) GS selection methods

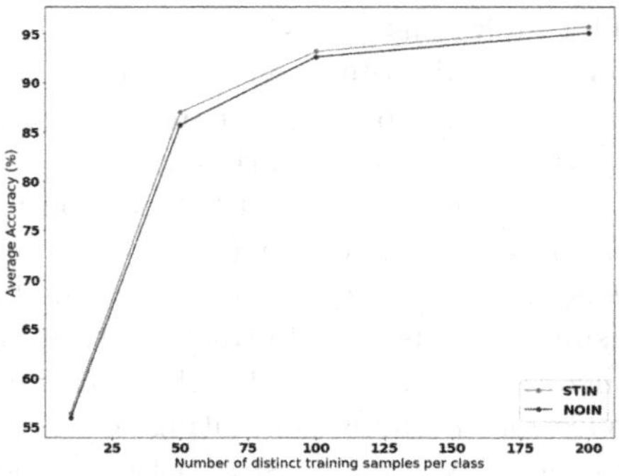

(b) STIN vs NOIN

Fig. 4. Classification Accuracies with different training dataset sizes using the MNIST dataset across (a) Different GS Selection Methods, (b) GS based method with IN (STIN) and without IN (NOIN)

shows the results of these experiments in terms of classification accuracy and inference time. As can be seen, on an average, the STIN outperforms the KNN approach in terms of accuracy. While the increase in accuracy seems marginal, the inference time taken by the STIN approach suggests it to be the clear winner. The average speedup factor of STIN over the KNN approach is around 74. This speedup can be attributed to the *parameterised* nature of STIN where the GSs are already distilled out *a priori* and are directly used as characteristics of the corresponding classes during inferencing. This saves time as there is no need to compare the test sample with each sample of a class during inference.

Table 1. Classification Accuracies and Inference Times using the MNIST dataset for KNN-based (KNN) and combined GS-IN based (STIN) methods (max accuracy and min time boldfaced)

Accuracy (%)		Inference time (seconds)	
KNN	STIN	KNN	STIN
87.57	**88.73**	2477.61	**33.45**
88.21	**88.49**	2722.31	**38.15**
87.0	**87.03**	2463.94	**33.11**
85.99	85.7	2492.50	**33.54**
87.71	**88.88**	2581.44	**33.15**

4.4 Comparison of STIN and Its Variants with Random Sampling Based Approaches for Multi-class Classification

In addition to RS, RMV and STIN, we introduced the following methods for experimentation:

- **STMV**: Multiple GS sets are found by using *Stimulations*. The inference is then drawn for a test sample based on *Majority Voting* by the GS sets. IN has not been used here.
- **STMVIN**: Using IN for each GS set along with STMV

In order to test the performance consistency of our proposed methods (STIN, STMV and STMVIN) against existing methods (RS and RMV), we conducted experiments across three different datasets: MNIST, FMNIST and KMNIST. Each of these datasets have different levels of complexity in terms of higher level features and extent of overlap among classes. For all the methods, the underlying SNN was trained separately using eight discrete training dataset sizes made by using 50, 75, 100, 250, 500, 1000, 3000 and 5000 distinct samples per class. These eight SNNs were then used individually for classification using each of the five methods, amounting to forty different experiments. Further to discourage any empirical analysis emerging from stochastic reasons, we conducted each of the experiments twenty times to find the average testing accuracies and inference times (leading to 800 experiments for each of the three datasets). Additionally, five GS sets were used for voting in RMV, STMV and STMVIN methods.

Table 2. Classification Accuracies and Inference Times with different training dataset sizes across different methods using the MNIST dataset (max accuracy and min time(s) boldfaced)

Samples/class	Accuracy (%)					Time (seconds)				
	RS	RMV	STIN	STMV	STMVIN	RS	RMV	STIN	STMV	STMVIN
50	83.30	85.73	**87.68**	86.26	**87.60**	**51.73**	263.26	**52.27**	260.75	266.18
75	87.43	90.99	**92.29**	91.51	**92.16**	**53.71**	264.158	**54.42**	261.96	265.59
100	90.07	92.63	**93.57**	93.01	**93.62**	**54.82**	263.84	**55.34**	263.51	267.57
250	93.42	94.50	**95.34**	94.61	**95.35**	**52.53**	264.55	**53.06**	266.44	268.13
500	95.10	95.69	**96.47**	95.89	**96.52**	**52.54**	264.00	**53.52**	261.2	268.23
1000	96.29	96.73	**97.21**	96.85	**97.21**	**53.01**	262.997	**53.55**	265.82	268.23
3000	97.02	97.60	**97.93**	97.66	**97.92**	**52.91**	261.28	**53.67**	258.57	264.13
5000	97.22	98.00	**98.151**	98.03	**98.154**	**52.55**	262.26	**52.76**	332.59	277.90

The accuracy and inference time (for all the test samples) obtained across the experiments using the three datasets are shown in Table 2 and Fig. 5, leading to the following significant observations:

- Accuracies across all the methods increase with increase in training dataset size. However, for a given dataset size, the accuracies generally decrease across datasets MNIST, FMNIST and KMNIST; in that order. This implies the relatively increasing complexities of these datasets.
- For all the datasets, our proposed methods (STIN, STMV, STMVIN) consistently outperform the existing methods (RS, RMV). This can be observed from Table 2 where the top two accuracy values for each dataset size (row)

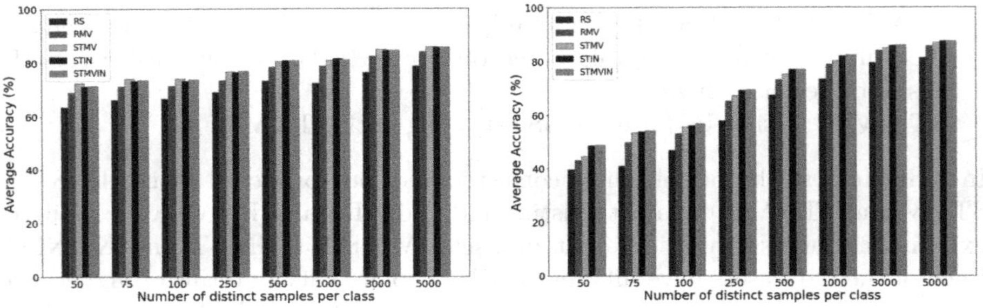

(a) FMNIST (b) KMNIST

Fig. 5. Classification Accuracies with different training dataset sizes across different methods using (a) FMNIST dataset (b) KMNIST dataset

of the MNIST dataset are boldfaced. Figure 5 also depicts the same trend for FMNIST and KMNIST datasets. This highlights the utility of the proposed mechanisms of *stimulations* (over random sampling [4,9]) and IN.
- STMVIN generally performs better than STIN because of the involvement of more than one GS sets. STMVIN also outperforms STMV due to the use of an IN by the former.
- STIN and RS take the least time (speedup of almost 5 over both RMV and STMVIN) as shown in Table 2. Experiments conducted using FMNIST and KMNIST datasets also showed similar trends. It may thus, be inferred that when time is a crucial factor, it is best to opt for STIN. STMVIN can be used otherwise, when accuracies are to be increased even further.
- In few cases, STIN manages to yield a higher accuracy than STMVIN (with lesser inference time). It can also be noted that STIN generally outperforms STMV. These observations strongly suggest the towering impact of the use of *stimulations* and the IN on classification performance as compared to the conventional majority voting mechanism.

4.5 Comparison over Multiple Metrics for Multi-class Classification

In addition to accuracy and inference time, multi-class classifiers can be analysed in detail by using precision, recall and F-score metrics. F-score is generally the preferred metric for comparing classifiers since it merges both precision and recall. Table 3 provides a class-wise comparison of these metrics for RMV, STIN and STMVIN methods for the MNIST dataset. The underlying SNN for all these methods was trained using 900 distinct samples per class from the training dataset. The F-score tends to be the highest for STMVIN in most cases (values boldfaced), highlighting its consistent performance gain, even at the class level.

Table 3. Class-wise comparison of Precision, Recall and F-score for different methods using the MNIST dataset (max values boldfaced for each class)

Class	F-score			Precision			Recall		
	RMV	STIN	STMVIN	RMV	STIN	STMVIN	RMV	STIN	STMVIN
0	**0.985**	**0.985**	**0.985**	**0.988**	0.985	0.985	0.981	**0.985**	**0.985**
1	0.988	**0.989**	0.988	**0.991**	0.987	0.986	0.984	**0.991**	**0.9912**
2	0.971	**0.976**	**0.976**	0.953	0.960	**0.961**	0.990	**0.992**	**0.992**
3	0.965	0.971	**0.972**	0.960	0.971	**0.972**	0.970	0.971	**0.972**
4	0.972	**0.976**	**0.976**	0.987	**0.988**	**0.988**	0.958	**0.965**	**0.965**
5	0.965	0.966	**0.9666**	**0.960**	0.955	0.957	0.971	**0.976**	**0.976**
6	**0.979**	**0.979**	**0.979**	**0.986**	0.983	0.983	0.972	**0.975**	**0.975**
7	0.967	0.972	**0.974**	**0.984**	0.980	0.981	0.951	0.965	**0.967**
8	0.951	**0.958**	**0.958**	0.928	**0.948**	**0.948**	**0.974**	0.968	0.968
9	0.939	0.944	**0.945**	0.949	0.960	**0.961**	0.929	**0.930**	**0.930**

5 Conclusions and Future Work

This paper proposes an approach to enhance the performance of an SNN for multi-class classification through an immuno-inspired method. The experimental analyses strongly support the proposal, showing a consistent increment in the accuracy with a speedup of upto five times in a dataset agnostic manner. By utilising an SNN as an affinity function for an Artificial Immune System, this work attempts to bridge both the domains. In addition, the proposed methods (STIN and STMVIN) provide a choice for higher accuracy upon investment of more time, thereby contributing to an *Anytime* algorithm [22].

We have used a multiplication based strategy to accommodate contribution by Immune Network during classification. Considering the promising preliminary results, this strategy needs to be investigated further which can also be enhanced by involving parallel *stimulations* and *suppressions*. Since SNNs can also tackle scalability towards new classes, we can widen the scope of our proposed methods in terms of adding new classes and samples. As the approach has been tested with small sized datasets, its usage can also be explored for low-resource devices.

References

1. Bromley, J., et al.: Signature verification using a "siamese" time delay neural network. Int. J. Pattern Recognit. Artif. Intell. **7**, 25 (1993)
2. Jindal, S., Gupta, G., Yadav, M., Sharma, M., Vig, L.: Siamese networks for chromosome classification. In: 2017 IEEE International Conference on Computer Vision Workshops (ICCVW), pp. 72–81 (2017)
3. Nanni, L., Brahnam, S., Lumini, A., Maguolo, G.: Animal sound classification using dissimilarity spaces. Appl. Sci. **10**(23), 8578 (2020). https://www.mdpi.com/2076-3417/10/23/8578

4. Hindy, H., Tachtatzis, C., Atkinson, R., Bayne, E., Bellekens, X.: Developing a siamese network for intrusion detection systems. In: Proceedings of the 1st Workshop on Machine Learning and Systems, EuroMLSys 2021, pp. 120–126. Association for Computing Machinery, New York (2021). https://doi.org/10.1145/3437984.3458842
5. Jiang, W., Zhang, L.: Edge-SiamNet and edge-TripleNet: new deep learning models for handwritten numeral recognition. IEICE Trans. Inf. Syst. **103**(3), 720–723 (2020)
6. Zhu, R., Gong, X., Hu, S., Wang, Y.: Power quality disturbances classification via fully-convolutional siamese network and k-nearest neighbor. Energies **12**(24), 4732 (2019). https://www.mdpi.com/1996-1073/12/24/4732
7. Wang, B., Wang, D.: Plant leaves classification: a few-shot learning method based on siamese network. IEEE Access **7**, 151754–151763 (2019)
8. Zhou, M., Tanimura, Y., Nakada, H.: One-shot learning using triplet network with kNN classifier. In: Ohsawa, Y., et al. (eds.) JSAI 2019. AISC, vol. 1128, pp. 227–235. Springer, Cham (2020). https://doi.org/10.1007/978-3-030-39878-1_21
9. Veal, C., et al.: Doing more with less: similarity neural nets and metrics for small class imbalanced data sets. In: Detection and Sensing of Mines, Explosive Objects, and Obscured Targets XXV, ser. Society of Photo-Optical Instrumentation Engineers (SPIE) Conference Series, vol. 11418, p. 1141802 (2020)
10. Jerne, N.K.: Towards a network theory of the immune system. Ann. Immunol. **125**(1–2), 373–389 (1974). https://pubmed.ncbi.nlm.nih.gov/4142565
11. Gregory, K., Zemel, R., Salakhutdinov, R.: Siamese neural networks for one-shot image recognition. In: ICML Deep Learning Workshop (2015)
12. Chopra, S., Hadsell, R., LeCun, Y.: Learning a similarity metric discriminatively, with application to face verification. In: 2005 IEEE Computer Society Conference on Computer Vision and Pattern Recognition (CVPR 2005), vol. 1, pp. 539–546 (2005)
13. Schroff, F., Kalenichenko, D., Philbin, J.: "Facenet: a unified embedding for face recognition and clustering. In: IEEE Conference on Computer Vision and Pattern Recognition (CVPR) 2015, pp. 815–823 (2015)
14. Chicco, D.: Siamese neural networks: an overview. In: Cartwright, H. (ed.) Artificial Neural Networks. MMB, vol. 2190, pp. 73–94. Springer, New York (2021). https://doi.org/10.1007/978-1-0716-0826-5_3
15. López, G.Q., Morales, L.A., Niño, L.F.: Immunological computation. In: Autoimmunity: From Bench to Bedside [Internet]. El Rosario University Press (2013). https://www.ncbi.nlm.nih.gov/books/NBK459484
16. LeCun, Y., Cortes, C.: MNIST handwritten digit database (2010). http://yann.lecun.com/exdb/mnist/
17. Xiao, H., Rasul, K., Vollgraf, R.: Fashion-MNIST: a Novel Image Dataset for Benchmarking Machine Learning Algorithms (2017). http://arxiv.org/abs/1708.07747
18. Clanuwat, T., Bober-Irizar, M., Kitamoto, A., Lamb, A., Yamamoto, K., Ha, D.: Deep Learning for Classical Japanese Literature (2018)
19. Project Jupyter (2022). https://jupyter.org. Accessed 31 May 2022
20. Keras: The Python Deep Learning API (2022). https://keras.io. Accessed 31 May 2022
21. TensorFlow (2022). https://www.tensorflow.org. Accessed 31 May 2022
22. Dean, T., Boddy, M.: An analysis of time-dependent planning. In: Proceedings of the Seventh AAAI National Conference on Artificial Intelligence, AAAI 1988, pp. 49–54. AAAI Press (1988)

PCMask: A Dual-Branch Self-supervised Medical Image Segmentation Method Using Pixel-Level Contrastive Learning and Masked Image Modeling

Yu Wang[1], Bo Liu[1,2(✉)], and Fugen Zhou[1,2]

[1] Image Processing Center, Beihang University, Beijing 100191, People's Republic of China
bo.liu@buaa.edu.cn

[2] Beijing Advanced Innovation Center for Biomedical Engineering, Beihang University, Beijing 100083, People's Republic of China

Abstract. Supervised deep learning methods have gained prevalence in various medical image segmentation tasks for the past few years, such as U-Net and its variants. However, most methods still need a large amount of annotation data for training, and the quality of annotation will also affect the performance of the model. To address this issue, we propose a novel self-supervised model named PCMask. Specifically, it is a self-supervised method with a dual branch suitable for pre-training the U-Net architecture with a ViT encoder. While the masked image modeling branch pre-trains the encoder through the reconstruction of masked tokens, the pixel-level contrastive branch utilizes a contrastive learning strategy to pre-train both the encoder and decoder. Its advantage lies in the introduction of the pixel-level contrastive learning strategy, which can improve the reconstruction of high-resolution features at different scales. We validate the effectiveness of the proposed framework by transferring the pre-trained UNETR backbone to two different datasets. Favorable results were obtained over existing methods.

Keywords: Medical image segmentation · Self-supervised learning · Contrastive learning · Masked image modeling

1 Introduction

Medical image segmentation, including lesion and organ segmentation, plays an essential role in clinical diagnosis and treatment planning. In recent years, vision transformers (ViTs) [1] have started a revolutionary trend in computer vision and medical image analysis due to their outstanding modeling capabilities. ViT-based segmentation networks are proposed, such as UNETR [2] and Swin UNETR [3], which achieved impressive results. However, most current works rely on a large amount of labeled data to achieve good performance, which is costly and challenging to obtain, especially for 3D medical images.

W. Q. Yan et al. (Eds.): IVCNZ 2022, LNCS 13836, pp. 501–510, 2023.
https://doi.org/10.1007/978-3-031-25825-1_36

To address this issue, self-supervised learning has been extensively studied as a practical way to learn better-generalized representations using a large amount of unlabeled data. And self-supervised methods for ViT-based architectures have been proposed recently, such as BEiT [4], MAE [5], and MaskFeat [6]. However, to the best of our knowledge, most of the existing self-supervised methods only focus on the feature extraction ability of the encoder, leading to representations that may be sub-optimal for downstream tasks such as object detection or semantic segmentation. The weights of the head networks or decoders need to be randomly initialized and trained from scratch. A recent work [7] has shown that only partial capability of the self-supervised methods is used for representation learning, and a large part of the capacity is wasted to cater to the pretext tasks. In the fields of semantic segmentation, especially medical image segmentation [8], the ability to reconstruct high-resolution feature maps is as important as the ability of feature encoding.

To address this and try to improve the performance of self-supervised segmentation for medical images, we propose a novel dual-branch self-supervised framework using **P**ixel-level **C**ontrastive learning and **Mask**ed image modeling to learn representation from unlabeled data (named PCMask). The proposed methods can improve the feature extraction ability of the encoder and constrain the representation extracted by the decoder with our pixel-level contrastive learning branch, promoting the pre-training quality of the whole encoder-decoder architecture.

Our main contributions can be summarized as follows:

1. We proposed a simple and effective self-supervised learning framework named PCMask, to learn general vision representations with a dual-branch architecture. It mainly focuses on medical image segmentation tasks and can be used to improve ViT-based segmentation.
2. We introduce a pixel-level contrastive learning strategy focusing on decoder structure and thus can achieve better decoding performance to reconstruct high-resolution features at different scales. We validate the effectiveness of the proposed framework by transferring the pre-trained UNETR [2] backbone to two different datasets.

2 Related Work

This section reviews deep learning-based self-supervised learning approaches in natural and medical images.

2.1 Self-supervised Learning in Natural Images

Early methods for self-supervised learning try to mine the information from unlabeled data by predicting input images' properties that are covariant to some destabilization, such as inpainting [9], decoupling [10], and colorization [11]. Afterward, contrastive learning-based self-supervised learning methods such as MoCo [12], SimCLR [13], and BYOL [14] were proposed to learn representations by enforcing features to be similar for matched pairs and dissimilar for unmatched pairs and showed better performance.

Besides the commonly-studied classification problems, some contrastive learning methods focused on detection or segmentation are also proposed to promote the performance on downstream tasks [15, 16]. Following the success of masked language modeling in the natural language process, transformer-based methods like masked image modeling [4, 5, 17] are proposed and achieve outstanding performance in downstream tasks.

Pretexted tasks are used for self-supervised learning at the early stage of visual representation learning, aiming to mine the information of unlabeled data by predicting input images' properties which are covariant to some destabilization, such as inpainting [9], decoupling [10], colorization [11]. As a relatively new pretext task, contrastive learning-based methods such as MoCo [12], SimCLR [13], and BYOL [14] also utilize pretext tasks to learn representations by enforcing features to be similar for matched pairs and dissimilar for unmatched pairs. Some contrastive learning methods focused on detection or segmentation are proposed to promote the performance of downstream tasks [15, 16]. Recently, following the success of masked language modeling in the natural language process, transformer-based methods like masked image modeling [4, 5, 17] have been proposed and achieved outstanding performance in downstream tasks.

2.2 Self-supervised Learning in Medical Images

The success of self-supervised in natural images has prompted researchers to apply self-supervised learning to medical image analysis. Pretext tasks are designed to learn representations, such as reconstructing corrupted images [18], solving jigsaw puzzles [19], solving Rubik's cube problems [20], and random rotation prediction [21]. Contrastive learning methods [22–25] are applied similarly to natural images, which constitute an essential step toward better self-supervised learning approaches to medical image analysis.

3 Method

In this section, we will first introduce the overall overview of the framework. Then we present the two branches of self-supervision respectively and explain their different roles in self-supervision pre-training.

3.1 Overview

The overview of the proposed PCMask architecture is illustrated in Fig. 1, which is composed of two branches, the pixel-level contrastive branch and the masked image modeling branch. A ViT-based medical image segmentation method called UNETR was used as the backbone. The encoder of the UNETR, which encodes semantic features, was shared in the two branches with the same weights.

Both our proposed pixel-level contrastive learning strategy and the masked image modeling task can help improve the ability of the encoder. The difference between the two branches is that the pixel-level contrastive branch adds pixel consistency constraints in the decoder, which could enhance the pixel-level encoder ability. In contrast, the masked image modeling branch only adds constraints to the reconstruction of the original image at the top of the decoder, which is helpless for the downstream segmentation task. Therefore, the decoder in the masked image modeling branch will be discarded in the fine-tuning stage. We used the decoder in the pixel-level contrastive learning branch when transferring to downstream tasks.

For the input for the PCMask, blocks with the size of 96 × 96 × 96 were randomly cropped from the original image. Different data augmentation strategies were applied to the blocks for the two branches. While strong augmentations (random regions shuffle and dropout) were applied to construct two different views of a sample for the pixel-level contrastive learning branch, only the aforementioned random crop of the blocks was applied to the input of the masked image modeling branch. The input patch size for the ViT encoder was set as 16 × 16 × 16.

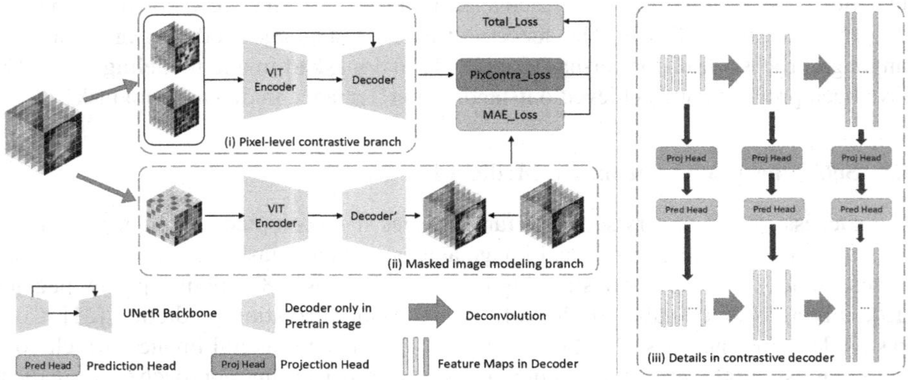

Fig. 1. Overview of our PCMask. (i) Sketch of the pixel-level contrastive learning branch. Here, the ViT encoder and decoder construct the UNETR backbone. (ii) Sketch of the masked image modeling branch. This ViT encoder shares the same structure and weights as that in the pixel-level contrast branch, and the decoder will be discarded in the fine-tuning stage. (iii) Details of the decoders of the pixel-level contrast branch. Pixel features are used to generate positive and negative pairs.

3.2 Masked Image Modeling Branch

To improve the performance of the encoder in our self-supervised framework, we proposed to utilize the masked image modeling branch, in which the ViT encoder reconstructs randomly masked tokens [26] by leveraging a few visible patches. Two hundred sixteen tokens were regularly extracted from the input block, and some of them were randomly masked and replaced with learnable embeddings. With the features extracted by the ViT encoder, a simple decoder with a series of Transformer blocks was used to reconstruct the missing patches.

The mask autoencoder has a solid ability to encode semantic features. It can extract the information of unlabeled images through the simple task of reconstructing the original tokens, thereby guiding the ViT encoder to encode image features better. On the other hand, we think that the simple decoder learns more about how to reconstruct the original image and does not help our segmentation task. Therefore, we discard this decoder when applying it to downstream tasks.

3.3 Pixel-Level Contrastive Learning Branch

Although the masked image modeling scheme can help the encoder learn to extract better semantic information, it focuses more on instance-level information. The segmentation task has high requirements on the capability of the backbone to encode and reconstruct pixel-level information. The pixel-level contrastive branch was proposed to address this, with the aim that the whole encoder-decoder architecture can learn more pixel-level information besides encoding the instance-level information.

Like most instance-level contrastive learning methods, our pixel-level contrastive learning branch starts by applying two different augmentations to the same block. What is different is that the two views v_1, v_2 have the same size and location as the original image, which is convenient for applying pixel consistency constraints. The two views will be fed into two UNETR backbones with different parameters θ_1, θ_2 respectively, with θ_1 being the regular parameters and θ_2 a momentum-updated moving average as in MoCo. With the generated feature maps of the decoder, pixel features were selected to generate positive and negative pairs according to their locations. If x_i and x_j are the pixel features in two views, which were encoded as queries q and keys k after the projection heads; a query should be similar to its matching key and dissimilar to others through a similarity function defined as

$$sim(q, k) = max(cos(q, k), 0) \tag{1}$$

The strategy was designed from the following considerations. Since the two views were in the same position when we cropped them from the original image, the high-dimensional features of the corresponding positions of the two views on the feature maps of the decoder should also be consistent. We should ensure that the corresponding labels of the two views are the same after the reconstruction of the segmentation result of the original input size.

Note that a projection head and a prediction head are used when computing the contrastive loss, just like BYOL. Besides, considering the label consistency and the consistency of prediction results between different views, we add contrastive constraints at different scales of feature maps in the decoder.

3.4 Loss Function

InfoNCE [27] is a commonly used contrastive loss in self-supervised learning, which is considered in MoCo as

$$L = -log \frac{exp(q \cdot k_+/\tau)}{\sum_{i=0}^{K} exp(q \cdot k_i/\tau)} \tag{2}$$

where q is the encoded query from one of the views; the sum is over one positive and K negative samples; k_+ is a single key in the dictionary that q matches. τ is a temperature hyper-parameter.

Specifically, our pixel-level contrastive loss is defined as

$$L_{PCL} = -log \frac{exp(sim(q, k_+)/\tau)}{exp(sim(q, k_+)/\tau) + \sum_{(q,k)\in\Omega-} exp(sim(q, k)/\tau)} \quad (3)$$

where (q, k) is the encoded query and key from the pixel-level feature of feature maps in the different scales of the decoder; the $sim()$ is a cosine similarity function; Ω is the pixel-level feature sets constructed by feature maps in the decoder.

The loss function in the masked image modeling branch is a simple $L2$ reconstruction loss between the original images and the reconstructed images, which is similar to MAE [5]. We used normalized pixels as the reconstruction target and computed the loss only on masked patches.

Finally, our total loss can be defined as

$$L_{total} = L_{PCL} + \alpha L_{MIM} \quad (4)$$

where L_{PCL} and L_{MIM} are defined as above; α is a multiplicative factor to balance these two loss functions.

4 Experiments and Results

4.1 Experimental Setup

We conduct experiments on two datasets consisting of different modalities. We ignore the labels in the pre-training stage and then finetune the model with different scales of labeled data.

The MMWHS [28] **dataset** was for the segmentation of seven cardiac sub-structures on cardiac CT, hosted in STACOM and MICCAI 2017 challenge. The included twenty labeled CT images are used in this work.

The MRA-18 dataset was collected by the authors, which comprised 18 MRAs with expert annotations for cerebrovascular segmentation.

Each of the datasets is split into a pre-training set X_{pre} and a test set X_{ts}. The fine-tuning set X_{ft} is a subset from X_{pre}. We pre-train our UNETR backbone on X_{pre}, and then fine-tune the whole network using a small number of labeled data from X_{ft}, and eventually, validate the performance of the fine-tuned network on X_{ts}. More details of our datasets are shown in Table 1.

Table 1. Data splitting in our experiments

Name	Modalities	Data splitting		
		Pre-train	Fine-tune	Test
MMWHS	CT	16	10%–40%	4
MRA-18	MRA	14	10%–40%	4

We use the AdamW [29] optimizer with a cosine learning rate scheduler [30]. The framework was implemented using PyTorch and MONAI [31] and trained with 2 NVIDIA GeForce 3090 GPUs.

Pre-training Stage. We used our proposed PCMask to pre-train the model. For data augmentation, we just employed some simple transforms such as random region dropout and intensity scaling. All of the CT images without labels in X_{pre} are used in the pre-training stage. All models were trained for a total of 500 epochs, and the best model checkpoints on the validation set were used for the fine-tuning stage.

Fine-tuning Stage. We used the same input size as the pre-training stage. In the fine-tuning stage, we discarded the decoder in the masked image modeling branch and only used the decoder obtained by pixel-level contrastive learning. Note that we will add an output layer according to the number of output labels for overall fine-tuning. In order to better verify the performance of the pre-training, we use different scales of the finetuning set to fine-tune the model.

4.2 Results

The performance of our proposed PCMask methods is presented in Table 2. PCMask was compared with the supervised method with randomly initialized weights and some advanced self-supervised methods such as Inpainting [9], SimCLR [13], and the self-supervised method proposed in MONAI [31] and MAE [5].

Inpainting applied reconstruction as the pretext task, aiming to reconstruct the color information of the masked pixels. Instance-level contrastive learning method was used in SimCLR through InfoNCE loss. The MONAI method extended a segmentation ViT encode to learn from self-supervised reconstruction tasks with various data augmentation and a contrastive loss. MAE used a mask-and-predict pretext task similar to our masked image modeling branch.

Results showed that our proposed self-supervised method achieved better results on a smaller fine-tuning set. When using all labeled samples to fine-tune the model, different methods showed similar results due to the limited pre-training data.

To investigate the efficacy of the proposed strategy, we compared the proposed self-supervised method with several variants on the two datasets, using the same UNETR architecture across all methods. Table 3 shows the results of different settings of pretext tasks, including inpainting [9], instance-level contrastive learning [13], pixel-level contrastive learning in the decoder, and masked image modeling. Results showed the combination of the two advanced pretext tasks by our dual branch architecture achieved better performance.

Table 2. Comparison of methods using different datasets. The dice similarity coefficient (DSC) was used to evaluate the segmentation performance. The supervised method only used X_{ft} for training without pre-training. All of the methods used the UNETR backbone.

Method	Dataset	10%	20%	40%	100%
Supervised	MMWHS	59.03	75.85	88.07	91.38
Inpainting [9]		61.51	76.25	88.65	91.45
SimCLR [13]		62.81	79.08	88.05	91.27
MONAI [31]		62.33	78.56	89.02	**91.54**
MAE [5]		62.06	77.22	88.10	91.51
Ours		**64.99**	**80.87**	**89.76**	91.49
Supervised	MRA-18	53.68	71.00	85.22	88.03
SimCLR [13]		53.96	74.03	86.55	**88.96**
MONAI [31]		54.20	74.95	86.95	88.21
Ours		**56.65**	**75.69**	**87.85**	88.82

Table 3. Results of different tasks. (i) R-Loss indicates the inpainting pretext task aiming to reconstruct the original image used in [9]. (ii) ICL-Loss indicates image-level contrastive task in the encoder. (iii) PCL-Loss indicates our proposed pixel-level contrast branch in the decoder. (iv) MIM-Loss indicates the masked image modeling branch. The proposed self-supervised pre-training provides better results than other methods.

Pretext tasks				DSC
R-loss	ICL-loss	PCL-loss	MIM-loss	
√				87.51
	√			88.01
√	√			87.97
			√	88.65
		√		88.46
		√	√	**88.99**

5 Conclusion

In this work, a simple and effective self-supervised pre-training method for the segmentation task was proposed, which was named PCMask. This method combines the advantages of generative pretext task and predictive pretext task with a dual branch structure, with each of them targeting different parts of the segmentation network. Experiments showed that the proposed method achieved better performance on the downstream semantic segmentation tasks by pre-training the entire encoder-decoder network.

Acknowledgment. This work was supported by the National Key R&D Program of China under Grant Numbers:2018YFA0704100 and 2018YFA0704101, the National Natural Science Foundation of China (61971443), and the Fundamental Research Funds for the Central Universities.

References

1. Dosovitskiy, A., Beyer, L., Kolesnikov, A., et al.: An image is worth 16 x 16 words: Transformers for image recognition at scale. arXiv preprint arXiv:2010.11929, (2020)
2. Hatamizadeh, A, Tang, Y., Nath, V., et al.: Unetr: transformers for 3d medical image segmentation. In: Proceedings of the IEEE/CVF Winter Conference on Applications of Computer Vision, pp. 574–584 (2022)
3. Tang, Y, Yang, D., Li, W., et al.: Self-supervised pre-training of swin transformers for 3d medical image analysis. arXiv preprint arXiv:2111.14791 (2021)
4. Bao, H., Dong, L., Wei, F.: Beit: BERT pre-training of image transformers. arXiv preprint arXiv:2106.08254 (2021)
5. He, K., Chen, X., Xie, S., et al.: Masked autoencoders are scalable vision learners. arXiv preprint arXiv:2111.06377 (2021)
6. Wei, C., Fan, H., Xie, S., et al.: Masked feature prediction for self-supervised visual pre-training. arXiv preprint arXiv:2112.09133 (2021)
7. Chen, X., Ding, M., Wang, X., et al.: Context autoencoder for self-supervised representation learning. arXiv preprint arXiv:2202.03026 (2022)
8. Ronneberger, O., Fischer, P., Brox, T.: U-net: Convolutional networks for biomedical image segmentation. In: Navab, N., Hornegger, J., Wells, W.M., Frangi, A.F. (eds.) MICCAI 2015. LNCS, vol. 9351, pp. 234–241. Springer, Cham (2015). https://doi.org/10.1007/978-3-319-24574-4_28
9. Pathak, D., Krahenbuhl, P., Donahue, J., et al.: Context encoders: feature learning by inpainting. In: Proceedings of the IEEE Conference on Computer Vision and Pattern Recognition, pp. 2536–2544 (2016)
10. Feng, Z., Xu, C., Tao, D.: Self-supervised representation learning by rotation feature decoupling. In: Proceedings of the IEEE/CVF Conference on Computer Vision and Pattern Recognition, pp. 10364–10374 (2019)
11. Larsson, G., Maire, M., Shakhnarovich, G.: Learning representations for automatic colorization. In: Leibe, B., Matas, J., Sebe, N., Welling, M. (eds.) ECCV 2016. LNCS, vol. 9908, pp. 577–593. Springer, Cham (2016). https://doi.org/10.1007/978-3-319-46493-0_35
12. He, K., Fan, H., Wu, Y., et al.: Momentum contrast for unsupervised visual representation learning. In: Proceedings of the IEEE/CVF Conference on Computer Vision and Pattern Recognition, pp. 9729–9738 (2020)
13. Chen, T., Kornblith, S., Norouzi M., et al.: A simple framework for contrastive learning of visual representations. In: International Conference on Machine Learning, PMLR, 2020, pp. 1597–1607 (2020)
14. Grill, J.B., Strub, F., Altché, F., et al.: Bootstrap your own latent: a new approach to self-supervised learning. arXiv preprint arXiv:2006.07733 (2020)
15. Xie, Z,. Lin, Y., Zhang, Z., et al.: Propagate yourself: exploring pixel-level consistency for unsupervised visual representation learning. In: Proceedings of the IEEE/CVF Conference on Computer Vision and Pattern Recognition, pp. 16684–16693 (2021)
16. Wang, X., Zhang, R., Shen, C., et al.: Dense contrastive learning for self-supervised visual pre-training. In: Proceedings of the IEEE/CVF Conference on Computer Vision and Pattern Recognition, pp. 3024–3033 (2021)

17. Shi, Y., Huang, Z., Feng, S., et al.: Masked label prediction: unified message passing model for semi-supervised classification. arXiv preprint arXiv:2009.03509 (2020)
18. Chen, L., Bentley, P., Mori, K., et al.: Self-supervised learning for medical image analysis using image context restoration. Med. Image Anal. **58**, 101539 (2019)
19. Noroozi, M., Favaro, P.: Unsupervised learning of visual representations by solving jigsaw puzzles. In: Leibe, B., Matas, J., Sebe, N., Welling, M. (eds.) ECCV 2016. LNCS, vol. 9910, pp. 69–84. Springer, Cham (2016). https://doi.org/10.1007/978-3-319-46466-4_5
20. Tao, X., Li, Y., Zhou, W., et al.: Revisiting Rubik's cube: self-supervised learning with volume-wise transformation for 3D medical image segmentation. In: International Conference on Medical Image Computing and Computer-Assisted Intervention, pp. 238–248 (2020).https://doi.org/10.1007/978-3-030-59719-1_24
21. Gidaris, S., Singh, P., Komodakis, N.: Unsupervised representation learning by predicting image rotations. arXiv preprint arXiv:1803.07728 (2018)
22. Taleb, A., Loetzsch, W., Danz, N., et al.: 3d self-supervised methods for medical imaging. Adv. Neural. Inf. Process. Syst. **33**, 18158–18172 (2020)
23. Zhou, H.Y., Yu, S., Bian, C., Hu, Y., Ma, K., Zheng, Y.: Comparing to learn: surpassing imagenet pretraining on radiographs by comparing image representations. In: Martel, A.L., Abolmaesumi, P., Stoyanov, D., Mateus, D., Zuluaga, M.A., Zhou, S.K., Racoceanu, D. (eds.) MICCAI 2020. LNCS, vol. 12261, pp. 398–407. Springer, Cham (2020). https://doi.org/10.1007/978-3-030-59710-8_39
24. Zhou, H.Y., Lu, C., Yang, S., et al.: Preservational learning improves self-supervised medical image models by reconstructing diverse contexts. In: Proceedings of the IEEE/CVF International Conference on Computer Vision, pp. 3499–3509 (2021)
25. Chaitanya, K., Erdil, E., Karani, N., et al.: Contrastive learning of global and local features for medical image segmentation with limited annotations. Adv. Neural. Inf. Process. Syst. **33**, 12546–12558 (2020)
26. Devlin, J., Chang, M.W., Lee, K., et al.: BERT: pre-training of deep bidirectional transformers for language understanding. arXiv preprint arXiv:1810.04805 (2018)
27. Van den Oord, A., Li, Y., Vinyals, O.: Representation learning with contrastive predictive coding. arXiv e-prints. arXiv: 1807.03748 (2018)
28. Wang, C., Smedby, Ö.: Automatic whole heart segmentation using deep learning and shape context. In: Pop, M., Sermesant, M., Jodoin, P.M., Lalande, A., Zhuang, X., Yang, G. (eds.) STACOM 2017. LNCS, vol. 10663, pp. 242–249. Springer, Cham (2018). https://doi.org/10.1007/978-3-319-75541-0_26
29. Loshchilov, I., Hutter, F.: Decoupled weight decay regularization. arXiv preprint arXiv:1711.05101 (2017)
30. Loshchilov, I., Hutter, F.: Sgdr: Stochastic gradient descent with warm restarts. arXiv preprint arXiv:1608.03983 (2016)
31. MONAI Consortium: MONAI: Medical Open Network for AI (Version 0.8.1) [Computer software] (2022). https://github.com/Project-MONAI/MONAI

Vision Transformer-Based Bark Image Recognition for Tree Identification

Towa Yamabe and Takeshi Saitoh(✉)

Kyushu Institute of Technology, 680–4 Kawazu, Iizuka, Fukuoka, Japan
saitoh@ai.kyutech.ac.jp

Abstract. Our group is studying tree species recognition using image processing technology. In the previous research, we proposed an image-based bark recognition using CNN. In this paper, we propose a method of recognizing bark image using Vision Transformer (ViT), which has attracted attention in the image recognition task in recent years. Four public datasets of NewBarkTex, TRUNK12, BarkNet1.0, and Bark-101, and a new dataset of 150 tree species originally collected, KyutechBark150, were used in the evaluation experiment. Several CNN models were used as comparison methods. As a result of the recognition experiment, the highest recognition accuracy of ViT was obtained in all the datasets. In addition, the trained model was visualized by t-SNE and attention map, and this paper shows that ViT is effective for bark image recognition.

Keywords: Bark texture · Tree species recognition · Vision transformer

1 Introduction

There is a need for computer-based tree species identification technology in the forestry, real estate, and education industries. This need is because many species are difficult for amateurs to identify. Against this background, we are working on tree species recognition using image processing technology. So far, we have proposed recognition methods for leaf image [16] and bark image [8,9]. This paper proposes a recognition method for bark images using Vision Transformer (ViT), attracting attention in image recognition tasks in recent years.

Flowers, leaves, bark, and fruits can identify trees. Flowers and fruits have brighter colors than leaves and bark, so these are easy to identify. However, the period during which they can be observed is limited. The leaves are simpler in shape than the flowers and are almost green. Although the leaves have deciduous seeds, they have the advantage that they can be photographed for a longer period than flowers and fruits, and the number of leaves is larger than the number of flowers, making it easier to photograph. However, as they grow, flowers, leaves, and fruits change in size, color, and shape. The bark has less color and may have moss. It is generally more difficult to identify than flowers, leaves, and fruits. However, the bark is easy to photograph since the trees grow from the ground. Thus, the bark is set as the recognition target in this paper.

W. Q. Yan et al. (Eds.): IVCNZ 2022, LNCS 13836, pp. 511–522, 2023.
https://doi.org/10.1007/978-3-031-25825-1_37

The bark texture is qualitatively described in pictorial books as smooth, lenticels, furrows, ridges, cracks, scales, and strips. Therefore, the classification of bark texture is considered a difficult problem task. In reports on standard texture analysis, bark and textures such as blocks and stones are discussed. This paper focuses only on the bark and is positioned as a difficult task to classify similar textures.

Previously, texture analysis methods were used for bark image recognition, but studies applying deep learning have shown significant improvements in accuracy in recent years. The following studies have been reported as an approach that does not use deep learning. Fiel et al. [5] adopted an approach using leaves and bark and proposed a method for recognizing bark using gray level co-occurrence matrix (GLCM) and wavelet coefficients as features and using bags of words and support vector machine (SVM). Porebski et al. [13] studied the classification problem of color texture and constructed the bark image dataset BarkTex as a benchmark. Haralick features and 3D color histogram features are defined and recognized by 1-NN. Boudra et al. [1] propose a statistical radial binary pattern (SRBP) and a method of recognition by SVM. Experiments are conducted using four datasets, AFF, TRUNK12, BarkNet1.0, and Bark-101. Ratajczak et al. [14] constructed the dataset Bark-101 and evaluated it using existing texture features and color histograms. Evaluation experiments are conducted using the BarkTex, AFF, and TRUNK12 datasets. Remes et al. [15] proposes a method using 2DSCAR and evaluates it using four datasets: AFF, BarkTex, TRUNK12, and BarkNet1.0.

On the other hand, Carpentier et al. [2] constructs a dataset BarkNet1.0 and proposes a method for recognizing it using ResNet, which is a type of CNN. Nanii et al. [12] use ResNet50, but the research subject compares various data expansion methods, and bark images are used for the evaluation. In our previous research [9], we proposed a method to automatically extract the bark region by applying semantic segmentation to the photographed image and recognizing it by CNN. Although it is evaluated using an original private small dataset, three models (Inception-v3, VGG16, ResNet-50) are used.

The contributions of this research are as follows.

- We constructed a new bark image dataset KyutechBark150, with more tree species than the conventional public datasets.
- We show that ViT has higher bark image recognition accuracy for multiple bark image datasets than other methods.

This paper is organized as follows. Section 2 describes the proposed method. Section 3 introduces the datasets used for the recognition experiment and shows the recognition experiment. Finally, Sect. 4 concludes this paper.

2 Proposed Method

This section describes the overview of the proposed method.

(a) Original natural images.

(b) Region image after applying semantic segmentation, and extracted ROI.

(c) Extracted ROIs

Fig. 1. ROI extraction process from original natural images. (Color figure online)

2.1 ROI Extraction

In this research, similar to the previous research [8,9], the tree images taken in the natural state, as shown in Fig. 1(a) are processed. The target tree is photographed in the center of the image, but other trees may be reflected behind it.

In order to extract the tree to be recognized from the captured image and extract the ROI, the ROI extraction process of the previous research [9] is applied.

1. First, we apply semantic segmentation to classify the tree region and other regions. In this paper, DeepLab v3+ [3] is applied. At this time, there are three classification classes: bark, moss, and background. Figure 1(b) shows the result of applying semantic segmentation to Fig. 1(a).
2. Multiple trees may appear in the captured image. Depending on the photography conditions, the tree in the center of the image is extracted as the target tree.
3. The target tree's left and right boundary pixels are detected for the region image resulting from semantic segmentation. A tree has branches, but the boundary between the tree and the background can be considered a straight line. Therefore, a spatial filter detects edge pixels from the region image. The

tree boundary is detected by applying RANSAC [6] to the edge pixels to remove outliers and detect two approximate straight lines.
4. The tilt angle of the tree is calculated using the left and right lines extracted in the previous step. The tilt angle is used to correct the tilt since the tree may appear tilted when the image is taken. After that, a square region is extracted from the tree as ROI.

The red rectangle shows the position of the extracted ROI in Fig. 1(b). The extracted ROI is shown in Fig. 1(c). This ROI is used as input data for the subsequent recognition process.

2.2 Recognition

Most of the previous studies on bark images have applied CNN. In this study, Vision Transformer (ViT) [4] is applied.

ViT is a model that applies transformers to image classification tasks. The contents of ViT are almost the same as the original transformer, but we have devised a way to handle images as input, just like natural language processing.

ViT divides an image into N patches of 16×16 size. Since a patch is a 3D data (height $\times$ width $\times$ number of channels), it cannot be handled directly by a transformer that handles language (2D). Therefore, linear projection is applied and converted into two-dimensional data after flattening. This allows each patch to be treated as a word-like token and input to the transformer. Here, the ROI size is changed to a multiple of 16 in advance.

3 Experiment

3.1 Datasets

This experiment conducted evaluation experiments using five datasets: NewBarkTex, TRUNK12, BarkNet1.0, Bark-101, and KyutechBark150. The last one is originally constructed in this paper.

NewBarkTex [13] consists of six species, Betula Pendula, Fagus Silvatica, Picea Abies, Pinus Silvestris, Quercus Robus, Robinia Pseudacacia, and the provided ROI is 64×64 [pixels] shown in Fig. 2(a). A total of 1,632 ROIs are contained, 272 ROIs of each type.

TRUNK12 [18][1] contains twelve tree species of Alder, Beech, Birch, Chestnut, Ginkgo Biloba, Hornbeam, Horse Chestnut, Linden, Oak, Oriental Plane, Pine, and Spruce. Several samples of the provided image are shown in Fig. 2(b). The image size is a rectangular area of $3,000 \times 4,000$ [pixels], including 30 to 45 samples per species. The total number of samples is 393.

[1] https://www.vicos.si/resources/trunk12/.

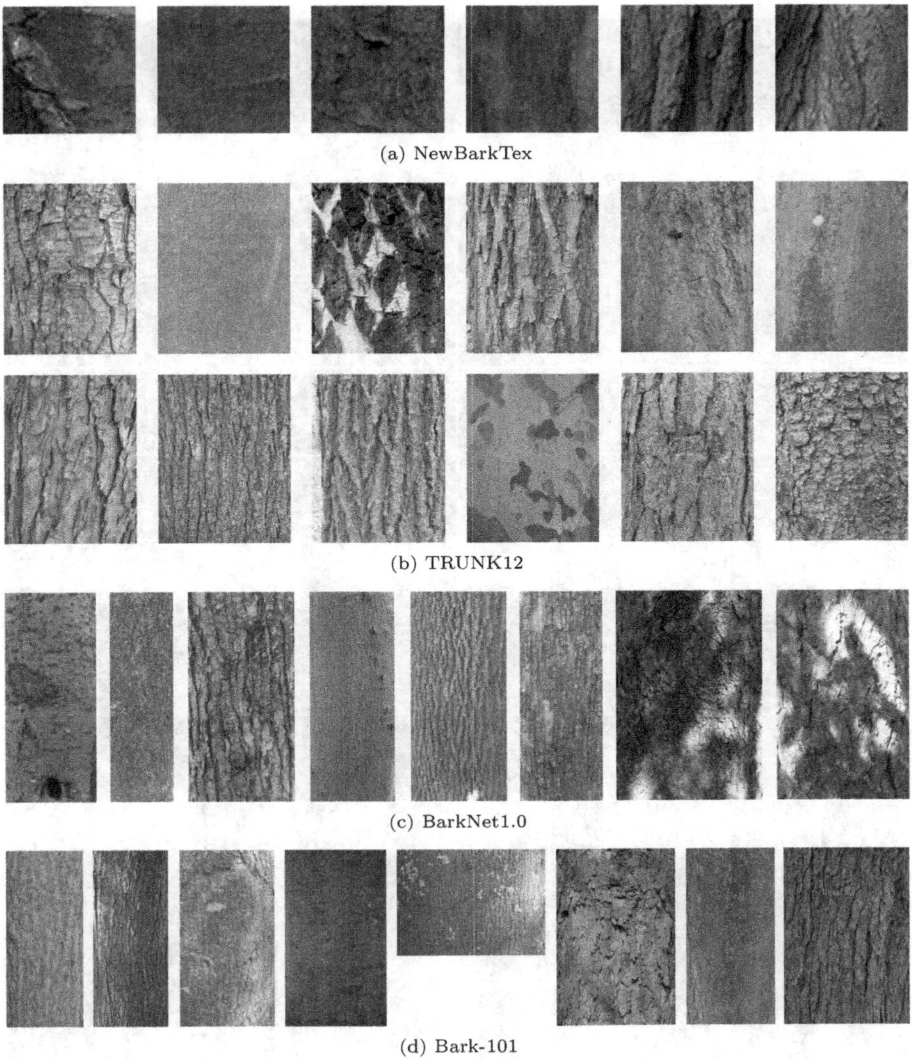

Fig. 2. Four publicly available datasets used in our experiment.

BarkNet1.0 [2] contains 21 species. Several samples are shown in Fig. 2(c). The image size is $(260 \times 370) \sim (2,984 \times 5,312)$ [pixels] and includes 64 to 2,724 images per species for a total of 20,988 images.

Bark-101 [14][2] contains 101 species. Several samples are shown in Fig. 2(d). The image size is $(86 \times 202) \sim (800 \times 804)$ [pixels] and includes 1 to 138 images per species, for a total of 2,586 images.

[2] http://eidolon.univ-lyon2.fr/~remi1/Bark-101/.

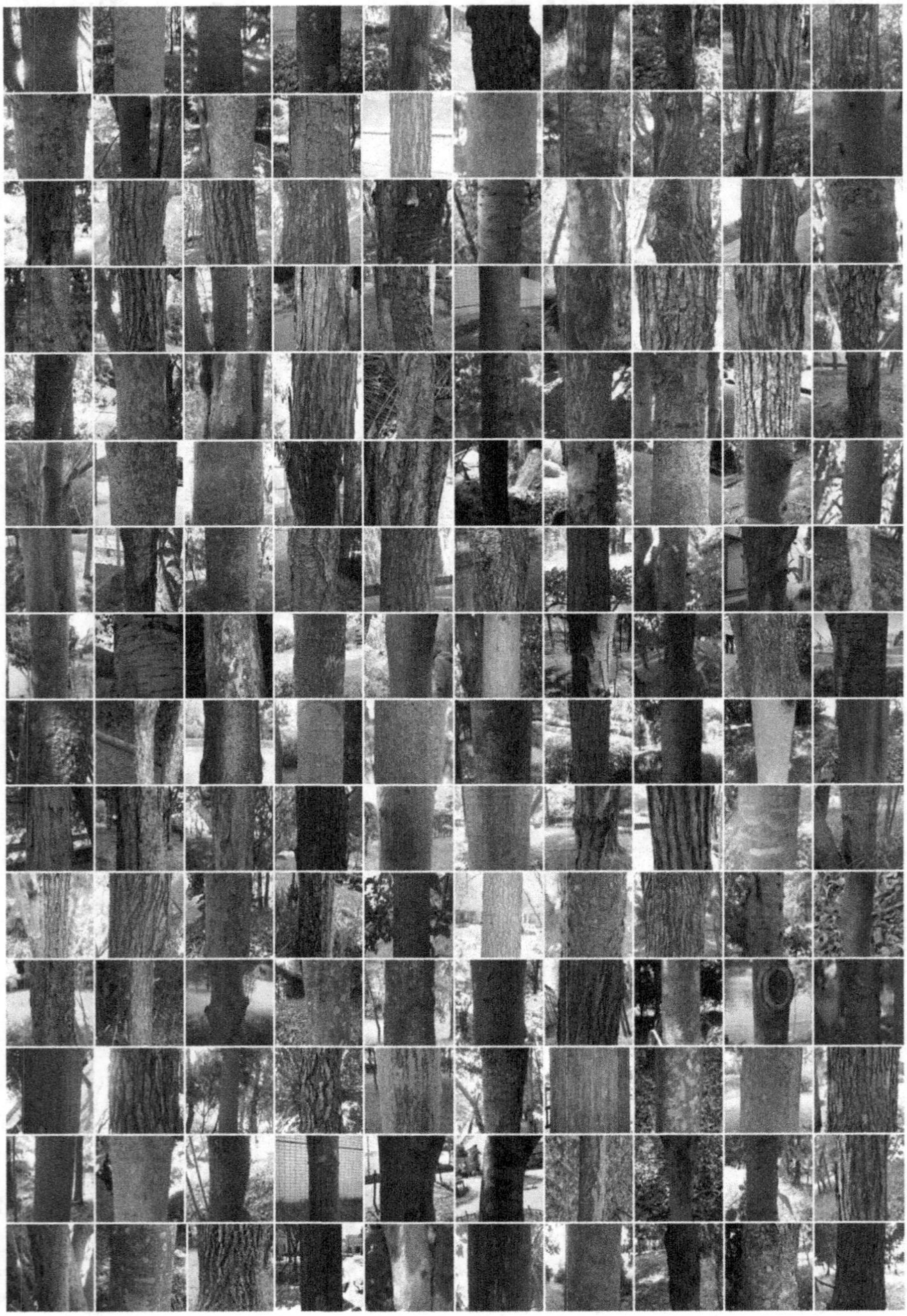

Fig. 3. Our collected dataset: Kyutech150.

KyutechBark150 is a dataset originally collected in this study. We photographed in the Kyushu area in Japan and photographed 30 to 85 trees per species. The total number of images is 6,171. Image sizes range from 640×480 to $4,608 \times 3,648$ [pixels]. Figure 3 shows the tree images of each sample of 150 species. Our dataset has more tree species than other public datasets. Here, we collected bark images under expert guidance.

3.2 Recognition Experiment

The datasets introduced above differ in the types of images provided. For ViT or CNN, a square ROI is used as input data. Therefore, extracting the square ROI from each dataset is necessary. Since NewBarkTex is provided with the square ROI, the provided data is used. For KyutechBark150, the method described in Sect. 2 should be applied. The other TRUNK12, BarkNet1.0, and Bark-101 are bark images without a background but have various aspect ratios and image sizes. For these, the square ROI was automatically extracted from the center of the provided bark image. Also, unlike other datasets, KyutechBark150 is a photographed image that includes the background. Generally, in bark image recognition, a bark image that does not include the background is used, but in this experiment, it is also considered to give the photographed image directly as input data.

The deep learning framework Keras was used to implement ViT and CNN. There are three ViT model sizes, Base/Large/Huge, but Base and Large were used in this experiment. Two types of patch sizes, 16 and 32, were used. The optimization method is Adam [10], the number of epochs is 50, the initial learning rate is 0.001, and the learning rate is multiplied by 0.1 every seven epochs. All ROIs were converted to 224×224 [pixels] for input to ViT. Typical CNN models of VGG16 [17], VGG19 [17], and ResNet50 [7] were used as comparative models. All ROIs were converted to 224×224 [pixels] for input to CNN. We experimentally adjusted the parameters according to each CNN model and dataset for the number of epochs and learning rate.

The experimental conditions and results for the five datasets are shown in Table 2. Here, the training and test data are specified for the data provided by NewBarkTex [13], and Bark-101 [14], but the rest of the datasets are not specified. Therefore, for these datasets, the training and test data are randomly divided at $4:1$, and the training and evaluation experiments tasks are performed three times each to obtain the average recognition rate. Here, the recognition rate is the total number of correctly identified bark images divided by the total number.

From Table 2, a higher recognition rate was obtained using ViT rather than CNN in all datasets. Regarding the four conditions of ViT, although there are

Table 1. Detail of datasets.

Dataset	NewBarkTex [13]	TRUNK12 [18]	BarkNet1.0 [2]
# of classes	6	12	23
# of trainings/class	136	24	51 ~ 240
# of tests/class	136	6	13 ~ 60
# of total images	1,632	393	23,000+
Image size [pixel]	64 × 64	2,400 × 2,400	224 × 224
Dataset	Bark-101 [14]	KyutechBark150	
		whole image	ROI
# of classes	101	150	
# of trainings/class	1 ~ 69	24	
# of tests/class	1 ~ 69	6	
# of total images	2,592	6,169	
Image size [pixel]	86 × 86 ~ 751 × 751	640 × 480 ~ 4,608 × 3,648	256 × 256

Table 2. Recognition results.

Dataset	NewBarkTex [13]	TRUNK12 [18]	BarkNet1.0 [2]	Bark-101 [14]	KyutechBark150	
Image type	ROI	ROI	ROI	ROI	Whole image	ROI
VGG16 [17]	0.901	0.935	0.646	0.384	0.653	0.566
VGG19 [17]	0.881	0.931	0.620	0.329	0.659	0.533
ResNet50 [7]	0.654	0.690	0.353	0.205	0.384	0.238
ViT-Base/16	0.960	**0.963**	0.812	**0.553**	**0.894**	**0.834**
ViT-Base/32	0.951	0.954	0.783	0.510	0.859	0.790
ViT-Large/16	0.956	0.940	**0.829**	0.499	0.886	0.826
ViT-Large/32	**0.966**	0.931	0.807	0.517	0.876	0.818

differences in accuracy depending on the dataset, a high recognition rate was obtained with ViT-Base/16.

Table 3 shows the comparison results between the conventional and proposed methods in the four public datasets except for our dataset. However, for datasets not used in each paper, "—" is entered. In the table, the * mark means that the voting process has been applied. Voting processing is processing to determine the final class by voting based on the output of multiple ROIs without determining the class with only one ROI. The maximum recognition accuracy was obtained in many models by using ViT. Deep BarkID has been reported to have higher recognition accuracy than our proposed method in BarkNet 1.0.

In KyutechBark150, we used the original image and extracted ROI for input data. The former obtained higher recognition accuracy from Table 2. In order to analyze this factor, we computed an attention map. The obtained attention

Table 3. Comparison results.

Dataset	NewBarkTex [13]	TRUNK12	BarkNet1.0 [2]	Bark-101 [14]
ResNet34+Multiple Crop [2]	—	—	0.9388	—
LCoLBP [14]	0.893	0.842	—	0.419
2DSCAR [15]	—	0.929	0.904	—
EnsDA_all [12]	—	—	0.913	—
Deep BarkID [19]	—	—	**0.9436***	—
sSRBP [1]	—	0.8804	0.6674	0.4888
ViT (ours)	**0.9657**	**0.9630**	0.8291 (0.9386*)	**0.5529**

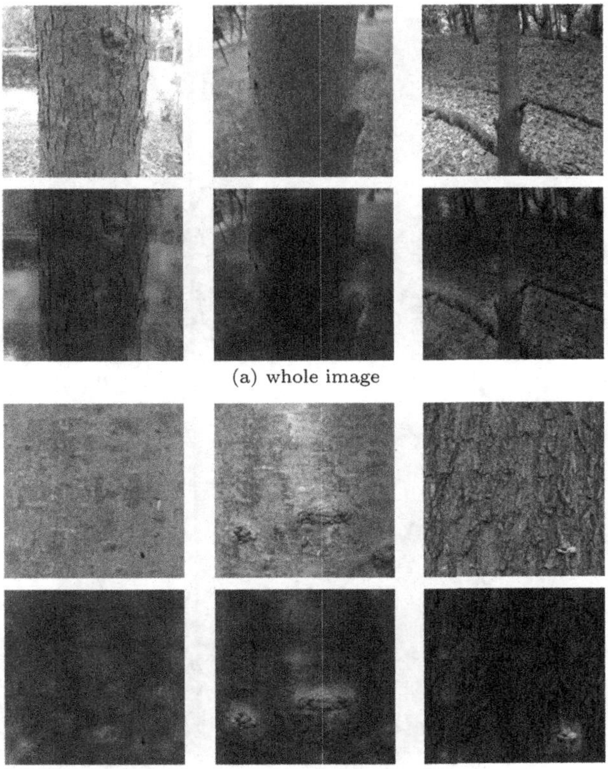

(a) whole image

(b) ROI

Fig. 4. Attention maps when ViT-Base/16 is applied to KyutechBark150.

maps are shown in Fig. 4. In the figure, the upper part is the input image, and the lower part is the image in which the attention map overlaps the input image. The brighter the pixel, the more important it is for recognition. By observing

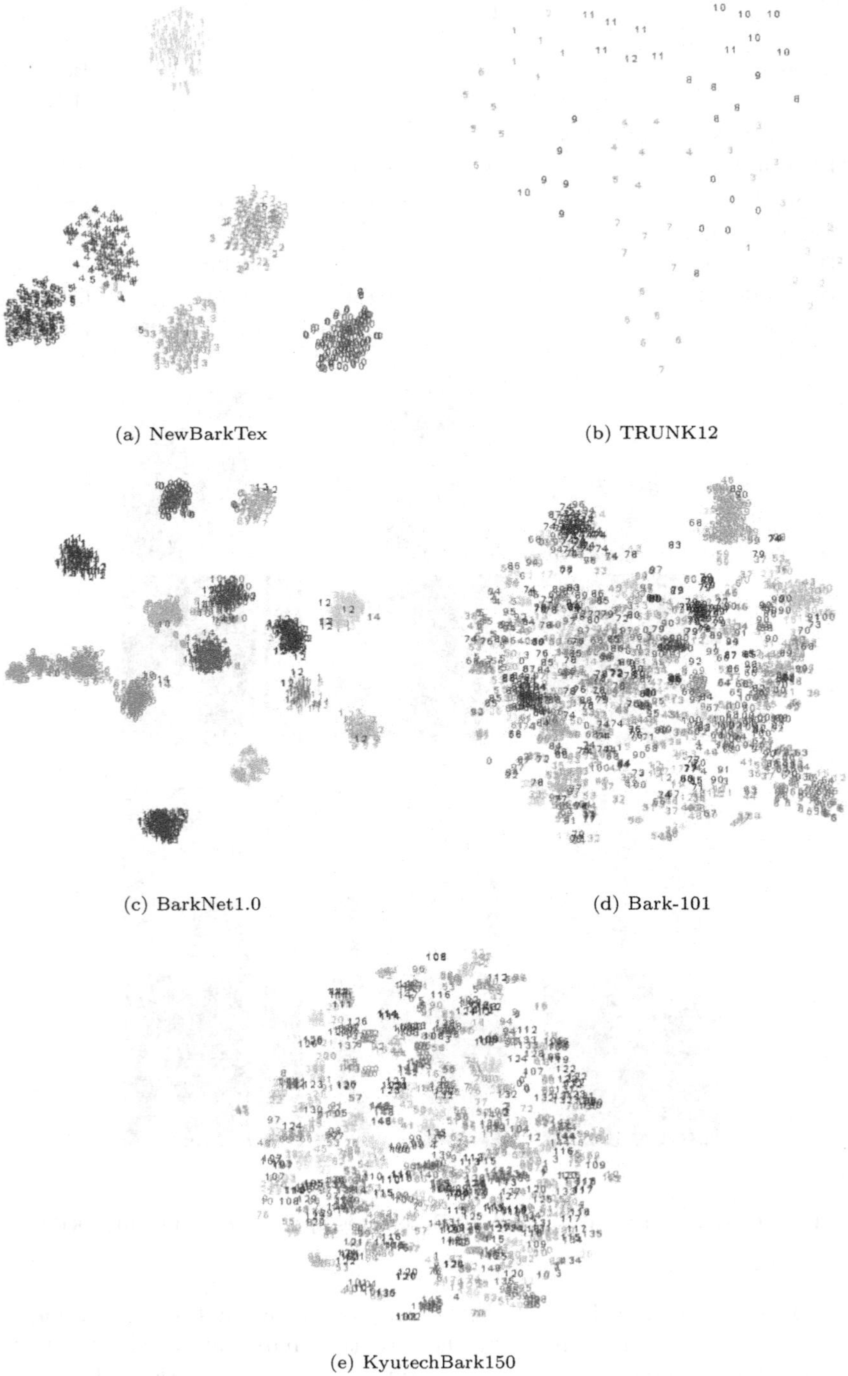

(a) NewBarkTex

(b) TRUNK12

(c) BarkNet1.0

(d) Bark-101

(e) KyutechBark150

Fig. 5. Visualization results by t-SNE (ViT-Base/16).

Fig. 4(a), the whole image in which the background is contained, the background or the boundary between the tree and background, is more focused than the tree area. There are many similar environments where the tree grows. It is guessed that the recognition accuracy was improved by using the background region compared to ROI alone. Focusing on the ROI in Fig. 4(b), it was confirmed that the uneven parts of the bark surface became bright, which was important for recognition.

In order to visualize the trained model, we applied visualization by t-SNE [11] using the trained model of ViT-Base/16 of five datasets. The result is shown in Fig. 5. Ince the recognition accuracy of Bark-101 is low, and the number of species is large, there needs to be cohesiveness for each tree species in the visualization results. However, it was confirmed that groups were formed for each tree species for the other four datasets. This result confirmed that ViT was trained correctly.

4 Conclusion

This paper proposed the ViT-based bark image recognition method for tree species identification. As a result of conducting comparative experiments by applying CNN and ViT to the originally constructed dataset and four public datasets, ViT obtained the highest recognition accuracy in all the datasets. Furthermore, visualization by t-SNE and attention map was performed to show the effectiveness of ViT.

Of the five datasets used in the recognition experiment, the recognition accuracy of Bark-101 is lower than that of the other datasets. It is because the number of training data is small. As a future task, we will further improve the recognition accuracy. The datasets were relatively small and may be overfitted. Therefore, we will apply data augmentation.

References

1. Boudra, S., Yahiaoui, I., Behloul, A.: A set of statistical radial binary patterns for tree species identification based on bark images. Multim. Tools Appl. **80**, 22373–22404 (2021). https://doi.org/10.1007/s11042-020-08874-x
2. Carpentier, M., Giguere, P., Gaudreault, J.: Tree species identification from bark images using convolutional neural networks. In: 2018 IEEE/RSJ International Conference on Intelligent Robots and Systems (IROS), pp. 1075–1081 (2018). https://doi.org/10.1109/IROS.2018.8593514
3. Chen, L.-C., Zhu, Y., Papandreou, G., Schroff, F., Adam, H.: Encoder-decoder with Atrous separable convolution for semantic image segmentation. In: Ferrari, V., Hebert, M., Sminchisescu, C., Weiss, Y. (eds.) ECCV 2018. LNCS, vol. 11211, pp. 833–851. Springer, Cham (2018). https://doi.org/10.1007/978-3-030-01234-2_49
4. Dosovitskiy, A., et al.: An image is worth 16x16 words: transformers for image recognition at scale. In: International Conference on Learning Representations (ICLR) (2021)
5. Fiel, S., Sablatnig, R.: Automated identification of tree species from images of the bark, leaves and needles. In: 16th Computer Vision Winter Workshop (2011)

6. Fischler, M.A., Bolles, R.C.: Random sample consensus: a paradigm for model fitting with applications to image analysis and automated cartography. Commun. ACM **24**(6), 381–395 (1981). https://doi.org/10.1145/358669.358692
7. He, K., Zhang, X., Ren, S., Sun, J.: Deep residual learning for image recognition. In: 2016 IEEE Conference on Computer Vision and Pattern Recognition (CVPR), pp. 770–778 (2016).. https://doi.org/10.1109/CVPR.2016.90
8. Ido, J., Saitoh, T.: CNN-based tree species identification from bark image. In: 10th International Conference on Graphics and Image Processing (ICGIP 2018). vol. 11069 (2019). https://doi.org/10.1117/12.2524213
9. Ido, J., Saitoh, T.: Automatic tree species identification from natural bark image. In: 11th International Conference on Graphics and Image Processing (ICGIP 2019). vol. 11373, pp. 29–34 (2020). https://doi.org/10.1117/12.2557187
10. Kingma, D., Ba, J.: Adam: a method for stochastic optimization. arXiv:1412.6980 (2014). 10.48550/arXiv. 1412.6980
11. van der Maaten, L., Hinton, G.: Visualizing data using t-SNE. J. Mach. Learn. Res. **9**, 2579–2605 (2008)
12. Nanni, L., Paci, M., Brahnam, S., Lumini, A.: Comparison of different image data augmentation approaches. J. Imaging **7**(12) (2021). https://doi.org/10.3390/jimaging7120254
13. Porebski, A., Vandenbroucke, N., Macaire, L., Hamad, D.: A new benchmark image test suite for evaluating colour texture classification schemes. Multim. Tools Appl. **70**, 543–556 (2014). https://doi.org/10.1007/s11042-013-1418-8
14. Ratajczak, R., Bertrand, S., Crispim-Junior, C., Tougne, L.: Efficient bark recognition in the wild. In: International Conference on Computer Vision Theory and Applications (VISAPP2019) (2019)
15. Remes, V., Haindl, M.: Bark recognition using novel rotationally invariant multispectral textural features. Pattern Recogn. Lett. **125**, 612–617 (2019). https://doi.org/10.1016/j.patrec.2019.06.027
16. Saitoh, T., Iwata, T., Wakisaka, K.: Okiraku search: Leaf images based visual tree search system. In: 14th IAPR International Conference on Machine Vision Applications (MVA), pp. 242–245 (2015). https://doi.org/10.1109/MVA.2015.7153176
17. Simonyan, K., Zisserman, A.: Very deep convolutional networks for large-scale image recognition. arXiv preprint arXiv:1409.1556 (2014). 10.48550/arXiv. 1409.1556
18. Švab, M.: Computer-vision-based tree trunk recognition. B.sc thesis, Fakulteta za računalništvo in informatiko, Univerza v Ljubljani (2014)
19. Wu, F., Gazo, R., Benes, B., Havia, E.: Deep barkid: a portable tree bark identification system by knowledge distillation. Eur. J. Forest Res. **140**, 1391–1399 (2021). https://doi.org/10.1007/s10342-021-01407-7

Author Index

Printed in the United States
by Baker & Taylor Publisher Services